Drilling Technology Handbook

Drilling Technology Handbook

Edited by **Annabelle Turner**

SYRAWOOD
PUBLISHING HOUSE
New York

Published by Syrawood Publishing House,
750 Third Avenue, 9th Floor,
New York, NY 10017, USA
www.syrawoodpublishinghouse.com

Drilling Technology Handbook
Edited by Annabelle Turner

International Standard Book Number: 978-1-68286-343-5 (Hardback)

Printed in China.

Contents

Preface

I am honored to present to you this unique book which encompasses the most up-to-date data in the field. I was extremely pleased to get this opportunity of editing the work of experts from across the globe. I have also written papers in this field and researched the various aspects revolving around the progress of the discipline. I have tried to unify my knowledge along with that of stalwarts from every corner of the world, to produce a text which not only benefits the readers but also facilitates the growth of the field.

Drilling technology has been remarkably useful in the study of geothermal sciences. Much has been found and understood about the earth's inner structure in the recent past owing to the progress of drilling and its scientific applications. This book contains some path-breaking studies in the field of drilling technology as it is a collective work of an eminent panel of internationally renowned scholars. Topics discussed in this book include instrumentation and observation, technological advances in drilling, earth sampling, mineral resources, etc. It strives to provide a fair idea about this discipline and develop a better understanding of the latest advances within this field. This book will serve as a reference to a broad spectrum of readers. It is highly recommended for students and researchers pursuing drilling technology and allied disciplines.

Finally, I would like to thank all the contributing authors for their valuable time and contributions. This book would not have been possible without their efforts. I would also like to thank my friends and family for their constant support.

Editor

Investigating Large Igneous Province Formation and Associated Paleoenvironmental Events: A White Paper for Scientific Drilling

by Clive R. Neal, Millard F. Coffin, Nicholas T. Arndt, Robert A. Duncan, Olav Eldholm, Elisabetta Erba, Cinzia Farnetani, J. Godfrey Fitton, Stephanie P. Ingle, Nao Ohkouchi, Michael R. Rampino, Marc K. Reichow, Stephen Self, and Yoshiyuki Tatsumi

Introduction

Earth's history has been punctuated over at least the last 3.5 billion years by massive volcanism on a scale unknown in the recent geological past. Largely unknown mechanical and dynamic processes, with unclear relationships to seafloor spreading and subduction, generated voluminous, predominately mafic magmas that were emplaced into the Earth's lithosphere. The resultant large igneous provinces (LIPs; Coffin and Eldholm, 1994; Ernst and Buchan, 2001; Bryan and Ernst, 2008) were at times accompanied by catastrophic environmental changes. The interaction of the LIP-associated mantle processes with the Earth's crust have produced a variety of surface expressions (Fig. 1a and 1b); the most common present-day examples are oceanic plateaus (e.g., Kerguelen/Broken Ridge, Ontong Java, Manihiki, Hikurangi, Shatsky), ocean basin flood basalts (e.g., Caribbean, Nauru), magma-dominated divergent continental margins (e.g., the North Atlantic), and continental flood basalts (e.g., Columbia River, Deccan Traps, Siberian Traps). Environmental effects associated with LIP formation include climate changes, mass and other extinctions, variations in ocean and atmospheric chemistry, and Oceanic Anoxic Events (OAEs). Therefore, the geodynamic processes in the mantle that produce LIPs have potentially profoundly affected the Earth's environment, particularly the biosphere and climate. The Integrated Ocean Drilling Program (IODP) affords unique opportunities to investigate LIPs and associated environmental effects, building upon results from the Ocean Drilling Program (ODP) and Deep Sea Drilling Project (DSDP) (Coffin et al., 2006). To this end, a workshop on LIPs, sponsored by IODP Management International (IODP-MI) and the Consortium for Ocean Leadership, was held at the University of Ulster in Coleraine, Northern Ireland, U.K. on 22–25 July 2007 (Coffin et al., 2007).

A multi-disciplinary group of eighty scientists representing academia, government, and industry from sixteen countries discussed strategies for advancing understanding of LIPs and associated environmental events using the three different IODP platforms and related technologies. During the workshop, which began with an examination of the United Nations Educational, Scientific, and Cultural Organization (UNESCO) World Heritage "Giant's Causeway" exposure of the North Atlantic LIP (Fig. 2), scientists investigating LIPs through field, laboratory, and modeling approaches shared their expertise. Specifically, outstanding problems related to LIP origin, emplacement, and environmental impacts were discussed in plenary and breakout sessions. The workshop achieved the following:

- Identified multidisciplinary, synergistic approaches to addressing outstanding Earth system problems associated with LIP science.
- Brought together scientists from widely different areas of expertise (e.g., petrology, paleontology, geodynamics, oceanography, geophysics, logging, volcanology, geochemistry, stratigraphy, tectonics, paleoceanography, paleomagnetics) to focus on how ocean drilling can address unresolved issues associated with the origin,

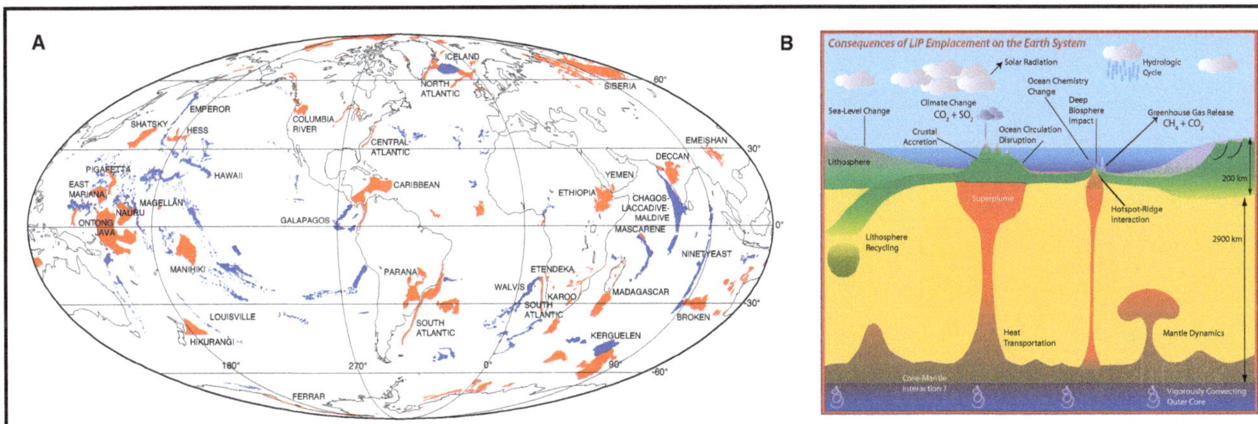

Figure 1. [A] Phanerozoic global LIP distribution. Red = LIPs (or portions thereof) generated by a transient "plume head"; Blue = LIPs (or portions thereof) generated by a persistent "plume tail". Taken from Coffin (2006). [B] A holistic representation of the process of LIP emplacement and associated environmental effects.

emplacement, and evolution of LIPs, as well as their impact on the environment.

- Exposed new scientists to the IODP and nurtured early career scientists, specifically with respect to how scientific drilling can advance understanding of LIPs.
- Enhanced cooperation between the IODP and the International Continental Scientific Drilling Program (ICDP).
- Explored partnerships among IODP, government, and industry.

One of the major outcomes of the workshop was to define multiple pathways to enhance our knowledge of LIPs through scientific drilling. These ranged from ancillary project letters, through individual expedition proposals, to a mission proposal.

This white paper highlights the major problems associated with LIPs that can be addressed through scientific drilling and related studies. It also highlights multidisciplinary approaches required to address such problems, as studies of LIPs encompass mantle geodynamics, emplacement processes, and environmental events affecting the lithosphere, hydrosphere, atmosphere, and biosphere.

Past Achievements

Scientific drilling has played a vital role in the exploration of LIPs, most importantly by providing the first, and in many cases the only, ground truth from the igneous basement of submarine LIPs. While major advances in LIP research involve holistic observational, experimental, and modeling studies involving a broad array of Earth science expertise, results from scientific drilling have been and will continue to be key components of interdisciplinary work. The first dedicated igneous basement sampling investigations of submarine LIPs, conducted during the ODP, concentrated on three provinces: the Ontong Java Plateau (Pacific Ocean), the Kerguelen Plateau/Broken Ridge (Indian Ocean), and the North Atlantic magma-dominated divergent continental margins (Fig. 1). Below, we highlight the major results of these investigations, which have laid the groundwork for the next major phase of discovery during the IODP.

Drilling results from ~120-Ma Ontong Java Plateau basement rocks are complemented by studies of obducted plateau rocks exposed in the Solomon Islands (Andrews et al., 1975; Kroenke et al., 1991; Berger et al., 1993; Neal et al., 1997; Mahoney et al., 2001; Fitton et al., 2004a, b). All Ontong Java basement rocks recovered to date are remarkably homogeneous tholeiitic basalts with minor variations in elemental and isotopic composition, and they were deposited in a submarine environment. Partial batch melting (≥30%) generated the basalts (Fitton and Godard, 2004; Herzberg, 2004), with melting and fractional crystallization at depths of <6 km (Sano and Yamashita, 2004). The lavas and their overlying sediment indicate relatively minor uplift accompanying emplacement and relatively minor subsidence since emplacement. Primarily on the basis of drilling results, multiple models—plume (e.g., Fitton et al., 2004a), bolide impact (Ingle and Coffin, 2004), and upwelling eclogite (Korenaga, 2005)—have been proposed for the origin of the Ontong Java Plateau.

Figure 2. Giant's Causeway (Ireland), part of the North Atlantic LIP. Note the stepped topography in the background, characteristic of flood basalt provinces.

Uppermost igneous basement of the Kerguelen Plateau/Broken Ridge is dominantly tholeiitic basalt erupted above sea level (Barron et al., 1989, 1991; Schlich et al., 1989; Wise et al., 1992; Coffin et al., 2000; Frey et al., 2003), and it shows two apparent peaks in magmatism at 119–110 Ma and 105–95 Ma (Coffin et al., 2002). Geochemical differences among tholeiitic basalts erupted at each site are attributable to varying proportions of components from the primary mantle source (plume?), depleted mid-ocean ridge basalt (MORB)-related asthenosphere, and continental lithosphere. Proterozoic-age zircon and monazite in clasts of garnet-biotite gneiss in a conglomerate intercalated with basalt at one drill site demonstrate the presence of fragments of continental crust in the Kerguelen Plateau, as inferred previously from geophysical (e.g., Operto and Charvis, 1995) and geochemical (e.g., Alibert, 1991) data. For the first time from an intra-oceanic LIP, alkalic lavas, rhyolite, and pyroclastic deposits were sampled. Flora and fauna preserved in sediment overlying igneous basement provide a long-term record of the plateau's subsidence, beginning with terrestrial and shallow marine deposition and continuing to deep water deposition.

Seaward-dipping reflector (SDR) wedges of the latest Paleocene-earliest Eocene North Atlantic LIP drilled off the British Isles during DSDP (Roberts et al., 1984), off Norway during ODP (Eldholm et al., 1987, 1989), and off SE Greenland during ODP (Duncan et al., 1996; Larsen et al., 1994, 1999) confirmed them to be thick series of subaerial lava flows covering large areas. Lavas on the landward side of the SDRs show geochemical and petrological evidence of contamination by continental crust, implying that they ascended through continental crust during early rifting,

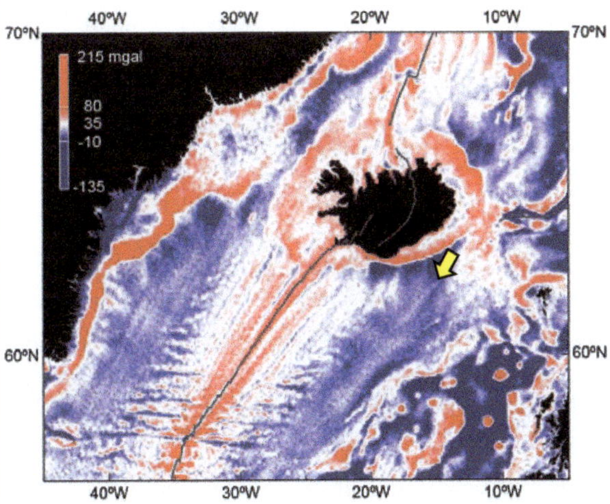

Figure 3. Gravity map of the North Atlantic Ocean showing prominent V-shaped ridges in the basement topography around the Reykjanes Ridge south of Iceland. These represent crustal thickness variations of ~2 km. The yellow arrow shows the location of thick sediment drifts deposited by northern component water flowing southwards over the Iceland-Faeroes ridge. Thinner drifts are present south of the Greenland-Iceland ridge.

whereas oceanward SDR lavas appear to have formed at a seafloor spreading center resembling Iceland. Drilling results from these margins indicate extreme magmatic productivity over a distance of at least 2000 km during continental rifting and breakup, with spatiotemporal influence of the Iceland plume during rifting, breakup, and early seafloor spreading (Saunders et al., 1998).

Mantle Geodynamics

In many respects, LIP magmatism remains quite enigmatic, yet its unique characteristics (e.g., volumes of erupted lava, time duration of volcanism, distinct geochemical composition with respect to MORBs, and frequency over geologic time) unequivocally highlight the importance of understanding the underlying physical, chemical, mechanical, and dynamic processes.

The mantle plume model, in which dynamic instabilities of a thermal boundary layer (e.g., the D'' zone at the base of the Earth's mantle) give rise to a mantle plume with a large head and a long-lived conduit or tail, provides a context for interpreting a wealth of surface observations related to LIPs and hotspot volcanism. And, if plumes do originate in the lowermost mantle, they may be the only way to 'sample' the deepest part of the Earth's mantle. Regardless of the origin of LIPs, studying them and hotspot volcanism provides unique and fundamental information about mantle dynamics, global heat and mass transfer in the mantle, time scales for 'storage' of geochemical heterogeneities and, ultimately, mixing efficiency of mantle convection.

Over the last decade, our views of mantle plumes have evolved considerably; the 'mushroom' shaped plume is certainly possible, but it is not *the* unique plume morphology. By moving beyond the simplistic assumption of a compositionally homogeneous mantle, where density differences are only due to temperature differences, numerical simulations and fluid dynamics laboratory experiments have found an astounding variety of plume sizes and shapes (Tackley, 1998; Davaille, 1999).

Although the nature of compositional heterogeneities in the Earth's mantle is still a matter of debate (e.g., recycled denser, eclogitic crust; early Earth crust; ancient mantle possibly interacting with the metallic core), many lines of seismological and mineral physics evidence suggesting the presence of chemically distinct, denser material in the lowermost mantle. Therefore, thermo-chemical plumes (i.e., their density contrast with respect to the surrounding mantle depends on both excess temperature and intrinsic composition) provide a new and exciting framework to relate deep mantle dynamics to LIP-forming events and pose a variety of new questions that can be addressed through scientific drilling.

Mantle Temperature Versus Mantle Fertility in LIP Formation

A primary question in understanding LIP formation is the extent to which melting anomalies reflect excess fertility in the mantle rather than excess mantle temperature. This issue lies at the heart of the current mantle plume debate. Mantle temperature can be addressed through the major-element composition of primitive LIP basalt via phase equilibria (Herzberg et al., 2007), and studies thereof suggest excess temperatures of <200°C. Mantle fertility is more difficult to investigate because its effects can be mimicked by lower degrees of mantle melting. Diachronous V-shaped ridges around the Reykjanes Ridge south of Iceland (Fig. 3) provide an example of how this problem might be addressed. These ridges reflect fluctuations in crustal thickness of ~2 km, and they are caused by pulses in magma productivity radiating outwards from the Iceland hotspot at ~20 cm yr^{1}. Drilling on the peaks and in the troughs along a transect away from the Reykjanes Ridge will allow basalt composition and hence mantle temperature estimates to be related to crustal thickness. Temperature fluctuations should result in an inverse correlation between gravity (a proxy for crustal thickness) and incompatible-element concentrations in basalt because these concentrations will be lower in larger-degree mantle melts. A direct correlation will indicate fertility pulses. A fluctuation in mantle temperature will cause a correlated fluctuation in water depth along the Greenland–Iceland–Faeroes ridge, affecting the flow of northern component water southwards from the North Atlantic (Wright and Miller, 1996; Jones et al., 2002; Poore et al., 2006). Sediment drifts south of the Iceland–Faeroes ridge (yellow arrow in Fig. 3) should record variations in this flow, and coring these will provide an independent proxy for temperature fluctuations in the Iceland hotspot. This synergy between mantle-dynamic and paleoceanographic objectives will be of considerable mutual benefit.

Thermo-Mechanical Plume-Lithosphere Interaction with Massive Magmatism

The amount of surface dynamic uplift induced by the arrival of a deep mantle plume is being widely debated. Pioneering models (e.g., Farnetani and Richards, 1994) suggested an unrealistically high uplift rate before volcanism, due to the following: (1) the assumption of purely thermal plumes, (2) poor representation of the lithosphere and the overlying crust, and (3) limited numerical resolution.

The plume uplift issue is currently being revisited using thermo-chemical plumes, which show a lower vertical velocity component than purely thermal plumes, and thus induce lower strain rates at the base of the lithosphere. Furthermore, the new generation of numerical models incorporates both a buoyant residual solid in the plume head resulting from partial melting, and surface exchanges of energy and mass between the ascending melts and surrounding rocks. For various plume buoyancy fluxes, lithospheric ages, and geodynamic settings, we can now calculate surface uplift/subsidence as a function of time and distance from the plume center. Model predictions should then be compared with geological observations obtained through drilling. It is important to emphasize that the geologically reconstructed time sequence of surface deformation needs to encompass a time interval extending from a few million years before to a few million years after the main phase of LIP construction.

Coupling of theoretical predictions with observations will lead to a self-consistent and coherent model of mantle plume dynamics and its thermo-mechanical effect on the overlying lithosphere. The periphery of the Ontong-Java Plateau (OJP) away from its convergent margin with the Solomon Islands may be ideal for this type of study. The OJP represents the world's most extensive LIP (equivalent to the size of Greenland or western Europe) that, on the basis of current knowledge, formed in ~5 myr at ~122 Ma. It is divided into a High Plateau containing the bathymetric high and a seismically slow cylindrical mantle root extending to ~300 km depth (Richardson et al., 2000; Klosko et al., 2001; Gomer and Okal, 2003) and an Eastern Salient (Fig. 4), with all drill sites encountering igneous basement being located on the High Plateau. Both isostatic (minimum) and dynamic (maximum) crustal uplift were significantly less for the OJP than for active hotspots today, and total subsidence is also anomalously less (Neal et al., 1997; Ito and Clift, 1998; Ito and Taira, 2000; Ingle and Coffin, 2004; Roberge et al., 2005) than that of any other known oceanic lithosphere (Parsons and Sclater, 1977; Coffin, 1992; Stein and Stein, 1992). However, to complicate this situation, ODP Leg 192 recovered a sequence of volcaniclastic sediments at Site 1184 on the Eastern Salient that represent subaerial eruptions (Thordarson, 2004); this is interesting because this site is not over the mantle root. The sequence was divided into several sub-units separated by wood horizons (Mahoney et al., 2001). Borehole heat flow measurements from the High Plateau and Eastern Salient will establish the thermal or chemical nature of the mantle root, and recovery of basement lava flows from the Eastern Salient will test whether volcanism was extensively subaerial in this area during formation of the OJP (consistent with plume theory), or whether the Site 1184 volcaniclastic sequence was produced by a late-stage volcanic edifice with the bulk of the Eastern Salient being constructed by submarine volcanism.

LIP Internal Architecture

Although drilling typically only scratches the surface of thick igneous basement, it is important to explore the internal structure of a LIP to quantify the relative volumes of extrusive versus intrusive magmatism (Cox 1992; Kerr et al., 1997). Better estimates of total melt volumes and compositions will help to bracket melting rates and to define melt transport/storage mechanisms between subsurface magma chamber(s) and the surface. Moreover, interactions between partial melts and the surrounding rocks determine the length- and time-scales over which LIP magmatism can change the temperature, the internal loads, and the state of stress of the pre-existing lithosphere and crust. LIPs emplaced in oceanic lithosphere (rather than the thicker and more compositionally complex continental lithosphere) are thus better suited to investigate the thermo-mechanical and compositional modifications induced by LIP magmatism in the lithosphere and the uppermost asthenosphere.

Moreover, we note that the current debate on the role of plumes in the formation of thick, *subduction-resisting* lithosphere in the Archean will benefit from a better characterization of the density and viscosity structure resulting from LIP volcanism. This may be difficult to achieve directly through scientific drilling, even with riser capability, although direct

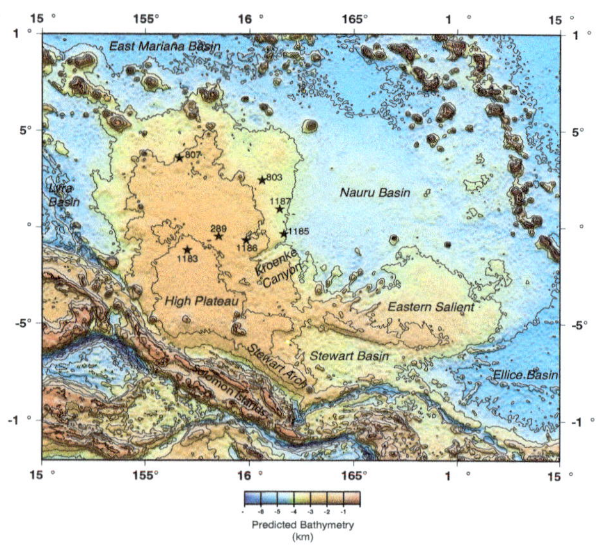

Figure 4. Bathymetry of the western Pacific Ocean including features of the Ontong Java Plateau and surrounding basins. Numbered stars represent DSDP Leg 30, ODP Leg 130, and ODP Leg 192 drill sites that penetrated and recovered igneous basement.

results of site survey work, done in preparation for drilling, can directly bear on this question.

Plumes and Superplumes: Exploring Interactions with Slabs and the Earth's Core

Although we generally consider plumes to be isolated 'bodies', they are part of mantle convection and are likely to interact with subduction processes and deep Earth dynamics. Plumes upwelling from the thermal boundary layer at the base of the Earth's mantle could enhance flushing of stagnant slabs into the lower mantle, and vice versa (Nakagawa and Tackley, 2005). During Cretaceous time, LIP and arc magmatism—which are surface manifestations of mantle upwelling and downwelling, respectively—occurred simultaneously (Reymer and Schubert, 1984; Larson, 1991; Eldholm and Coffin, 2000). Decoding the time relation between magmatism at LIPs and at convergent margins is, therefore, key to understanding the primary cause of mantle convection.

Recently, Burke and Torsvik (2004) restored twenty-five LIPs of the past 200 myr to their eruption sites using a new global paleomagnetic reference model. Ninety percent of the LIPs, when erupted, lay above low-velocity seismic-shear-wave regions of the D'' zone, as indicated in current tomographic models, suggesting that the deep mantle beneath the Central Pacific and Africa may represent a long-lived source region for plumes. Better characterization of LIPs will therefore help to understand the nature of 'superplumes', their stability over time (≥200 myr?), and their potential to sample distinct geochemical reservoirs. By using 2-D numerical simulations, Farnetani and Samuel (2005) have shown the complex internal dynamics of thermo-chemical plumes, leading to the possible coexistence of different types of plumes and superplumes (Fig. 5).

Finally, the accumulation of cold slabs and the upwelling of hot material induce spatial and temporal variations in heat flow at the core-mantle boundary, affecting the outer core

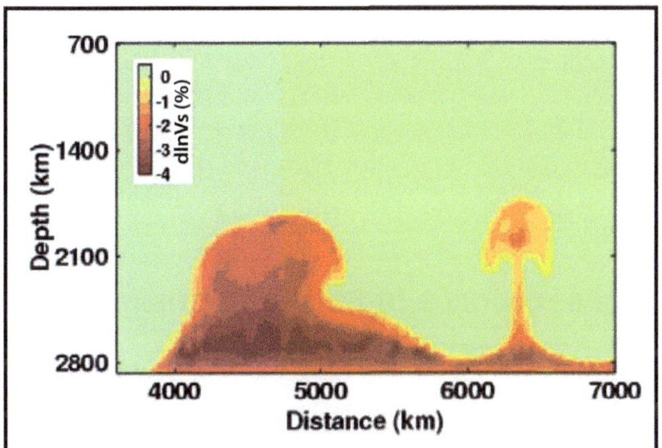

Figure 5. 2-D model showing the coexistence of plumes and superplumes in time and space (Farnetani and Samuel, 2005). Colors correspond to calculated seismic velocity anomalies (see inset key for quantitation).

convection and the frequency of polarity reversals (e.g., low (high) core heat flux, infrequent (frequent) polarity reversals). Courtillot and Olson (2007) showed that three magnetic superchrons preceded the largest Phanerozoic mass extinctions (Cretaceous-Tertiary, Triassic-Jurassic, Permo-Triassic), which are associated with major flood basalt events. These authors suggest that thermal instabilities in the D'' layer may increase heat flow from the core and trigger the end of a magnetic superchron. Documenting the timing of LIP magmatism and superchron events will provide key information on dynamic linkages between the core and mantle.

Geodynamics and Tectonic Setting

LIPs are emplaced along active plate boundaries (e.g., magma-dominated divergent continental margins) and in intraplate settings (e.g., continental flood basalts). The tectonic setting of emplacement for most oceanic plateaus and ocean basin flood basalts, however, is not completely understood. A key question is whether upwelling mantle can erode the lithosphere sufficiently to instigate formation of a divergent plate boundary (Hill, 1991; Davies, 1994). While this appears to be the case for the North Atlantic volcanic province, where extension and seafloor spreading relocated from the Labrador Sea to the incipient northernmost North Atlantic coincident with massive North Atlantic volcanic province magmatism (Srivastava and Tapscott, 1986), the situation is significantly less clear for other magma-dominated divergent continental margins (e.g., Northwest Australia and Kerguelen/Antarctica/India/Australia). To gain a thorough understanding of relationships among mantle geodynamics, tectonics, and basaltic magmatism, we need to investigate more than a single example of magma-dominated divergent margins by scientific ocean drilling.

In marked contrast to the North Atlantic example, the Northwest Australian margin is segmented, and igneous rock volumes vary considerably along strike, without clear evidence for a related mantle plume (Mutter et al., 1988; Hopper et al., 1992; Symonds et al., 1998; Planke et al., 2000). This makes the margin a strong candidate to test the edge-driven/small-scale convection hypothesis (Mutter et al., 1988; King and Anderson, 1998; Korenaga, 2004) for generating excessive magma by drilling a margin transect across multiple seaward-dipping reflection wedges, the Wallaby Plateau, and normal oceanic crust. The geochemistry, petrology, and geochronology of the recovered rocks will yield melting conditions, mantle reservoir type, extent of continental contamination, and the spatiotemporal evolution of the magma source.

The Kerguelen Plateau/Broken Ridge appears to have begun forming in the nascent eastern Indian Ocean at least 10 myr after breakup among Antarctica, India, and southwestern Australia (Coffin et al., 2002; Gaina et al., 2007). Despite these conjugate continental margins exhibiting some characteristics of excessive magmatism during breakup (e.g., SDR wedges; Stagg et al., 2007) and contemporaneous continental basaltic volcanism (Storey et al., 1992; Frey et al., 1996), the lack of physical and therefore geochronological continuity between these margins and the Kerguelen Plateau/Broken Ridge reveals yet another variation from the North Atlantic example of simultaneous continental breakup and LIP formation. Drilling the Early Cretaceous SDR wedges of the conjugate southwest Indian Ocean margins and the oldest portions of the Kerguelen Plateau to determine geochemistry, petrology, and geochronology will address critical questions involving relationships between geodynamics and tectonic setting.

Emplacement

Unravelling the emplacement of LIPs is pivotal in understanding their significance in the formation of the Earth's crust as well as any potential environmental consequences. Our understanding of these processes remains quite limited, and basic questions—such as whether or not the centers of these eruptions occur along fissures—are still matters for debate. Onset of volcanism can be determined for most continental LIPs, although cessation of activity is more problematic to determine due to erosion of uppermost lavas. Oceanic LIPs provide a unique opportunity to address these issues as they are generally better preserved, and consequently can provide a more complete picture of LIP formation. The major issue concerns timing of LIP generation and emplacement (i.e., whether the event was short- or long-lived). We observe examples of both in almost wholly submarine LIPs and in subaerially emplaced LIPs that have been partly rifted and submerged. This overarching question encompasses several missing links in our present understanding that are immediately addressable by scientific drilling.

Duration of Emplacement

Many LIPs appear to have at least one major pulse of volcanism when the most voluminous lava sequence was erupted. Identifying the duration of the full sequence of lavas enables identification of the most voluminous intervals. Knowing the time span of the most voluminous interval(s) for each LIP is crucial for determining the mechanism of emplacement and the potential environmental impact. Furthermore, sampling and dating the lavas allow questions concerning episodicity (about which we know little in major LIPs) to be addressed. For example, was there more than one interval of voluminous volcanism associated with the LIP's formation? As well as from direct sampling of LIP lavas, much can be learned about the onset and cessation of LIP volcanism from environmental indicators contained in sedimentary sections in older basins adjacent to the LIPs and (potentially) worldwide. These include shifts in isotopic and trace metal content of the sediment that can be linked to the arrival of the first LIP lavas onto the seafloor and/or (perhaps) volcanic degassing to the atmosphere. If the sedimentary sequence yields a high-resolution record, this can provide a highly sensitive time indicator of the onset and cessation of LIP activity, directly correlating the LIP with any contemporaneous environmental events.

Style and Timing of LIP Eruptions

Many eruptions occur during the lifespan of a LIP. For almost all LIPs, the total volumes of lavas are not accurately known, let alone those contained in each eruptive package (or other volcanic deposits such as hyaloclastites). What is also unknown is whether each package was produced by short, intense eruptive episodes, or more pulse-like and protracted ones. Critical for addressing environmental recovery is accurate determination of intervals between eruptions. Drilling into LIPs will provide samples of sediment between lavas to address this (e.g., ODP sites 642 and possibly 807; Larson and Erba, 1999). The probability of finding chronological and environmental markers (e.g., carbonaceous or siliceous sediment deposition) between flow units is much higher in the oceanic environment than on the continents. Erosion during emplacement of submarine LIPs is expected to be minimal; hence, it is more likely to recover flows erupted in sequence. Furthermore, the final (uppermost) lavas are commonly not preserved in continental flood basalts, but truly submarine LIPs offer a much greater chance of obtaining the full sequence of lava products, or at least the last erupted lavas.

If these two fundamental problems can be addressed, we will be able to answer questions such as:

1) When was the big pulse or biggest pulses of volcanism during a LIP's lifespan?
2) Does this time correlate globally with sudden environmental events?
3) What possible role(s) does LIP emplacement have on these sudden events?

Timing of emplacement is critical for determining magma fluxes and, ultimately, the mantle processes controlling magma production. Is magma production in plume heads controlled by 'bottom-up' or 'top-down' processes? To better understand the possible effects of LIP volcanism being causal for brief but severe environmental crises, it is necessary to determine eruption rates, and this, in turn, will be useful to model the transport and emplacement distances of lava flows.

Consequences of Shallow Intrusions

Another topical and critical question concerns LIP-related sill emplacement into sediments leading to gas release into the ambient environment; this has occurred on land and on the ocean floor (Svensen et al., 2004). Would submarine gas or fluid release directly affect the oceanic environment (Fig. 6)? To better examine relationships between sills and lava flows, drilling in specific areas can address the timing of the intrusion of the sills, whether they were the feeders for the flood lavas (see section Duration of Emplacement above) and plumbing systems for LIPs. If the sills and their interactions with sediment cause environmental changes, then it is vital to establish the timing of sill and LIP lava emplacement. Further to the consideration of sills, we know little about magma reservoirs for LIPs, including their locations and dimensions. Are the chambers deep or shallow? Perhaps in oceanic crust, more primitive magmatic material can be erupted at the surface that would otherwise be stalled in continental crust, and a study of this may provide clues to the nature of the initial magmatism related to LIPs.

Relationships Between Subaerial and Submarine LIP Emplacement

What are the differences between subaerial and submarine emplacement of LIP lavas, and thus the similarities or differences in LIP architecture between the continental and oceanic realms? In the ocean, very large flow fields may be possible underwater, and perhaps few barriers halt the spread of lava flows. We might thus predict widespread lava fields covering whole provinces, but is this realistic? What do LIP super-eruption products look like when they are emplaced under water? Are the component lava bodies of oceanic flow fields similar in dimension to those in continental flood basalt provinces, with flow unit thicknesses of 100 m or greater? What is the overall architecture of an oceanic plateau sequence? It appears that their dominantly shallow dipping, low angle slopes would favor the inflation of lava flows (Self et al., 1998).

Lava age distributions are also needed to assess how LIPs are constructed. Even though younger, post-major pulse lavas cover most oceanic LIPs, it may be possible to find widespread main-pulse lavas at the edges of LIPs. Using the architecture of better-known subaerial LIPs (Fig. 7), we can assess where best to drill to intersect long lava flows in submarine LIPs.

An important aspect of subaerial versus submarine emplacement of LIPs is the nature of seaward-dipping reflectors (SDRs). These are highly significant components of several LIPs (such as the North Atlantic and Kerguelen) and not observed in continental LIPs. They are most likely related to the tectono-magmatic setting, but what do they represent? Are they exclusively subaerially emplaced, but subsided parts of a LIP, or might they also form in marine environments? Some geophysical observations suggest the latter, but confirmation by ocean drilling is required. To understand the nature of SDRs, it is crucial to drill full sequences to determine if these were emplaced as deep marine sheet flows, perhaps infilling a subsiding rift basin.

LIP-Generated Felsic Magmas

Felsic volcanic rocks are typically much higher in potassium than their mafic counterparts and thus contain ideal minerals for Ar-Ar dating. Drilling on the Kerguelen

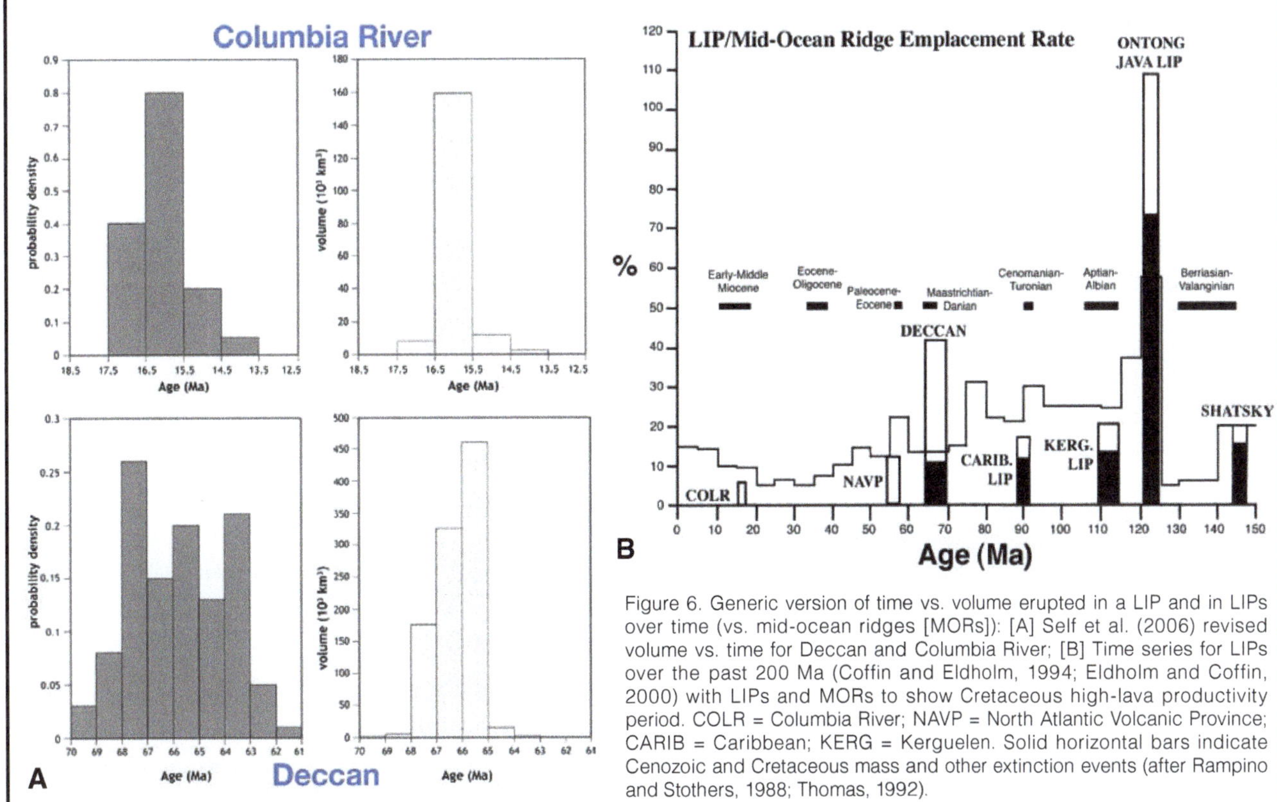

Figure 6. Generic version of time vs. volume erupted in a LIP and in LIPs over time (vs. mid-ocean ridges [MORs]): [A] Self et al. (2006) revised volume vs. time for Deccan and Columbia River; [B] Time series for LIPs over the past 200 Ma (Coffin and Eldholm, 1994; Eldholm and Coffin, 2000) with LIPs and MORs to show Cretaceous high-lava productivity period. COLR = Columbia River; NAVP = North Atlantic Volcanic Province; CARIB = Caribbean; KERG = Kerguelen. Solid horizontal bars indicate Cenozoic and Cretaceous mass and other extinction events (after Rampino and Stothers, 1988; Thomas, 1992).

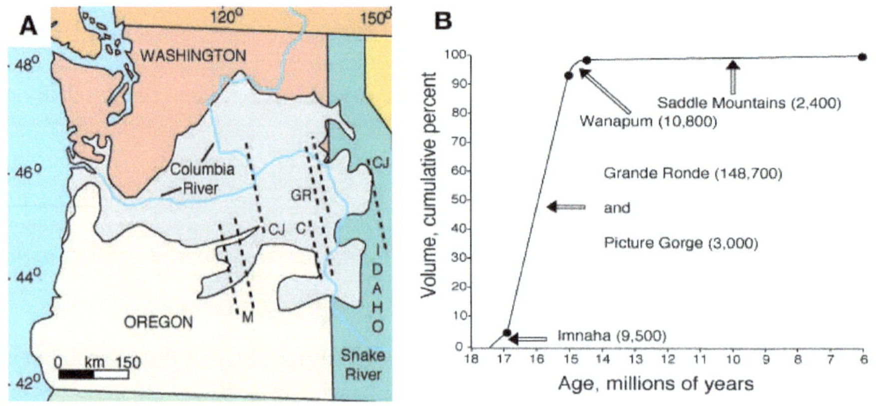

Figure 7. [A] Map of the Columbia Plateau, showing the limits of the Columbia River flood basalts and the feeder dike systems (dashed lines). CJ indicates the longitudinal boundaries of the Chief Joseph dike swarms that fed the Clarkston (Imnaha, Grande Ronde, and Wanapum) and Saddle Mountain basalts. The Grande Ronde (GR) and Cornucopia (C) swarms are concentrations within the CJ. M indicates the Monument dike swarm that fed the Picture Gorge basalt. [B] Ages and estimated volumes of the major eruptive units of the Columbia River basalts. This figure illustrates that surficial sampling of a LIP can yield information on its history of formation.

Plateau (ODP Leg 183) recovered significant amounts of a wide variety of felsic volcaniclastics and lavas (Coffin et al., 2000). Other oceanic LIPs, such as Hess Rise in the Pacific (DSDP Leg 62), contain discrete intervals of felsic lavas. Many, if not all, continental LIPs contain felsic components in the main-series lavas (Bryan et al., 2002), but not from the Deccan and the Columbia River provinces. Eruptions of these evolved magmas can be explosive (e.g., the Ethiopian ignimbrites) and represent valuable stratigraphic and geochronologic markers if they can be correlated to felsic ash fall deposits in deep sea sediment (Peate et al., 2003). Drilling into felsic volcanic products of LIP volcanism, on both the LIP proper and in neighboring sedimentary sections, will help anchor the timing of eruptions via chronostratigraphy.

Environmental Consequences

Connections between large historic basaltic eruptions and perturbations of the global environment are well documented. Classic examples include the eruptions of Laki, Iceland in 1783, and Eldgjá, Iceland in 934, both of which were followed in subsequent years by historically cold summers in the northern hemisphere (Stothers, 1998). In contrast, LIPs represent much larger outpourings of basaltic magmas over much longer time scales, and recent data (Coffin and Eldholm, 1994; Kerr, 1998; Larson and Erba, 1999; Leckie et al., 2002; Snow et al., 2005; Kuroda et al., 2007) increasingly suggest temporal correlations between LIP formation and significant oceanographic, biotic, and climatic events, the most severe of which are OAEs. Greenhouse gases released from LIP magmas, as well as sediment-magma interactions and ocean chemistry changes associated with the LIP emplacement, may have had major effects on the global environment (Fig. 1b). Critical in understanding the impact LIP magmatism may have had on the environment is determining LIP volume and, more importantly, magma eruption rates. Scientific drilling has a vital role to play in identifying the mechanisms and quantifying the timescale and magnitude of effects, and interdisciplinary, synergistic collaborations among scientists from a variety of disciplines are required.

Atmospheric Impacts of LIP Emplacement

LIPs may impact the atmosphere, oceans, and biosphere by rapidly releasing huge amounts of magmatic volatiles (CO_2, SO_2, etc.) or volatiles (CO_2, CH_4) from intruded sediments (e.g., carbonates, organic-rich shales, evaporites) (Fig. 8). Directly or indirectly, they may cause changes in the atmosphere/ocean system that lead to perturbations of atmosphere/ocean chemistry, circulation, ecology, and biological productivity (Self et al., 2006). This was especially true in Cretaceous and Early Tertiary time, when the atmospheric CO_2 content of the atmosphere was three to seven times higher than that of today (Fig. 9), and perhaps more susceptible to short-term perturbations in ocean/atmosphere dynamics and their ensuing effects on life. Furthermore, recent compilations suggest that a sudden sea level rise (~60 m; Miller et al., 2005) was associated with the Paleocene-Eocene Thermal Maximum (PETM) event (Kennett and Stott, 1991; Bains et al., 1999; Svensen et al., 2004), which in turn was concurrent with the formation of the North Atlantic Volcanic Province (Eldholm and Thomas, 1993).

Oceanic Anoxic Events

Episodes of complete depletion of oxygen below surface levels in the Earth's oceans, known as OAEs (Schlanger and Jenkyns, 1976), represent the most momentous environmental changes in the ocean of the past 250 million years, and bear some similarities to the less impactive PETM event. Examples of linked LIP emplacement and major environmental/biological crises include (Fig. 10): Oceanic Anoxic Event (OAE)-1a in the early Aptian (~122 Ma) and the Ontong Java Plateau (~120 Ma), Manihiki Plateau (~120 Ma) and the Kerguelen Plateau (~119 Ma); OAE-2 around the Cenomanian-Turonian boundary (~94 Ma) and the Caribbean-Colombian flood basalts (~92–94 Ma); and the PETM (55 Ma) and the North Atlantic Volcanic Province (~55 Ma).

There are two possible causal relationships between LIPs and OAEs: 1) the instigation of oceanic anoxia through global warming by greenhouse gases (e.g., CO_2, CH_4) that leads to ocean stagnation, which in turn induces anoxia in deep/intermediate depths of the ocean; and 2) submarine volcanic eruptions and associated massive hydrothermal release of trace metals into the global ocean instigating black shale events (Snow et al., 2005). Furthermore, the connection between LIPs and marine biotic changes (including some extinctions) has been ascribed to acidification of seawater by adding CO_2 and SO_2 (Coffin and Eldholm, 1994; Kerr, 1998). However, these ideas require critical evaluation.

Exploring Linkages between LIP Emplacements and Environmental Events

To evaluate linkages between LIP emplacements and environmental events requires information on multiple topics:

1) Accurate and precise timing, duration, and magnitude of LIP magmatic activity and environmental/biotic perturbations;
2) Volatile fluxes related to LIP emplacement events (i.e., LIP magma degassing as well as those potentially derived from magma-sediment interactions) and the extent of gas release into the ocean/atmosphere environment;
3) Hydrothermal fluid compositions (especially trace metals) and fluxes related to LIP emplacement events, and the extent of their release into the ocean environment;
4) Eruptive environments (e.g., deep/shallow submarine; subaerial) and sediment-magma interactions, as well as the overall tectonic settings, of LIP events;

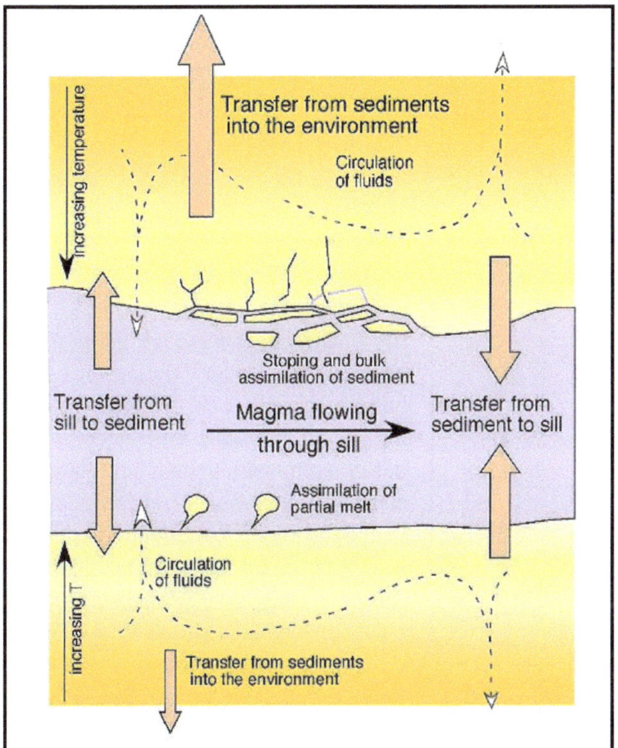

Figure 8. Hypothetical representation of sill-sediment interaction and the release of gases liberated from the sedimentary wallrock.

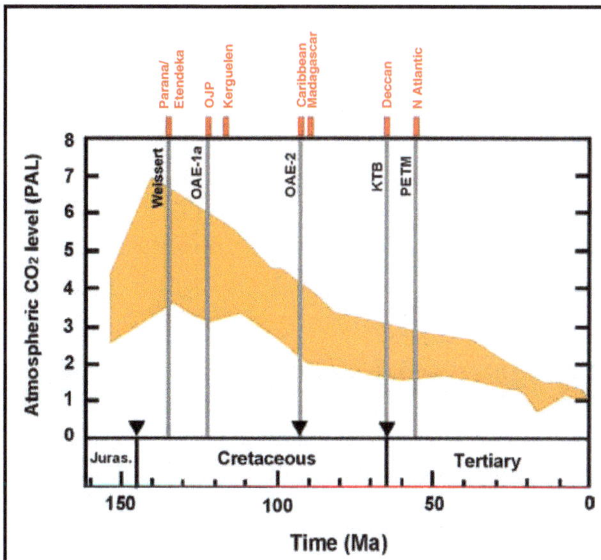

Figure 9. Estimated atmospheric CO_2 concentrations during the last 150 myr (Berner and Kothavala, 2001; Tajika, 1998). Shaded areas indicate global oceanic anoxic events (OAE-1a and OAE-2), Cretaceous-Tertiary boundary (KTB), and Paleocene-Eocene Thermal Maximum (PETM). Mass extinctions events (Sepkoski, 1996) are shown as black inverted triangles.

5) Specific paleoenvironmental conditions prior to, at the onset of, and during times of LIP emplacements (i.e., to define changes in atmospheric and oceanic physical and chemical properties and circulation).

Temporal coincidence between LIPs and environmental crises is clear for a few examples (e.g., North Atlantic LIP, Deccan Traps, Siberian Traps), but temporal correlations between others need further evaluation.

Sediment and volcanic rock recovered by the IODP in various oceanic basins and at different paleolatitudes will provide opportunities to resolve these issues in the following ways:

- Obtaining complete, high-resolution sedimentary records from critical ocean environments (e.g., the Mesozoic pelagic/deep, shallow, and atoll; sediment from the Pacific, Indian, Arctic, and Southern oceans; and high paleo-latitudinal sites are critically required).
- Obtaining syn-sedimentary sections within or adjacent to individual LIPs to estimate the potential hydrothermal fluid release and gas release through sediment-magma interactions.
- Targeted drilling to bracket the duration, peak, and volume of magmatic activity (e.g., tectonic windows and feather edges of LIP basement, and syn-sedimentary sections).
- Bracketing the chronology/duration of the environmental events through recovery of sedimentary sections containing carbonates at locations permanently above the carbonate compensation depth, or CCD (e.g., Magellan Rise).

New Approaches to Drilling of LIPs

Global understanding of LIPs will benefit greatly from new approaches, with drilling as a key investigative tool. Such new approaches, detailed below, include strengthening collaboration with industry; more integrative, multidisciplinary studies; development and deployment of new technologies; and coordinated IODP-ICDP investigations.

Industry Connections to Drill LIPs

Drilling of LIPs is of interest to the hydrocarbon industry to understand the fundamental processes, in time and space, involved in LIP development and evolution. In particular, interest focuses on how these processes influence the thermal, structural, depositional, and vertical motion histories of adjacent sedimentary basins (i.e., how LIPs influence the generation, maturation, and migration of hydrocarbons). Among the various LIP categories, the primary industry interest is magma-dominated divergent continental margins and, to a lesser degree, oceanic plateaus underlain, in whole or in part, by continental crust. Specific LIPs (e.g., the Norwegian, northwest Australian, and South Atlantic margins) are in regions of interest to the hydrocarbon industry, and ties with industry should be developed to the maximum extent possible.

Collaborative IODP–industry LIP investigations could take the form of industry–academia consortia established to address topics of interest to both industry and members of the IODP community using any of the three IODP platforms. Industry contributions to such consortia could include site surveying and seismic processing/reprocessing of existing seismic data using sophisticated techniques, as well as financial support for planning, execution, and interpretation phases of drilling programs.

Joint IODP–industry drilling ventures should address highly ranked scientific objectives, and the resulting data should be public domain, or offered to the community after a brief participant-exclusive period. Joint, dedicated LIP scientific drilling ventures provide a means to expand opportunities for scientific drilling while maintaining IODP scientific integrity.

An Integrated Approach to Drilling LIPs

LIP science will be advanced in five key areas by drilling, with objectives and potential drilling sites outlined below.

Obtaining deep sections within multiple LIPs to examine magmatic (and therefore mantle source) variability through time. This will require offset drilling along a rifted LIP margin or into a deep erosional feature within a LIP. Potential locations include (a) conjugate rifted margins of the Kerguelen Plateau and Broken Ridge, Indian Ocean; (b) conjugate rifted margins of the Hikurangi and Manihiki plateaus, Pacific Ocean; (c) Danger Islands Troughs of the Manihiki Plateau, Pacific Ocean; (d) proposed conjugate rifted margins of the Ontong Java Plateau bordering the Stewart Basin, Pacific Ocean; (d) Kroenke Canyon of the Ontong Java Plateau, Pacific Ocean; and (e) flanks of the TAMU Massif on the Shatsky Rise, Pacific Ocean. Note that as one of the oldest oceanic plateaus, Shatsky Rise is an important drilling target for increasing our understanding of the processes that form LIPs as well as how they evolve over time (e.g., subsidence history, secondary volcanism), although basalts from the feature are characterized by MORB-type isotopic signatures (Mahoney et al., 2005).

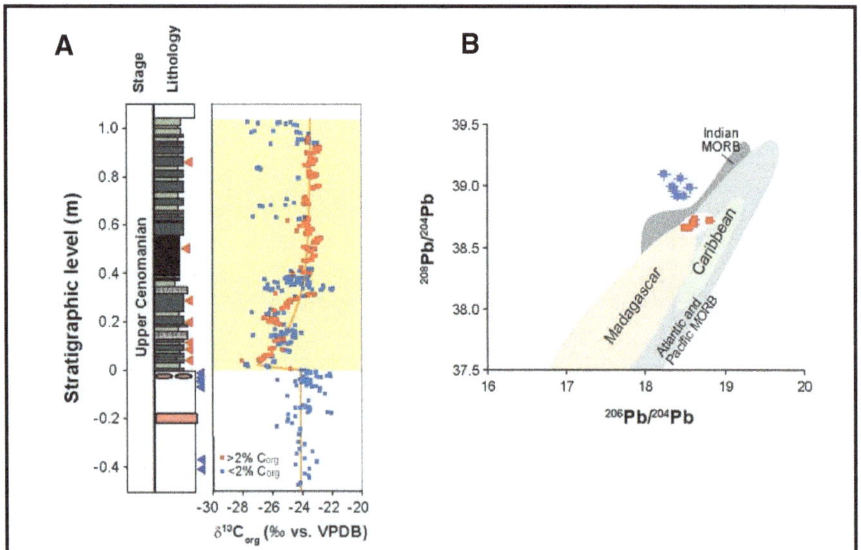

Figure 10. [A] A high-resolution profile of carbon isotopic composition of total organic matter from OAE-2 section from central Italy (Bonarelli Event, colored interval). Data from organic-rich (>2% total organic carbon content) and -poor (<2%) sediment are indicated by red and blue symbols, respectively. The profile indicates negative excursion at the base of the OAE-2. [B] A cross-plot of $^{206}Pb/^{204}Pb$ vs. $^{208}Pb/^{204}Pb$ in the Bonarelli (red symbols) and underlying Cenomanian limestone (blue symbols). For comparison, Pb isotopic compositions of basaltic rock from Caribbean (88–95 Ma), Madagascar flood basalts (88 Ma), and from MORB (present) from the Atlantic, Pacific, and Indian oceans are also shown. Both figures are referred from Kuroda et al. (2007).

Defining the nature of melting anomalies (i.e., compositional vs. thermal) that produce LIPs. Understanding the underlying mechanics and dynamics of melting anomalies can be tested where basalt composition can be related to crustal thickness or where there is evidence for anomalous mantle beneath a LIP, such as (a) diachronous V-shaped ridges around the Reykjanes Ridge south of Iceland (Fig. 3) in the North Atlantic (see section Mantle Geodynamics above for rationale); and (b) Ontong Java's High Plateau, underlain by a 300-km-deep "root" of seismically anomalous mantle that has been postulated to represent the fossil plume head of the OJP (Richardson et al., 2000; Klosko et al., 2001; Gomer and Okal, 2003). By determining heat flow from drill holes above the interpreted fossil plume head as well as away from it, the nature of the melting anomaly can be tested, as numerical models suggest that it should retain a detectable thermal signature.

Defining precise durations of oceanic LIP events. Two obvious ways to bracket LIP events are to (a) drill through the oldest and youngest eruptive sequences of a LIP, and (b) core a syn-LIP sedimentary sequence in an older proximal basin. Pursuing option (a) is not feasible by drilling through the entire eruptive sequence of a LIP, but if the age final eruption can be determined, the age of the start of LIP formation can be approximated by drilling through the lava flow sequence at the feather (distal) edge of a LIP. Option (b) will be pursued by drilling syn-LIP sedimentary sequences and analyzing for both age-dateable ash layers and chemical anomalies that are related to LIP formation and/or ash layers (taking into account the varying residence times in ocean water of different elements). Ideally, both options will be pursued. The feather edge of most oceanic LIPs may be drilled providing that it is distal to any volcanic vents and is not tectonic in nature. Examples of syn-LIP sedimentary sections in proximal basins include the following:

Kerguelen Plateau: Perth Basin off SW Australia, Enderby Basin and Princess Elizabeth Trough between Kerguelen and Antarctica;

Deccan Traps: Western and Northern Somali Basin west of Seychelles, where Deccan basalts crop out;

Agulhas Plateau: Transkei Basin between the Agulhas and Mozambique plateaus;

North Atlantic Volcanic Province: central and northern North Atlantic, ideally recovering sections through the Paleocene-Eocene Thermal Maximum event;

Shatsky Rise: Northwest Pacific Basin west of Shatsky Rise;

Ontong Java Plateau: Nauru Basin west of, and East Mariana and Pigafetta basins north of, the Ontong Java Plateau.

Moreover, the ~145-Ma Magellan Rise has a carbonate, chert, and black shale (OAE) section (Winterer et al., 1973) encompassing the formations of the Ontong Java, Manihiki, and Hikurangi plateaus. Similarly, the crests of Late Jurassic and Early Cretaceous seamounts in the western Pacific preserve syn-sedimentary sections deposited above the CCD (i.e., carbonate sediments).

Defining modes of eruption-constant effusion over several million years or several large pulse events over the same time interval. This will be achieved by a) age dating of discrete ash layers in syn-sedimentary sections, b) drilling through the feather edge of a LIP reached by only the largest flows—test for age progression, and c) drilling through basement reflections interpreted to represent alternating thin and thick flows (Inoue et al., 2008).

Establishing relationships among oceanic LIPs, OAEs, and other major environmental changes (e.g., ocean acidification and fertilization). Late Jurassic and Cretaceous OAEs are known to be approximately synchronous with LIP events (Jones and Jenkyns, 2001). Syn-sedimentary sections containing OAE, and bounding intervals are critical for analyses of elemental and isotopic variations associated with OAEs. These data can be compared with similar data from synchronous LIPs. Recovery of OAE intervals at multiple locations around an oceanic LIP allows directionality of fluxes to be evaluated. Knowledge of the duration of the LIP event is required for these studies (see above).

Technology and LIP Drilling

Advances in drilling technology will improve our understanding of LIP origin, emplacement, and environmental impacts dramatically. Specifically, technologies that should either be developed or implemented by the IODP that will advance our understanding of LIPs significantly include the following:

Enhanced recovery of syn-sedimentary sections, especially those with alternating hard-soft (e.g., chert-chalk) layers. To date, recovery of intercalated hard/soft sediment from the Pacific and Indian oceans has been exceedingly difficult during the DSDP, ODP, and IODP, and poor core recovery precludes recovery of important syn-sedimentary sections.

Sidewall coring, important for recovering soft sediment from alternating hard-soft layers. OAEs encompass a maximum of 150 cm of vertical section, although recovery of underlying and overlying sediment is necessary as well for biostratigraphic dating each OAE.

Oriented cores. Linking hotspots to LIPs, especially in the Pacific, is hampered because unoriented cores yield only paleolatitude information. Oriented cores are critical for determining sediment magnetostratigraphy in low latitudes, investigating geomagnetic field behavior, studying plate motions, and establishing flow directions of lavas.

Riser drilling in >2500 m of water. This will open opportunities for drilling through the feather edges of LIPs to the basement or sediment beneath, thereby bracketing the durations of LIP events.

Collaborations between IODP and ICDP

LIPs are equally well manifested in the oceans as on land. Questions such as the nature of the mantle source of volcanic plateaus, the ascent of magma from source to surface, magma fluxes through the life of a plateau, and the impact of LIP emplacement on global climate are best treated by a joint IODP/ICDP approach. The lessons learned from drilling on variably well exposed continental flood basalts can then be applied to the study of their more spectacular, but less accessible, oceanic counterparts. Proxies of climate change, measured in oceanic and continental sequences, provide information on the contrasting impact of the wholly submarine emplacement of an oceanic plateau and the open-to-the-atmosphere impact of continental volcanism. A key target of ICDP drilling should be the sill complexes that presumably underlie most LIPs. These complexes, relatively inaccessible in ocean basins, are important for four reasons. 1) They are an important element in the magmatic plumbing of each LIP. 2) Volatile-release at sill-sediment contacts contributes greatly to climate impact. 3) Valuable deposits of Ni, Cu, and Pt-group elements are located in these sills. 4) Intrusions in sedimentary basins influence the maturation of petroleum deposits and complicate exploration for such deposits. An understanding of the sill complexes, therefore, has important economic implications, in both continental and oceanic settings. Most importantly, unique and promising opportunities exist for combined IODP/ICDP drilling of the same LIP (e.g., onshore and offshore sections of the North Atlantic Volcanic Province; the Parana-Etendeka flood basalts (South Atlantic); the Deccan Traps-Seychelles Bank dikes (Indian Ocean); *in situ* and obducted (Caribbean, Central and South America) Caribbean flood basalts; Alpha Ridge and the High Arctic LIP; and the Ontong Java Plateau and obducted sections thereof in the Solomon Islands (Pacific Ocean).

Conclusions

The LIPs workshop was highly successful in showing that studying LIPs requires an integrated approach involving mantle geodynamics, plume modeling, petrology, environmental impacts, paleoceanography, physical volcanology, micropaleontology, geophysics, and tectonics. The workshop also concluded that oceanic LIPs must be studied in concert with their continental counterparts to better understand emplacement mechanism and environmental effects of their emplacement. A number of conceptual drilling targets and prospective regions were identified. In addition, areas where technology development was needed were highlighted, as were potential LIP-focused IODP-industry and ICDP/IODP collaborations. The result of this workshop will allow focused IODP LIP drilling proposals to be developed.

Acknowledgements

The authors are grateful for input to this paper by the workshop participants as well as contributors who did not attend the workshop. We thank the IODP and the Consortium for Ocean Leadership for funding. Paul Lyle contributed greatly to the logistics of the workshop and led the field trip to the Causeway Coast featuring outcrops of the North Atlantic Volcanic Province. We appreciate the reviews of this paper by Jan Behrmann, Christian Koeberl, and Hans Christian Larsen. We thank Kelly Kryc, Therese Lowe, and Charna Meth for organizing the workshop.

References

Alibert, C., 1991. Mineralogy and geochemistry of a basalt from Site 738: Implications for the tectonic history of the southernmost part of the Kerguelen Plateau. *In* Barron, J., Larsen, B., et al., *Proc. ODP, Sci. Res., 119*. College Station, Texas (Ocean Drilling Program), 293–298.

Andrews, J.F., Packham, G., et al., 1975. Initial Reports of the Deep Sea Drilling Project, 30, Washington, DC (U.S. Government Printing Office), 753 pp.

Bains, S., Corfield R.M., and Norris R.D., 1999. Mechanisms of climate warming at the end of the Paleocene. *Science*, 285:724–727.

Barron, J., Larsen, B., et al., 1989. *Proc. ODP, Init. Rep., 119*. College Station, Texas (Ocean Drilling Program), 942 pp.

Berger, W.H., Kroenke, L.W., Mayer, L.A., et al., 1993. *Proc. ODP, Sci. Res., 130*. College Station, Texas (Ocean Drilling Program), 867 pp.

Berner, R.A. and Kothavala Z., 2001. GEOCARB III: a revised model of atmospheric CO2 over Phanerozoic time. *Am. J. Sci.*, 301:182–204.

Bryan, S.E., and Ernst, R.E., 2008. Revised definition of Large Igneous Provinces (LIPs). *Earth-Science Rev.*, 86:175–202.

Bryan, S.E, Riley, T.R., Jerram, D.A., Leat, P.T., and Stephens, C.J., 2002. Silicic volcanism: an under-valued component of large igneous provinces and volcanic rifted margins. *In* Menzies, M.A., Klemperer, S.L., Ebinger, C.J., and Baker, J. (Eds.), *Magmatic Rifted Margins. Geological Society of America Special Paper*, 362:99–118.

Burke, K., and Torsvik, T.H., 2004. Derivation of Large Igneous Provinces of the past 200 million years from long-term heterogeneities in the deep mantle. *Earth Planet. Sci. Lett.*, 227:531–538.

Coffin, M.F., 1992. Emplacement and subsidence of Indian Ocean plateaus and submarine ridges. *In* Duncan, R.A., Rea, D.K., Kidd, R.B., von Rad, U., and Weissel, J.K. (Eds.), *Synthesis of Results from Scientific Drilling in the Indian Ocean, Geophysical Monograph* 70, Washington, DC (American Geophysical Union), 115–125.

Coffin, M.F., and Eldholm, O., 1994. Large igneous provinces: Crustal structure, dimensions, and external consequences. *Rev. Geophys.*, 32:1–36.

Coffin, M.F., Duncan, R.A., Eldholm, O., Fitton, J.G., Frey, F.A., Larsen, H.C., Mahoney, J.J., Saunders, A.D., Schlich, R., and Wallace, P.J., 2006. Large igneous provinces and scientific ocean drilling: Status quo and a look ahead. *Oceanography*, 19:150–160.

Coffin, M.F., Frey, F.A., Wallace, P.J., and the ODP Leg 183 Shipboard Scientific Party, 2000. Kerguelen Plateau – Broken Ridge: A large igneous province. *Proc.ODP*. College Station, Texas (Ocean Drilling Program), [CD-ROM].

Coffin, M.F., Pringle, M.S., Duncan, R.A., Gladczenko, T.P., Storey, M., Müller, R.D., and Gahagan, L.A., 2002. Kerguelen hotspot magma output since 130 Ma. *J. Petrol.*, 43:1121–1139.

Coffin, M.F., Neal, C.R., Duncan, R.A., Eldholm, O., Erba, E., Farnetani, C., Fitton, J.G., Ingle, S.P., Ohkouchi, N., Rampino, M.R., Reichow, M., Self, S., and Tatsumi, Y., 2007. Large igneous province workshop, *Eos*, 88:505, doi:10.1029/2007EO470009.

Courtillot, V., and Olson, P., 2007. Mantle plumes link magnetic superchrons to phanerozoic mass depletion events. *Earth Planet. Sci. Lett.*, 260:495–504.

Cox, K.G., 1992. Continental magmatic underplating. *Phil. Trans. Roy. Soc. Lond. A*, 342:155–166.

Davaille, A., 1999. Simultaneous generation of hotspots and superswells by convection in a heterogeneous planetary mantle. *Nature*, 402:756–760.

Davies, G.F., 1994. Thermomechanical erosion of the lithosphere by mantle plumes. *J. Geophys. Res.*, 99:15709–15722.

Duncan, R.A., Larsen, H.C., Allan, J.F., et al., 1996. *Proc. ODP, Init. Rep., 163*. College Station, Texas (Ocean Drilling Program), 279 pp.

Eldholm, O., and Coffin, M.F., 2000. Large Igneous Provinces and plate tectonics. *In* Richards, M., Gordon, R., and van der Hilst, R. (Eds.), *The History and Dynamics of Global Plate Motion, Geophysical Monogr.*, 121, Washington, DC (American Geophysical Union), 309–326.

Eldholm, O., and Thomas, E., 1993. Environmental impact of volcanic margin formation. *Earth Planet. Sci. Lett.*, 117:319–329.

Eldholm, O., Thiede, J., Taylor, E., et al., 1987. *Proc. ODP, Init. Rep., 104*. College Station, Texas (Ocean Drilling Program), 783 pp.

Eldholm, O., Thiede, J., and Taylor, E., 1989. Evolution of the Vøring volcanic margin. *In* Eldholm, O., Thiede, J., Taylor, E., et al. (Eds.), *Proc. ODP, Sci. Res., 104*. College Station, Texas (Ocean Drilling Program), 1033–1065.

Ernst, R.E., and Buchan, K.L., 2001. Large mafic magmatic events through time and links to mantle plume heads. *In* Ernst, R.E., and Buchan, K.L. (Eds.), *Mantle Plumes: Their Identification through Time. Geol. Soc. Am. Special Paper*, 352:483–575.

Farnetani, C.G., and Richards, M.A., 1994. Numerical investigation of the mantle plume initiation model for flood basalt events. *J. Geophys. Res.*, 99:13813–13883.

Farnetani, C.G., and Samuel, H., 2005. Beyond the thermal plume paradigm. *Geophys. Res. Lett.*, 32:L07311, doi:10.1029/2005GL022360.

Fitton, J.G., and Godard, M., 2004. Origin and evolution of magmas on the Ontong Java Plateau. *In* Fitton, J.G., Mahoney, J.J., Wallace, P.J., and Saunders, A.D. (Eds.), *Origin and Evolution of the Ontong Java Plateau. Special Publications*, 229, London, (Geological Society), 151–178.

Fitton, J.G., Mahoney, J.J., Wallace, P.J., and Saunders, A.D., 2004a. *Origin and Evolution of the Ontong Java Plateau. Special Publications*, 229, London (Geological Society), 374 pp.

Fitton, J.G., Mahoney, J.J., Wallace, P.J., and Saunders, A.D., 2004b. *Proc. ODP, Sci. Res.*, 192, College Station, Texas (Ocean Drilling Program), [CD-ROM].

Frey, F.A., Coffin, M.F., Wallace, P.J., and Quilty, P.J., 2003. *Proc. ODP, Sci. Res.*, 183. College Station, Texas (Ocean Drilling Program), [CD-ROM].

Frey, F.A., McNaughton, H.J., Nelson, D.R., Delaeter, J.R., and Duncan, R.A., 1996. Petrogenesis of the Bunbury basalt, western Australia: interaction between the Kerguelen plume and Gondwana lithosphere? *Earth Planet. Sci. Lett.*, 176:73–89.

Gaina, C., Müller, R.D., Brown, B., Ishihara, T., and Ivanov, S., 2007. Breakup and early seafloor spreading between India and Antarctica. *Geophys. J. Int.*, 170:151–169, doi:10.1111/j.1365-246X.2007.03450.x

Gomer, B.M., and Okal, E.A., 2003. Multiple-ScS probing of the Ontong-Java Plateau. *Phys. Earth Planet. Int.*, 138:317–331.

Herzberg, C., 2004. Partial melting below the Ontong Java Plateau. *In* Fitton, J.G., Mahoney, J.J., Wallace, P.J., and Saunders, A.D. (Eds.), *Origin and Evolution of the Ontong Java Plateau. Special Publications*, 229. London (Geological Society), 179–183.

Herzberg, C., Asimow, P.D., Arndt, N., Niu, Y., Lesher, C.M., Fitton, J.G., Cheadle, M.J., and Saunders, A.D., 2007. Temperatures in ambient mantle and plumes: Constraints from basalts, picrites and komatiites. *Geochem. Geophys. Geosyst.*, 8: Q02006, doi: 10.1029/2006GC001390.

Hill, R.I., 1991. Starting plumes and continental break-up. *Earth Planet. Sci. Lett.*, 104:398–416.

Hopper, J.R., Mutter, J.C., Larson, R.L., Mutter, C.Z., and Northwest Australia Study Group, 1992. Magmatism and rift margin evolution: evidence from northwest Australia. *Geology*, 20:853–857.

Ingle, S.P., and Coffin, M.F., 2004. Impact origin for the Greater Ontong Java Plateau? *Earth Planet. Sci. Lett.*, 218:123–134.

Inoue, H., Coffin, M.F., Nakamura, Y., Mochizuki, K., and Kroenke, L.W., 2008. Intrabasement reflections of the Ontong Java Plateau: Implications for Plateau construction. *Geochem. Geophys. Geosyst.*, 9:Q04014, doi:10.1029/2007GC001780.

Ito, G., and Clift, P., 1998. Subsidence and growth of Pacific Cretaceous plateaus. *Earth Planet. Sci. Lett.*, 161:85–100.

Ito, G., and Taira, A., 2000. Compensation of the Ontong Java Plateau by surface and subsurface loading. *J. Geophys. Res.*, 105:11171–11183.

Jones, C.E., and Jenkyns, H.C., 2003. Seawater strontium isotopes, oceanic anoxic events, and seafloor hydrothermal activity. *Amer. J. Sci.*, 111:112–149.

Jones, S.M., White, N., and Maclennan, J., 2002. V-shaped ridges around Iceland: Implications for spatial and temporal patterns of mantle convection. *Geochem. Geophys. Geosyst.*, 3(10):1059, doi:10.1029/2002GC000361.

Kennett, J.P., and Stott, L.D., 1991. Abrupt deep-sea warming, palaeoceanographic changes, and benthic extinctions at the end of the Palaeocene. *Nature*, 353:225–229.

Kerr, A.C., 1998. Oceanic plateau formation: a cause of mass extinction and black shale deposition around the Cenomanian-Turonian boundary? *J. Geol. Soc. Lond.*, 155:619–626.

Kerr, A.C., Tarney, J., Marriner, G.F., Nivia, A., and Saunders, A., 1997. The Caribbean-Colombian Cretaceous igneous province: the internal anatomy of an oceanic plateau. *In* Mahoney, J.J., and Coffin, M.F. (Eds.), *Large Igneous Provinces: Continental, Oceanic, and Planetary Flood Volcanism. Geophysical Monograph, 100*. Washington, DC (American Geophysical Union), 123–144.

King, S.D., and Anderson, D.L., 1998. Edge-driven convection. *Earth Planet. Sci. Lett.*, 160:289–296.

Klosko, E.R., Russo, R.M., Okal, E.A., and Richardson, W.P., 2001.

Evidence for a rheologically strong chemical mantle root beneath the Ontong-Java Plateau. *Earth Planet. Sci. Lett.*, 186:347–361.

Korenaga, J., 2004. Mantle mixing and continental breakup magmatism. *Earth Planet Sci. Lett.*, 218:463–473, doi:10.1016/S0012-821X(03)00674-5.

Korenaga, J., 2005. Why did not the Ontong Java Plateau form subaerially? *Earth Planet. Sci. Lett.*, 234:385–399.

Kroenke, L.W., Berger, W.H., Janecek, T.R., et al., 1991. *Proc. ODP, Init. Rep.*, 130. College Station, Texas (Ocean Drilling Program), 1240 pp.

Kuroda, J., Ogawa, N.O., Tanimizu, M., Coffin, M.F., Tokuyama, H., Kitazato, H., and Ohkouchi, N., 2007. Contemporaneous massive subaerial volcanism and late Cretaceous Oceanic Anoxic Event 2. *Earth Planet. Sci. Lett.*, 256:211–223.

Larsen, H.C., Duncan, R.A., Allan, J.F., and Brooks, K., 1999. *Proc. ODP, Sci. Res.*, 163. College Station, Texas (Ocean Drilling Program), 173 pp.

Larsen, H.C., Saunders, A.D., Clift, P.D., et al., 1994. *Proc. ODP, Init. Rep.*, 152. College Station, Texas (Ocean Drilling Program), 977 pp.

Larson, R.L., 1991. Latest pulse of Earth: evidence for a mid-Cretaceous superplume. *Geology*, 19:547–550.

Larson, R.L., and Erba, E., 1999. Onset of the mid-Cretaceous greenhouse in the Barremian-Aptian events: Igneous events and the biological, sedimentary, and geochemical responses. *Paleoceanogr.*, 14:663–678.

Leckie, R.M., Bralower, T.J., and Cashman, R., 2002. Oceanic anoxic events and plankton evolution: Biotic response to tectonic forcing during the mid-Cretaceous. *Paleoceanogr.*, 17(3):10.1029/2001PA000623.

Mahoney, J.J., and Coffin, M.F., 1997. *Large Igneous Provinces: Continental, Oceanic, and Planetary Flood Volcanism. American Geophysical Union Geophysical Monograph 100.* Washington, DC (American Geophysical Union), 438 pp.

Mahoney, J.J., Duncan, R.A., Tejada, M.L.G., Sager, W.W., and Bralower, T.J., 2005. Jurassic-Cretaceous boundary age and mid-ocean-ridge-type mantle source for Shatsky Rise. *Geology*, 33:185–188, doi:10.1130/G21378.1.

Mahoney, J.J, Fitton, J.G., and Wallace, P.J. et al., 2001. Basement drilling of the Ontong Java Plateau, Sites 1183–1187, 8 September–7 November, 2000. *Proc. ODP Init. Rep.*, 192. College Station, Texas (Ocean Drilling Program), 75p.

Miller, K.G., Kominz, M.A., Browning, J.V., Wright, J.D., Mountain, G.S., Katz, M.E., Sugarman, P.J., Cramer, B.S., Christie-Blick, N., and Pekar, S.F., 2005. The Phanerozoic record of global sea-level change. *Science*, 310:1293–1298.

Mutter, J.C., Buck, W.R., and Zehnder, C.M., 1988. Convective partial melting: 1. A model for the formation of thick basaltic sequences during the initiation of spreading. *J. Geophys. Res.*, 93:1031–1048.

Nakagawa, T., and Tackley, P.J., 2005. The interaction between the post-perovskite phase change and a thermo-chemical boundary layer near the core-mantle boundary. *Earth Planet. Sci. Lett.*, 238:204–216.

Neal, C.R., Mahoney, J.J., Kroenke, L.W., Duncan, R.A., and Petterson, M.G., 1997. The Ontong Java Plateau. *In* Mahoney, J.J., and Coffin, M.F. (Eds.), *Large Igneous Provinces: Continental, Oceanic, and Planetary Flood Volcanism. Geophysical Monograph, 100.* Washington, DC (American Geophysical Union), 183–216.

Operto, S., and Charvis, P., 1995. Kerguelen Plateau: a volcanic passive margin fragment? *Geology*, 23:137-140, doi:10.1130/0091-7613(1995)023<0137:KPAVPM>2.3.CO;2

Parsons, B., and Sclater, J.G., 1977. An analysis of the variation of ocean floor bathymetry and heat flow with age. *J. Geophys. Res.*, 82:803–827.

Peate, I.U., Baker, J.A., Kent, A.J.R, Al-Kadasi, M., Al-Subbary, A., Ayalew, D., and Menzies, M., 2003. Correlation of Indian Ocean tephra to individual Oligocene silicic eruptions from Afro-Arabian flood volcanism. *Earth Planet. Sci. Lett.*, 211:311–327.

Planke, S., Symonds P.A., Alvestad E., and Skogseid, J., 2000. Seismic volcanostratigraphy of large-volume basaltic extrusive complexes on rifted margins. *J. Geophys. Res.*, 105:19335–19351, doi:10.1029/1999JB900005.

Poore, H.R., Samworth, R., White, N.J., Jones, S.M., and McCave, I. N., 2006. Neogene overflow of Northern Component Water at the Greenland-Scotland Ridge. *Geochem. Geophys. Geosyst.*, 7:Q06010, doi:10.1029/2005GC001085.

Rampino, M.R., and Stothers, R.B., 1988. Flood basalt volcanism during the past 250 million years. *Science*, 241:663–668.

Reymer, A., and Schubert, G., 1984. Phanerozoic addition rates to the continental crust and crustal growth. *Tectonics*, 3:63–77.

Richardson, W.P., Okal, E.A., and van der Lee, S., 2000. Rayleigh-wave tomography of the Ontong Java Plateau. *Phys. Earth Planet. Int.*, 118:29–51.

Roberge, J., Wallace, P., White, R.V., and Coffin, M.F., 2005. Anomalous subsidence of the Ontong Java Plateau inferred from CO2 contents of submarine basaltic glasses. *Geology*, 33:501–504.

Roberts, D.G., Schnitker, D., et al., 1984. *Initial Reports, Deep Sea Drilling Project, Leg 81*, Washington, D.C. (U.S. Government Printing Office), 923 pp.

Sano, T., and Yamashita, S., 2004. Experimental petrology of basement lavas from Ocean Drilling Program Leg 192: implications for differentiation processes in Ontong Java Plateau magmas. *In* Fitton, J.G., Mahoney, J.J., Wallace, P.J., and Saunders, A.D. (Eds.), *Origin and Evolution of the Ontong Java Plateau. Special Publications, 229.* London, (Geological Society), 185–218.

Saunders, A.D., Larsen, H.C., and Wise, S.W., Jr., 1998. *Proc. ODP, Sci. Res.*, 152. College Station, Texas (Ocean Drilling Program), 554 pp.

Schlanger, S.O., and Jenkyns, H.C., 1976. Cretaceous anoxic events: causes and consequences. *Geologie en Mijnbouw*, 55: 179-184.

Schlich, R., Wise, S.W., Jr., et al., 1989. *Proc. ODP, Init. Rep.*, 120. College Station, Texas (Ocean Drilling Program), 648 pp.

Self, S., Keszthelyi, L., and Thordarson, T., 1998. The importance of pahoehoe. *Ann. Rev. Earth Planet. Sci.*, 26:81–110.

Self, S., Widdowson, M., Thordarson, T., and Jay, A.E., 2006. Volatile fluxes during flood basalt eruptions and potential effects on the global environment: A Deccan perspective. *Earth Planet. Sci. Lett.*, 248:517–531.

Sepkoski, J.J., 1996. Patterns of Phanerozoic extinction: a perspective from global data bases. *In* Walliser, O.H. (Ed.), *Global Events and Event Stratigraphy.* Berlin (Springer), 35–51.

Snow, L.J., Duncan, R.A., and Bralower, T.J., 2005. Trace element abundances in the Rock Canyon Anticline, Pueblo, Colorado, marine sedimentary section and their relationship to Caribbean plateau construction and oxygen anoxic event 2. *Paleoceanogr,* 20:PA3005, doi:10.1029/2004PA001093.

Srivastava, S.P., and Tapscott, C.R., 1986. Plate kinematics of the North Atlantic. *In* Tucholke, B.E., Vogt, P.R. (Eds.), The Geology of North America. *The Western Atlantic Region, DNAG Series vol. M.* Boulder, Colo. (Geological Society of America), ,379–404.

Stagg, H.M.J., Colwell, J.B., Borissova, I., Ishihara, T., and Bernardel, G., 2006. The Bruce Rise Area, East Antarctica: formation of a continental margin near the Greater India-Australia-Antarctica triple junction. *Terra Antartica,* 13:13–22.

Stein, C.A., and Stein, S., 1992. A model for the global variation in oceanic depth and heat flow with lithospheric age. *Nature,* 359:123–129.

Storey, M.S., Kent, R., Saunders, A.D., Salters, V.J., Hergt, J., Whitechurch, H., Sevigny, J.H., Thirlwall, M.F., Leat, P., Ghose, N.C., and Gifford, M., 1992. Lower Cretaceous volcanic rocks on continental margins and their relationship to the Kerguelen Plateau. *In* Wise, S.W., Jr., Schlich, R., et al. (Eds.), *Proc. ODP, Sci. Res., 120.* College Station, Texas (Ocean Drilling Program), 33–53.

Stothers, R.B., 1998. Far reach of the tenth century Eldgjá eruption, Iceland. *Climatic Change,* 39:715–726.

Svensen, H., Planke, S., Malthe-Sorenssen, A., Jamtveit, B., Myklebust, R., Eidem, T.R., and Rey, S.S., 2004. Release of methane from a volcanic basin as a mechanism for initial Eocene global warming. *Nature,* 429:542–545.

Symonds, P.A., Planke, S., Frey, Ø., and Skogseid, J., 1998. Volcanic evolution of the Western Australian continental margin and its implications for basin development. *In* Purcell, P.G.R.R. (Ed.), *The Sedimentary Basins of Western Australia 2: Proceedings of the PESA Symposium.* Perth, Australia (Petroleum Exploration Society of Australia), 33–54.

Tackley, P.J., 1998. Three-dimensional simulations of mantle convection with a thermo-chemical basal boundary layer: D''? *In* Gurnis, M. (Ed.), *The Core-mantle Boundary Region, Geophys. Monogr. Ser., 28.* Washington, DC (American Geological Union), 231–253.

Tajika, E., 1998. Climate change during the last 150 million years: reconstruction from a carbon cycle model. *Earth Planet. Sci. Lett.,* 160:695–707.

Thomas, E., 1992. Cenozoic deep-sea circulation: Evidence from deep-sea foraminifera. *In* Kennet, J.P., and Warnke, D. (Eds.), *The Antarctic Paleoenvironment: A Perspective on Global Change, Antarct. Res. Ser. 56.* Washington, D.C. (American Geological Union), 141–165.

Thordarson, T., 2004. Accretionary-lapilli-bearing pyroclastic rocks at ODP Leg 192 Site 1184: A record of subaerial phreatomagmatic eruptions on the Ontong Java Plateau. *In* Fitton, J.G., et al., (Eds.), *Origin and Evolution of the Ontong Java Plateau. Spec. Pub. 229.* London (Geological Society), 275–306.

Winterer, E.L., Ewing, J.I., et al., 1973. *Init.Rep. Deep Sea Drill. Proj.* Washington, DC (U.S. Government Printing Office), 17:1–931.

Wise, S.W., Jr., Schlich, R., et al., 1992. *Proc. ODP, Sci. Res.,* 120. College Station, Texas (Ocean Drilling Program), 1155 pp.

Wright, J.D., and Miller, K., 1996. Control of North Atlantic Deep Water circulation by the Greenland-Scotland Ridge. *Paleoceanogr.,* 11:157–170.

Authors

Clive R. Neal, Department of Civil Engineering and Geological Sciences, 156 Fitzpatrick Hall, University of Notre Dame, Notre Dame, Ind., 46556, U.S.A., e-mail: neal.1@nd.edu.

Millard F. Coffin, National Oceanography Centre, Southampton, University of Southampton Waterfront Campus, European Way, Southampton SO14 3ZH, U.K.

Nicholas T. Arndt, Laboratoire de Géodynamique des Chaînes Alpines, Université de Grenoble, Grenoble, France.

Robert A. Duncan, College of Oceanic and Atmospheric Sciences, 104 COAS Admin Bldg., Oregon State University, Corvallis, Ore. 97331, U.S.A.

Olav Eldholm, Department of Earth Science, The University of Bergen, Allegaten 41, N-5007, Bergen, Norway.

Elisabetta Erba, Dipartimento di Scienze della Terra, University of Milan, via Mangiagalli, 34, 20133 Milano, Italy.

Cinzia Farnetani, Laboratoire de Dynamique des Fluides Géologiques, Institut de Physique du Globe, Boite 89, 4 pl. Jussieu, 75252 Paris, France.

J. Godfrey Fitton, School of Geosciences, 210 Grant Institute, University of Edinburgh, The King's Buildings, West Mains Road, Edinburgh EH9 3JW, U.K.

Stephanie P. Ingle, National Oceanography Centre, Southampton, University of Southampton Waterfront Campus, European Way, Southampton SO14 3ZH, U.K.

Nao Ohkouchi, Institute for Research on Earth Evolution, Japan Agency for Marine-Earth Science Technology, 2-15 Natsushima-Cho, Yokosuka, 237-0061, Japan.

Michael R. Rampino, Earth & Environmental Science Program, New York University, 1009 Silver Center, 100 Washington Square East, New York, N.Y. 10003, U.S.A.

Marc K. Reichow, Department of Geology, University of Leicester, University Road, Leicester, LE1 7RH, U.K.

Stephen Self, Department of Earth Sciences, The Open University, Milton Keynes MK7 6BJ, U.K.

Yoshiyuki Tatsumi, Institute for Research on Earth Evolution, Japan Agency for Marine-Earth Science Technology, Natsushima-cho 2-15, Yokosuka 237-0061, Japan.

Drilling to Decipher Long-Term Sea-Level Changes and Effects—A Joint Consortium for Ocean Leadership, ICDP, IODP, DOSECC, and Chevron Workshop

by Craig S. Fulthorpe, Kenneth G. Miller, André W. Droxler, Stephen P. Hesselbo, Gilbert F. Camoin, and Michelle A. Kominz

Introduction

One of the most societally relevant objectives of the Earth sciences is to understand the history and impact of global sea-level (eustatic) fluctuations at different timescales. Over a third of the world's population lives within 100 km of a coastline. One-tenth of the global population and thirteen percent of the world's urban population live in coastal areas that lie within just 10 m above sea level (the Low Elevation Coastal Zone or LECZ), which covers only two percent of the world's land area (McGranahan et al., 2007). Reconstruction of global mean sea level since 1870 indicates a twentieth century rate of sea-level rise of 1.7 ± 0.3 mm yr^{-1} and a significant acceleration of sea-level rise of 0.013 ± 0.006 mm yr^{-2} (Church and White, 2006), in part due to anthropogenic influences. Satellite observations in the last decade show that the rates have increased since 1993 to 3.3 ± 0.4 mm yr^{-1} (Cazenave and Nerem, 2004). Remote-sensing data suggest that ice sheets currently contribute little to sea-level rise. Best estimates are that sea level could rise by as much as 50 cm in the next 100 years (IPCC, 2007). However, dynamical instabilities in response to climate warming may cause faster ice-mass loss (Cazenave, 2006). Rahmstorf et al (2007) show that sea-level observations are tracking at the high end of the IPCC estimates and conclude that 80 cm, and perhaps >1 m, is the most likely global rise by 2100. In some of the most heavily populated areas (e.g., the U.S. Atlantic seaboard) relative sea-level rise exceeds 4 mm yr^{-1} (Psuty and Collins, 1996) due to combined effects of global sea-level rise and subsidence. While such rates are gradual on a human timescale, the geological record shows that they can increase rapidly and dramatically (e.g., >2 m in a century; Fairbanks, 1989; Bard et al., 1990); in addition, the retreat of shorelines can be erratic and rapid even under conditions of moderate global rises of sea level.

The geologic record provides an opportunity to quantify the timing, amplitudes, rates, mechanisms/controls, and effects (stratigraphic response) of eustatic change (Figs. 1 and 2). This information, in turn, provides a baseline for predicting future global sea-level changes and assessing anthropogenic influences. In order to understand the effects of potential future eustatic trends, it is vital to document how the Earth system has operated during past abrupt climate changes (e.g., the last and penultimate deglaciations) and under past conditions of extreme climate forcing, and to constrain the eustatic response to elevated CO_2 levels. For example, determining how sea level varied in response to past intervals of global warming—e.g., marine isotope chrons 5e (Thompson and Goldstein, 2005), 11 (Droxler et al., 2003), 31 (Scherer et al., 2008); "mid " Pliocene warmth (Draut et al., 2003), the middle Miocene climate optimum, the early Eocene (Zachos et al., 2001), and the Late Cretaceous (Abreu et al., 1998; Miller et al., 2005 a, b; Bornemann et al., 2008)—will provide a means to evaluate the eustatic impact of future climate trends. Understanding how processes/mechanisms yield specific eustatic responses will therefore improve our understanding of the societal impact of the resulting sea-level changes. Furthermore, understanding how process interactions produce the preserved stratigraphy of beds and sequences is fundamental to deciphering the long-term geologic and climatic history recorded by sediments in a variety of marine sedimentary basins. These environments are also economically and strategically important—testing predictive sequence models has a proven potential for identifying oil and gas resources and for ground water/pollution remediation issues. Such research also helps to achieve the long-sought goal of predicting margin lithologies in the absence of drilling, a concept pioneered by the Exxon group (Vail and Mitchum, 1977). Finally, constraining the history of sea-level change provides data of direct use to researchers in other disciplines because of the relationships between

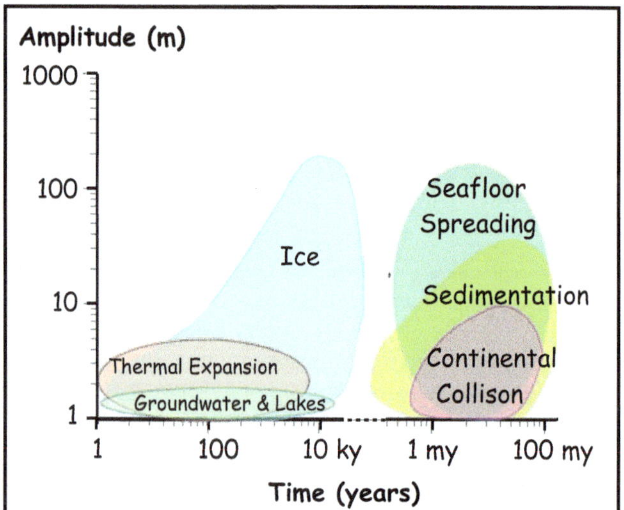

Figure 1. The mechanisms that generate eustatic change operate on different timescales and generate different magnitudes of sea-level change. These variations in conjunction with proxy data may be used to determine the causal mechanisms of eustatic change. (Modified from Miller et al., 2005a.)

eustasy and ice-sheet growth and decay, nutrients and ocean productivity, carbon storage, and ocean chemistry.

The challenge is considerable because eustatic effects are complexly intertwined with processes of basin subsidence and sediment supply (Cloetingh et al., 1985; Karner, 1986; Posamentier et al., 1988; Christie-Blick et al., 1990; Reynolds et al., 1991; Christie-Blick and Driscoll, 1995; Kominz et al., 1998; Kominz and Pekar, 2001). Extracting the eustatic signal requires integrated onshore/offshore drilling transects involving global retrieval of cores representing multiple timeframes and depositional settings, including siliciclastic, carbonate, and mixed systems (Fig. 3). Fundamental to the approaches recommended by our workshop are as follows: 1) to enhance our understanding of eustatic timing, amplitudes, rates, and stratigraphic response during the icehouse period, when glacioeustasy is known to be the principal eustatic mechanism, and; 2) to begin an aggressive program to understand the mechanisms responsible for greenhouse eustasy and how they relate to climatic trends and stratigraphic response.

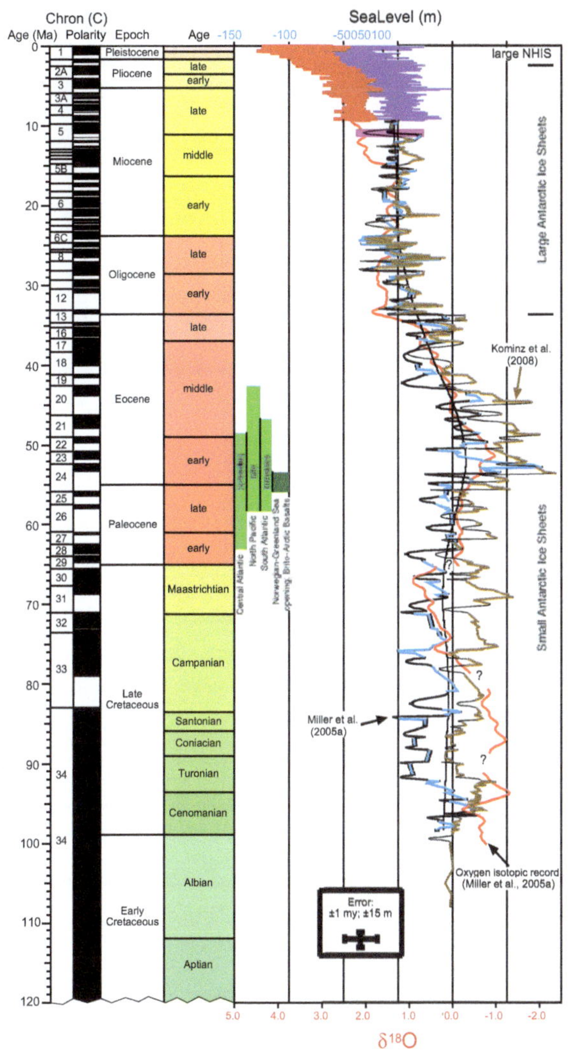

Figure 2. Global sea level (light blue) for the interval 7–100 Ma derived by back-stripping five New Jersey coastal plain core holes (Miller et al., 2005a). Revised back-stripped sea-level curve (brown) based on a total of eleven New Jersey core holes (Kominz et al., 2008). Global sea level (purple) for the interval 0–7 Ma derived from $\delta^{18}O$, shown in red for a benthic foraminiferal $\delta^{18}O$ synthesis from 0–100 Ma with the scale on the bottom axis (in parts per thousand, Miller et al., 2005a). The Miller et al. (2005a) back-stripped sea-level curve was smoothed with a 21-point Gaussian convolution filter to generate the smooth black curve. The pink box at 11 Ma is a sea-level estimate derived from the Marion Plateau (John et al., 2004). Light green boxes indicate times of spreading rate increases on various ocean ridges (Cande and Kent, 1992). Dark green box indicates the opening of the Norwegian-Greenland Sea and concomitant extrusion of basalts (Modified from Browning et al., 2008).

Salt Lake City Workshop

Various groups related to the Ocean Drilling Program (ODP) have developed strategies for studying eustasy on orbital (>19 kyr) and longer timescales (Imbrie et al., 1987; Watkins and Mountain, 1990; JOIDES, 1992). These strategies have begun to be implemented with drilling transects across the New Jersey margin (ODP Legs 150, 150X, 174A, and 174AX), the Bahamas (Leg 166 and mission-specific platform sites) and a targeted sea-level amplitude experiment on the Marion Plateau, Northeast Australia (Leg 194). However, an effective, coordinated strategy requires that additional margin transects be drilled. In addition, it has been fifteen years since the last of these groups; the Sea-Level Working Group (JOIDES, 1992), discussed goals and strategies of sea-level research. Recent drilling advances, including the use of mission-specific platforms (MSP) offshore and joint onshore-offshore drilling (e.g., IODP Expedition 313), together with new views on the roles of tectonics and sediment dynamics, required that the scientific community reevaluate the fundamental assumptions of sea-level studies.

As a follow-up to the SEALAIX Symposium ("Sea-Level Changes: Records, Processes and Modeling", September 2006, Presqu'île de Giens, France), an international workshop of more than fifty participants was held in Salt Lake City, Utah (8–10 October 2007) sponsored by Consortium for Ocean Leadership (formerly Joint Oceanographic Institutions), the International Continental Scientific Drilling Program (ICDP), the Integrated Ocean Drilling Program (IODP), Drilling, Observation and Sampling of the Earth's Continental Crust (DOSECC), and Chevron. The purposes of the workshop were 1) to review results of ODP and early IODP drilling for sea-level objectives; 2) to reevaluate principles and strategies for constraining genetic links between eustatic change and Earth system processes and for defining the relative roles of eustasy versus local processes in building the stratigraphic record; and 3) to identify possible geographical areas and time-intervals for future IODP drilling transects. Presentations about IODP, ICDP, and DOSECC were followed by keynote scientific talks and a series of short, three-minute presentations by participants. Breakout groups subsequently focused on the relationship between recorded sea-level cyclicity and eustatic mechanisms through time, and on deciphering the stratigraphic response to eustasy through a sedimentary process approach in both

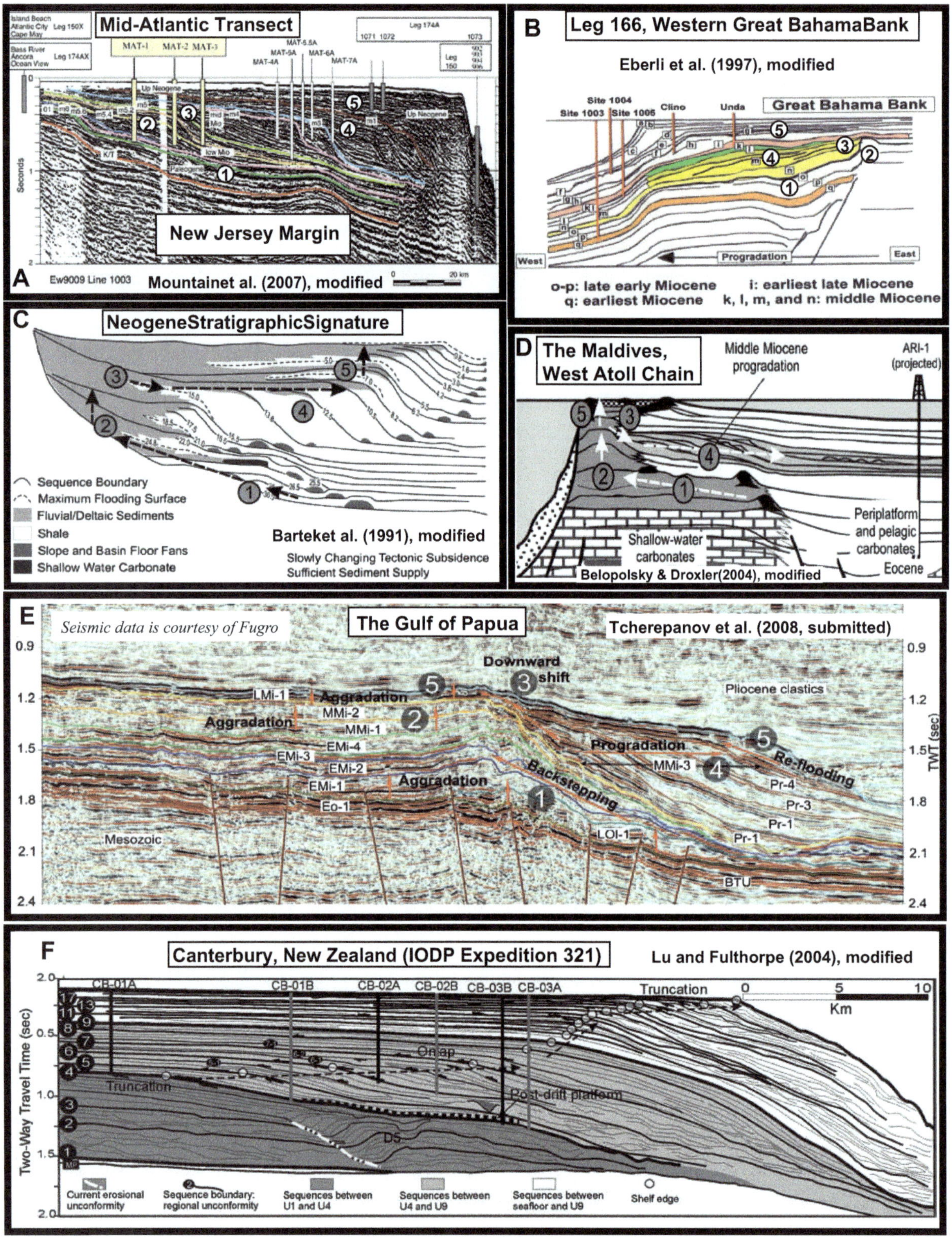

Figure 3. [A] Middle Atlantic Transect (MAT) off New Jersey, showing drillsites targeting a Paleogene-Pleistocene prograding clinoform succession. MAT has been drilled on the slope (ODP Leg 150; sites not shown) and shelf (Leg 174A; sites 1071–1073 along strike from this profile), as well as on the Coastal Plain (Legs 150X and 174AX). Inner shelf drilling (MAT 1-3) is planned as a Mission-Specific Platform IODP Expedition 213. [B] Line drawing of interpreted Great Bahama Bank sequences drilled during ODP Leg 166 (Sites 1003–1005) and the Bahamas Drilling Project (sites Clino and Unda). [C] The stratigraphic signature of the Neogene represented by: 1) late Oligocene-early Miocene aggradation, backstepping and partial drowning; 2) late early Miocene-early middle Miocene vertical growth or aggradation; 3) earliest middle Miocene downward shift of deposition; 4) late middle Miocene systematic lateral growth (progradation); and 5) late Miocene-early Pliocene re-flooding and aggradation (Bartek et al., 1991; Tcherepanov et al., 2008). [D] The Neogene stratigraphic signature along the West Maldives Inner Sea carbonate margin. [E] Neogene stratigraphic signature in the Gulf of Papua. [F] A future sea-level transect: line drawing of interpreted sequences, offshore Canterbury Basin, New Zealand, showing proposed IODP sites, scheduled for drilling as IODP Expedition 317.

icehouse and greenhouse worlds. Breakout Group Two was further subdivided into siliciclastic and carbonate groups. In addition to identifying scientific questions and objectives, the groups were also asked to consider drilling program design, potential target areas, and technology requirements (onshore and offshore).

Objective 1: Determining Eustatic Mechanisms

Understanding the mechanisms that drive eustatic change requires knowledge of the timing, amplitudes, and rates of global sea-level change (Fig. 1). It also requires information on climate and paleoceanography, mainly derived from proxy records (Fig. 4), and tectonic mechanisms that control the volume of the oceans. In turn, such quantification of eustatic change will contribute to other areas of the Earth sciences by helping to constrain such processes as ice-sheet growth and decay, ocean temperatures, carbon burial, and inorganic carbon precipitation in carbonates (Fig. 4) as well as global tectonism (Harrison, 1990). Sea level is important for the study of tectonic processes, because it is the datum against which vertical tectonic movements are measured.

1) Refining timing, amplitudes, and mechanisms of icehouse (Oligocene-Recent) eustatic change. ODP results to date have demonstrated that global sea-level changes over the past 42 myr can be explained, in part, by growth and decay of continental ice sheets (glacioeustasy; Miller et al., 1996; Eberli et al., 1997; Eberli, 2000). Such drilling has principally addressed the timing of sea-level change and has also determined that sequence boundaries indeed represent timelines as predicted in the sequence stratigraphic model (Eberli et al., 1997; Betzler et al., 2000).

However, our understanding of how climate change influences sea level, even during this "icehouse" period of large ice sheets, is incomplete. In particular, there are still uncertainties surrounding the hierarchy of eustatic and sequence periodicities, and particularly the origins of sequences with durations of >1 myr, which do not appear to conform to long-period (1.2 myr and 2.4 myr) astronomical variations (Miller et al., 2005a). It is surprising that modulation by the 1.2-myr-long tilt cycle is not a dominant periodicity in icehouse sea-level records, because it has been shown that the short 41-kyr tilt cycle dominates the ice-volume record of the past 34+ my (Zachos et al., 2001). The 2.4-myr very long eccentricity cycle dominates carbon isotopic records throughout the Cretaceous to Cenozoic through its effects on the carbon system, which might be expected to be influenced by sea-level changes. Spectral analysis of the Miller et al. (2005a) sea-level records shows that variations occur with an as-yet-unexplained, persistent 3-myr beat that may be either an interference between the 1.2 myr and 2.4 myr cycles or be an artifact of an undersampled sea-level signal. This intriguing relationship bears investigation because the million-year-scale sea-level signal can be shown to be a composite of 41-kyr tilt cycles, at least for the icehouse world (Miller et al., 2005a).

Moreover, sea-level amplitudes during this period have not yet

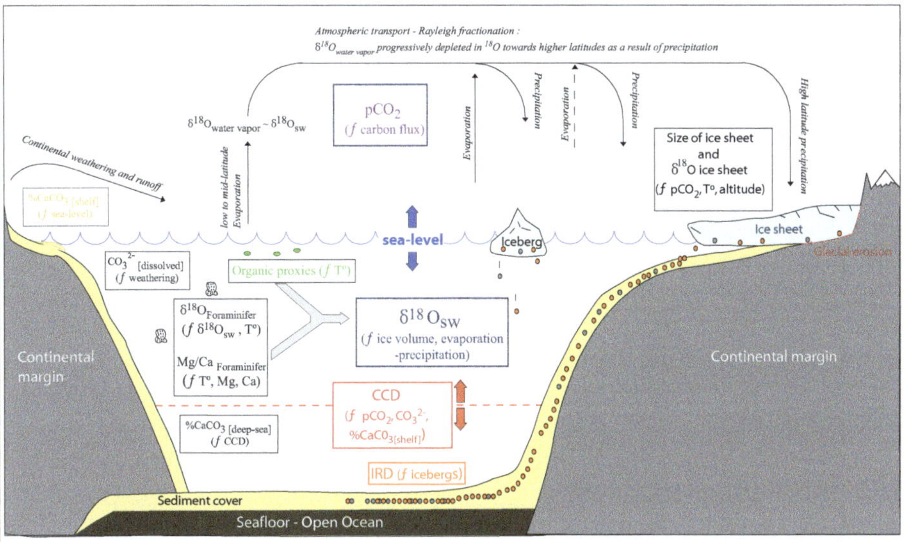

Figure 4. Details are given below.

Schematic figure illustrating how deep-sea geochemical records can be used to understand mechanisms of past eustatic changes by analogy to the modern ocean. During atmospheric transport from low to high latitudes water vapor becomes progressively more depleted in $\delta^{18}O$, and ice sheets have a very negative $\delta^{18}O$ signature. The isotopic composition of high latitude ice sheets is a function of the magnitude of isotopic fractionation within the hydrological cycle, which in turn is dependent on pCO_2 and temperature, and could vary over geological timescales. Consequently, the $\delta^{18}O$ composition of seawater ($\delta^{18}O_{SW}$) is largely a function (f) of ice volume and regional evaporation and precipitation processes. Reconstructing $\delta^{18}O_{S}w$ in different ocean basins will highlight times of eustatic changes due to ice-volume fluctuations, as well as provide a record of the timing and amplitude of these changes. $\delta^{18}O_{SW}$ can be derived by combining the isotopic composition of foraminiferal calcite ("$\delta^{18}O_{Foraminifer}$") with independent temperature proxies (e.g., Mg/Ca for deep-water temperature, and TEX86 and alkenones for surface water temperatures). Subsequently, correlation of excursions in $\delta^{18}O_{SW}$ to more positive values with independent evidence of sea-level change can be taken as support for the operation of a glacio-eustatic mechanism. Open ocean sites can also provide more indirect evidences of the relative role of glacio-eustasy through geological time. The presence of ice rafted debris (IRD) in open ocean sediments indicates iceberg transport, and thus a significant volume of ice at sea level along continental margins. The waxing and waning of ice sheets is a function of high-latitude temperatures and atmospheric pCO2, which also impact the position of the carbonate compensation depth (CCD). Fluctuations in the CCD are recorded as variable carbonate contents (%$CaCO_3$) within deep-sea sediments and could be used as indirect evidence of glacial/interglacial alternations, as CCD is sensitive to changes in carbonate burial on the shelf. Finally, eustatic variations control the area of shelf submerged, thus indirectly impacting the type of rocks subjected to continental weathering, the amount of nutrients and carbonate ions delivered to the coastal ocean, and the area available for carbon burial on continental margins. Some isotope systems (Os, Nd, and Sr) are available as proxies of continental weathering. The weathering processes ultimately have feedbacks on the carbon cycle, climate, and glacioeustasy.

been adequately constrained. One approach for determining eustatic amplitudes that has been applied with success to the New Jersey margin involves combining sequence stratigraphic and back-stripping analyses (Fig. 5; Kominz and Pekar, 2001; 2002; Pekar and Kominz, 2001). The resulting sea-level curve (Fig. 2; Browning et al., 2008) represents the best current estimate, but it is still incomplete because lowstand sediments were not recovered, introducing uncertainty to estimated amplitudes. Possibly as a result, the Miocene part of the New Jersey sea-level curve does not appear to correspond as well to the globally recognized stratigraphic signature of the Neogene as other eustatic curves (Fig. 6; Bartek et al., 1991). Furthermore, the New Jersey curve also differs from $\delta^{18}O$ records that have been corrected for paleotemperature and are therefore an improved record of ice-volume fluctuations (Billups and Schrag, 2002; Lear et al., 2004), and which do correspond well to the stratigraphic signature of the Neogene (Fig. 6).

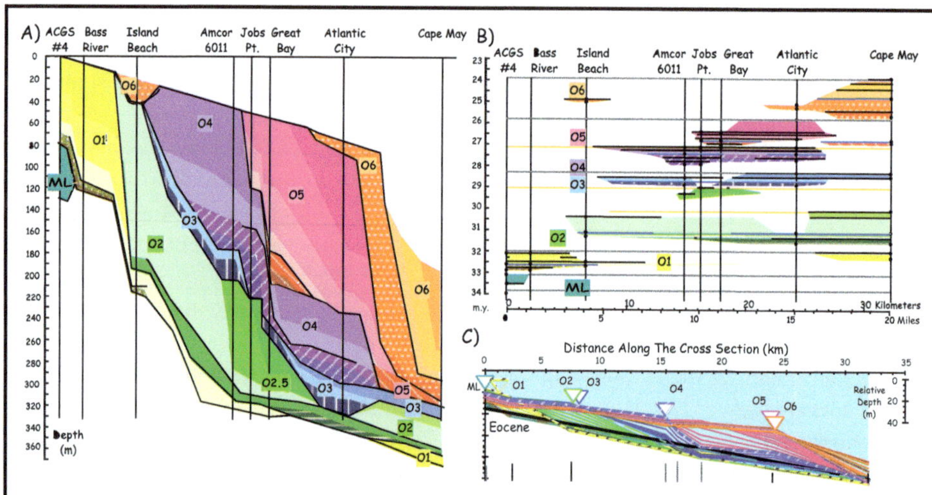

Figure 5. In order to determine sea-level change from a marginal marine setting, we recommend at least two-dimensional sequence stratigraphic back-stripping. An example of some of the data required for two-dimensional sequence stratigraphic back-stripping from Kominz and Pekar (2001, 2002) is shown for illustration. [A] Chronostratigraphic chart for New Jersey coastal plain Oligocene sequences. Solid colors represent highstand systems tracts, while lowstand systems tracts are depicted by patterned colors. Vertical lines show well control. [B] Sequence model for New Jersey coastal plain Oligocene sequences. Patterns and lines as described above. [C] Geometry of horizons identified in A and B after performing geohistory analysis. Sequences may be identified by colors, which reflect those in A and B, and the labeled offlap break points (inverted triangles) of the final horizon of each sequence. (Modified from Kominz and Pekar, 2002.)

Finally, estimates of the amplitudes of eustatic change from one-dimensional back-stripping at one location (e.g., New Jersey; Miller et al., 2005a; Kominz et al., 2008; Browning et al., 2008) requires supplemental application of this procedure to strata on distant continents (e.g., Carter et al., 1991). Future scientific drilling must therefore include additional drilling for icehouse eustatic objectives. Ideally, two- and three-dimensional back-stripping procedures would improve amplitude estimates (Kominz and Pekar, 2001). These approaches require good regional seismic coverage and a well-constrained, regional, sequence stratigraphic framework, including data that can only be obtained from cores.

2) Challenging the paradigm of a stable, ice-free "greenhouse" climate. Though we are beginning to unlock the mysteries of icehouse sea-level changes, our understanding of eustatic change during the preceding "greenhouse" world of the Triassic to early Eocene is controversial. For example, the Late Cretaceous has been reconstructed as a greenhouse world with warm polar climates (Bice et al., 2006), and most studies have assumed the absence of polar ice sheets (e.g., Huber et al., 1995). However, the work of Exxon Production Research Company (Vail and Mitchum, 1977; Haq et al., 1987) and more recent publications (Van Sickel et al., 2004; Miller et al., 2003, 2005a, b; Bornemann et al., 2008) have indicated large (tens of meters), short-period (<1 myr) Late Cretaceous global sea-level (eustatic) fluctuations (Fig. 2). In addi-

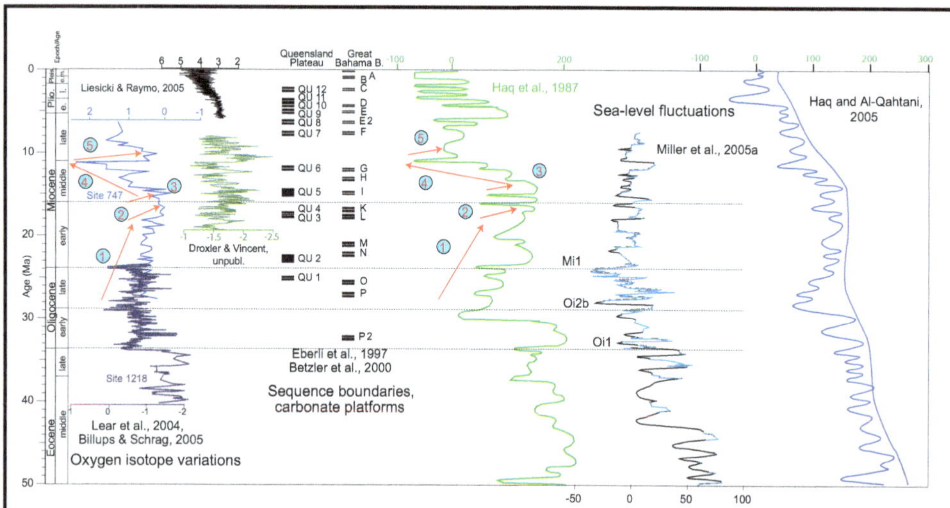

Figure 6. Oxygen isotopic and eustatic curves showing correspondence to the stratigraphic signature of the Neogene (red arrows and numbers 1–5; see Figure 3). The Miller et al. (2005a) eustatic curve from the New Jersey margin does not conform precisely to the global signature, particularly in the middle and late Miocene.

tion, second- (~10 my), third- (1–5 my) and fourth- (~0.5 my) order sequences can apparently be correlated widely between tectonically active and passive regions (e.g., Western Interior Seaway, Europe and India; Gale et al., 2002) suggesting eustatic control. Glacioeustasy is the only known mechanism for producing such large, rapid eustatic changes (Donovan and Jones, 1979; Fig. 1).

There are two solutions to this enigma: eustatic mechanisms are not fully understood, or there were ice sheets throughout much of the Triassic to early Eocene (Stoll and Schrag, 1996, 2000; Abreu et al., 1998; Miller et al., 2003, 2005a,b; Bornemann et al., 2008). ODP and ICDP drilling onshore New Jersey (Leg 174AX) have provided a detailed record of Cretaceous to early Eocene sequences. This record quantifies high amplitudes and rates of eustatic change (>25 m in <1 myr) in the Late Cretaceous to Eocene greenhouse world. Based on this sea-level history, Miller et al. (2003, 2005a, b) have proposed that ice sheets existed for geologically short intervals (i.e., lasting ~100 ky) during the Late Cretaceous-Eocene. This view can be reconciled with previous assumptions of an ice-free Greenhouse World. Eustatic changes on the 10^6 yr scale were typically ~15–30 m in the Late Cretaceous-Eocene (ca. 100–33.8 Ma), suggesting growth and decay of small- to medium-sized (10–15 x 10^6 km^3) ephemeral Antarctic ice sheets (Miller et al., 2005a, b).

However, although such indirect evidence for ephemeral ice sheets is growing, there is, as yet, no physical evidence for Late Cretaceous to early Eocene ice sheets. A particular difficulty is that other data indicate warm global temperatures for much of this interval—for example, very warm Albian-Santonian sea surface temperatures in the tropical Atlantic (Forster et al., 2007). There is therefore a need for additional high-resolution stratigraphic records from the greenhouse period.

Objective 2: Defining Stratigraphic Responses

The stratal geometries that define sedimentary sequences worldwide (Mitchum et al., 1977; Haq et al., 1987) result from a complex interplay of processes acting in three dimensions. Eustasy competes with climatic and paleoceanographic variations, tectonism, rates and modes of sediment supply, and submarine current activity to influence base level and shoreline position and, hence, stratal formation and preservation. Understanding margin sedimentation, therefore, requires evaluation of multiple processes (including eustasy) at various temporal and spatial scales (Nittrouer and Kravitz, 1995). However, predictive models of the distribution of sediments within unconformity-bounded sequences are based on assumptions about the importance of relative sea-level change (Posamentier et al., 1988; Vail et al., 1991) that have yet to be adequately tested.

Nevertheless, various industry and academic studies have established that unconformity-bounded sequences are indeed the building blocks of the stratigraphic record (see summary in Christie-Blick and Driscoll, 1995), as first proposed by Vail and Mitchum (1977), and that they can occur in predictable patterns (Fig. 3). For example, the geometric signature of stratigraphic sequences along continental margins for the last 30 Ma involves (Fig. 3C–E) (1) late Oligocene (Chattian) and early Miocene (Aquitanian) aggradation, back-stepping and drowning; (2) late early Miocene (Burdigalian) and earliest middle Miocene (early Langhian) aggradation; (3) earliest middle Miocene (late Langhian) downward shift of deposition; (4) middle Miocene (Serravallian) progradation; and (5) two stacked flooding and aggradational episodes in the late Miocene (Tortonian) and early Pliocene (Zanclean) separated by a late Miocene (Messinian) downward shift of deposition (Bartek et al., 1991, Tcherepanov et al., 2008a and b). Although this pattern is widespread and is observed globally in both siliciclastics and carbonates (Fig. 3; Bartek et al., 1991; Tcherepanov et al., 2008a and b), the heart of this section, the early and middle Miocene, has not yet been ground-truthed by drilling. The fundamental assumptions and predictive capabilities of sequence models can only be tested by drilling on shallow continental shelves where (3-D) sedimentary geometries are constrained by seismic data (e.g., Kominz and Pekar, 2002).

Siliciclastic Margins. Because of the complex interplay of forcing mechanisms responsible for the stratigraphic record, stratigraphic response must be defined in a diversity of time periods and settings, both tectonic and sedimentary. Siliciclastic sediments are excellent sea-level markers because both highly sensitive indicators of shoreline position, and they are globally widespread. However, it is essential to define the specific sedimentary processes (depositional, transportational, and erosional) responsible for the stratigraphic record and to distinguish the responses of these processes to eustasy from their responses to local forcing. This process-based approach must be a component of future drilling-based sea-level research.

The stratigraphic record comprises both surfaces and intervening sedimentary units. In offshore work, surfaces are often defined initially using seismic reflection profiles and later calibrated by coring (Fig. 3). However, only coring can provide the lithofacies and biofacies of the intervening units. Sequence stratigraphic models of such units (Posamentier et al., 1988; Van Wagoner et al., 1988; Vail et al., 1991) are based on simple assumptions about how facies respond to relative sea-level changes. However, the real world is rendered more complex by the additional influence of local forcing and the three-dimensionality of sequence architectures. Future drilling to investigate the stratigraphic response to eustasy must therefore evaluate the contributions of tectonism and sediment supply. In addition, geometrical variations must be constrained by pre-drilling seismic surveys.

Carbonate Platforms and Margins. Carbonates are excellent sea-level markers because carbonate facies are depth-dependent owing to the importance of sunlight to many carbonate-secreting organisms (Eberli et al., 1997; Camoin et al., 2007a). The relationship of these systems to the carbon cycle allows direct correlation of climatic and eustatic signals (Lear et al., 2004). In addition, multiple dating techniques are available for carbonates (including ^{14}C, U/Th, Sr, U/Pb, biostratigraphy, and magnetostratigraphy). These enable examination of a wide range of frequencies of sea-level change, from millennial scale to tens of millions of years.

Continental margin transects (Fig. 3) have the advantage that their stratigraphic architectures are well constrained by seismic data. However, they are complemented by tropical reefs and atolls, which provide the most reliable geological estimates of relative sea level by dating "fossil sunshine" (e.g., shallow dwelling corals). The study of coral reefs is of crucial importance in attempts to resolve the rates of millennial-scale changes in sea level, to clarify the mechanisms that drive glacial-interglacial cycles, and to constrain geophysical models. Coral reefs provide unparalleled records of sea-level amplitudes, particularly for the middle to late Pleistocene. For example, drilling reefs in Barbados has provided a precise estimate for the last eustatic lowstand (120 ± 5 m below present at 18 ka; Fairbanks, 1989; Bard et al., 1990; Peltier and Fairbanks, 2006). Shallow-water drilling of coral reefs remains challenging due to recovery problems, but is necessary to allow study of recent high-resolution climate changes, and it contributes to estimates of the future behavior of the Earth system on societal timescales. This approach was employed during IODP Expedition 310 off Tahiti (Camoin et al., 2007a, b).

Strategies

1) *A focus on both icehouse and greenhouse objectives.*

2) *Drilling transect approach.* This approach, which was tried and tested on the New Jersey Margin and Great Bahama Bank, with additional IODP drilling planned off New Zealand (Fig. 3), must be enhanced and extended by:

- Integration of onshore (e.g., ICDP, DOSECC) and offshore (IODP, DOSECC) drilling. The record of icehouse eustasy is best preserved offshore (e.g., on continental margins), but the older, greenhouse, record tends to be preserved and drillable beneath coastal plains or in onshore basins (e.g., the Western Interior Seaway). Onshore drilling is therefore expected to play an increasingly important role in sea-level studies.

- Drilling of sufficient boreholes, including multiple transects where necessary, and incorporation of sufficient seismic control to constrain three-dimensional stratigraphic architecture.

- Maximizing core recovery by using appropriate drilling technology (e.g., casing, mud) and platforms (e.g., Mission Specific Platforms [MSPs]), and by adapting coring strategies as needed, e.g., by using diamond coring or short advances of the XCB.

3) *Recognizing the value of addressing tectonically active settings.* For example, the stratigraphic expression of sea-level change in active foreland basins in the U.S. and Canadian western interior basins is superlative, though the stratigraphic records incorporate the effects of both eustasy and tectonism.

4) *Incorporating a focus on high-resolution (10^3–10^5 yr) glacial-interglacial cycles* (e.g., the last 130 kyr). Examination of margins with high stratigraphic resolution will allow evaluation of the interaction of eustasy and other processes (e.g., Papua New Guinea; Jorry et al., 2008), and integration with process-oriented modeling (e.g., physical and mathematical modeling done as part of the Margins and Intermargins Initiatives).

5) *Coordination with drilling operations designed to address other objectives.* Sea-level studies can benefit greatly from the results of research into, for example, paleoclimate, carbon cycling, and ice-sheet dynamics (Fig. 4). Conversely, these research programs will also gain necessary insights from a well constrained eustatic history.

Future Work

Future IODP drilling for sea-level objectives includes IODP Expedition 313, New Jersey inner shelf drilling scheduled for summer 2009 and IODP Expedition 317, Canterbury Basin, New Zealand, scheduled for November 2008–January 2009. Great Barrier Reef drilling is tentatively planned for 2009. These planned expeditions, and existing IODP proposals (e.g., Maldives, North West Australian Shelf, Gulf of Mexico - Southern Bank, Belize margin, Gulf of Papua), all address icehouse objectives. Such drilling is indeed vital, in particular to constrain icehouse eustatic amplitudes and to calibrate the stratigraphic signature of the Neogene (Bartek et al., 1991). However, the next phase of sea-level studies must include greenhouse objectives. We therefore encourage proponents to prepare and submit sea-level proposals for both offshore and onshore drilling focusing on the greenhouse world.

Acknowledgements

We thank all of the workshop participants for their contributions in support of scientific drilling for sea-level objectives. Particular thanks are due to breakout group chairs Andy Gale, John Jaeger, Michelle Kominz, Rick Sarg, and Bill Thompson. Charna Meth and Julie Farver of the

Consortium for Ocean Leadership provided excellent support before and during the workshop. Dave Zur of DOSECC also provided logistical support. Bryan Bracken of Chevron led an outstanding field trip to Book Cliffs before the workshop, ably assisted by Sunday Shephard and Peter Sixsmith. We express our sincere appreciation for the financial support of the Consortium for Ocean Leadership, ICDP, IODP, DOSECC, and Chevron. This joint support enabled us to make the workshop a truly international meeting.

References

Abreu, V.S., Hardenbol, J., Haddad, G., Baum, G.R., Droxler, A.W., and Vail, P.R., 1998. Oxygen isotope synthesis: A Cretaceous ice-house? *In* Graciansky, P.-C., Hardenbol, J., Jacquin, T., and Vail, P.R. (Eds.), *Mesozoic and Cenozoic Sequence Stratigraphy of the European Basins.* Tulsa, Okla. (SEPM (Society for Sedimentary Geology)) Spec. Pub. 60:75–80.

Bard, E., Hamelin, B., and Fairbanks, R.G., 1990. U-Th ages obtained by mass spectrometry in corals from Barbados: Sea level during the past 130,000 years. *Nature,* 346:456–458, doi: 10.1038/346456a0.

Bartek, L.R., Vail, P.R., Anderson, J.B., Emmet, P.A., and Wu, S., 1991. Effect of Cenozoic ice sheet fluctuations in Antarctica on the stratigraphic signature of the Neogene. *J. Geophys. Res.,* 96:6753–6778, doi:10.1029/90JB02528.

Belopolsky, A.V., and Droxler, A.W., 2004. Seismic expressions of prograding carbonate bank margins: Middle Miocene progradation in the Maldives, Indian Ocean. *AAPG Mem.,* 81:267–290.

Betzler, C., Kroon, D., and Reijmer, J.J.G., 2000. Synchroneity of major late Neogene sea-level fluctuations and paleoceanographically controlled changes as recorded by two carbonate platforms. *Paleoceanogr.,* 15:722–730, doi:10.1029/1999PA000481.

Bice, K.L., Birgel, D., Meyers, P.A., Dahl, K.A., Hinrichs, K., and Norris, R.D., 2006. A multiple proxy and model study of Cretaceous upper ocean temperatures and atmospheric CO_2 concentrations. *Paleoceanogr.,* 21:PA2002, doi: 10.1029/2005PA001203.

Billups, K., and Schrag, D.P., 2002. Paleotemperature and ice-volume of the past 27 Myr revisited with paired Mg/Ca and 18O/16O measurements on benthic foraminifera. *Paleoceanogr.,* 17(1):26–37, doi:10.1029/2000PA000567.

Bornemann, A., Norris, R.D., Friedrich, O., Beckmann, B., Schouten, S., Damsté, J.S.S., Vogel, J., Hofmann, P., and Wagner, T., 2008. Isotopic evidence for glaciation during the Cretaceous super greenhouse. *Science,* 319:189–192, doi:10.1126/science.1148777.

Browning, J.V., Miller, K.G., Sugarman, P.J., Kominz, M.A., McLaughlin, P.P., and Kulpecz, A.A., 2008. A 100 million year record of sequences, sedimentary facies and sea-level change from Ocean Drilling Program onshore core holes, U.S. Mid-Atlantic coastal plain. *Basin Res.,* in press, doi:10.1111/j.1365-2117.2008.00360.x

Camoin, G.F., Iryu, Y., McInroy, D., and the Expedition 310 scientists, 2007a. *Proc. IODP, 310.* College Station, Texas (Integrated Ocean Drilling Program - Management International, Inc.).

Camoin, G.F., Iryu, Y., McInroy, D., and the Expedition 310 scientists, 2007b. IODP Expedition 310 reconstructs sea-level, climatic and environmental changes in the South Pacific during the last deglaciation. *Sci. Drill.,* 5:4–12.

Cande, S.C., and Kent, D.V., 1992. A new geomagnetic polarity time scale for the Late Cretaceous and Cenozoic. *J. Geophys. Res.,* 97:13917–13951, doi:10.1029/92JB01202.

Carter, R.M., Abbott, S.T., Fulthorpe, C.S., Haywick, D.W., and Henderson, R.A., 1991. Application of global sea-level and sequence stratigraphic models in southern hemisphere Neogene strata from New Zealand. *In* MacDonald, D. (Ed.), *Sea Level and Active Plate Margins, International Association of Sedimentologists Spec. Pub.* 12:41–65.

Cazenave, A., 2006. How fast are the ice sheets melting? *Science,* 314:1250–1252, doi:10.1126/science.1133325.

Cazenave, A., and Nerem, R.S., 2004. Present-day sea-level change: Observations and causes. *Rev. Geophys., 42*:RG3001, doi: 10.1029/2003RG000139.

Christie-Blick, N., and Driscoll, N.W., 1995. Sequence stratigraphy. *Annu. Rev. Earth Planet. Sci.,* 23:451–478, doi:10.1146/annurev.ea.23.050195.002315.

Christie-Blick, N., Mountain, G.S., and Miller, K.G., 1990. Seismic stratigraphic record of sea-level change. *In* Revelle, R. (Ed.), *Sea-level Change,* Washington, DC (National Academy Press), 116–140.

Church, J.A., and White, N.J., 2006. A 20th century acceleration in global sea-level rise. *Geophys. Res. Lett.,* 33:L01602, doi:10.1029/2005GL024826.

Cloetingh, S., McQueen, H., and Lambeck, K., 1985. On a tectonic mechanism for regional sea-level variations. *Earth Planet. Sci. Lett.,* 75:157–166, doi:10.1016/0012-821X(85)90098-6.

Donovan, D.T., and Jones, E.J.W., 1979. Causes of world-wide changes in sea level. *J. Geol. Soc. London,* 136:187–192, doi:10.1144/gsjgs.136.2.0187.

Draut, A.E., Raymo, M.E., McManus, J.F., and Oppo, D.W., 2003. Climate stability during the Pliocene warm period. *Paleoceanogr.,* 18(4):1078, doi:10.1029/2003PA000889.

Droxler, A., Poore, R., and Burckle, L., 2003. Earth's climate and orbital eccentricity: The marine isotope stage 11 question. *Geophysical Monograph Series, 137.*

Eberli, G.P., 2000. The record of Neogene sea-level changes in the prograding carbonates along the Bahamas transects—166 syntheses. *In* Swart, P.K., Eberli, G.P., Malone, M.J., and Sarg, J.F., (Eds.), *Proc. ODP Sci. Results, 166.* College Station, Texas (Ocean Drilling Program), 167–177.

Eberli, G., Swart, P., Malone, M., and Shipboard Scientific Party, 1997. *Leg 166 Preliminary Report,* College Station, Texas (Ocean Drilling Program).

Fairbanks, R.G., 1989. Glacio-eustatic record 0-16,000 years before present: Influence of glacial melting rates on Younger Dryas event and deep ocean circulation. *Nature,* 34:637–642, doi:10.1038/342637a0.

Forster, A., Schouten, S., Baas, M., and Sinninghe Damsté, J.S., 2007. Mid-Cretaceous (Albian–Santonian) sea surface temperature record of the tropical Atlantic Ocean. *Geology,* 35:919–922; doi: 10.1130/G23874A.1

Gale, A.S., Hardenbol, J., Hathway, B., Kennedy, W.J., Young, J.R., and Phansalkar, V., 2002. Global correlation of Cenomanian

(Upper Cretaceous) sequences: Evidence for Milankovitch control on sea level. *Geology*, 30:291–294; doi: 10.1130/0091-7613(2002)030<0291:GCOCUC>2.0.CO;2

Haq, B.U., and Al-Qahtani, A.M., 2005. Phanerozoic cycles of sea-level change on the Arabian Platform. *GeoArabia*, 10:127–160.

Haq, B.U., Hardenbol, J., and Vail, P.R., 1987. Chronology of fluctuating sea levels since the Triassic (250 million years ago to present). *Science*, 235:1156–1167.

Harrison, C.G.A., 1990. Long-term eustasy and epeirogeny in continents. *In* Revelle, R. (Ed.), *Sea-level Change*, Washington, DC (National Academy Press), 141–158.

Huber, B.T., Hodell, D.A., and Hamilton, C.P., 1995. Mid to Late Cretaceous climate of the southern high latitudes: stable isotopic evidence for minimal equator-to-pole thermal gradients. *Geol. Soc. Am. Bull.*, 107:1164–1191, doi:10.1130/0016-7606(1995)107<1164:MLCCOT>2.3.CO;2.

Imbrie, J. et al., 1987. Scientific goals of an Ocean Drilling Program designed to investigate changes in the global environment. *In* Report of the Second Conference on Scientific Ocean Drilling (COSOD II), Joint Oceanographic Institutions for Deep Earth Sampling, 15–46.

Intergovernmental Panel on Climate Change (IPCC), 2007. Climate change 2007—Synthesis report. Summary for policymakers. Available at http://www.ipcc.ch/pdf/assessment-report/ar4/syr/ar4_syr_spm.pdf.

John, C.M., Karner, G.D., and Mutti, M., 2004. Delta ^{18}O and Marion Plateau back-stripping: combining two approaches to constrain late middle Miocene eustatic amplitude. *Geology*, 32:829–832, doi:10.1130/G20580.1.

JOIDES, 1992. Sea-Level Working Group (SLWG) Report. Loutit, T. S. (Ed.), *JOIDES Journal*, 18(3):28–36.

Jorry, S.J., Droxler, A.W., Mallarino, G., Dickens, G.R., Bentley, S.J., Peterson, L.C., and Opdyke, B., 2008. Bundled turbidite deposition in the Central Pandora Trough (Gulf of Papua) since Last Glacial Maximum: Linking sediment nature and accumulation to sea-level fluctuations at millennial time-scale. *J. Geophys. Res.*, 113:F01S19, doi:10.1029/2006JF000649.

Karner, G.D., 1986. Effects of lithospheric in-plane stress on sedimentary basin stratigraphy. *Tectonics*, 5:573–588, doi:10.1029/TC005i004p00573.

Kominz, M.A., and Pekar, S.F., 2001. Oligocene eustasy from two-dimensional sequence stratigraphic backstripping. *Geol. Soc. Am. Bull.*, 113:291–304, doi:10.1130/0016-7606(2001)113<0291:OEFTDS>2.0.CO;2.

Kominz, M.A., and Pekar, S.F., 2002. Sequence stratigraphy and eustatic sea level, *Proceedings 22nd Annual GCSSEPM Foundation Bob F. Perkins Research Conference*, "Sequence Stratigraphic Models for Exploration and Production: Evolving Methodology, Emerging Models and Application Case Histories", 8–11 August 2002, Houston, Texas, 349–365.

Kominz, M.A., Miller, K.G., and Browning, J.V., 1998. Cenozoic sea-level estimates from back-stripping. *Geology*, 26:311–314.

Kominz, M., Browning, J.V., Miller, K.G., Sugarman, P.J., Misintseva, S. and, Scotese, C.R., 2008. Late Cretaceous to Miocene sea-level estimates from the New Jersey and Delaware Coastal Plain core holes: An error analysis. *Basin Research*, in press.

Lear, C.H., Rosenthal, Y., Coxall, H.K., and Wilson, P.A., 2004. Late Eocene to early Miocene ice sheet dynamics and the global carbon cycle. *Paleoceanogr.*, 19(4):PA4015, doi:10.1029/2004PA001039.

Liesicki, L.E., and Raymo, M.E., 2005. A Pliocene-Pleistocene stack of 57 globally distributed d^{18} O records. Paleoceanography, 20, PA1003, doi:10.1029/2004PA001071.

Lu, H., and Fulthorpe, C.S., 2004. Controls on sequence stratigraphy of a middle-Miocene to Recent, current-swept, passive margin: offshore Canterbury Basin, New Zealand. *Geol. Soc. Am. Bull.*, 116:1345–1366, doi:10.1130/B2525401.1.

McGranahan, G., Balk, D., and Anderson, B., 2007. The rising tide: assessing the risks of climate change and human settlements in low elevation coastal zones. *Environment and Urbanization* 19:17–37, doi:10.1177/0956247807076960.

Miller, K.G., Kominz, M.A., Browning, J.V., Wright, J.D., Mountain, G.S., Katz, M.E., Sugarman, P.J., Cramer, B.S., Christie-Blick, N., and Pekar, S.F., 2005a. The Phanerozoic record of global sea-level change. *Science*, 310:1293–1298, doi:10.1126/science.1116412.

Miller, K.G., Mountain, G.S., the Leg 150 Shipboard Party, and Members of the New Jersey Coastal Plain Drilling Project, 1996. Drilling and dating New Jersey Oligocene-Miocene sequences: ice volume, global sea level, and Exxon records. *Science*, 271:1092–1094, doi:10.1126/science.271.5252.1092.

Miller, K.G., Sugarman, P.J., Browning, J.V., Kominz, M.A., Hernandez, J.C., Olsson, R.K., Wright, J.D., Feigenson, M.D., and Van Sickel, W., 2003. A chronology of Late Cretaceous sequences and sea-level history: Glacioeustasy during the Greenhouse World. *Geology*, 31:585–588, doi:10.1130/0091-7613(2003)031<0585:LCCOLR>2.0.CO;2.

Miller, K.G., Wright, J.D., and Browning, J.V., 2005b. Visions of ice sheets in a greenhouse world. *In* Paytan, A., and De La Rocha, C. (Eds.), *Ocean Chemistry Throughout the Phanerozoic, Marine Geology*, Special Issue, 217:215–231.

Mitchum, R.M., Jr., Vail, P.R., and Thompson, S., 1977. Seismic stratigraphy and global changes of sea level, part 2: the depositional sequence as a basic unit for stratigraphic analysis. *In* Payton, C.E., (Ed.), *Seismic Stratigraphy – Applications to Hydrocarbon Exploration, AAPG Memoir* 26:53–62.

Mountain, G.S., Burger, R.L., Delius, H., Fulthorpe, C.S., Austin, J.A., Jr., Goldberg, D.S., Steckler, M.S., McHugh, C.M., Miller, K.G., Monteverde, D.H., Orange, D.L., and Pratson, L.F., 2007. The long-term stratigraphic record on continental margins. *In* Nittrouer, C.A., Austin, J.A., Jr., Field, M.E., Kravitz, J.H., Syvitski, J.P.M., and Wiberg, P.L. (Eds.), *Continental Margin Sedimentation: From Sediment Transport to Sequence Stratigraphy*, International Association of Sedimentologists Special Publication 37:381–458.

Nittrouer, C.A., and Kravitz, J.H., 1995. Integrated continental margin research to benefit ocean and earth sciences. *Eos 76*:121, 124, 126.

Pekar, S.F., and Kominz, M.A., 2001. Two-dimensional paleoslope modeling: A new method for estimating water depths for benthic Foraminiferal biofacies and paleo shelf margins. *J. Sed. Res.*, 71:608–620, 2001, doi:10.1306/100600710608.

Peltier, W.R. and Fairbanks, R.G., 2006. Global glacial ice volume and

Last Glacial Maximum duration from an extended Barbados sea-level record. *Quat. Sci. Rev.*, 25:3322–3337, doi:10.1016/j.quascirev.2006.04.010.

Posamentier, H.W., Jervey, M.T., and Vail, P.R., 1988. Eustatic controls on clastic deposition I -- Conceptual framework. *In* Wilgus, C.K., Hastings, B.S., Kendall, C.G.St.C., Posamentier, H.W., Ross, C.A., and Van Wagoner, J.C. (Eds.), *Sea-Level Changes: An Integrated Approach*, SEPM Spec. Publ. 42:109–124.

Psuty, N.P., and Collins, D., 1996. Sea-level rise: A white paper on the measurements of sea-level rise in New Jersey and a perspective on the implications for management. *Coastal Hazard Management Report*, Office of Land and Water Planning, New Jersey Department of Environmental Protection.

Rahmstorf, S., Cazenave, A., Church, J.A., Hansen J.E., Keeling, R.F., Parker, D.E., and Somerville, R.C.J., **2007.** Recent climate observations compared to projections. *Science*, 316:709, doi:10.1126/science.1136843.

Reynolds, D.J., Steckler, M.S., and Coakley, B.J., 1991. The role of the sediment load in sequence stratigraphy: The influence of flexural isostasy and compaction. *J. Geophys. Res.*, 96:6931–6949, doi:10.1029/90JB01914.

Scherer, R.P., Bohaty, S.M., Dunbar, R.B., Esper, O., Flores, J.A., Gersonde, R., Harwood, D.M., Roberts, A.P., and Taviani, M., 2008. Antarctic records of precession-paced insolation-driven warming during early Pleistocene Marine Isotope Stage 31. *Geophys. Res. Lett.*, 35:L03505, doi:10.1029/2007GL032254.

Stoll, H., and Schrag, D.P., 2000. High resolution stable isotope records from the upper Cretaceous of Italy and Spain: glacial episodes in a greenhouse planer? *Geol. Soc. Am. Bull.*, 112:308–319, doi:10.1130/0016-7606(2000)112<0308:HRSIRF>2.3.CO;2.

Stoll, H.M., and Schrag, D.P., 1996. Evidence for glacial control of rapid sea-level changes in the early Cretaceous. *Science*, 272:1771–1774, doi:10.1126/science.272.5269.1771.

Tcherepanov, E.N., Droxler, A.W., Lapointe, P., Dickens, G.R., Bentley, S.J., Beaufort, L., Peterson, L.C., Daniell, J., and Opdyke, B.N., 2008a. Neogene evolution of the mixed carbonate-siliciclastic system in the Gulf of Papua, Papua New Guinea. *J. Geophys. Res.*, 11:F01S21, doi:10.1029/2006JF000684.

Tcherepanov, E.N., Droxler, A.W., Lapointe, P., and Mohn, K., 2008b. Carbonate seismic stratigraphy of the Gulf of Papua mixed depositional system: Neogene stratigraphic signature and eustatic control. Basin Res., 20(2): 185–209, doi:10.1111/j.1365-2117.2008.00364.x

Thompson, W.G., and Goldstein, S.L., 2005. Coral dating that corrects for exchange of uranium decay products with sea water reveals that sea level varied repeatedly by up to 30 meters during recent glaciations. *Science*, 308:401–404.

Vail, P.R., and Mitchum, R.M., Jr., 1977. Seismic stratigraphy and global changes of sea level, Part 1: Overview. *In* Payton, C.E. (Ed.), *Seismic Stratigraphy - Applications to Hydrocarbon Exploration. Am. Assoc. Petrol. Geol. Memoir* 26:51–52.

Vail, P.R., Audemard, F., Bowman, S.A., Eisner, P.N., and Perez-Cruz, C., 1991. The stratigraphic signatures of tectonics, eustasy and sedimentology—an overview. *In* Einsele, G., Ricken, W., and Seilacher, A. (Eds.), *Cycles and events in stratigraphy*, Berlin (Springer-Verlag), 617–659.

Van Sickel, W.A., Kominz, M.A, Miller, K.G., and Browning, J.V., 2004. Late Cretaceous and Cenozoic sea-level estimates back-stripping analysis of borehole data, onshore New Jersey. *Basin Res.*, 16:451–465, doi:10.1111/j.1365-2117.2004.00242.x.

Van Wagoner, J.C., Posamentier, H.W., Mitchum, R.M., Vail, P.R., Sarg, J.F., Loutit, T.S., and Hardenbol, J., 1988. An overview of the fundamentals of sequence stratigraphy and key definitions. *In* Wilgus, C.K., Hastings, B.S., Kendall, C.G.St.C., Posamentier, H.W., Ross, C.A., and Van Wagoner, J.C. (Eds.), *Sea-level Changes: An Integrated Approach. SEPM Spec. Publ.* 42:39–45.

Watkins, J.S., and Mountain, G.S., 1990. Role of ODP drilling in the investigation of global changes in sea level. *JOI-USSAC Workshop Rept.*, El Paso, Texas, 24–26 October 1988, 70 p.

Zachos, J., Pagani, M., Sloan, L., Thomas, E., and Billups, K., 2001. Trends, rhythms, and aberrations in global climate 65 Ma to Present. *Science*, 292:686–693, doi:10.1126/science.1059412.

Authors

Craig S. Fulthorpe, The University of Texas at Austin Institute for Geophysics, John A. and Katherine G. Jackson School of Geosciences, J.J. Pickle Research Campus, Bldg. 196 (ROC), 10100 Burnet Road (R2200), Austin, Texas 78758-4445, U.S.A., e-mail: craig@ig.utexas.edu.

Kenneth G. Miller, Department of Earth and Planetary Sciences, Wright Labs, Rutgers, The State University of New Jersey, 610 Taylor Road, Piscataway, N.J. 08854-8066, U.S.A.

André W. Droxler, Department of Earth Science, Rice University, P.O. Box 1892, Houston, Texas 77251-1892, U.S.A.

Stephen P. Hesselbo, Department of Earth Sciences, University of Oxford, Parks Road, Oxford OX1 3PR, U.K.

Gilbert F. Camoin, CEREGE UMR 6635 CNRS, BP 80, Europôle Méditerranéen de l'Arbois, F-13545 Aix-en-Provence cedex 4, France.

Michelle A. Kominz, Department of Geosciences, Western Michigan University, 1133 Rood Hall, 1903 West Michigan Avenue, Kalamazoo, Mich., 49008, U.S.A.

3

ANDRILL's Success During the 4th International Polar Year

by Fabio Florindo, David Harwood, Richard Levy, and SMS Project Science Team

Introduction

One of the scientific programs of the Fourth International Polar Year (Allison et al., 2007; www.ipy.org), the ANDRILL (ANtarctic geological DRILLing) Program demonstrated ability to recover high quality marine and glacimarine sedimentary drill cores from high latitude ice-covered areas. ANDRILL's inaugural 2006 and 2007 drilling seasons resulted in the two deepest drill holes on the Antarctic continental margin, recovering 2,400 meters of high-quality and nearly continuous sediment core. A chief scientific objective of this collaborative effort of scientists, engineers, technicians, students, educators, drillers, and support personnel from Germany, Italy, New Zealand, and the United States is the recovery of sedimentary archives from which past climatic and environmental changes in the southern high latitudes can be reconstructed. More than 120 individuals have been involved in each of the two drilling projects, eighty of whom worked in Antarctica during each austral summer season.

Two Successful Antarctic Seasons

Completed in early December 2007, the AND-2A drillcore of the Southern McMurdo Sound (SMS) Project (Figs. 1 and 2) recovered 98% of the 1138.54 m of sedimentary rock penetrated, demonstrating that the ANDRILL drilling system (Figs. 1, 3, 4, and 5) is capable of consistently recovering high quality cores. The AND-2A drillcore is exceeded in depth only by ANDRILL's first drill hole AND-1B of the McMurdo Ice Shelf (MIS) Project (Fig. 2), which reached a total depth of 1284.87 m with similar success of high core recovery. Capable of operating in a range of environmental settings, the MIS Project drilled from an 84-m-thick ice shelf platform in 943 meters of water, and the SMS Project operated on the surface of 8.4 m of multi-year sea-ice over a 384-m water column. Alex Pyne, the Drilling Science Manager, merits recognition for the concept, design, and successful integration of elements of the ANDRILL drilling system.

Both of the ANDRILL projects reached their scientific targets. The MIS Project, led by Tim Naish (NZ) and Ross Powell (U.S.), recovered a record of a dynamic cryosphere of the last 13 million years of glacial and climatic variation of the West Antarctic Ice Sheet and Ross Sea region. The SMS Project (led by the authors of this report) completed drilling operations in late 2007 after recovering an expanded 600-m-thick section of the target interval that recorded a history of ice-proximal, shallow marine paleoenvironmental variation during the middle Miocene. This interval has long been held as a fundamental step in development of the Cenozoic cryosphere—interpretations of deep-sea oxygen isotope records suggest the middle Miocene encompassed a change from a period of warm climatic optima, approximately 17.5 Ma to 14.5 Ma, to the onset of major cooling between c. 14 Ma to 13 Ma, and the formation of a quasi-permanent ice sheet on East Antarctica.

In addition to the 600-m-thick middle Miocene interval (800–223 mbsf), the AND-2A drill core also recovered an expanded lower Miocene section (1138.54 up to c. 800 mbsf) through an interval previously recovered during the Cape Roberts Project and an upper Pliocene to Recent interval (223 mbsf to 0.0 mbsf) that is thinner but correlative to parts of the Upper Neogene section recovered by the ANDRILL MIS Project in drill core AND-1B. Abundant volcanic clasts and tephra layers in the core, also provide the materials to document the 20-million-year evolution of the McMurdo Volcanic Province including several previously unknown explosive volcanic events.

Figure 1. Panoramic image of the ANDRILL drilling rig and science laboratories in Southern McMurdo Sound during late 2007, operating from a 8.4-meter-thick sea-ice platform. Transantarctic Mountains are visible in the background.

Figure 2. Satellite image of the McMurdo Sound region, including location of the Southern McMurdo Sound (SMS) and McMurdo Ice Shelf (MIS) drill sites.

Other results of the SMS Project include (1) a nearly continuous downhole logging operation, including deployment of a range of tools and a borehole televiewer, to match the excellent core physical (MSCL) and chemical (XRF scanning) properties data collected on-ice; (2) the first Antarctic *in situ* stress measurements from hydrofracture experiments conducted near the bottom of the borehole; (3) successful reconstruction and orientation of the core, using physical features, borehole tele-imaging, and a core orienting tool; (4) a robust chronostratigraphic framework, developed through integrated diatom biostratigraphy, magnetostratigraphy, and radiometric dating of volcanic materials, which now provides age control for the drill hole and the network of seismic lines in the western Ross Sea; (5) a vertical seismic profile study with three-component data; (6) the richest Cenozoic macropaleontological resource in East Antarctica, with more than fifty productive marine horizons; and (7) a record of terrestrial and marine temperature variations from a variety of climate proxies.

Future Activities and Planning

Programs like ANDRILL can help constraining uncertainties about the future behavior of Antarctic ice sheets and resultant sea-level change. These stratigraphic records will be used to determine the behavior of ancient ice sheets and to better understand the factors driving past ice sheet, ice shelf, and sea-ice growth and decay. This knowledge will enhance our understanding of Antarctica's potential responses to future global climate change. With this in mind, the ANDRILL scientific and operations teams continue to plan for future scientific progress using the ANDRILL system through field-based site surveys, scientific planning, and technological developments. ANDRILL's capabilities are expanding to operate from an ice shelf platform several hundred meters thick and moving at a rate of more than two meters per day. The ANDRILL teams are also assessing the feasibility of reentering a drill hole, following relocation of the drilling rig and drill site science facilities. In the meantime, the science team members involved in the MIS and SMS projects are actively studying the drill cores and reporting initial results (Harwood et al., 2006, 2008; Naish, et al., 2007a, 2007b, 2008). These results are vital to SCAR's (Scientific Committee on Antarctic Research) ACE (Antarctic Climate Evolution) program (www.ace.scar.org), whose objectives are to integrate geological and paleoclimatic data into climate and ice sheet models to constrain estimates of Cenozoic ice volume variability, and terrestrial and marine paleotemperatures.

In support of the 4th IPY's focus on education and outreach objectives (http://www.ipy.org), ANDRILL is also engaging and training the next generation of Antarctic geoscientists and educators through exciting and collaborative international research and is taking polar science adventure into classrooms and homes through a stimulating and diverse education and outreach program (http://andrill.org/iceberg).

Acknowledgements

The ANDRILL project is a multinational collaboration between the Antarctic Programs of Germany, Italy, New Zealand, and the United States. Antarctica New Zealand is the project operator, and developed the drilling system in collaboration with Alex Pyne at Victoria University of Wellington and Webster Drilling and Exploration Ltd. Antarctica New Zealand supported the drilling team at Scott Base, and Raytheon Polar Services supported the Science team at McMurdo Station and the Crary Science and Engineering Center. Scientific support was provided by the ANDRILL Science Management Office, University of

Figure 3. ANDRILL sediment cores were flown by helicopter daily from the drillsite to the Crary Science and Engineering Center at McMurdo Station. Photo by Lucia Simion.

Figure 4. Diamond-impregnated drilling bits enable the ANDRILL drilling system to recover high-quality core, with up to 98% core recovery through glacial and glacimarine sediments.

Figure 5. Drilling operations under the ANDRILL drill-rig canopy.

Nebraska-Lincoln. Scientific studies are jointly supported by the U.S. National Science Foundation, NZ Foundation for Research, the Italian Antarctic Research Program, the German Science Foundation, and the Alfred Wegener Institute. For more information, please visit the ANDRILL website at http://andrill.org.

References

Allison, I., Béland, M., Alverson, K., Bell, R., Carlson, D., Danell, K., Ellis-Evans, C., Fahrbach, E., Fanta, E., Fujii, Y., Glaser, G., Goldfarb, L., Hovelsrud, G., Huber, J., Kotlyakov, V., Krupnik, I., Lopez-Martinez, J., Mohr, T., Qin, D., Rachold, V., Rapley, C., Rogne, O., Sarukhanian, E., Summerhayes, C., and Xiao, C., 2007. The scope of science for the International Polar Year 2007-2008. *WMO/TD-No.1364*, Geneva (World Meteorological Organization), 1–79.

Harwood, D.M., Levy, R.H., Cowie, J., Florindo, F., Naish, T., Powell, R.D., and Pyne, A., 2006. Deep drilling with the ANDRILL program in Antarctica. *Sci. Drill.*, 3:43–45.

Harwood, D.M., Florindo, F., Talarico, F., Levy, R.H., and SMS Science Team, 2008. Initial results of ANDRILL's Southern McMurdo Sound Project drill core AND-2A: early Miocene to Recent paleoclimate and geological history of the Victoria Land Basin, Antarctica. *Geophys. Res. Abstr.*, 10: EGU2008-A-12320.

Naish, T., Powell, R. D., Levy, R. H., Florindo, F., Harwood, D. M., Kuhn, G., Niessen, F., Talarico, F., and Wilson, G. S., 2007a. A record of Antarctic climate and ice sheet history recovered. *Eos Trans. AGU*, 88(50):557–558, doi:10.1029/2007EO500001.

Naish, T.R., Powell, R.D., and Levy, R.H., 2007b. Initial science results from AND-1B, ANDRILL McMurdo Ice Shelf Project, Antarctica. *Terra Antartica*, 14(3):111–328.

Naish, T.R., Powell, R.D., Barrett, P.J., Levy, R.H., Henry, S., Wilson, G.S., Krissek, L.A., Niessen, F., Pompilio, M., Ross, J., Scherer, R., Talarico, F., Pyne, A., and the ANDRILL-MIS Science Team, 2008. Late Neogene climate history of the Ross Embayment from the AND-1B drill core: culmination of three decades of Antarctic margin drilling. *In* Cooper, A. K. (Ed.), *Antarctica: A Keystone in a Changing World. Proceedings of the 10th International Symposium on Antarctic Earth Sciences.* Washington, DC (The National Academies Press), 71–82.

Authors

Fabio Florindo, Istituto Nazionale di Geofisica e Vulcanologia, Via di Vigna Murata, 605, 00143 Rome, Italy, e-mail: florindo@ingv.it

David Harwood and **Richard Levy**, ANDRILL Science Management Office, and Department of Geosciences, University of Nebraska-Lincoln, 126 Bessey Hall, Lincoln, Neb. 68588-0341, U.S.A., e-mail: dharwood1@unl.edu and rlevy2@unl.edu

*The names and affiliations of those **SMS Project Science Team** members engaged during the SMS Project's initial core characterization phase can be found at the Web site: http://andrill.org/projects/sms/team.html

Related Web Links

http://www.ipy.org
http://andrill.org
http://andrill.org/iceberg
http://www.ace.scar.org

Photo Credits

Fig. 1: Photo by Simon Nielsen
Fig. 2: Satellite image courtesy of Steve Fischbein
Fig. 4: Photo courtesy of ANDRILL
Figs. 3 and 5: Photo by Lucia Simion

4

The CO_2SINK Boreholes for Geological Storage Testing

by Bernhard Prevedel, Lothar Wohlgemuth, Jan Henninges, Kai Krüger, Ben Norden, Andrea Förster, and the CO_2SINK Drilling Group

Introduction

Europe's first onshore scientific carbon dioxide storage testing project CO_2SINK (CO_2 Storage by Injection into a Natural saline aquifer at Ketzin) is performed in a saline aquifer in NE Germany. The major objectives of CO_2SINK are the advancement of the science and practical processes for underground storage of carbon dioxide, and the provision of operational field results to aid in the development of standards for CO_2 geological storage. Three boreholes (one injection well and two observation wells) have been drilled in 2007, each to a depth of about 800 m. The wells are completed as "smart" wells containing a variety of permanent downhole sensing equipment, which has proven its functionality during its baseline surveys. The injection of CO_2 is scheduled for spring 2008 and is intended to last up to two years to allow for monitoring of migration and fate of the injected gas through a combination of downhole monitoring with surface geophysical surveys. This report summarizes well design, drilling, coring, and completion operations.

Since the publication of the Intergovernmental Panel on Climate Change Report (IPCC, 2005), carbon dioxide capture and storage, including the underground injection of CO_2 through boreholes, became a viable option to mitigate atmospheric CO_2 release. One of the major goals for the immediate future is to investigate the operational aspects of CO_2 storage and whether the risks of storage can be successfully managed.

CO_2SINK is the first European research and development project on *in situ* testing of geological storage of CO_2 in an onshore saline aquifer (Förster et al., 2006). Key objectives of the project are to advance understanding of and develop practical processes for underground storage of CO_2, gain operational field experience to aid in developing a harmonized regulatory framework and standards for CO_2 geological storage, and build confidence towards future set in "projects of that kind".

The CO_2SINK site is located near the town Ketzin to the west of Berlin, Germany (Fig. 1). The plan is to inject into a saline aquifer over a period of two years a volume of approximately 60,000 t of CO_2. For this purpose, one vertical injection well (Ktzi-201) and two vertical observation wells (Ktzi-200 and Ktzi-202) were drilled at a distance of 50 m to 100 m from each other (Fig. 1). All three wells are equipped with downhole instrumentation to monitor the migration of the injected CO_2 and to complement the planned surface geophysical surveys. The injection of CO_2 will be interrupted at times for repeated downhole seismics (VSP, MSP), cross-hole seismic experiments, and downhole geoelectrics.

The preparatory phase for CO_2 injection started in April 2004 with a comprehensive geological site characterization and a baseline fluid monitoring (Förster et al., 2006). This was followed by a baseline 3-D seismic survey (Juhlin et al., 2007) and the development of a drilling and completion concept (Fig. 2) allowing for monitoring during CO_2 injection and storage observation.

Geological Background

The CO_2SINK site is located in the Northeast German Basin (NEGB), a subbasin of the Central European Basin System. The sedimentary succession in the NEGB is several kilometers thick containing geological formations of Permian to Quaternary age, comprising abundant deep saline aquifers. The CO_2 will be injected into the Stuttgart Formation (lower portion, Fig. 3) of Triassic (Middle Keuper) age, into the southern flank of a gently dipping double anticline.

The 80-m-thick target formation rests at about 630–710 m depth at a temperature of about 38°C. The formation is made up of siltstones and sandstones interbedded by mudstones deposited in a fluvial environment. The reservoir is in sandstone channels as well as levee and crevasse splay deposits. These channel-(string)-facies rocks alternate with muddy

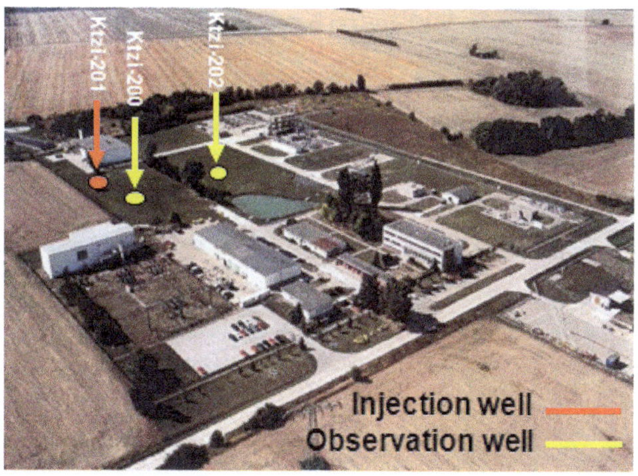

Figure 1. Location of boreholes at Ketzin industrial park.

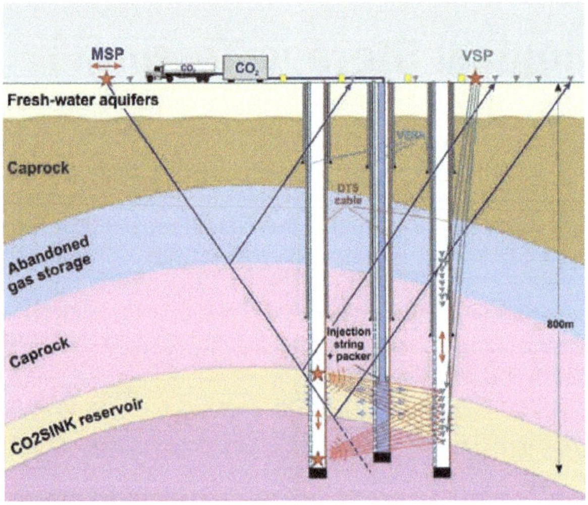

Figure 2. Schematic concept of Ketzin geology, drilling, and geophysical monitoring including moving source seismic profiling (MSP), vertical seismic profiling (VSP), cross-hole seismics, surface 3-D seismics, and surface and cross-hole geoelectrics using a permanently installed Vertical Electrical Resistivity Array (VERA), and Distributed Temperature Sensing (DTS) (from Förster et al., 2006).

flood-plain-facies rocks of poor reservoir quality. A geostatistical approach applied to the reservoir architecture (Frykman et al., 2006) pointed towards variable dimensions of the sandstone bodies and was supported by continuous wavelet transforms on 3-D seismic data (Kazemeini et al., 2008).

The Stuttgart Formation is underlain by the Grabfeld Formation (Middle Keuper), which is a thin-bedded mudstone succession with interbedded marlstone, marly dolomite and thin anhydrite or gypsum beds deposited in a clay/mud-sulfate playa depositional environment (Fig. 3; Beutler and Nitsch, 2005). The immediate caprock of the Stuttgart Formation, the Weser Formation (Middle Keuper), also is of continental playa type, consisting mainly of fine-grained clastics such as clayey and sandy siltstone that alternate with thin-bedded lacustrine sediments, like carbonates, and evaporites (Beutler and Nitsch, 2005). The high clay-mineral content and the observed pore-space geometry of these rocks attest sealing properties appropriate for CO_2 capture (Förster et al., 2007). The Weser Formation is overlain by the Arnstadt Formation (Middle Keuper), again of lacustrine character (mud/clay-carbonate playa; Beutler and Nitsch, 2005) with similar sealing properties. The two caprock formations immediately overlying the Stuttgart Formation are about 210 m thick (Fig. 3).

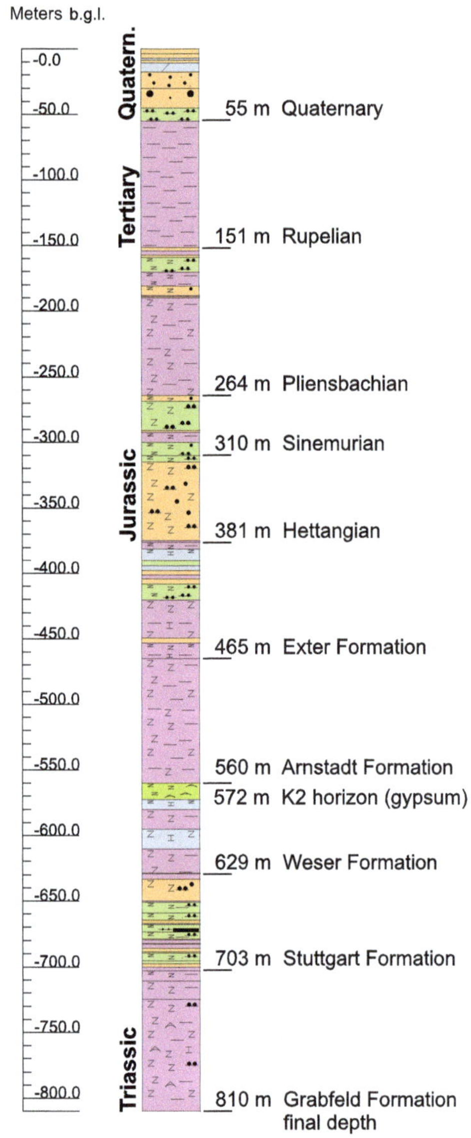

Figure 3. Condensed geological profile of the Ktzi 200/2007 borehole. Lithological color code: mudstone (magenta), siltstone (green), sandstone (yellow), anhydrite (light blue).

Borehole Design

All three wells were designed with the same casing layout, including stainless production casings equipped with pre-perforated sand filters in the reservoir section and wired on the outside with a fiber-optical cable, a multi-conductor copper cable, and a PU-heating cable to surface (Table 1). The reservoir casing section is externally coated with a fiber-

Table 1. Casing Schemes

	Depth [m]	Diameter [inch]	[mm]	[lb/ft]	Quality	Connection
Stand pipe	30	24	610	125.5	4140	welded
Conductor	150	18 5/8	473	87.5	X56	Buttress-BTC
Reserve Casing	ca.340	13 3/8	340	54.5	K-55	Buttress-BTC
Intermediate	590	9 5/8	244	36	K-55	Buttress-BTC
Production String	800	5 1/2	140	20	13Cr80 (outside coating)	VAM Top
Injection String	680	3 1/2	89	9.3	C-95 (inside coating)	TS-8

glass resin wrap for electrical insulation. A staged cementation program was planned around the application of newly developed swellable elastomer packer and stage cementation downhole tools. This technology was preferred over perforation work that would have caused unmanageable risks of potential damage of the outside casing cables.

The 200-m core sections for detailed reservoir and sealing property investigations were recovered with a 6" x 4" wire-line coring system using polycrystalline diamond compact (PDC) core bits. The 6 1/4" core hole sections were enlarged to 8 1/2", and the wells finally deepened below the reservoir zone to accommodate sufficient sensor spacing for installation of behind-casing sensor arrays.

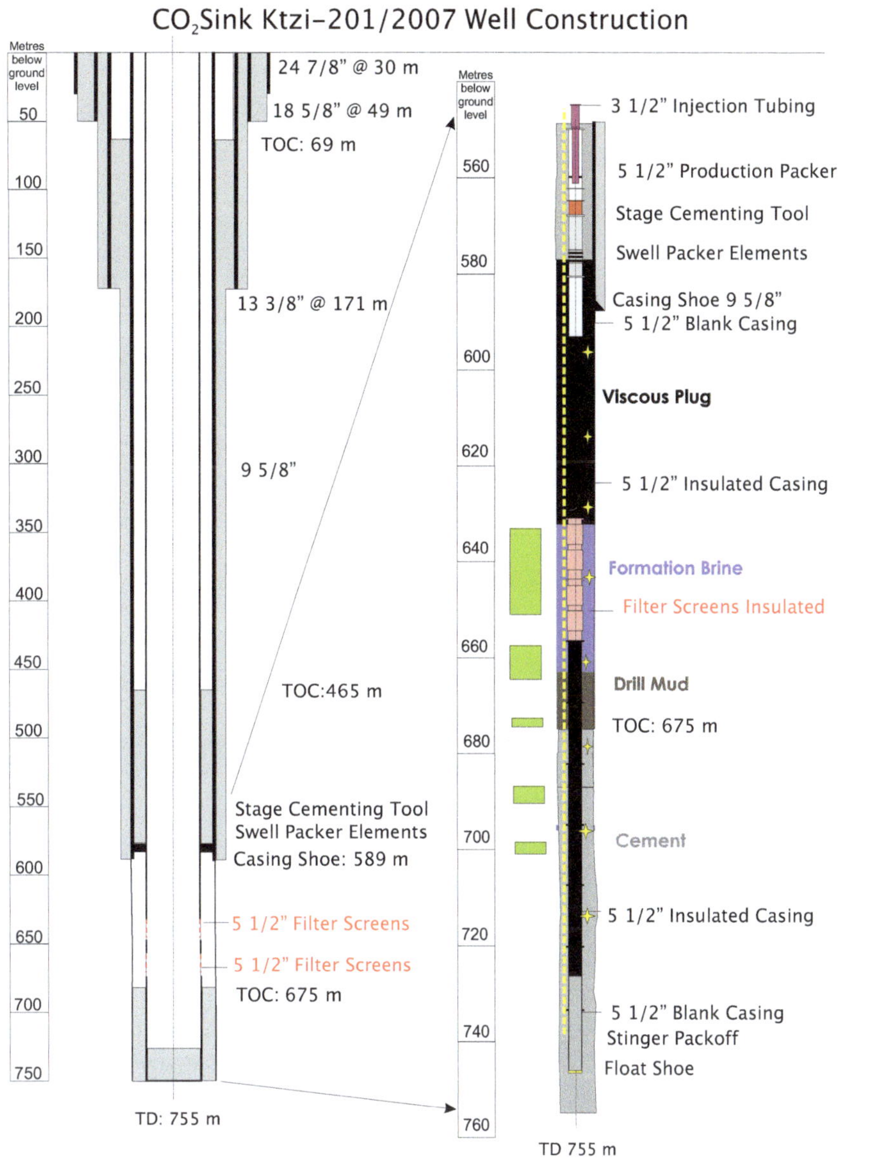

Figure 4. Drilling design and well completion of the Ktzi 201/2007 borehole. Yellow line indicates DTS and ERT cables with location of ERT electrodes (yellow pluses). Sandstone reservoir intervals are shown in green.

Drilling and Completion Operations

Constructing three wells close to each other and with such a dense sensor and cable population requires detailed planning. For this purpose, high-end oilfield QHSE (Quality, Health, Safety, Environment) management tools were applied, such as "drill well on paper" (DWOP), hazardous operation identification, repeated incident reporting, post job analysis, and risk management.

Drill site construction started in December 2006, and the drilling operation commenced on 13 March 2007 with the mobilization of a truck-mounted and top-drive equipped rotary drill rig. All the Ketzin wells were drilled with a shale inhibited KCl-water-based mud system, with the exception of the top-hole section in the fresh-water aquifers, where a K_2CO_3-water-based system was required by the authorities. Both drill muds were conditioned at 1.05–1.16 gcm^{-3} density. In order to avoid potential risks from environmental hazards, the project further implemented a "shallow gas" procedure in this well section to avoid spills when the wells would encounter high pressurized shallow gas from the past gas storage activity. For this purpose, the top-hole section of the first borehole was pre-drilled with a blow-out preventer/diverter/gas-flare installation on the rig to capture and control unexpected and sudden shallow gas influxes. As no stranded shallow gas was encountered during drilling (as also confirmed by reconnaissance wire-line logging and surface seismic processing), this pilot drilling was consequently skipped for the second and third well. Casing (18 5/8") running and cementation with stinger to surface were performed in all three wells without problems.

In the following 12 1/4" sections, the wells penetrated the Jurassic aquifer systems in which under-balanced pressure regimes were supposed. All wells encountered a minimum of three loss circulation zones between 366 m and 591 m with cumulative mud losses of 550 m^3. The addition of medium- to coarse-grained shell grit to the mud cured the loss of circulation and brought the wells safe to the 9 5/8" casing depth between 588 m and 600 m.

The lower part of Weser Formation and the entire Stuttgart reservoir section were cored with a specially designed $CaCO_3$-water/polymer drilling mud (1.1 g cm^{-3}). In the first well, a total of 100 m core was drilled in thirty-nine core runs, and an average recovery of 97% was achieved. In the second well 80 meters of core was retrieved in thirty-one runs (100% recovery). In the third well only the top 18 m of the Stuttgart Formation was cored with the same excellent performance. The 6 1/4" core hole section was then enlarged to 8 1/2", and the wells finally deepened below the reservoir into the Grabfeld Formation.

Stainless steel 5 1/2" production casings (Fig. 4) were installed and cemented in all wells with sensors and cables on the outside. The cables were terminated and fed pressure tight at the wellhead to the outside through the drilling spool below the casing slips. The cement selected in all casing cementations was standard class-G with fresh water and no additives (SG = 1.98 kg L^{-1}), with the exception of the plug cementation, for which a specially designed CO_2-resistant class-G salt cement was selected.

The CO_2 injection well was completed with a gas-tight and internally coated production tubing, including a permanent production packer above the injection horizon, a fiber-optic pressure and temperature mandrel/gauge arrangement above the packer and a wire-line-retrievable subsurface safety valve at 50 m depth below the well head. The optical cables and hydraulic safety valve actuation lines were clamped to the outside of the production tubing and fed pressure tight to the outside at the tubing hanger adaptor below the Christmas tree gate valves.

Permanent Downhole Sensors for Monitoring of CO_2

Geophysical monitoring techniques are applied in CO_2SINK to delineate the migration and saturation of injected CO_2 (Fig. 2). The injection well and the two observation wells are equipped with state-of-the-art as well as newly developed geophysical sensors. The data from this permanent downhole monitoring will be interpreted in combination with data from periodic seismic monitoring (VSP, MSP, and cross-hole seismics) and periodic fluid sampling and well logging (Reservoir Saturation Tool).

The following permanent components were installed in the boreholes for scientific monitoring:

- a fiber-optic-sensor cable loop for Distributed Temperature Sensing (DTS; all wells)
- a two-line electrical heater cable (Ktzi 201/2007, Ktzi 202/2007)
- a Vertical Electrical Resistivity Array (VERA) consisting of fifteen toroidal steel electrodes, 15-line surface connection cable (all wells)
- fiber-optic pressure/temperature (P/T) sensor, fiber-optic surface connection cable (at injection string only).

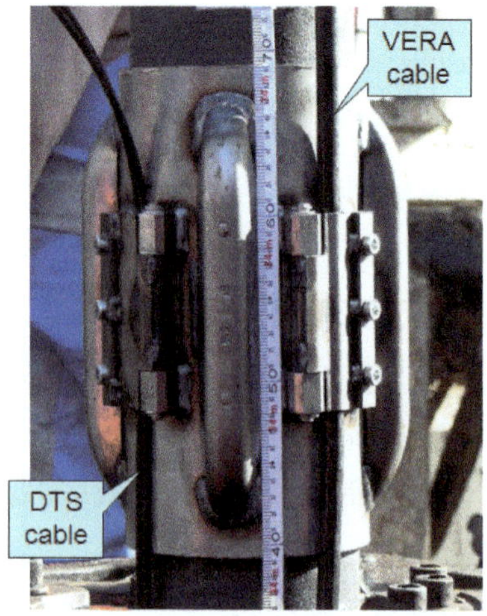

Figure 5. Centralizer attached to casing string with DTS (left) and VERA cables (right).

Using the DTS technology, quasi-continuous temperature profiles can be measured on-line along the entire length of the wells with high temporal and spatial resolution (Förster et al., 1997; Büttner and Huenges, 2003). The permanent installation of DTS sensors behind the casing (Hancock et al., 2005; Henninges et al., 2005) offers the advantage of full access to the well during technical operations, which, for example, allows control of the process of casing cementation (Henninges and Brandt, 2007). The borehole temperature data will primarily serve in the delineation of physical properties and of the state of the injected CO_2. To enhance the thermal signal and improve the monitoring of brine and CO_2 transport, successive thermal perturbation experiments (Freifeld et al., 2006) will be performed, using the electrical heater cable installed adjacent to the DTS cables. VERA provides data on the CO_2 saturation employing the Electrical Resistivity Tomography (ERT) method. Each of the VERA arrays covers an interval of about 140 m centered in the injection horizon and consisting of fifteen electrodes spaced at about 10-m intervals. The P/T-sensor installed at the bottom of the injection string above the packer system will continuously monitor the downhole pressure and temperature changes during injection. Data will be transferred via optical fiber attached to the injection string.

The inclusion of the permanent downhole sensors into the well completion required a selection of suitable completion components and procedures. Custom-made casing centralizers were used for outside-casing installation of sensor cables, for centralization of the casing inside the borehole, and for protection of cables from mechanical damage during installation (Fig. 5). The 8 1/2" borehole diameter in the lower reservoir sections allowed for sufficient clearance within the annular space between casing and borehole wall and thus for a safe installation of the downhole sensors. Within the 140-m zone, where the VERA electrodes are placed, the steel casing was electrically insulated outside using a fiberglass coating.

After an on-site installation test had been conducted, the installation of the DTS and VERA cables (Fig. 5) and electrodes in the Ktzi 200, 201, and 202 wells was performed on 5 May, 5 July, and 18 August 2007. After careful installation operations of up to 18–24 h duration, the cables were guided into the substructure of the drill rig, and the casing was cemented.

The DTS monitoring allowed online monitoring and control of the cementing operations and provided valuable information about the positions of the cemented sections during the setting of the cement. This information was verified by subsequent industry-standard cement-bond logs. The installation of monitoring tools was finished by feeding the cables into the casing spool at the wellhead, which was subsequently pressure-sealed using a stuffing box. Preliminary tests of VERA have shown that all electrodes and cables are fully functional.

Field Laboratory

The CO_2SINK field laboratory comprised core-cleaning and core-sealing facilities, a full core imager, and a Geotek gamma-ray density core logger. The field lab was designed to record and describe a high core-run volume within a short handling time to quickly generate the litholog for the drilled boreholes and to identify the reservoir section. This procedure was necessary in order to proceed rapidly with decision making on the selection of the borehole intervals completed with filter casings through which the CO_2 would be injected into the formation or monitored.

In the preparation for unconsolidated sandstone in the Stuttgart Formation, coring was performed with PVC liners in 3-m liner intervals. At the drill rig, liners were cut after orientation marking into 1-m sections, and the cut surface geologically described was sealed before being transferred to the field lab for analyses. Sections containing sandstone were shipped preserved in liners to a commercial laboratory for "hot-shot" poro-perm analysis. Reservoir sandstone intervals (Fig. 6) with porosities on the order of 20%–25%, together with requirements for permanent ERT sensor arrangement on the casing, guided the depths at which the wells were completed with filter screens for CO_2 injection and monitoring.

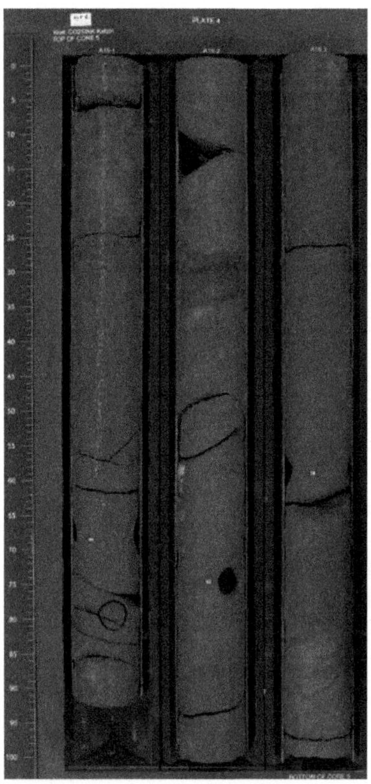

Figure 6. Core image of reservoir sandstone showing cross-bedding.

The geological description of core started with the sections of well-cemented mudstone after its cleaning with synthetic formation water, reorientation, and scanning unrolled using an optical core scanner. Later, the "hot-shot" reservoir sections were included. From the geological core and cutting descriptions and interpreted petrophysical well logs, stratigraphic-lithologic logs (Fig. 3) were finally generated for all three CO_2SINK wells to refine the geological model. For example, the stratigraphic-lithologic logs were used to calibrate the 3-D seismic time sections (Juhlin et al., 2007). Petrographical and mineralogical studies and geochemical analyses from reservoir and caprock were performed to characterize the Ketzin site on micro-scale as a basis for fluid-rock-alteration modeling.

Outlook

CO_2SINK is the first project that extensively uses behind-casing installations for a study of the CO_2 injection and storage process in a geological medium. In this regard, CO_2SINK differs from other scientific projects of CO_2 test storage, such as the Frio experiment in Texas (Hovorka et al., 2006), the Nagaoka experiment in Japan (Kikuta et al., 2004), the field test in the West Pearl Queen Reservoir in New Mexico (Pawar et al., 2006), and the Otway Basin pilot project in Australia (Dodds et al., 2006).

It is envisaged that the extensive set of data generated by cross-correlation of seismic surface monitoring, well-logging and monitoring, and simulations, will allow for verification of *a priori* scenarios of storage/migration of fluids. Emphasis, for example, will be given to the observation of non-isothermal effects in the storage formation during injection as described by Kopp et al. (2006). This type of effect also can occur during leakage from a storage reservoir along a fracture zone as numerically investigated by Pruess (2005). Thus, the observations in progress will contribute to a sound understanding of the thermodynamic processes of CO_2 injection at well-scale as well as in the short and longer term the processes during CO_2 storage at larger scale.

Acknowledgements

We would like to thank all partners of the CO_2SINK project for their continued support and contributions that helped to finish the three wells in a healthy and environmentally safe manner. Special thanks go to Shell International and StatoilHydro for their most valuable advice and operational support during the planning and drilling phase in Ketzin. Furthermore, thanks go to VNG AG for letting us use their site at Ketzin and for their local logistical support on site. The CO_2SINK project receives its funding from the European Commission (Sixth Framework Program, FP6) and two German ministries, the Federal Ministry of Economics and Technology (CO_2-Reduction-Technologies for fossil fuelled power plants, COORETEC Program), and the Federal Ministry of Education and Research (Geotechnologien Program).

References

Beutler, G., and Nitsch, E., 2005. Paläographischer Überblick. *In* Beutler, G. (Ed.), *Stratigraphie von Deutschland IV, Keuper.* Stuttgart (Courier Forschung-Institut Senckenberg, 253), 15–30.

Büttner, G., and Huenges, E., 2003. The heat transfer in the region of the Mauna Kea (Hawaii) – constraints from borehole temperature measurements and coupled thermo-hydraulic modeling. *Tectonophysics*, 371:23–40.

Dodds, K., Sherlock, D., Urosevic, M., Etheridge, D., de Vries, D., and Sharma, S., 2006. Developing a monitoring and verification scheme for a pilot project, Otway Basin, Australia. *Proceedings GHGT-8 Conference*, Trondheim, Norway, CD-ROM.

Förster, A., Norden, B., Zinck-Jørgensen, K., Frykman, P., Kulenkampff, J., Spangenberg, E., Erzinger, J., Zimmer, M., Kopp, J., Borm, G., Juhlin, C., Cosma, C.-G., and Hurter, S., 2006. Baseline characterization of the CO_2SINK geological storage site at Ketzin, Germany. *Environmental Geosciences*, 13:145–161.

Förster, A., Schrötter, J., Merriam, D.F., and Blackwell, D.D., 1997. Application of optical-fibre temperature logging; an example in a sedimentary environment. *Geophysics*, 62(4): 1107–1113.

Förster, A., Springer, N., Beutler, G., Luckert, J., Norden, B., and Lindgren, H., 2007. The mudstone-dominated caprock system of the CO_2-storage site at Ketzin, Germany. *Proceedings of the 2007 AAPG Annual Convention and Exhibition*, Long Beach, Calif., U.S.A., CD-ROM.

Freifeld, B.M., Walker, J., Doughty, C., Kryder, L., Gilmore, K., Finsterle, S., and Sampson, J., 2006. Evidence of rapid localized groundwater transport in volcanic tuffs beneath Yucca Mountain, Nevada. *Eos Trans. AGU*, 87:52:H43A-0480.

Frykman, P., Zinck-Jørgensen, K., Bech, N., Norden, B., Förster, A., and Larsen, M., 2006. Site characterization of fluvial, incised valley deposits. *Proceedings CO2SC Symposium*, Lawrence Berkeley National Laboratory, Berkeley, Calif., U.S.A., 121–123.

Hancock, S., Collett, T.S., Dallimore, S.R., Satoh, T., Inoue, T., Huenges, E., Henninges, J., and Weatherill, B., 2005. Overview of thermal-stimulation production-test results for the JAPEX/JNOC/GSC Mallik 5L-38 gas hydrate production research well. *In* Dallimore, S.R., and Collett, T.S. (Eds.), *Scientific Results from the Mallik 2002 Gas Hydrate Production Research Well Program, Mackenzie Delta, Northwest Territories, Canada, Geological Survey of Canada, GSC Bulletin*, 585:CD-ROM.

Henninges, J., and Brandt, W., 2007. Evaluation of cement integrity using distributed temperature sensing. *Proceedings Engine Workshop 4 "Drilling cost effectiveness and feasibility of high-temperature drilling"*, ISOR, Reykjavik, Iceland, 41p.

Henninges, J., Schrötter, J., Erbas, K., and Huenges, E., 2005. Temperature field of the Mallik gas hydrate occurrence – implications on phase changes and thermal properties. *In* Dallimore, S.R., and Collett, T.S. (Eds.), *Scientific Results from the Mallik 2002 Gas Hydrate Production Research Well Program, Mackenzie Delta, Northwest Territories, Canada, Geological Survey of Canada, GSC Bulletin*, 585:CD-ROM.

Hovorka, S.D., Benson, S.M., Doughty, C.K., Freifeld, B.M., Sakurai, S., Daley, T.M., Kharaka, Y.K., Holtz, M.H., Trautz, R.C., Nance, H.S., Myer, L.R., and Knauss, K.G., 2006. Measuring permanence of CO_2 storage in saline formations – The Frio experiment. *Environmental Geosciences*, 13:103–119.

IPCC, 2005. *IPCC Special Report on Carbon Dioxide Capture and Storage.* Prepared by Working Group III of the IPCC [Metz, B., Davidson, O., de Coninck, H.C., Loos, M., and Meyer, L.A. (Eds.)]. Cambridge (Cambridge University Press), 442 pp.

Juhlin, C., Giese, R., Zinck-Jørgensen, K., Cosma, C., Kazemeini, H., Juhojuntti, N., Lüth, S., Norden, B., and Förster, A., 2007. 3D baseline seismics at Ketzin, Germany: the CO_2SINK project. *Geophysics*, 72(5):B121–B132.

Kazemeini, H., Juhlin, C., Zinck-Jørgensen, K., and Norden, B., 2008. Application of continuous wavelet transform on seismic data for mapping of channel deposits and gas detection at the CO_2SINK site, Ketzin, Germany. *Geophysical Prospecting*, accepted.

Kikuta, K., Hongo, S., Tanase, D., and Ohsumi, T., 2004. Field test of CO_2 injection in Nagaoka, Japan. *Proceedings GHGT-7 Conference*, Vancouver, Canada, CD-ROM.

Kopp, A., Bielinski, A., Ebigbo, A., Class, H., and Helmig, R., 2006. Numerical investigation of temperature effects during injection of carbon dioxide into brine. *Proceedings GHGT-8 Conference*, Trondheim, Norway, CD-ROM.

Pawar, R.J., Warpinski, N.R., Lorenz, J.C., Benson, R.D., Grigg, R.B., Stubbs, B.A., Stauffer, P.H., Krumhansl, J.P., and Cooper, S.P., 2006. Overview of a CO_2 sequestration field test in the West Pearl Queen reservoir, New Mexico. *Environmental Geosciences*, 13(3):163–180.

Pruess, K., 2005. Numerical simulations show potential for strong nonisothermal effects during fluid leakage from a geologic disposal reservoir for CO_2. *In* Faybishenko, A., Witherspoon, P.A., and Gale, J. (Eds.), *Dynamics of Fluids and Transport in Fractured Rock, Geophysical Monograph Series 162*, Washington, DC (American Geophysical Union), 81–89.

Authors

Bernhard Prevedel, Lothar Wohlgemuth, Jan Henninges, Kai Krüger, Ben Norden, Andrea Förster, GeoForschungsZentrum Potsdam, Telegrafenberg, D-14473 Potsdam, Germany, e-mail: prevedel@gfz-potsdam.de, and **CO_2SINK Drilling Group:** Ronald Conze, Knut Behrends, Erik Danckwardt, Jörg Erzinger, Tor Harald Hanssen, Jochem Kück, Dana Laaß, Andre Mol, Fabian Möller, Peter Pilz, Mathias Poser, Manfred Rinke, Cornelia Schmidt-Hattenberger, Birgit Schöbel, Jörg Schrötter, Hartmut Schütt, Gerardo Stapel, Thomas Wöhrl, Hilke Würdemann, and Martin Zimmer.

Related Web Links

http://www.co2sink.org/
http://www.icdp-online.org/
http://www.gfz-potsdam.de/

Photo Credits

Fig. 1. VNG - Verbundnetz Gas AG, Leipzig, Germany
Fig. 5. Silvio Mielitz, GeoForschungsZentrum Potsdam, Germany

The First *D/V Chikyu* IODP Operations: Successful Logging and Coring During NanTroSEIZE Stage 1 Expeditions

Shin'ichi Kuramoto, Daniel Curewitz, Moe Kyaw Thu, Hideki Masago, and the Exp. 314, 315, and 316 Science Parties

The Nankai Trough Seismogenic Zone Experiment (NanTroSEIZE) is a multi-expedition IODP drilling project aimed at drilling, coring, logging, and instrumenting the seismogenic zone of an active subduction margin , in a region thought to generate megathrust earthquakes of magnitude >8.0 on the moment-magnitude scale (Tobin and Kinoshita, 2006). The Nankai Trough, offshore of the Kii Peninsula, Honshu, Japan (Fig. 1) was chosen as the location for this project based on a number of scientific drilling proposals to IODP. These reviewed existing drilling data in the region, the long-term historical and recent record of great earthquakes, the social and societal relevance of the area, and the accessibility of the seismogenic zone to present drilling technology. The first stage of this multi-stage project was intended to accomplish a broad characterization of the shallow geology, geophysics, physical properties, heat flow, and fluid flow in a transect across the downgoing Philippine Sea Plate, the toe of the Nankai accretionary prism, the megasplay fault zone region on the continental slope, and the Kumano Basin that lies between the accretionary prism and the Kii Peninsula, on the continental shelf (Fig. 2).

Between September 2007 and February 2008, IODP Expeditions 314, 315, and 316 were carried out in order to complete Stage 1 of the NanTroSEIZE; operations included Logging While Drilling (LWD), coring, and downhole measurement at eight sites (thirty-three holes) in the Nankai Trough accretionary prism. On 21 September 2007, *Chikyu* left Shingu port, Wakayama Prefecture, Japan, for her first scientific voyage as an IODP drilling platform. The start of NanTroSEIZE operations took place seven years after the first scientific proposals were submitted (see Kimura et al., 2003; Kinoshita et al., 2003; Suyehiro et al., 2003; Underwood et al., 2003; Screaton et al., 2005), and marked the culmination of thousands of hours of preparation and planning by a Project Management Team (PMT), the IODP Science Advisory Structure (SAS), and the CDEX engineers, technicians, and marine workers.

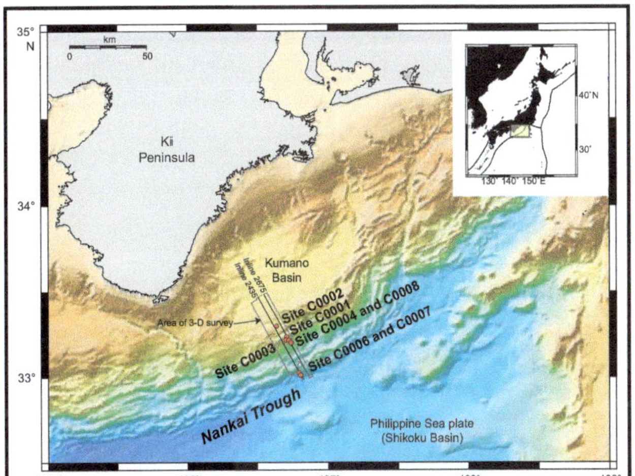

Figure 1. Location of drilled sites during NanTroSEIZE Project Stage 1 in the Nankai Trough off Kumano. Gray box = location of the 2006 three-dimensional seismic survey, black lines = 3-D inlines 2435 and 2675 (see Figure 2), red circles = location of drilling sites. Inset shows location of Nankai Trough.

After 138 days of continuous operations, Stage 1 operations were completed on 5 February 2008. The overviews of each operation as well as the overall accomplishments of Stage 1 are described below. Preliminary reports for each of the expeditions have been published and posted on the Web (http://www.iodp.org/scientific-publications/).

Expedition 314: LWD Transect

Expedition 314, NanTroSEIZE Stage 1 LWD Transect, was planned to obtain a comprehensive suite of geophysical logs and other downhole measurements at six primary sites along a transect: two on the incoming plate, two through major active thrust faults, and two pilot holes for deeper riser drilling that also address scientific targets in the splay fault thrust sheet and the Kumano forearc basin and underlying prism.

Due to the difficulties experienced during attempted wireline logging and relatively higher success during limited LWD operations in previous accretionary prism drilling within ODP, Expedition 314 was entirely dedicated to the LWD effort. LWD tool selection focused on maximizing scientific returns from the tools and the time needed for deployment. The LWD tools used during Expedition 314 were state-of-the-art industrial tools, and this expedition marked the first ODP/IODP use of check shot interval velocity, and sonic velocity, and density/porosity measurements while drilling tools that have only been used on very limited occasions.

The operational plan included one-third of the total days (22.5 days out of 57 days) for contingencies (typhoons, mechanical downtime) and casing operations. Six primary and two contingency sites were planned for the expedition with the drilling order determined by scientific priority and operational difficulty. The location of planned sites and the order of drilling changed after the team faced various challenges. Allotted contingency days were consumed by expected mechanical downtime caused by the strong Kuroshio Current flowing through the area, by difficult drilling conditions, and by unforeseen mechanical and operational events.

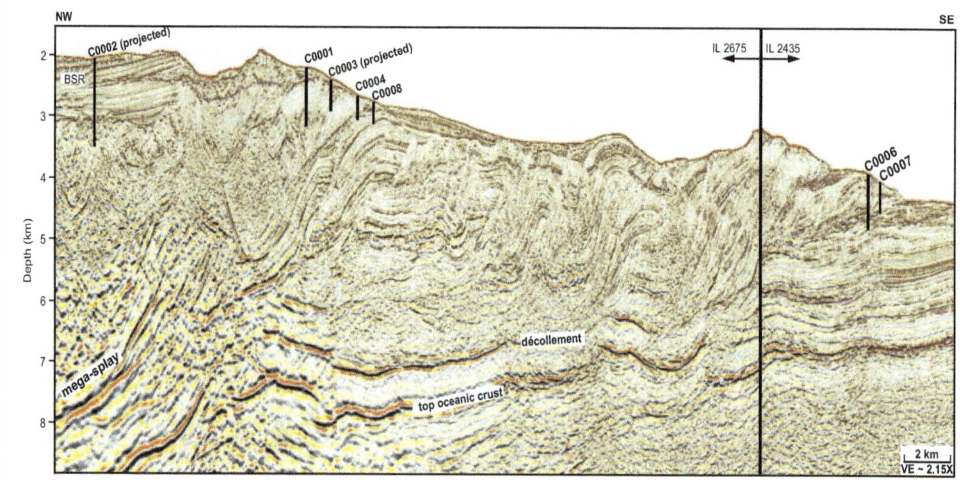

Figure 2. NW-SE slice through the NanTroSEIZE 3-D seismic survey (composite of 3-D Inlines 2435 and 2675) showing the projected locations of the sites, cored and logged during Stage 1. (Moore et al., 2007)

During the eight-week LWD campaign, various measurements were successfully completed at four of the twelve total drilled holes, with LWD data coverage ranging from 400 mbsf to 1400 mbsf (Fig. 3). Tool failure, loss of the Bottom Hole Assembly (BHA) due to borehole collapse, and pilot hole drilling (required for operational safety) consumed the remaining drilling time and accounted for the eight holes not fully logged. Logging data were initially analyzed by the shipboard scientists during the expedition in order to make the LWD data available to the scientific parties of the subsequent NanTroSEIZE Stage 1 coring expeditions (Exp. 315, 316). During the expedition, the shipboard science party worked in four groups: log characterization and lithologic interpretation, physical properties, structural geology and geomechanics, and core-log-seismic integration using various applications for log and seismic data analyses.

The principal results of Expedition 314 reveal a wealth of information about the accretionary prism, and provide a strong foundation for future planning of NanTroSEIZE, analysis of cores, downhole measurements, and observatory monitoring data. Additionally, these data provide a starting point for addressing questions raised about current models of the margin, including the composition and physical characteristics of the stacked thrust sheets comprising the prism, the distribution and character of the faults bounding the structural blocks and thrust sheets, information about the stress state in different areas across the margin, the distribution and occurrence of gas hydrates, and the characteristics of the sediments deposited in both the ancient, transported rocks and the more recent sediments draping the margin. Drilling of Sites C0001 and C0002 will also provide pilot hole information important in preparation for the planned deep-riser sites for later stages of NanTroSEIZE drilling.

Figure 3. LWD drilling operations during Expedition 314. [A] Real time monitoring in the lab area; [B] LWD toolstring on the rig floor; [C] Driller's house operations.

Expedition 315: Megasplay Riser Pilot

IODP Expedition 315 took place between 16 November and 18 December 2007, planned as a geotechnical and scientific pilot study for future riser drilling of the megasplay fault (Fig. 2). Borehole LWD data obtained during Expedition 314 was available for use during Expedition 315; this proved extremely beneficial in planning for drilling, coring, and sampling of the scheduled operations.

The operational plan for the expedition was to drill and core to 1000 mbsf at site C0001, a site considered for future riser drilling, then install riser top-hole casing in the latter half of the expedition. However, the Kuroshio Current was deemed to be running too strong (3–5 knots) to safely install the 36-inch and 20-inch casing strings required for the riser pilot. Drilling and operation plans were changed, and Expedition 315 was devoted to coring and downhole temperature measurement at two sites, C0001 and C0002 (Figs. 1 and 2).

Coring at C0001 was conducted with the foreknowledge that the borehole conditions might deteriorate quickly at or near 500 mbsf, in the same formation that proved so difficult during Expedition 314 and where caving and borehole collapse resulted in the loss of the LWD BHA. Coring commenced with HPCS (hydraulic piston coring system) from the seafloor, and obtained good core recovery down to 230 mbsf, at which point coring was conducted using the ESCS (extended shoe coring system). Core recovery by HPCS was good (75%–90%); however, core quality using the ESCS was quite poor, and RCB (rotary core barrel) coring then commenced. Using the results of LWD, we expected to encounter a zone of drilling difficulty and borehole instability around 450–600 mbsf. Accordingly, the strategy was to wash down to and through this formation and try to core the formations below that. We succeeded in drilling down to the target depth; however, caving and borehole collapse began nearly immediately, precluding attempts to core below the unstable region.

Because riser top-hole casing operations were postponed due to the intense current, Site C0002 in the Kumano Basin was chosen for coring, and it was decided that the deeper section was the most critical target. RCB coring began after drilling down to 475 mbsf, and sixty-six cores were collected from that point down to the total depth. We penetrated the basal unconformity of the Kumano basin at 922 mbsf and continued more than 100 m into accretionary prism materials with moderate recovery. As borehole conditions steadily worsened deeper in the accretionary prism, the hole was abandoned at 1057 mbsf. Two short coring operations using the HPCS system were conducted between the seafloor and 204 mbsf for both geotechnical investigations and for scientific data collection.

As the first coring expedition for the *Chikyu*, Expedition 315 provided valuable scientific and geotechnical data that will be critical in understanding the Nankai accretionary prism off the Kii Peninsula. In addition, the expedition proved to be invaluable in terms of improving drilling, management, and laboratory performance under harsh weather, strong current, and unfavorable operational conditions. The drilling conditions encountered at intermediate depths at C0001 and C0002 provide important constraints on riserless drilling operations in the fractured and faulted formations that comprise much of the shallow prism, and they will provide data necessary to plan future drilling, coring, and casing operations. Critical scientific data was acquired from the upper part of the accretionary prism at both sites, including geological (lithology, structure, and age), geotechnical (physical properties, temperature) and geochemical information for the accretionary prism materials (Fig. 4). These data will help planning for the engineering and scientific aspects of future riser drilling.

Expedition 316: Shallow Megasplay and Frontal Thrusts

IODP Expedition 316 drilled several sites in a transect across the outer part of the Nankai accretionary prism. The aims of the expedition were to characterize the sedimentology, stratigraphy, and physical properties of the slope sediments, the accretionary prism rocks, and the underthrust material, and to sample and understand the character of fault rocks within fault zones controlling the geometry and evolution of the accretionary prism.

We targeted two major structures: the frontal thrust (sites C0004 and C0008) and the shallow portion of the megasplay fault zone (sites C0006 and C0007) near its intersection with the seafloor (Figs. 1 and 2). Comprehensive sampling and measurement both in boreholes and on core materials were carried out, including downhole temperature measurements, interstitial water sampling, microbiological sampling, gas sampling, and a wide array of sampling for chemistry, mineralogy, and physical properties. More than 1300 meters of core were recovered using HPCS, ESCS, and RCB. More than 5000 samples were taken from the cores for shore-based investigation, and many thousands of continuous and discrete measurements were carried out using the array of laboratory tools on board the *Chikyu*.

We successfully recovered a wide array of fault rocks from fractured rocks to breccia to fault gouge, sedimentary materials ranging from fine clay and mud to siltstone and sand, and we

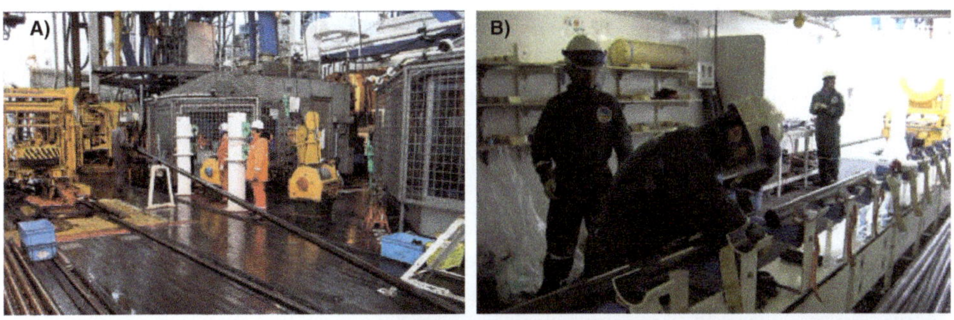

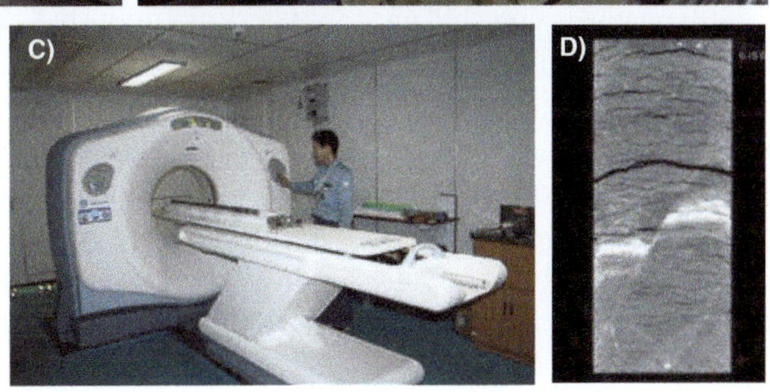

Figure 4. Coring and laboratory operations during Expedition 315. [A] Core barrel being laid out on the drill floor prior to removal of the collected core. [B] Laboratory technicians cutting a core into sections and collecting safety and time-sensitive samples prior to delivery of core section to the core lab. [C] Core section being run through the X-ray CT scanner. [D] X-ray CT image of a small fault zone cutting across interbedded mud (dark areas) and sandy ash (bright areas).

managed to recover extremely coarse grained materials from paleo-trench-axis channels. Materials sampled include recent slope apron/slope basin deposits (including several mass transport complexes that may shed light on the periodicity of slope failure), ancient accretionary prism rocks, and material that has been over-ridden during thrust faulting events (Fig. 5). Analyses of these wide-ranging data sets will shed new light on the evolution, structure, and architecture of the Nankai accretionary prism off the Kii Peninsula.

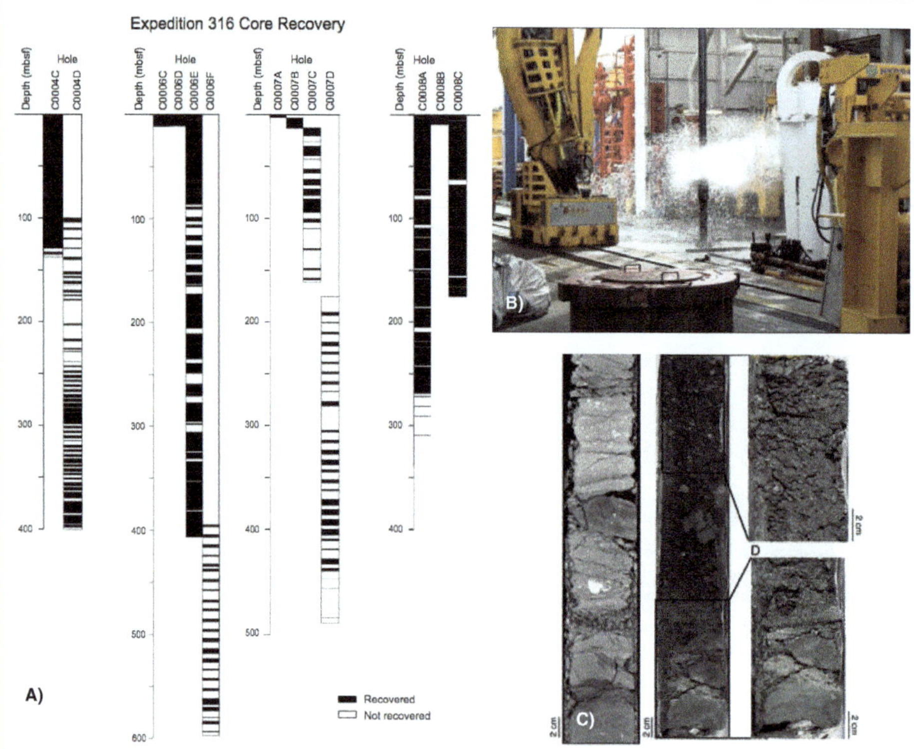

Figure 5. Coring results, coring operations and laboratory work during Expedition 316. [A] Core recovery during Exp. 316. [B] Drill pipe being disconnected immediately prior to removing a core. [C] Composite showing CT image, split core surface, and detail photos of fault zone material.

Overview and Primary Accomplishments

Chikyu accomplished an LWD and two coring expeditions in a transect across the Nankai Trough accretionary prism, gathering large volumes of data, cores, and samples that will help understand the structure and mechanics of the shallow accretionary prism, and that will aid in planning and carrying out future deep drilling to the shallow part of the lying 6–7-km-deep seismogenic zone.

Expedition 314 (LWD Transect) collected 4274 m of LWD logs and conducted 2285.5 m of MWD logs from pilot holes. Expedition 315 (Megasplay Riser Pilot) penetrated 1287 m and recovered 808 m of core. Expedition 316 (Megasplay and Frontal Thrusts) penetrated 2103 m and recovered 1340 m of core. Total drilling length was approximately 12,800 m, and total attempted coring was about 3400 m with an average core recovery rate of about 65%. All cores have been archived for preservation, future description, and sampling at the Kochi Core Center (KCC), one of the three IODP core repositories.

During the course of these expeditions, *Chikyu* was able to test and refine operational techniques and structures that are well-established components of the ODP/IODP operational paradigm, but that are still relatively untested on this newest platform for scientific ocean drilling. These include severe, real-world tests of the Dynamic Positioning system, drilling, coring, and laboratory systems, expedition management, safety monitoring, and onboard operational safety and communications infrastructure. In addition, *Chikyu* was able to pioneer the use of new tools and equipment, including the scientific use of never-before-used LWD tools, testing of new coring systems and techniques, testing of new downhole temperature measurement tools, and extensive use of X-ray Computed Tomography (XCT, Fig. 6) on whole core sections. The use of XCT scanning in particular proved to be an exciting and revolutionary addition to the scientific drilling, as it allows inspection of core before any destructive measurement or sampling is carried out, provides real-time guidelines for sampling and handling before the core is split, and allows the identification of critical intervals for special handling. In addition, the use of XCT imaging to construct pseudo-density logs (using the "CT number" which is a semi-quantitative proxy for bulk density), to measure faults and fractures, and to identify unconformities and fine-scale details provided exciting results on board and opened a new avenue for extensive post-expedition research.

While shipboard and shore-based scientists from the NanTroSEIZE Stage 1 expeditions have already begun analyzing the huge volume of data taken from LWD and core measurements, studies for future deep riser drilling, in terms of geomechanical modeling and borehole stability, are underway in CDEX in collaboration with industry and research specialists.

An immense amount of work has been required to prepare the *Chikyu* for IODP science operations. While it is clear that there is a lot of further effort required to develop the fullest potential of the vessel and its laboratories, when we take into account the successes, problems, and challenges experienced during NanTroSEIZE Stage 1, the results of these three expeditions give us great confidence in the future of IODP operations using the *Chikyu*.

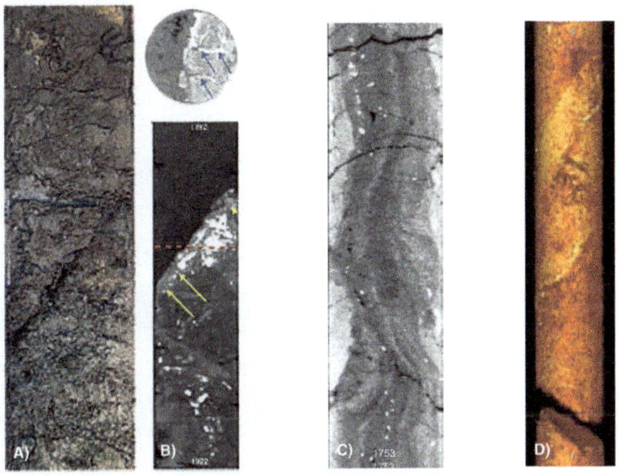

Figure 6. Usage and results from the X-ray computed tomography analyses. [A] Selected X-ray CT images. [B] The relationships that can be seen between the core photos and the CT images; Yellow and blue arrows indicate the position of the clear discontinuity seen in both core and CT image, with bright areas corresponding to mineralized material. The red line in the along-core CT image shows the position of the slice in the upper right. Spatial and angular relationships can be determined with high precision using core orientation. [C] X-ray CT imagery can be used to evaluate coring induced or drilling induced deformation or disruption of the core sample. This image shows flow-in of fluidized mud that occurred during core recovery of a piston core. [D] False color imagery based on CT-number (density contrasts related to pore spaces, water content, and mineral chemistry) can be used to highlight structural features such as this set of normal faults (cutting the core from top right to bottom left in this image). In all cases, cores are 2.5 inches across.

Acknowledgements

We would like to thank the Japanese Ministry of Education, Culture, Sports, Science and Technology (MEXT) for funding *Chikyu*, and to the host of staff, technical workers, managers, and operations personnel in all the various departments and agencies at JAMSTEC, at CDEX, at TAMU, at the ESO, and within the global science community for the work, advice, efforts, and support required to begin this scientific endeavor with such success. Great thanks are owed to the crew of the *Chikyu*, the drilling crew and operational teams, the helicopter pilots, shorebased support staff and supply vessel crews, the technical and design teams who helped construct and test the vessel prior to IODP operations, to the science parties and laboratory staff, and to the enormous number of technicians, scientists, designers, programmers, and engineers who have contributed so much of their time, expertise, and advice to the preparation and execution of the first successful IODP operations of the *Chikyu*.

References

Kimura, G., Tobin, H., and the NanTroSEIZE working group, 2003. NanTroSEIZE: The Nankai Trough Seismogenic Zone Experiment Complex Drilling Project. IODP-603 CDP3, revised proposal received 1 October 2003.

Kinoshita, M., Brown, K., Saffer, D., Henry, P., Chester, F., Goto, T., Gulick, S., Hirono, T., Ito, H., Kato, A., Kimura, G., Kopf, A., Moore, G., Moore, J.C., Nakamura, Y., Park, J.-O., Saito, S., Schwartz, S., Shinohara, M., Stephen, R., Tobin, H., Ujiie, K., Tsunogai, U., and Yamano, M., 2003. NanTroSEIZE Drilling and Observatory Phase 2. Mechanical and Hydrologic State of Mega-Splay Faults: Implications for Seismogenic Faulting and Tsunami Generation, IODP 603B-Full 2, revised proposal received October 2003.

Moore, G.F., Bangs, N.L., Taira, A., Kuramoto S., Pangborn, E., and Tobin H., 2007. Three-dimensional splay fault geometry and implications for Tsunami generation. *Science.*, 318(5853):1128–1131, doi:10.1126/science.1147195

Screaton, E., Underwood, M., Saffer, D., Wang, K., Wheat, G., Obana, K., Moore, G., Brown, K., and Ashi, J., 2005. The Nankai Trough Seismogenic Zone Experiment: Observatory Science at the Reference Sites. IODP 603D-Full2, revised proposal received April 2005.

Suyehiro, K., Tobin, H., Araki, E., Bilek, S., Goto, T., Henry, P., Kimura, G., Kato, A., Kinoshita, M., Marone, C., Moore, G., Moore, J. C., Saffer, D., Sakaguchi, A., Shinohara, M., Stephen, R., Tsutsumi, A., Ujiie, K., and Wang, K., 2003. NanTroSEIZE Drilling and Observatory Phase 3: A Window into the Seismogenic Zone. IODP 603C-Full, revised proposal received October 2003.

Tobin, H.J., and Kinoshita, M., 2006. Investigations of seismogenesis at the Nankai Trough,Japan. *IODP Sci. Prosp., NanTroSEIZE Stage 1.* doi:10.2204/iodp.sp.nantroseize1.2006. Available online at http://publications.iodp.org/scientific_prospectus/NanTroSEIZE_stage1/.

Underwood, M., Ashi, J., Soh, W., Morgan, J., Saito, S., Saffer, D., Screaton, E., Kinoshita, M., Moore, G., Kastner, M., Bilek, S., and Ujiie, K., 2003. NanTroSEIZE Reference Sites: Sampling and Measuring Inputs to the Seismogenic Zone, IODP 603A-Full2, revised proposal received April 2003.

Authors

Shin'ichi Kuramoto, Daniel Curewitz, Moe Kyaw Thu, Hideki Masago, JAMSTEC, Yokohama Institute of Earth Sciences, 3173-25 Showamachi, Kanazawa, Yokohama 236-0001 Japan, e-mail: s.kuramoto@jamstec.go.jp; daniel@jamstec.go.jp; moe@jamstec.go.jp; masagoh@jamstec.go.jp.
and the Exp. 314, 315, and 316 Science Parties

Related Web Link

http://www.iodp.org/scientific-publications/

Figure and Photo Credits

All photos courtesy of IODP and CDEX.

6

Estimation of Minimum Principal Stress from an Extended Leak-off Test Onboard the *Chikyu* Drilling Vessel and Suggestions for Future Test Procedures

by Weiren Lin, Koji Yamamoto, Hisao Ito, Hideki Masago, and Yoshihisa Kawamura

Introduction

To understand the physics of faulting and rupture propagation for the great M8-class Nankai earthquakes that recur approximately every 100 years, a comprehensive drilling project is underway: the Nankai Trough Seismogenic Zone Experiment (NanTroSEIZE; Tobin and Kinoshita, 2007), which is part of the Integrated Ocean Drilling Program (IODP). Stress levels along seismogenic faults must be known in order to understand processes controlling the timing, energetics, and extent of earthquake ruptures. For scientific drilling projects such as NanTroSEIZE, it is very important to determine the *in situ* stress state at the decollement and the mega splay fault in the Nankai Trough.

Preliminary experiments to determine the orientations and magnitudes of principal stresses in the Nankai Trough were undertaken during the NanTroSEIZE Stage 1 expeditions using borehole image analysis (stress-induced breakouts and tensile fractures; Kinoshita et al., 2008) and indirect, core-based methods such as anelastic strain recovery (ASR; Lin et al. 2006). These experiments will provide necessary and important information about *in situ* stress. However, to improve reliability and reduce experimental uncertainties in these stress determinations, it is necessary to have direct *in situ* measurements of stress magnitudes—in particular, the minimum principal stress—at depth. These direct measurements are best obtained using methods involving the initiation and propagation of hydraulic fractures at depth, such as the traditional hydraulic fracturing test, a leak-off test (LOT), or an extended leak-off test (XLOT, sometimes ELOT) (Zoback et al., 2003). In the present paper, we aim to show that with the advent of the riser drilling vessel *Chikyu*, the XLOT is applicable and effective in deep scientific ocean drilling projects.

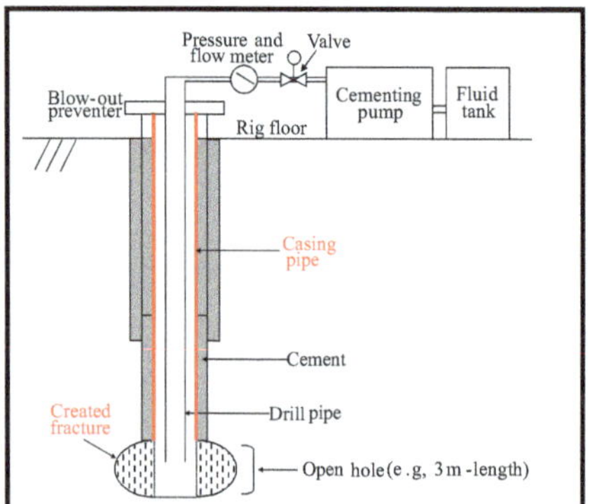

Figure 1. Schematic borehole configuration during a leak-off test (LOT) or extended leak-off test (XLOT; after Yamamoto, 2003)

During previous ODP expeditions and non-riser IODP expeditions, LOT or XLOT (which are sometimes used to determine drilling parameters such as optimal mud density) have not been conducted because the borehole was open to the seafloor. Thus, it has been impossible to pressurize a short interval of open hole below the casing as needed to conduct a LOT or XLOT (see below) without utilizing time-consuming and frequently unreliable drill-pipe-deployed packers. In contrast, the new drilling vessel *Chikyu* provides a riser-drilling capability that allows pressuring the entire casing string with drilling mud immediately after the casing is cemented in place. Therefore, NanTroSEIZE Stage 2 will present the first opportunity for a scientific ocean drilling program to use LOT or XLOT procedures without using a packer, providing direct information on the *in situ* magnitude of the minimum principal stress at minimal cost and risk.

In this study we will demonstrate the feasibility of using LOT and XLOT data acquired during the new riser-drilling program to determine stress magnitude. We will first describe LOT and XLOT procedures, and then use an XLOT data set that was acquired during the 2006 Shimokita shakedown cruise of the *Chikyu* drilling vessel to estimate the magnitude of minimum principal stress. We then recommend what we believe to be the optimum procedures for implementation of LOT–XLOT for determination of stress magnitude during future *Chikyu* riser-drilling programs.

Description of the Tests

A LOT is a pumping pressure test carried out immediately below newly set casing in a borehole (Fig. 1). It is similar to other pumping pressure tests known as the pressure integrity test, formation integrity test, or casing-shoe integrity test. Each of these tests has a different target pumping pressure. The LOT technique was originally developed in the oil industry to assess the "fracture gradient" of the formation (i.e., the maximum borehole pressure that can be applied without mud loss) and to determine optimal drilling parameters such as mud density (Kunze and Steiger, 1991). The LOT procedures are relatively simple. An XLOT is a more complex test with extended pressurizing procedures, as described in detail below. In future riser-drilling by *Chikyu*, it may be possible to regularly implement LOT or XLOT at each casing shoe immediately after casing has been run and cemented.

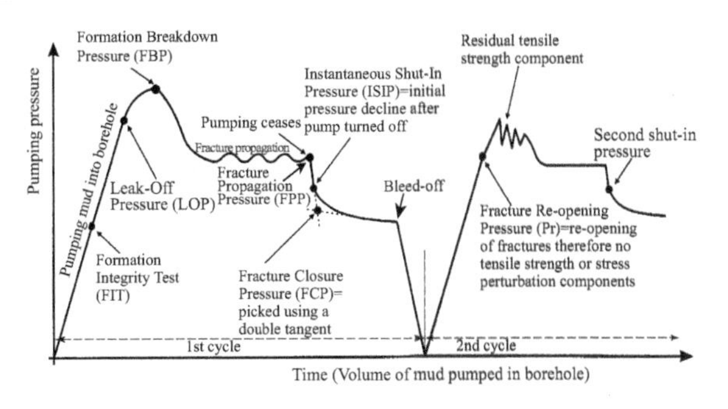

Figure 2. Idealized relationship between pumping pressure and time or volume of injected fluid during an XLOT (after White et al., 2002).

LOT and, in particular, XLOT procedures have been successfully and widely used to estimate the magnitude of minimum *in situ* horizontal stress (Addis et al., 1998; White et al., 2002; Yamamoto, 2003), mainly for the practical purpose of determining borehole stability during drilling operations. These data can be used for another important application—that is, to obtain *in situ* stress information that can be used in scientific objectives. In a similar case in which high borehole temperatures precluded use of a packer Hickman et al. (1998) conducted this kind of test to obtain *in situ* stress magnitude.

To carry out LOT or XLOT after setting casing and cementing, a short length (several meters) of extra open hole is drilled below the casing shoe. The casing shoe is then pressurized by drilling fluid delivered through drill pipe from a cementing pump set on the rig floor of the drilling vessel. The pressure at the casing shoe is equal to the sum of the hydrostatic pressure of the drilling fluid column and the ship-board pumping pressure. Figure 2 shows an idealized pumping pressure curve for XLOT (White et al., 2002).

Initially, pumping fluid into the borehole results in volumetric compression of the drilling mud column and elastic expansion of the casing string plus rock around the borehole. As the pressure in the borehole increases, the leak-off pressure (LOP) is reached when the relationship between pressure increase and volume of fluid pumped deviates from linear. This occurs when fluid begins to diffuse into the formation at a more rapid rate as the rock begins to dilate (Fig. 2). Generally, a LOT is a test that finishes immediately after LOP is reached.

An XLOT is an extended version of a LOT, but it is also similar to the hydraulic fracturing test used for stress measurement. During an XLOT, pumping continues beyond the LOP point until the pressure peaks at formation breakdown pressure (FBP). This creates a new fracture in the borehole wall. Pumping is then continued for a few more minutes, or until several hundred liters of fluid have been injected, to ensure stable fracture propagation into the undisturbed rock formation. The pumping pressure then stabilizes to an approximately constant level, which is called the fracture propagation pressure (FPP). Pumping then ceases (known as "shut-in"). The instantaneous shut-in pressure (ISIP) is defined as the point where the steep pressure decreases after shut-in deviates from a straight line. From our perspective, the most important pressure parameter is the fracture closure pressure (FCP), which occurs when the newly created fractures closes again. FCP is determined by the intersection of two tangents to the pressure versus mud volume curve (Fig. 2). The value of FCP represents the minimum principal stress (Yamamoto, 2003), because the stress in the formation and the pressure of fluid that remains in the fractures have reached a state of mechanical equilibrium. White et al. (2002) collected high-quality XLOT data and showed that both FCP and ISIP provide better estimates of minimum principal stress than LOP, although the difference in the values of LOP and ISIP was small in their study. In addition, ISIP is visually easier to determine than FCP. To end the test, the valve in rig floor is opened, and some of the fluid in the borehole flows back into the fluid tank (known as "bleed-off").

To confirm the pressure values obtained from the initial XLOT, a second pressurization cycle is warranted (Fig. 2). Because a fracture has been created by the first execution of XLOT, in the second cycle the pressure at the time of re-opening of the fracture corresponds approximately to the FPP of the first cycle. In general, it is advisable to conduct additional pressurization cycles beyond the second cycle in order to confirm that stable values of FCP and ISIP have been obtained.

An Extended Leak-off Onboard the *Chikyu*

During the Shimokita shakedown cruise (6 August to 26 October 2006), an XLOT was conducted onboard the *Chikyu*. The test was carried out at a depth of 525 meters below seafloor (mbsf) in 1180 m water depth; fluid density (seawater) was 1.030 g·cm^{-3}, and the injection flow rate was 0.5 bbl·min^{-1} (about 80 L·min^{-1}). Pressure and flow rate were recorded at the surface, using a sample rate of 5 min^{-1}. The resolution of the pressure measurements was 1 psi (about 7kPa) its accuracy is less than ±37 psi (about ±259 kPa). Because the main objectives of the first drilling operation test of the *Chikyu* during the Shimokita shakedown cruise were confirming basic drilling procedures, pure sea water was used, and rough measurement conditions were adopted for the preliminary XLOT. At the Shimokita site, core samples were retrieved only to a depth of 365 mbsf. However, the lithology at the XLOT depth was identified from cuttings analysis as volcanic tuff.

The fluid pumping rate was constant, and pumping was stopped immediately after formation breakdown (Fig. 3). About 400 liters (2.5 bbl) of seawater was injected into a length of about 3 m of uncased borehole for about 6 min, thus creating a fracture in the borehole wall. After shut-in, pressure was monitored for about 14 min and then released (bleed-off). Although two cycles were tried, only a data set of the first cycle was successfully obtained in this test.

The processes of formation breakdown and stable fracture propagation were not clearly evident in this test (compare Figs. 2 and 3). Moreover, the pressure versus time curve was not smooth, owing to the large data sampling interval during the pumping and monitoring processes and the relatively poor accuracy of the rig-floor pressure recorders. Thus, it was hard to pick the FCP with any confidence, as this requires that two tangents be drawn to the pressure decay curve. Instead, we estimate that the magnitude of the minimum principal stress lies between the pressure at the moment the pumps were turned off, which should be a close upper bound to the ISIP since we are conducting the test with low-viscosity sea water, and our estimated value for the FCP, obtained as best we could using a bi-linear tangent approach (Fig. 3). In this manner, we estimate that the magnitude of the minimum principle stress is 18.3–18.5 MPa. For comparison, we estimated the magnitude of vertical stress at the test depth from the density of the formation. An average formation density of 1.5 g·cm^{-3} from 0 mbsf to 365 mbsf was determined from the density profile of core samples retrieved during the Shimokita cruise. We assumed that the average density for the interval 365–525 mbsf was 1.8 g·cm^{-3}; therefore, the vertical stress was estimated to be approximately 20 MPa. Thus, the magnitudes of the minimum principal stress from the XLOT and the vertical stresses are close to one another, suggesting that we either measured the vertical stress with the XLOT or that we measured the minimum horizontal stress and are in a transitional strike-slip to reverse faulting environment. Since we were not able to determine the attitude of the hydraulic fracture in the test interval, we cannot ascertain which of these two possibilities is correct. Considering the many past applications of XLOT, both in continental scientific drilling projects and in industry oil fields (Kunze and Steiger, 1991; Lund and Zoback, 1999), we suggest that, although it is not a perfect and universally used technique, XLOT can provide data that are both valuable and practical for estimating the magnitude of minimum principal stress (Nelson et al., 2007).

XLOT Procedures for Stress Estimation

The XLOT procedure that we suggest for determination of stress magnitudes during future riser-drilling programs conducted onboard *Chikyu* is shown in Fig. 4. This procedure has several advantages over the types of tests often conducted following borehole completion. First, the XLOT procedure is superior to the LOT procedure. It can be difficult to obtain reliable estimates of minimum principal stress by using only the value of LOP, which is the only stress-related parameter obtained by the LOT procedure. Second, we suggest that implementation of multiple XLOT cycles (at least 3 cycles) will provide more reliable results than the LOT or XLOT procedure alone. The first cycle (equivalent to the previously mentioned casing-shoe integrity test) uses a lower maximum injection pressure than the predicted LOP and is designed to estimate the permeability of the formation, determine whether there are pre-existing fracture(s) and weakness(s), and check the effectiveness of cementing. The second cycle is a standard XLOT procedure, and the third cycle is a repetition of the second cycle to confirm the diagnostic pressure values obtained from the previous XLOT.

It is also important to record a high accuracy, closely sampled data set to avoid some of the difficulties in accurately picking test parameters, discussed in the example presented above. Data monitoring and recording details should include pumping pressure, the volume of fluid injected, and the volume of fluid returned to the fluid tank during bleed-off. We think this recording is quite easy. It is also important that the density of the fluid being injected is well known so that the hydrostatic pressure at the casing shoe under *in situ* pressure and temperature conditions can be calculated; alternatively, down-hole pressure recording at the casing shoe can be employed (using a wireline or memory tool) to measure directly pressure at the casing shoe.

The procedures that we suggest (Fig. 4) and describe in detail below are similar to those conducted in deep onshore wells (Yamamoto, 2003).

(1) In the first (LOT) cycle, drilling fluid is pumped into the borehole at a constant flow rate (e.g., 0.5 bbl·min^{-1}, or about 80 L·min^{-1}); pumping stops before the expected LOP, and the well is shut-in for 5–10 min. The pressure decline during the very early stage of shut-in reflects the decay of viscous pressure losses in the surface plumbing and drill pipe, and the pressure change during the later stage of shut-in is controlled by the permeability of the formation. If the pressure decline in the late stage of shut-in is large and does not stabilize, the leak-off of fluid might be attributed to the existence of natural fractures or to ineffective cementing. If the casing shoe is too permeable, then the

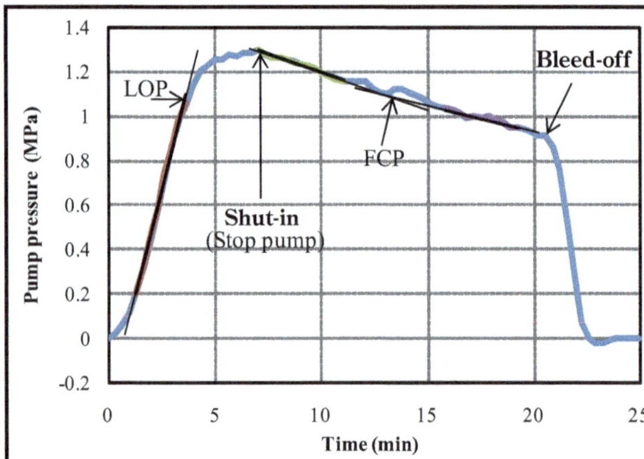

Figure 3. Pumping pressure at drilling rig level versus elapsed time during XLOT carried out on board the riser vessel *Chikyu*.

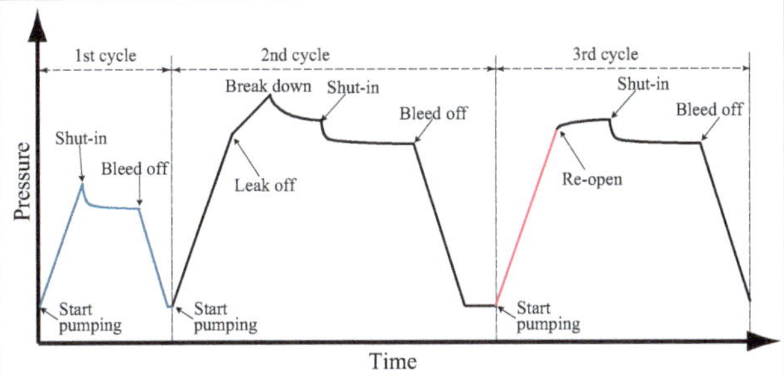

Figure 4. Suggested procedures for conducting XLOT to determine the magnitude of the minimum principal stress.

second and third test cycles are unnecessary, as a reliable measure of the minimum principal stress will not be possible.

(2) In the second injection cycle, pumping continues for at least 1 min beyond formation breakdown, and the well is then shut-in. If formation breakdown is not achieved but pressure decreases during pumping (indicating fracture propagation, perhaps from a pre-existing fracture), then pumping should continue until the volume of fluid injected reaches at least several barrels (e.g., 3 bbl, or about 450 L) and the well is shut in.

(3) The well then remains shut-in while pressure is monitored for at least 10 min or until the pressure ceases to decay. The well is then bled off.

(4) To evaluate the pressure versus volume curve during bleed-off, flow-back volume is monitored with a flow meter. The curve shown in Fig. 5 is an idealized relation between pumping pressure and volume, and it indicates the total amount of fluid lost into the formation (or through other system leaks) during the test. Raane et al. (2006) also mentioned that pump-in/flow-back test appears to give a robust estimate of the minimum principal stress.

(5) The third cycle repeats steps 2–4 and allows comparison of the pressure parameters obtained during the second cycle.

(6) Comparison of the pressure decline curves of the third and second cycles provides information about the state of the borehole. For example, if the pressure decline after shut-in during the third cycle is comparable to that observed in earlier cycles, then the cement bond has not been damaged, and with the test interval permeability has not been significantly affected.

(7) If required, a fourth cycle of pumping can be undertaken to investigate borehole integrity, including the extent of formation permeability during the test. In this case, the casing shoe is again pressurized to the maximum pressure of the first cycle. The well is then shut in, and the pressure and fluid volume monitored. Comparison of the pressure build-up rate (pressure versus volume) during injection and the pressure decline after shut-in during this cycle with those of the first cycle will show whether or not borehole integrity has been compromised.

There may be concern that the new fracture created during the XLOT has affected casing-shoe integrity. In general, casing-shoe integrity is maintained if appropriate drilling fluid (mud) has been used (Morita et al., 1997).

Calculation of minimum principal stress by using LOT–XLOT data depends on the assumption that a new fracture is created in a plane perpendicular to the minimum principal stress by the pumping pressure and that pre-existing fracture(s), weakness(s), anisotropy, and heterogeneity of the formation have no significant influences. Therefore, knowing with certainty the attitude of the new fracture produced is very helpful to determine direction of the minimum principal stress. For this purpose, Fullbore Formation Microimager (FMI) and/or Ultrasonic Borehole Image (UBI) logs or impression packer before and after the test can be conducted to acquire borehole images in cases of hydraulic fracturing which is conducted not at the borehole bottom (casing shoe). However, it should be difficult in case of an XLOT before the test because its test interval is too short to allow installing of FMI- or UBI-type logs. Additionally, in many cases the resolutions of FMI or UBI images are too low to see a hydraulic fracture. Also, given the low probability of success, it is hard to justify the expense and rig time for running a log to image a 1–3 m section of borehole.

The minimum principal stress determinate by an XLOT is equivalent to minimum principal horizontal stress in normal and strike-slip faulting environments; and hydraulic fracture is induced in a vertical plane. In contrast, in reverse faulting environments the minimum principal stress is equivalent to vertical stress; the fracture is formed in a horizontal plane. In general, it is difficult to identify if the minimum principal stress is vertical or horizontal stress without knowing attitude of hydraulic fracture induced. Only in cases where the minimum principal stress from an XLOT is significantly lower in magnitude than the calculated vertical stress, can the minimum principal stress be identified as the minimum horizontal stress.

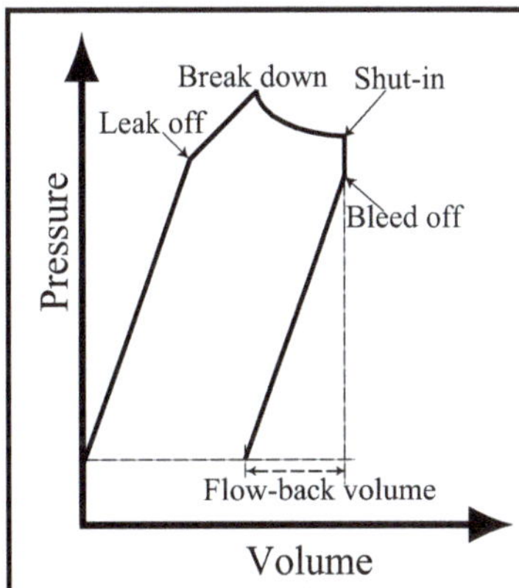

Figure 5. Relationship between pumping pressure and injected volume corresponding to the 2nd cycle in Fig. 4.

A drawback of the XLOT procedure that we have recommended is that it cannot be used to determine the magnitude of maximum principal stress, which is also difficult to determine using the standard hydraulic fracturing test (Ito et al., 2007).

The magnitude of the maximum principal stress in deep wells is best practically determined through an integrated analysis of borehole breakouts and tensile fractures from image logs, rock strength, and the minimum principal horizontal stress from the XLOT, as discussed, for example, in Zoback et al. (2003). However this integrated analysis has several problems which should be solved in the near future, such as rock strength problem (Haimson and Chang, 2002) and the effect of fluid compressibility and compliance of the test system (Raaen et al., 2006). In the near future, it is preferable and hopeful that more reliable and robust *in situ* stress measurements will be developed and applied onboard the *Chikyu*.

Summary

Investigation of *in situ* stress at depth is a necessary and important outcome of IODP drilling programs such as NanTroSEIZE. Fortunately, the availability of the new research vessel *Chikyu* means that LOT and XLOT procedures can be readily undertaken during future riser-drilling programs; these will yield important information about *in situ* stress magnitude as well as providing some of the data needed for drilling operations (e.g., borehole stability analysis). We used data from the 2006 *Chikyu* Shimokita shakedown cruise to demonstrate the feasibility of using XLOT data to determine the magnitude of the *in situ* minimum principal stress at depth. The procedures that we have recommended for the application of XLOT to determine stress magnitude during future riser-drilling programs of the *Chikyu* represent the most important outcome of this work.

Acknowledgements

We thank the *Chikyu* drilling operation team for allowing us to use XLOT data acquired by them during the Shimokita cruise. We gratefully acknowledge Stephen Hickman for his careful reviewing and many constructive comments which greatly improved the manuscript. This work was partly supported by Grant-in-Aid for Scientific Research (C: 19540453) of the Japan Society for the Promotion of Science.

References

Addis, M.A., Hanssen, T.H., Yassir, N., Willoughby, D.R., and Enever, J., 1998. A comparison of leak-off test and extended leak-off test data for stress estimation, SPE/ISRM 47235. *Proc. SPE/ISRM Eurock 98*, Volume 2, Trondheim, Norway, 8–10 July 1998, pp. 131–140.

Haimson, B.C., and Chang, C., 2002. True triaxial strength of the KTB amphibolite under borehole wall conditions and its use to estimate the maximum horizontal *in situ* stress. *J. Geophys. Res.*, 107(B10):2257, doi:10.1029/2001JB000647.

Hickman, S., Zoback, M.D., and Benoit, R., 1998. Tectonic controls on fault-zone permeability in a geothermal reservoir at Dixie Valley, Nevada. *In* Holt, R.M., et al. (Eds.), *Rock Mechanics in Petroleum Engineering*, vol. 1, Richardson, Texas (Society of Petroleum Engineers), 79–86.

Ito, T., Omura, K., and Ito, H., 2007. BABHY – A new strategy of hydrofracturing for deep stress measurements. *Sci. Drill., Special Issue*, 1:113–116.

Kinoshita, M., Tobin, H., and Moe, T., 2008. Preliminary results from NanTroSEIZE IODP Expedition 314, LWD Transect across the Nankai Trough off Kumano. Japan. Geoscience Union Meeting 2008, Chiba, Japan, 25–30 May 2008, U054-004.

Kunze, K.R., and Steiger, R.P., 1991. Extended leak-off tests to measure *in situ* stress during drilling. *In* Roegiers, J.-C. (Ed.), *Rock Mechanics as a Multidisciplinary Science*. Rotterdam (Balkema) 35–44.

Lin, W., Kwasniewski, M., Imamura, T., and Matsuki, K., 2006. Determination of three-dimensional *in situ* stresses from anelastic strain recovery measurement of cores at great depth. *Tectonophysics*, 426:221–238, doi: 10.1016/j.tecto.2006.02.019.

Lund, B., and Zoback, M.D., 1999. Orientation and magnitude of *in situ* stress to 6.5 km depth in the Baltic Shield. *Int. J. Rock Mech. Min. Sci.*, 36:169–190.

Morita, N., Fuh, G.-F., and Boyd, P.A., 1997. Safety of casing-shoe test and casing-shoe integrity after testing. *SPE Drilling & Completion*, 12:266–274.

Nelson, E.J., Chipperfield, S.T., Hillis, R.R., Gilbert, J., McGowen, J., and Mildren, S.D., 2007. The relationship between closure pressures from fluid injection tests and the minimum principal stress in strong rocks. *Int. J. Rock Mech. Min. Sci.*, 44:787–801.

Raaen, A.M., Horsrud, P., Kjorhold, H., and Okland, D., 2006. Improved routine estimation of the minimum horizontal stress component from extended leak-off tests. *Int. J. Rock Mech. Min. Sci.*, 43:37–48.

Tobin, H., and Kinoshita, M., 2007. The IODP Nankai Trough Seismogenic Zone Experiment, *Sci. Drill., Special Issue*, 1:39–41.

White, A.J., Traugott, M.O., and Swarbrick, R.E., 2002. The use of leak-off tests as means of predicting minimum *in situ* stress. *Pet. Geosci.*, 8:189–193.

Yamamoto, K., 2003. Implementation of the extended leak-off test in deep wells in Japan. *In* Sugawara, K. et al. (Eds.), *Proceedings of the Third International Symposium on Rock Stress Rs, Kumamoto '03*, Rotterdam (Balkema), 225–229.

Zoback, M.D., Barton, C.A., Brudy, M., Castillo, D.A., Finkbeiner, T., Grollimund, B.R., Moos, D.B., Peska, P., Ward, C.D., and Wiprut, D.J., 2003. Determination of stress orientation and magnitude in deep wells. *Int. J. Rock Mech. Min. Sci.*, 40:1049–1076.

Authors

Weiren Lin, Kochi Institute for Core Sample Research, Japan Agency for Marine-Earth Science and Technology, Nankoku 783-8502, Japan, e-mail: lin@jamstec.go.jp
Koji Yamamoto, Technology Research Center, Japan Oil, Gas and Metals National Corporation, Chiba, Japan
Hisao Ito, Hideki Masago, and Yoshihisa Kawamura, Center for Deep Earth Exploration, Japan Agency for Marine-Earth Science and Technology, Yokohama, Japan

7

Ultra-Deepwater Riserless Mud Circulation with Dual Gradient Drilling

by Greg Myers

Introduction

Drilling deep holes in very deep water presents the offshore drilling community with major wellbore stability challenges that are typically mitigated through the circulation of dense drilling mud to prevent hole collapse and to remove drilling debris ("cuttings"). This is normally accomplished through the application of a riser system (Fig. 1); however, riser lengths are presently limited to use in water depths of around 3047 m. In the scientific ocean drilling realm, we have been very successful in drilling relatively shallow holes (<1500 m) in water depths greater than 3657 m, a range we call "hyper-deep". Drilling in these extreme water depths requires the use of the "riserless" drilling technique (Fig. 1A) which is not constrained by the length limitations of a riser system ("riser").

The new riser-capable drilling vessel *Chikyu* has enabled the Integrated Ocean Drilling Program (IODP) to drill deep holes in water depths up to 2500 m. Scientific objectives in greater water depths with borehole penetrations deeper than 2000 m still remain a major challenge, because IODP does not have effective options for drilling in these environments. A technological solution to improve the drilling of deep holes in ultra-deep water with riserless drilling equipment may now be emerging through a joint effort between IODP-Management International (IODP-MI) and the energy industry. The technology, known as Riserless Mud Recovery (RMR), has been developed and commercialized by the Norwegian firm, AGR Drilling Services (http://www.agr.com/). The RMR system allows for the drilling of the upper section of a borehole using the dual gradient drilling technique (Fig. 1B). The AGR system is presently operating in water depths up to 457 m. This article describes this emerging drilling technology and outlines joint efforts by IODP-MI and the energy industry to modify the system and deploy it in water depths up to 3657 m.

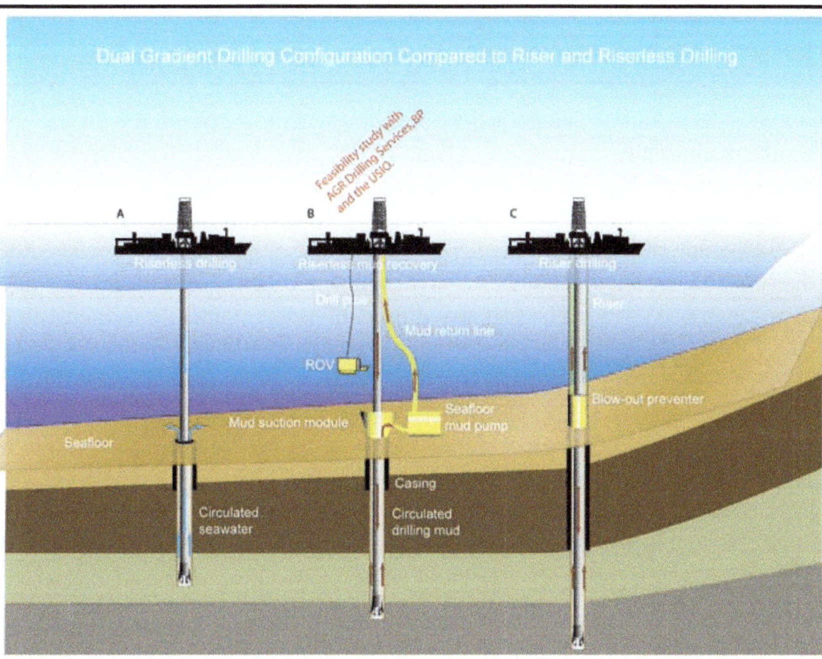

Figure 1. Dual gradient drilling (DGD) equipment may provide the technology needed to drill deep holes in deepwater. Riser drilling will still yield the deepest holes and riserless drilling will provide access to the deepest waters. Riserless [A] and riser [C] drilling configurations are depicted with a version of dual gradient drilling equipment provided by AGR Drilling services, named Riserless Mud Recovery (RMR) system [B].

Summation of Offshore Drilling Problem – Well Control and Costs

Whether a drill site is located in terrestrial or marine environments, several problems are common to rotary drilling operations. Drilling cuttings must be removed from the borehole, drilling fluid density must be managed to keep the borehole open without fracturing the formation, circulation must be maintained even when drilling fluid is lost to the formation, and pressure must be contained if over-pressured strata or gas are encountered (Weddle and Schubert, 2000). These problems are exacerbated for offshore wells due to hydrostatic pressure constraints, especially in the upper section of the borehole where the sedimentary overburden is reduced and/or increased pore pressures may be encountered (Fig. 2).

The result of this dynamic is a narrowing of the drilling "window" between the formation pore pressure and formation fracture pressure (Smith et al., 1999). As water depth increases, the "window" between the pore pressure and fracture pressure becomes even narrower as a result of the increasing influence of the weight of the fluid column.

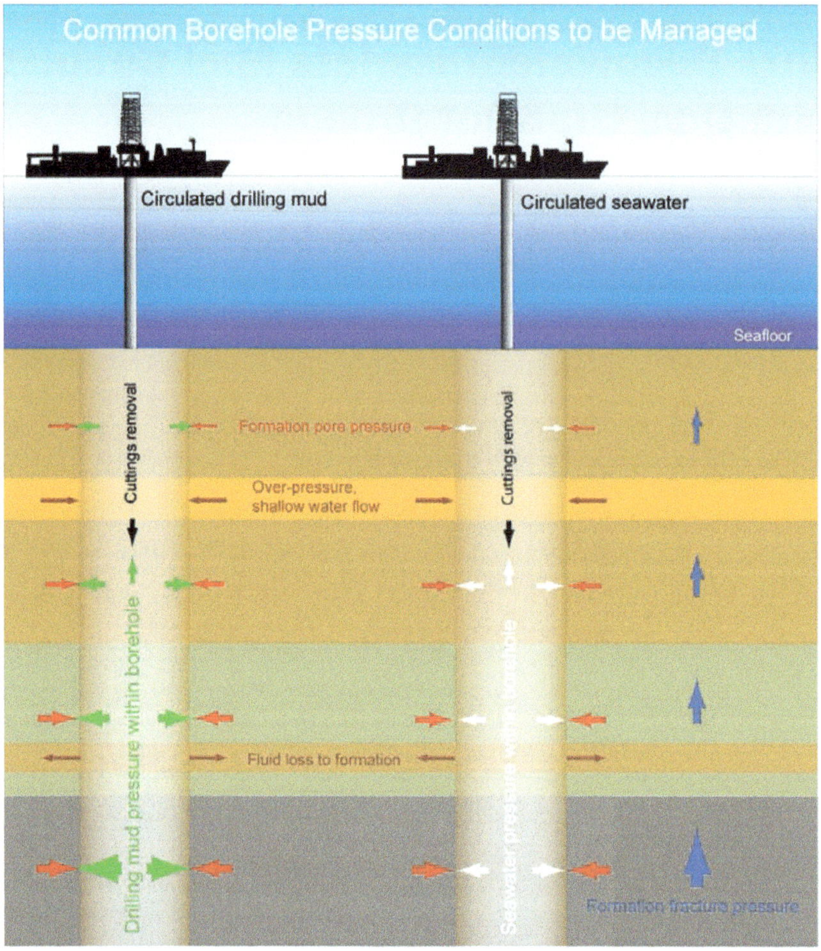

Figure 2. The pressure conditions within an offshore borehole are complex and must be precisely managed to achieve the target depth. Carefully selected drilling fluids provide the pressure needed to resist the formation pore pressure thereby keeping the borehole open without applying too much pressure and fracturing the formation. This optimal pressure zone between formation pore pressure and formation fracture pressure is known as the drilling "window". When drilling with seawater, the formation pore pressure overcomes the borehole fluid pressure leading to borehole collapse.

Riserless drilling primarily uses seawater, rather than drilling mud, to manage the borehole because drilling mud cannot be recirculated. Seawater has a significantly lower density than drilling mud, thus the boreholes are more likely to collapse with increasing hole depth as the pore pressure exceeds the hydrostatic pressure exerted by the seawater. For deeper drilling, drilling mud with regulated density must be continuously circulated in order to keep the borehole from collapsing and to remove cuttings from the borehole. The widely accepted standard equipment to establish continuous circulation is a marine drilling riser, which provides a conduit for returning the mud and cuttings to the drill rig. Current riser depth capabilities, however, only extend to just over 3047 m and are extremely expensive. Scientific ocean drilling and the energy industry have ultra-deepwater targets of interest beyond the reach of current risers. Research is underway to design risers of even greater depth potential using a combination of metallic and composite materials. However, the stresses due to vibration of these long riser strings and the well control problems associated with hydrostatic pressure of the mud pose significant operational problems. The use of alternate techniques, such as riserless mud recovery and dual gradient drilling, could therefore provide an attractive option for both IODP and industry to drill deep holes in deep water settings where current capabilities are either insufficient or not economic.

Dual Gradient Drilling Application for IODP

RMR provides a dual gradient drilling setup of the well, while capturing the drilling fluid and returning it to the drillship (Fig. 3). The term "dual gradient" implies two hydrostatic gradients: 1) the seawater gradient that begins at the sea surface, and 2) the drilling mud gradient that begins at the seafloor. Conventional drilling has only one pressure gradient for both seawater and mud that originates at the sea surface (Schubert, et al., 1999). Because dual gradient drilling has much less hydrostatic "head" associated with the drilling mud in the borehole, drilling fluids can be properly weighted, allowing drilling to be more easily contained within the "window" between the formation pore pressure (Fig. 3, red line) and the formation fracture pressure (blue line), thereby avoiding wellbore instability.

Riser drilling and dual gradient drilling provide different options for controlling borehole conditions while drilling. The primary difference is the point where the hydrostatic head of the drilling mud begins accumulating. Another way of visualizing dual gradient drilling is to imagine the drilling equipment positioned on the seafloor. In this scenario, the weight of the fluid column from the sea surface down to the seafloor is eliminated, thereby providing a larger drilling "window" between the formation pore pressure and formation fracture pressures, resulting in increased well bore pressure management.

In addition to well control, the RMR system also helps reduce the costs associated with deepwater drilling. First, the system eliminates the "pump and dump" drilling strategy, where mud that is sometimes utilized for well control in riserless drilling is directly discharged to the seafloor. Collecting and reusing drilling mud significantly reduces the amount of mud required to install an offshore well. Second, as the number of casing strings can be reduced, the upfront costs for this material can be significantly reduced. Thus, the RMR system offers a complementary solution to the capabilities of the *Chikyu* and the standard riserless capabilities of the *JOIDES Resolution*. *Chikyu* provides much deeper borehole penetration with blow-out prevention capabilities, but it has a limited water depth range. The current *JOIDES Resolution* provides hyper-deepwater capabilities but with limited borehole penetration. A drilling platform successfully equipped with RMR and dual gradient capabilities would fill in this technology gap in a cost-effective manner, allowing pursuit of deep drilling targets in water depths up to 3657 m.

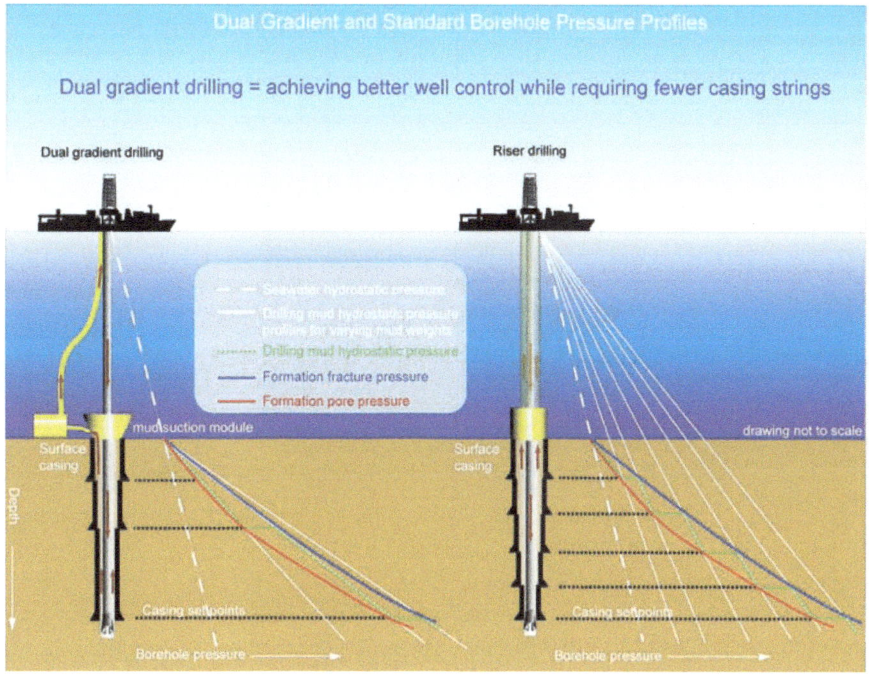

Figure 3. The dual gradient drilling approach establishes the drilling mud hydrostatic gradient at the seafloor which allows for the use of heavier drilling muds with an advantageous pressure increase with depth. The use of optimized drilling mud provides better well control and fewer casing points by keeping the drilling fluid within the drilling "window" for more depths within the borehole.

AGR Drilling Services has been providing RMR dual gradient drilling equipment to the energy industry for a number of years. The system has been successfully utilized in gas fields in Sakhalin (Brown et al., 2007) and the Caspian Sea. The equipment is mobilized and installed on the drill rig in modular sea-freight containers. Returning the drilling mud back to the drill rig for conditioning and recirculation requires seafloor mud suction equipment (Figs. 4 and 5), seafloor mud pumps, and a mud return line extending from the seafloor equipment back to the drill rig. In shallower water depths below 1524 m, this line can be a flexible, large-diameter hose. Beyond 1524 m, the mud return line will be fashioned from steel pipe, such as a drill pipe. The mud return line is deployed down to the seafloor and then secured as the pipe exits the rig's moon pool. A remotely operated vehicle (ROV) is utilized to provide visual inspection of the seafloor equipment setup. Following the equipment setup, the drill string is run to the seafloor, and the hole is spudded. The ROV system is deployed continuously to monitor the mud recovery system during drilling.

To investigate the feasibility of utilizing the RMR system on the *JOIDES Resolution* to 3657 m, a small team was formed by IODP-MI with members from the IODP United States Implementing Organization (USIO), AGR Drilling Services, and British Petroleum (BP). This team submitted afeasibility and planning proposal to DeepStar, an industry technology development consortium consisting of eight energy companies focused on advancing the technologies to meet deepwater and ultra-deepwater business needs. DeepStar provides a forum to execute deepwater technology development projects and leverage the financial and technical resources of the deepwater industry (www.deepstar.org).

The study, planned for completion by the end of 2008, will investigate the system requirements for several different well configurations and water depths ranging from 1523 m to 3657 m, depths of interest to both the energy industry and the scientific community. This study includes determining the feasibility studies of modifying the deepwater AGR

Figure 4. The AGR riserless mud recovery subsea equipment consists of a mud suction module (see above) and seafloor mud pumps. The mud is collected in the mud suction module and returned to the drillship via the seafloor mud pumps (image: courtesy of John Thorogood).

Figure 5. The AGR riserless mud recovery system deployed on an oil and gas exploration vessel.

system to operate in up to 3657 m of water and a feasibility study to determine the suitability of the *JOIDES Resolution* to deploy the AGR RMR system. Should the feasibility study show that the RMR system can be deployed from the *JOIDES Resolution*, IODP-MI will work with the USIO, AGR, and BP to secure funds for a Gulf of Mexico deployment within two years to test the system capabilities in actual field operations.

Conclusions and Applications

If RMR and dual gradient drilling prove feasible for use on the *JOIDES Resolution*, IODP will be in a much improved position to address drilling problems (Larsen and Kushiro, 1997) that consistently plagued the successful completion of highly ranked programs such as achieving deep crustal and even upper mantle objectives. And, it could open new areas of study in drilling deeper targets in over-pressured regions by providing a continuous supply of weighted mud needed to suppress shallow water flows (Myers et al., 2007). The modified AGR Drilling Services system could be leased as needed and deployed from the *JOIDES Resolution*, *Chikyu*, or another suitable platform as required to meet the scientific objectives of a given project.

Acknowledgements

The concepts and ideas presented in this article represent the amalgamation of previous work and informal discussions with many engineers from industry and scientific drilling circles. I wish to acknowledge all who have helped developed this concept on any level, in particular, Pierre Beynet and Warren Winters at BP, Tom Janecek of IODP-MI and Tom Williams helped turn these thoughts into an engineering initiative. Special thanks to Chris Haver and Jim Chitwood of the DeepStar consortium for hearing our proposal and seeing the value in symbiotic industry and scientific engineering efforts.

References

Brown, J.D., Urvant, V.V., Thorogood, J.L., and Rolland, N.L., 2007. Deployment of a riserless mud recovery system offshore Sakhalin Island. Presentation 105212, SPE/IADC Drilling Conference and Exhibition, 20–22 February 2007, Amsterdam, The Netherlands.

Larsen, H.C., and Kushiro, I., 1997. Report of the Conference on Cooperative Ocean Riser Drilling (CONCORD). 22–24 July 1997, Tokyo, Japan. http://www.odplegacy.org/PDF/Admin/Long_Range/CONCORD.pdf.

Myers, G., Winker, C., Dugan, B., Moore, C., M., Sawyer, D., Flemings, P., and Iturrino, G., 2007. Ursa Basin explorers shine new light on shallow water flow. *Offshore Engineer*, September 2007, pp. 88–93.

Schubert, J.J., Seland, S., Johansen, T.J., and Juvkam-Wold, H.C., 1999. Greater kick tolerance and fewer casing strings make dual gradient drilling a winner. IADC Well Control Conference of the Americas, 25–26 August 1999, Houston, Texas.

Smith, K.L., Gault, A.D., Witt, D.E., Peterman, C., Tangedahl, M., Weddle, C.E., Juvkam-Wold, H.C., and Schubert, J.J., 1999. Subsea mudlift drilling joint industry project achieving dual gradient drilling technology. *World Oil, Deepwater Technology (supplement)*, August 1999. http://www.worldoil.com/Magazine/MAGAZINE_DETAIL.asp?ART_ID=3539&MONTH_YEAR=Aug-1999.

Weddle, C.E., and Schubert, J.J., 2000. Factors to consider in dual gradient well control. IADC Well Control Conference and Exhibition, 6–7 December 2000, Houston, Texas.

Author

Greg Myers, IODP-MI Engineering and Operations Manager, IODP-MI Headquarters, 815 Connecticut Avenue, NW, Suite 210, Washington, DC 20006 U.S.A., e-mail: gmyers@iodp.org.

Related Web Links

http://www.agr.com/
http://www.deepstar.org

Photo Credits

Figs. 4 and 5: courtesy of John Thorogood (Drilling Global Consultant LLP)

8

Magnetic Susceptibility as a Tool for Investigating Igneous Rocks—Experience from IODP Expedition 304

by Roger C. Searle

Introduction

Continuous measurements of magnetic susceptibility have been commonly used on Ocean Drilling Program (ODP) and Integrated Ocean Drilling Program (IODP) expeditions to study minor lithological variations (for example, those related to climatic cycles) in sedimentary rocks, but they have been less frequently used on igneous rocks, although important post-cruise studies have utilized them (e.g., Ildefonse and Pezard, 2001). Here I report its use (and that of the closely related electrical conductivity) on IODP Expedition 304 to examine igneous crustal rocks. Expedition 304/305 targeted the Atlantis Massif, an oceanic core complex on the Mid-Atlantic Ridge, and recovered a suite of igneous rocks comprising mainly gabbros, troctolites, and some diabases (Blackman et al., 2006; Ildefonse et al., 2006, 2007; IODP Expeditions 304 and 305 Scientists, 2005). Shipboard measurements (on *D/V JOIDES Resolution*) of physical properties were made to characterize lithological units and alteration products, to correlate cored material with down-hole logging data, and to interpret broader-scale geophysical data.

Shipboard Measurements

Magnetic susceptibility, k, is a dimensionless measure of the degree to which material can be magnetized in an external magnetic field:

$$k = M/H$$

where M is the magnetization induced in the material by an external field of strength H. Magnetic susceptibility is sensitive to variations in the type and concentration of magnetic grains in rocks and is thus an indicator of compositional variations.

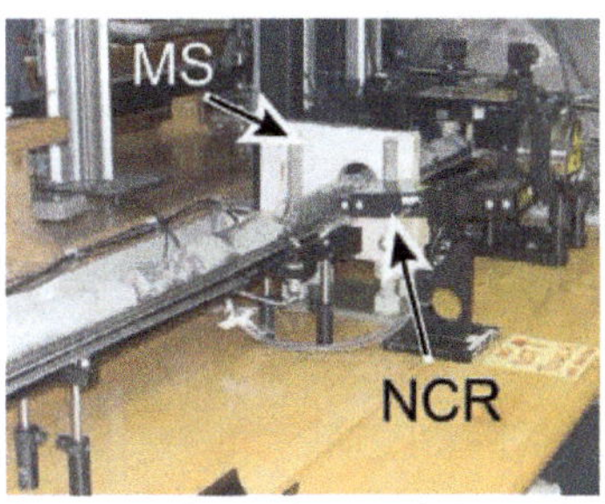

Figure 1. View of part of the multi-sensor track on *JOIDES Resolution* during Expedition 304, showing the MS and NCR sensors.

After recovery, cores were allowed to come to approximately room temperature (22°–25°C), then magnetic susceptibility (MS) and non-contacting electrical resistivity (NCR) were measured following standard IODP procedures on whole core in split liner in the multi-sensor track (Blum, 1997, Fig. 1).

MS was measured inductively at 2-cm intervals down core, using a model MS2C Bartington susceptibility meter, which has an 8-cm loop and operates at 0.565 kHz with a field intensity of 80 A/m (Bartington Instruments, 1995). The instrument is constructed so that for a core of diameter 65 mm, the recorded value is the absolute volume susceptibility. The diameter of the Expedition 304 cores was always smaller (~5.5–6.0 cm), so I report results in Instrument Units (IU), which under the conditions given above approximate to dimensionless Système International (SI) units × 10^{-5}. Because measured susceptibility depends on sample volume, measurements on pieces shorter than ~8 cm will be underestimated; such samples were flagged in my interpretations. A further complication is that the Bartington MS2C sensor currently has a maximum range of 10^4 IU; all readings greater than this lose the most significant digit, so that the signal appears to fall discontinuously to a low value, or "wrap around". As a result, intervals where k values appear to approach 10^4 IU and then fall rapidly should be examined and used with care. However, there is a potential solution to this as described below.

NCR was measured every 2 cm down core using a non-contacting inductive instrument, purpose-built for the MST by Geotek Limited (http://www.geotek.co.uk/site/index.php). Instrument output (in volts) is approximately inversely proportional to resistivity; the precise relationship was determined at the start of Expedition 304 by measuring brine samples of known salinity, though data in the IODP database are in uncalibrated voltages. The instrument is rated to

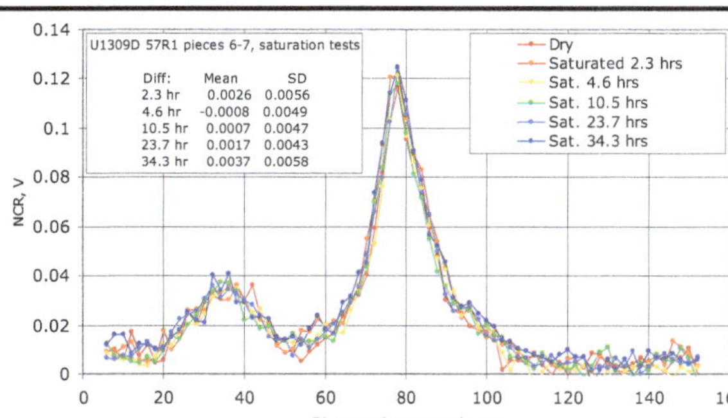

Figure 2. Results of repeated measurements of NCR on U1309D 57R-1 pieces 6 and 7, following drying and then saturation with seawater for periods from 2.3 h to 34.3 h.

measure resistivity in the range 0.1 to 10 ohm-meters. I obtained apparently useful measurements to >100 ohm-m, but these high values are poorly calibrated. It is often more useful and intuitive to consider the reciprocal of resistivity, which is conductivity, measured in siemen per meter (S m^{-1}), and values are presented as such here. Geotek estimates the spatial resolution of the NCR as approximately 2 cm down core (http://www.geotek.co.uk/site/scripts/module.php?webSubSectionID=31). My calibration showed, however, that the sensor has to be between 4 cm and 8 cm from the end of the sample before the full resistivity was measured. Thus, resistivity measured on pieces shorter than ~10 cm will be overestimated, and, as with MS, these were flagged.

The Origin of Conductivity and Susceptibility in Igneous Rocks

Electrical conductivity in rocks occurs by one of two mechanisms: ionic conductivity (in which the drift of ions through conducting pore-water carries the current) and electronic conductivity (in which electrons travel through conducting solid minerals). Most Expedition 304 rocks have low porosity (<3%), but the connectivity of pore spaces is critical in determining ionic conductivity. Normally, resistivity was measured on the MST about 2 h after the core came on board (the delay being the time required for core curation), and in this time the core can lose a significant proportion of its pore water. I ran a test on two long pieces (U1309D-57R-1 Pieces 6 and 7) from Hole 1309D Unit 137 (olivine-bearing gabbro), which was very conductive. After storage on board for several days allowed them to dry out, the working and archive halves were put together and held in place by elastic bands, and then were measured dry and after saturation in seawater for periods from 2.3 h to 34.3 h. Little variation was seen, suggesting that ionic conductivity was not the dominant mechanism (Fig. 2). Similar results were obtained from several other sections with varying lithology.

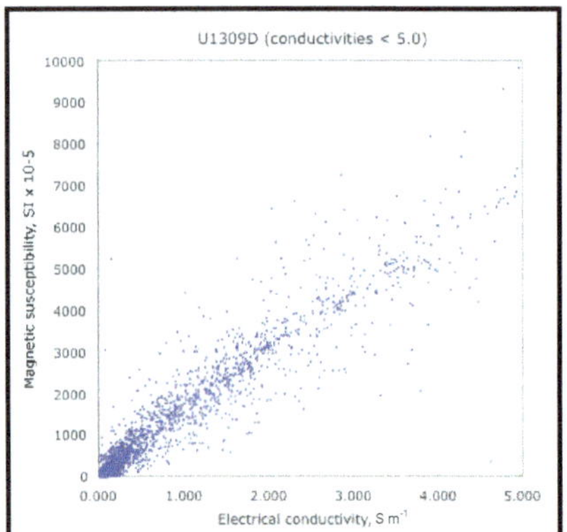

Figure 3. Plot of susceptibility versus conductivity for all values of conductivity <5 S m^{-1} from Hole U1309D measured on Expedition 304.

Early in Expedition 304 electrical conductivity and magnetic susceptibility seemed to be closely correlated, independent of lithology (Fig. 3). This strongly suggests that the same minerals are responsible for both the conductivity and susceptibility. A check of common minerals that exhibit high conductivity and high susceptibility produced the results shown in Table 1. Pyrrhotite, ilmenite, and magnetite all have high mean susceptibility and potentially high conductivity, and are relatively common in igneous rocks. Pyrrhotite was rare in the rocks recovered during Expedition 304, and I suspect that magnetite is the dominant mineral, as either a primary component or one produced during serpentinization.

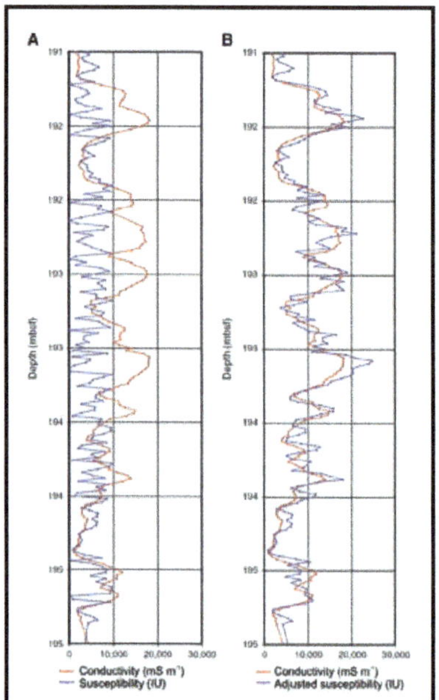

Figure 4. Plots of conductivity (red) and susceptibility (blue) against depth for sections U1309D 35R-1 to 36R-1 (unit 88, oxide gabbros). [A], logged values; [B] susceptibility augmented by integral amounts (1 or 2) of 10,000 IU to compensate for limited range of the installed susceptibility sensor. Note the good correlation with conductivity following this correction.

The strong correlation of conductivity and susceptibility offers a way of checking and perhaps correcting the wrap-around of strong susceptibility signals. Figure 4a shows measured susceptibility plotted alongside conductivity for sections U1309D 35R-1 to 36R-1 (unit 88, oxide gabbros). Some parts, particularly between 194 and 195 mbsf, correlate very well, while others do not. However, by adding 10,000 IU, or occasionally 20,000 IU, to the logged value, a much improved correlation is seen (Fig. 4b). While this manual correction can be applied in some places, it is tedious, and the number of wrap-around can be ambiguous; clearly, there is need for a susceptibility meter with extended range. Results here suggest it should be increased to at least 30,000 IU and probably to 50,000 IU.

Applications

During Expedition 304, MS sometimes showed variations that are not immediately apparent in on-board lithological

Table 1. Common minerals with high electrical conductivity or magnetic susceptibility (Telford et al., 1990)

Mineral	Susceptibility mean, IU	Susceptibility range, IU	Conductivity mean, Sm^{-1}	Conductivity range, Sm^{-1}
Pyrrhotite	1.5	0.006–1.6	104	$20.0–1.5\times10^{5}$
Ilmenite	1.9	0.3–3.8		$20.0\times10^{-2}–1.0\times10^{3}$
Magnetite	6.3	$5.0\times10^{-6}–5.7\times10^{3}$		$1.8\times10^{-4}–2.0\times10^{5}$
Pyrite	0.0015		3.3	

descriptions. For example, Figure 5 shows the susceptibility logged for U1309B Unit 62, which appears to be a fairly uniform oxide diabase (Blackman, 2006), but it exhibits variations from essentially zero to over 5,000 IU (approximately 0.05 SI) over distances of ~1m. This observation spawned the hypothesis that the variations might be related to flow phenomena and perhaps grain size variations at the edges of the unit. On-board examination of photomicrographs suggested that the intervals with low susceptibility might be places where magnetite had been extensively altered to lower susceptibility ilmenite (R. Frost, pers. comm., 2004, 2007), and led to a program of post-cruise research, which has shown that the susceptibility actually correlates quite well with observed proportion of oxides in thin section.

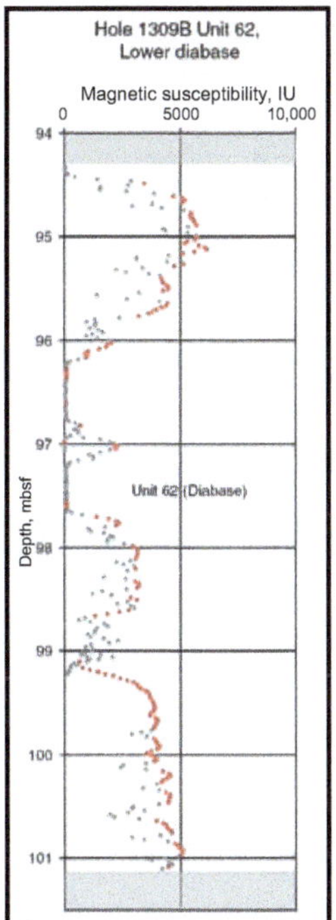

Figure 5. Susceptibility versus depth for Hole U1309B Unit 62, logged as a uniform oxide diabase. Red: MS measurements made >8 cm from the ends of pieces, which are considered reliable. Gray: points measured <8 cm from the ends of pieces, which are probably underestimates of true susceptibility.

Following this early success in using MS to identify potential lithological variations, we instituted the practice of routinely including MS in the visual core description sheets (so-called "barrel plots") produced on board (Fig. 6). These have already proved valuable for re-surveying the Expedition 304/305 cores post-cruise (B. John, pers. comm., 2007), and for investigations of the formation mechanism of serpentine from olivine-rich troctolites (R. Frost, pers. comm., 2007).

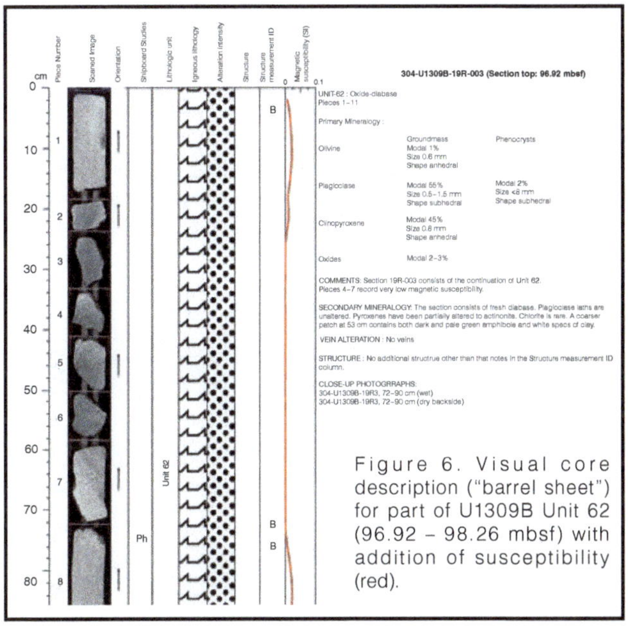

Figure 6. Visual core description ("barrel sheet") for part of U1309B Unit 62 (96.92 - 98.26 mbsf) with addition of susceptibility (red).

Acknowledgements

I am indebted to the help of the Expedition 304/305 shipboard scientists, ship's crews, and coring staff. Special thanks go to Physical Properties Marine Laboratory Specialist for Expedition 304, Heather Paul, who expertly guided this neophyte through the complexities of the on-board Physical Properties Laboratory. My participation in Expedition 304 was funded by the Natural Environment Research Council's thematic programme UKIODP.

References

Bartington Instruments, 1995. Operation Manual for MS2 Magnetic Susceptibility System, Oxford, U.K.

Blackman, D.K., Ildefonse, B., John, B.E., Ohara, Y., Miller, D.J., MacLeod, C.J., and Expedition 304/305 Scientists, 2006. Oceanic Core Complex Formation, Atlantis Massif: Expeditions 304 and 305 of the riserless drilling platform from and to Ponta Delgada, Azores (Portugal), Sites U1309–U1311, 17 November 2004–7 January 2005, and from and to Ponta Delgada, Azores (Portugal), Site U1309, 7 January–2 March 2005. *Proc. IODP, 304/305,* doi:10.2204/iodp.proc.304305.2006.

Blum, P., 1997. Physical properties handbook: A guide to the shipboard measurement of physical properties of deep-sea cores. *ODP Tech. Note,* 26, doi:10.2973/odp.tn.26.1997.

Ildefonse, B., and Pezard, P. A., 2001. Electrical properties of slow-spreading ridge gabbros from ODP Site 735, Southwest. *Indian Ridge. Tectonophysics,* 330:69–92, doi:10.1016/S0040-1951(00)00220-1.

Ildefonse, B., Blackman, D., John, B.E., Ohara, Y., Miller, D.J., MacLeod, C.J., and IODP Expeditions 304/305 Scientists, 2006. IODP Expeditions 304 & 305 characterize the lithology, structure, and alteration of an oceanic core complex. *Sci. Drill.,* 3:4–11.

Ildefonse, B., Blackman, D., John, B.E., Ohara, Y., Miller, D.J., MacLeod, C.J., and IODP Expeditions 304/305 Science Party, 2007. Oceanic core complexes and crustal accretion at slow-spreading ridges. *Geology,* 35(7):623–626, doi: 10.1130/G23531A.1.

IODP Expeditions 304 and 305 Scientists, 2005. IODP Expeditions 304 and 305: Oceanic Core Complex Formation, Atlantis Massif. *Sci. Drill.,* 1(1):28–31, doi:10.2204/iodp.sd.1.05.2005.

Telford, W.M., Geldart, L.P., and Sheriff, R.E., 1990. *Applied Geophysics.* Cambridge (Cambridge University Press).

Author

Roger C. Searle, Department of Earth Sciences, Durham University, Durham, DH1 3LE, U.K., e-mail: r.c.searle@durham.ac.uk.

Related Web Links

http://www.geotek.co.uk/site/index.php
http://www.geotek.co.uk/site/scripts/module.php?webSubSectionID=31

9

Marine Impacts and Environmental Consequences - Drilling of the Mjølnir Structure, the Barents Sea

by Henning Dypvik, Philippe Claeys, Alex Deutsch, Frank T. Kyte, Takafumi Matsui, and Morten Smelror

Introduction

In September 2007, thirty-three scientists attended an international workshop in Longyearbyen (Svalbard, Norway) to discuss impacts of extraterrestrial bodies into marine environment and to prepare for the drilling of the 142-Ma-old Mjølnir impact structure in the Barents Sea (Fig. 1; Gudlaugsson, 1993; Dypvik et al., 1996, Tsikalas et al., 1998). A field trip visited the ejecta layer in the Janusfjellet Mountain in Isfjorden, just outside Longyearbyen (Fig. 2).

The workshop focused on two topics: 1) mechanisms of marine impact cratering including ejecta formation and distribution, geothermal reactions, and the formation of tsunami, and 2) environmental effects of marine impacts. Both topics are highly relevant to the Mjølnir event and the geological evolution of the Arctic, as well as to the biological changes at the Jurassic-Cretaceous boundary. Against this background were a) concrete drilling targets formulated, b) plans outlined for compiling data from existing geological and geophysical surveys as the basis for Integrated Ocean Drilling Program (IODP) and International Continental Scientific Drilling Program (ICDP) drilling proposals, and c) a steering group and science teams established for compiling old and new material as a foundation for the development of drilling proposal.

Scientific Background

Asteroid and comet impacts are now recognized as an important and regular geological process releasing vast amounts of energy and resulting in near instantaneous increase in temperature and pressure, structural deformation, and redistribution of target materials. It is presently accepted that impacts, especially those in a marine environment, have very important influences on the development of the Earth. However, detailed knowledge of the geological and physical aspects of the impact process itself, as well as its environmental and biological consequences, is still limited. This is mainly due to the fact that a large majority of the ~170 currently known impact craters on the Earth and their ejecta deposits are rather poorly preserved. Only twenty-five of these craters represent marine impacts, and very few of those have remained submerged with a potential for preservation of the original structure (Dypvik and Jansa, 2003). No completely retained marine crater has been investigated in detail yet, while in the last years ICDP land coring projects in the Chicxulub, Bosumtwi, and Chesapeake Bay impact structures were of great scientific gain.

One of the best preserved known impact craters on Earth is the Mjølnir impact structure. It was discovered by seismic data during petroleum exploration in the Barents Sea but never sampled by coring. An extensive geophysical database has been collected over the Barents Sea, and more than sixty petroleum exploration wells have been drilled, particularly along basin margins and on structural highs. In addition, many shallow drill holes on sub-cropping sedimentary sequences have been drilled in the more central and remote areas of the Barents Sea. See also the Norwegian Petroleum Directorate (NPD; http://www.npd.no).

The Mjølnir Structure is 40 km in diameter and is located at the Bjarmeland Platform in the central Barents Sea (Fig. 1), beneath 350 m of water. Its elevated central high (Fig. 3) is

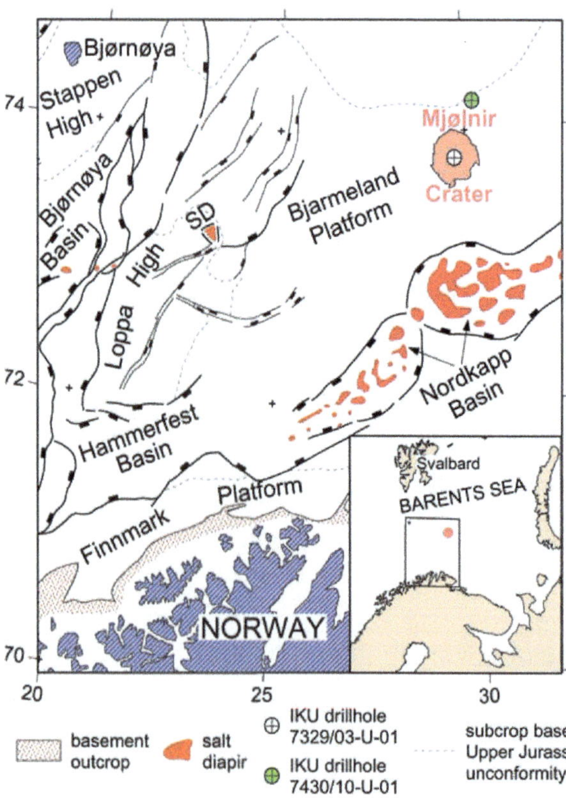

Figure 1. A tectonic map of the Bjarmeland Platform area in the Barents Sea, with a regional inset showing the Mjølnir and Svalbard position. The Mjølnir impact crater location and key shallow drillholes are shown in the main map.

covered by ~50 m of younger sediments. Geophysical, geological, and mineralogical data unequivocally substantiate the origin of the structure by an impact event into a sedimentary platform with 300–500 m paleo-water depth (e.g., Dypvik et al., 1996; Smelror et al., 2001; Sandbakken et al., 2005). The impact has been dated at about 142 Ma (Dypvik et al., 1996), very close to the Jurassic-Cretaceous boundary. At this time, the platform comprised upper Paleozoic strata, mainly carbonates and evaporates, overlain by 4–5 km of thick Mesozoic siliciclastic marine sediments (Dallmann, 1999).

The Workshop Program

The workshop included the following topics:

(1) Review the science behind marine impacts and the Mjølnir project. The state of knowledge, and ongoing geological and geophysical investigations in the Arctic realm, the Barents Sea, and Mjølnir were outlined by specialists of Arctic geology and members of the Mjølnir research group.
(2) Review of petroleum exploration drilling in the Barents Sea was presented by one representative from the NPD and representatives from Norsk Hydro and Statoil (now StatoilHydro). Drilling experts from ICDP, Drilling, Observation and Sampling of the Earths Continental Crust (DOSECC) and IODP presented different drilling options.
(3) Scientific goals and drilling strategies for the Mjølnir. A plenary session was followed by discussions in two break-out groups, whose recommendations are summarized below.
(4) An excursion was organized to the site of possible Mjølnir ejecta at the mountain Janusfjellet in Isfjorden (Fig. 2).

The Workshop Outcome

Deep wells in the Mjølnir impact structure would be of great interest to the international scientific community, in order to study the shock propagation, collapse, and re-sedimentation of the >6-km-thick sedimentary succession. Coring through this succession will make structural analysis and detailed understanding of crater generation and deformation possible and help constrain numerical modeling. However, the costs for deep coring in the harsh environments of the Barents Sea makes it unrealistic to raise funding for such operations in the foreseeable future.

Figure 2. The sun over Isfjorden and the ejecta locality at Janusfjellet.

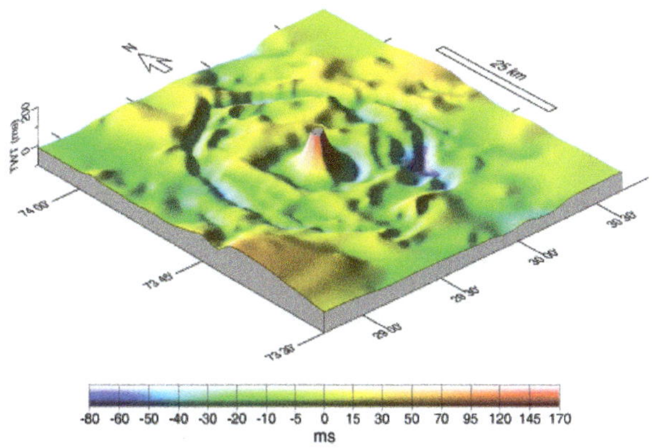

Figure 3. The seismic model of Mjølnir structure.

One of the great scientific advantages with the Mjølnir impact crater is the clear correlation between the crater and its very well preserved ejecta found in shallow drillings in the Barents Sea and on land (Svalbard and possibly Siberia; Dypvik et al., 2004; 2006). During a large part of late Jurassic and early Cretaceous, the Barents Sea region formed an epicontinental sea dominated by anoxic sedimentation of black, organic-rich clays. The Mjølnir bolide impacted into these sediments, and the crater and portions of the ejecta localities were buried and have remained buried under sediments and water since its formation. Those ejecta localities are well-preserved and accessible by shallow drilling (e.g., Bugge et al., 2002). It is one of the few places on the Earth where such important relations can be studied in detail. This is clearly of great importance for understanding the crater and ejecta formation, including the study the environmental consequences of marine impacts (Dypvik et al., 2006; Smelror et al., 2002). We will use Mjølnir as a type locality to study ejecta generation and distribution and possible relationships between the impact and biotic evolution. Mjølnir ejecta may even serve as a Boreal-Tethyan stratigraphic marker and could be useful in correlation of these two distinct provinces near the poorly understood Jurassic-Cretaceous boundary (Smelror et al., 2001; 2002). Further research could greatly expand our initial knowledge on tsunami generation and formation, impact ignitions of hydrocarbons in the target area, fires and subsequent soot precipitation. Calculations show that organic matter equivalent to a year's oil production of one Norwegian Shelf giant field (about thirty million std. m^3 oil in place) was burned during the first twenty minutes of the Mjølnir event (Dypvik et al., 2008).

The development of the Mjølnir research program should be carried out in full cooperation with the NPD and in close contact with the oil industry active in the region (e.g., StatoilHydro, ENI), making use of their extensive geophysical database and deep wells. A two-step drilling project was recommended:

Step 1. Drilling of five to six, up to 300-m-deep core holes in 350–400 m water depth around the Mjølnir structure to map and understand ejecta formation and distribution, coupled with *in situ* disturbance of sediments due to seismic and shock waves, or erosion by displaced water near the crater. Analysis of the cored material will be accompanied by sophisticated simulation models (Shuvalov and Dypvik, 2004) of the formation and deposition of ejecta in a marine environment.

Step 2. Drilling of one or two deep holes within the central moat to understand the inner structure of a large crater. At this point, however, the cost of such a project possibly requiring riser drilling is difficult to assess.

Future Plans

An international steering group (the authors of this paper) was established and charged with producing a draft project proposed by the end of 2008. The steering group will also be responsible for compiling the final drilling proposals to IODP, ICDP, and the Norwegian Research Council (NFR) by spring 2009. For further information on the Mjølnir drilling project, please contact the authors or visit http://mjoelnir.icdp-online.org/.

Acknowledgements

The workshop was kindly supported by ICDP, ESF-The Magellan Program, Statoil, Norsk Hydro, University Center on Svalbard, and University of Oslo.

References

Bugge, T., Elvebakk, G., Fanavoll, S., Mangerud, G., Smelror, M., Weiss, H.M., Gjelberg, J., Kristensen, S.E., and Nilsen, K., 2002. Shallow stratigraphic drilling applied in hydrocarbon exploration of the Nordkapp Basin, Barents Sea. *Mar. Pet. Geol.*, 19:13–37, doi: 10.1016/S0264-8172(01)00051-4.

Dallmann, W.K., 1999. *Lithostratigraphic Lexicon of Svalbard*. Tromsø, Norway (Norwegian Polar Institute), 318 pp.

Dypvik, H. and Jansa, L., 2003. Sedimentary signatures and processes during marine bolide impacts: a review. *Sed. Geol.*, 161:309–337, doi: 10.1016/S0037-0738(03)00135-0.

Dypvik, H., Gudlaugsson, S.T., Tsikalas, F., Attrep, M., Jr., Ferrell, R.E., Jr, Krinsley, D.H., Mørk, A., Faleide, J.-I., and Nagy, J., 1996. The Mjølnir structure – An impact crater in the Barents Sea. *Geology*, 24:779–782, doi:10.1130/0091-7613(1996)024<0779:MLSAIC>2.3.CO;2.

Dypvik, H., Smelror, M., Sandbakken, P.T., Salvigsen, O.T., and Kalleson, E., 2006. Traces of the marine Mjølnir impact event. *Palaeogeogr. Palaeoclimatol. Palaeoecol.*, 241: 621–634.

Dypvik, H., Mørk, A., Smelror, M., Sandbakken, P.T., Tsikalas, F., Vigran, J.O., Bremer, G.M.A., Nagy, J., Gabrielsen, R.H., Faleide, J.I., Bahiru, G.M., and Weiss, H.M., 2004. Impact breccia and ejecta from the Mjølnir crater in the Barents Sea—The Ragnarok Formation and Sindre Bed. *Nor. J. Geol.*, 84 (3), 143–167.

Dypvik, H., Wolbach, W., Shuvalov, V., and Weaver, S., 2008. Did the Mjølnir asteroid impact ignite Barents Sea hydrocarbon source rocks? *In* Evans, K., Horton, J.W., Jr., King, D.T., Jr., and Morrow, J.R. (Eds.) *The Sedimentary Record of Meteorite Impacts: Geological Society of America Special Paper, 437*: 65–72, Boulder, Colo. (Geological Society of America)

Gudlaugsson, S.T., 1993. Large impact crater in the Barents Sea. *Geology*, 21:291–294, doi:10.1130/0091-7613(1993)021<0291:LICITB>2.3.CO;2.

Sandbakken, P.T., Lagenhorst, F., and Dypvik, H., 2005. Shock metamorphism of quartz at the submarine Mjølnir impact crater, Barents Sea. *Meteor. Planet. Sci.*, 40:1363–1375.

Shuvalov, V. and Dypvik, H., 2004. Ejecta formation and crater development of the Mjølnir impact. *Met. Planet. Sci.*, 39(3):467–478.

Smelror, M., Dypvik, H., and Mørk, A., 2002. Phytoplankton blooms in the Jurassic-Cretaceous boundary beds of the Barents Sea possibly induced by the Mjølnir impact. *In* Buffetaut, E., and Koeberl, C. (Eds.), *Geological and Biological Effects of Impact Events. Impact Studies, vol. 1*, Heidelberg, (Springer), 69–81.

Smelror, M., Kelley, S., Dypvik, H., Mørk, A., Nagy, J., and Tsikalas, F., 2001. Mjølnir (Barents Sea) meteorite impact ejecta offers a Boreal Jurassic-Cretaceous boundary marker. *Newsletters in Stratigraphy*, 38:129–140.

Tsikalas, F., Gudlaugsson, S.T., and Faleide, J.-I., 1998. The anatomy of a buried complex impact structure: the Mjølnir Structure, Barents Sea. *J. Geophys. Res.*, 103:30469–30483.

Authors

Philippe Claeys, Department of Geology, Vrije Universiteit Brussel, Pleinlaan 2, B-1050 Brussels, Belgium, e-mail: phclaeys@vub.ac.be.

Alex Deutsch, Institut für Planetologie, Westfaelische Wilhelms-Universität Münster Wilhelm-Klemm-Str. 10, D-48149 Münster, Germany.

Henning Dypvik (coordinator), Department of Geosciences, University of Oslo, P.O. Box 1047, No 0316 Oslo, Norway.

Frank T. Kyte, Institute of Geophysics and Planetary Physics, UCLA University of California, Los Angeles, Calif. 90095-1567, U.S.A.

Takafumi Matsui Department of Complexity Science and Engineering, University of Tokyo, Tokyo, Japan.

Morten Smelror, Geological Survey of Norway, No 7491 Trondheim, Norway.

Related Web Links

http://mjoelnir.icdp-online.org/
http://www.npd.no

Photo Credits

Figs. 1 and 3: Filippos Tsikalas, University of Oslo and ENI, Norway
Fig. 2: P. Claeys, A. Deutsch.

10

The IODP Nankai Trough Seismogenic Zone Experiment

by Harold Tobin and Masataka Kinoshita

Introduction

The Integrated Ocean Drilling Program's (IODP) Nankai Trough Seismogenic Zone Experiment (NanTroSEIZE) attempts to drill into, sample, and instrument the updip portion of the Nankai subduction megathrust asperity. Access to the interior of active faults in which *in situ* processes can be monitored and fresh fault zone materials can be sampled is of fundamental importance to the understanding of earthquake mechanics. Great subduction earthquakes (with their accompanying tsunamis) represent one of the greatest natural hazards on the planet. Accordingly, drilling into and instrumenting an active interplate seismogenic zone was identified as a very high priority in the IODP Initial Science Plan, 2001. Through a decade-long series of international workshops, a consensus was made that the Nankai Trough subduction zone was an ideal place to attempt drilling the seismogenic plate interface. The first stage of NanTroSEIZE drilling operations is now scheduled for the late summer of 2007. It involves parallel deployment of the new U.S. Scientific Ocean Drilling Vessel (SODV) and the riser vessel *Chikyu*.

The fundamental goal of the NanTroSEIZE science plan (Tobin and Kinoshita, 2006; Kinoshita et al., 2006) is the creation of a distributed observatory spanning the updip limit of seismogenic and tsunamigenic behaviors at a location where great subduction earthquakes occur, allowing us to observe the hydrogeologic behavior of subduction megathrusts and the aseismic-to-seismic transition of the megathrust system. This will involve drilling of key elements of the active plate boundary system at several locations off the Kii Peninsula of Japan from the shallow onset of the plate interface to depths where earthquakes occur (Figs. 1 and 2). At this location, the plate interface and active megasplay faults implicated in causing tsunami are accessible to drilling in the region of coseismic rupture in the 1944 Tonankai magnitude 8.0 earthquake. The most ambitious objective is to access and instrument the Nankai plate interface within the seismogenic zone. The science plan entails sampling and long-term instrumentation of (a) the inputs to the subduction conveyor belt, (b) faults that splay from the plate interface to the surface that may accommodate a major portion of coseismic and tsunamigenic slip, and (c) the main plate interface at depths of ~4–6 km.

In addition to NanTroSEIZE, there is a burgeoning interest in active fault drilling, represented by the San Andreas Fault Observatory at Depth (SAFOD), Taiwan Chelungpu Fault Drilling Project (TCDP), Corinth Rift Laboratory (CRL), Nojima Fault Drilling Project, and other active projects on land and at sea (Reches and Ito, 2007). This is taking place in the context of rapidly growing research efforts on the mechanics and dynamics of faulting processes that integrate rock mechanics, seismology, geodesy, frictional physics, and fluid-fault interactions. Despite recent advances, there is at present no unified theory of fault slip to account for earthquake nucleation and propagation, nor to explain the mechanisms of strain across the spectrum of observed deformation rates ranging from seconds to years. Consequently, the question of whether precursor signals exist for major earthquakes, even in theory, remains under discussion. Progress on these topics is severely limited by a lack of information on

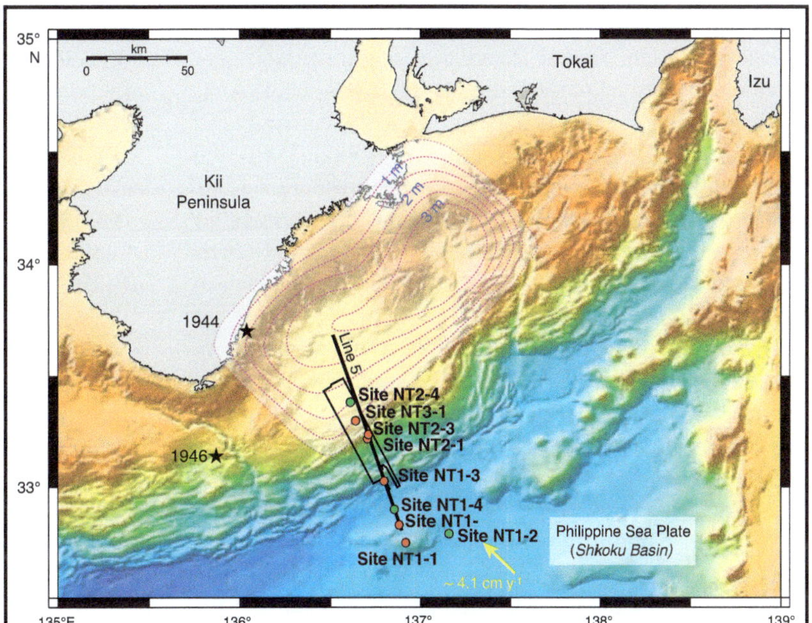

Figure 1. Bathymetric map of Kumano Basin region, with Kikuchi et al. (2003) inversion of the co-seismic rupture area of the 1944 Tonankai earthquake (gray contours), 3-D multichannel seismic survey area shown (bold black outline), and location of planned drill sites identified. Deep-penetration Multi Channel Seismic (MCS) Line 5 (thick black line) is shown in Fig. 2 with drill sites projected onto it.

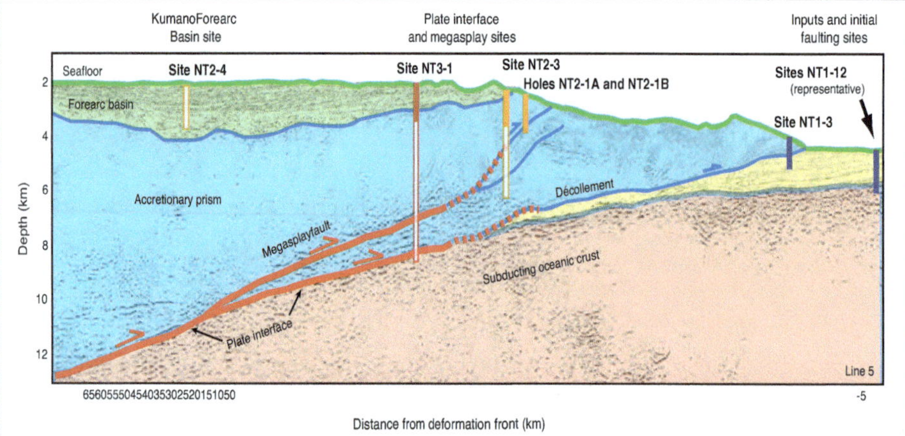

Figure 2. Prestack depth-migrated seismic line 5 of Park et al. (2002), with locations of planned drill sites. Riserless sites planned for Stage 1 drilling are shown in solid color, and riser-drilling sites are shown in outline. Faults labeled with red arrows are inferred possible co-seismic slip zones during 1944 event on basal decollement and mega-splay fault. Vertical exaggeration = 2x.

ambient conditions and mechanical properties of active faults at depth. Extant rheological models for how faults behave depend on specific physical properties at the fault interface and in the surrounding rock volume. Coefficients of friction, permeability, pore-fluid pressure, state of stress, and elastic stiffness are examples of such parameters that can best (or only) be measured through drilling and through geophysical sensing of the surrounding volume.

Geological and Geophysical Background

Subduction zones like the Nankai Trough, on which great earthquakes (M>8.0) occur, are especially favorable for study because the entire width (dip extent) of the seismogenic zone ruptures in each great event. Thus, the future rupture area is perhaps more predictable than for smaller earthquakes. The Nankai Trough region is among the best-studied subduction zones in the world. It has a 1300-year historical record of recurring and typically tsunamigenic great earthquakes, including the 1944 Tonankai (M8.1) and 1946 Nankaido (M8.3) earthquakes (Ando, 1975; Hori et al., 2004). The rupture area and zone of tsunami generation for the 1944 event are now reasonably well understood (Baba and Cummins, 2005; Ichinose et al., 2003). Land-based geodetic studies suggest that the plate boundary thrust here is strongly locked (Miyazaki and Heki, 2001). Similarly, the relatively low level of microseismicity near the updip limits of the earthquakes in the 1940s (Obana et al., 2004) implies significant interseismic strain accumulation on the megathrust; however, recent observations of very low frequency (VLF) earthquake event swarms apparently taking place within the accretionary prism in the drilling area (Obara and Ito, 2005) demonstrate that interseismic strain is not confined to slow elastic strain accumulation, but also is released as slow events.

In the Kumano Basin, the seismogenic zone lies ~6000 m beneath the seafloor (Nakanishi et al., 2002). Slip inversion studies suggest that only here did past coseismic ruptures clearly extend shallow enough for drilling (Baba and Commins, 2005; Ichinose et al., 2003), and, consequently, an updip zone of large slip (asperity) has been identified and targeted. Coseismic slip during events like the 1944 Tonankai earthquake likely occurred on the megasplay fault rather than on the décollement beneath it, though slip on either plane is permissible given the available data. Therefore, the megasplay fault is a primary drilling target equal in importance to the basal décollement zone.

Overall Project Scientific Objectives

Conditions for stable versus unstable sliding—which define seismic versus aseismic behavior—have long been the subject of research and debate, as has the frictional strength of likely fault zone material. Fault zone composition, consolidation state, normal stress magnitude, pore-fluid pressure, and strain rate may affect the transition from aseismic to seismic slip (Saffer and Marone, 2003). NanTroSEIZE will sample fault rocks over a range of pressure and temperature (P-T) conditions across the aseismic-seismogenic transition; the composition of faults and fluids and associated pore pressure and state of stress and will address partitioning of strain spatially between the décollement and splay faults. NanTroSEIZE will also install borehole observatories to provide *in situ* monitoring of these critical parameters (by monitoring of seismicity, strain, tilt, pressure, and temperature) over time, and test whether interseismic variations or detectable precursory phenomena exist prior to great subduction earthquakes.

The overarching hypotheses to be addressed are as follows:

- Systematic, progressive material and state changes control the onset of seismogenic behavior on subduction thrusts.
- Subduction zone megathrusts are weak faults.
- Within the seismogenic zone, relative plate motion is primarily accommodated by coseismic frictional slip in a concentrated zone.
- Physical properties, chemistry, and state of the fault zone change systematically with time throughout the earthquake cycle.
- The mega splay (out-of-sequence thrust; OOST) thrust fault system slips in discrete events, which may include tsunamigenic slip during great earthquakes.

Wiring the Seismogenic Fault

There are several borehole observatories installed on land and at sea. Over 700 on-land borehole observatory stations were installed at NIED (National Institute for Earth Science and Disaster Prevention) around Japanese islands. Basically, stations consist of broadband seismometers and tiltmeters. Comparable seismometer arrays have been deployed at the SAFOD, Nojima, and TCDP sites. Araki et al. (2004) installed a tiltmeter-strainmeter-broadband seismometer system in a seafloor borehole at the forearc of the Japan Trench.

The NanTroSEIZE observatory will be constructed based on all the experience obtained through the above mentioned installations. Since the temperature at 6 km below sea floor will exceed 150°C, we anticipate serious difficulties with any instrumentation of the seismogenic megathrust zone. Since good coupling of the instruments to the formation is essential, we need to develop a good clamping method and a good cementing method between the open hole and casing. Significantly tilted holes will prevent instruments from residing within measurable ranges. Wellhead seals are required for safety and environmental reasons, and this reduces the number of feedthroughs of electric and hydraulic conductors across the wellhead. We plan to develop a sub-seafloor A/D conversion and telemetry system. Finally, borehole data will be connected to a currently developed seafloor cable network off the Kumano area.

The Nankai Trough area poses a big challenge for ultra-deep drilling, but it is hoped that the outcome will be much larger than the investment for this project.

References

Ando, M., 1975. Source mechanisms and tectonic significance of historical earthquakes along the Nankai Trough, Japan. *Tectonophysics,* 27(2):119–140, doi:10.1016/0040-1951(75)90102-X.

Araki, E., Shinohara, M., Sacks, S., Linde, A., Kanazawa, T., Shiobara, H., Mikada, H., and Suyehiro, K., 2004. Improvement of seismic observation in the ocean by use of seafloor boreholes. *Bull. Seismol. Soc. Am.,* 94:678–690, doi:10.1785/0120020088.

Baba, T. and Cummins, P.R., 2005. Contiguous rupture areas of two Nankai Trough earthquakes revealed by high resolution tsunami waveform inversion. *Geophys. Res. Lett.,* 32(8): L08305, doi:10.1029/2004GL022320.

Hori, T., Kato, N., Hirahara, K., Baba, T., and Kaneda, Y., 2004. A numerical simulation of earthquake cycles along the Nankai Trough in southwest Japan: lateral variation in frictional property due to the slab geometry controls the nucleation position. *Earth Planet. Sci. Lett.,* 228(3–4):215–226. doi:10.1016/j.epsl.2004.09.033.

Ichinose, G.A., Thio, H.K., Somerville, P.G., Sato, T., and Ishii, T., 2003. Rupture process of the 1944 Tonankai earthquake (Ms 8.1) from the inversion of teleseismic and regional seismograms. *J. Geophys. Res.,* 108(B10):2497, doi:10.1029/2003JB002393.

Kinoshita, M., Moore, G., von Huene, R., Tobin, H., and Ranero, C.R., 2006. The seismogenic zone experiment. *Oceanogr.,* 19:28–38, 2006.

Miyazaki, S. and Heki, K., 2001. Crustal velocity field of southwest Japan: subduction and arc-arc collision. *J. Geophys. Res.,* 106(B3):4305–4326, doi:10.1029/2000JB90031.

Nakanishi, A., Takahashi, N., Park, J.-O., Miura, S., Kodaira, S., Kaneda, Y., Hirata, N., Iwasaki, T., and Nakamura, M., 2002. Crustal structure across the coseismic rupture zone of the 1944 Tonankai earthquake, the central Nankai Trough seismogenic zone. *J. Geophys. Res.,* 107(B1):2007, doi:10.1029/2001JB000424.

Obana, K., Kodaira, S., and Kaneda, Y., 2004. Microseismicity around rupture area of the 1944 Tonankai earthquake from ocean bottom seismograph observations. *Earth Planet. Sci. Lett.,* 222(2):561–572. doi:10.1016/j.epsl.2004.02.032.

Obara, K. and Ito, Y., 2005. Very low frequency earthquake excited by the 2004 off the Kii Peninsula earthquake: a dynamic deformation process in the large accretionary prism. *Earth Planets Space,* 57:321–326.

Park, J.-O., Tsuru, T., Kodaira, S., Cummins, P.R., and Kaneda, Y., 2002. Splay fault branching along the Nankai subduction zone. *Science,* 297(5584):1157-1160. doi:10.1126/science.1074111.

Reches, Z. and Ito, H., 2007. Scientific drilling of active faults: Past and future., *In* Harms, U., Koeberl, C., and Zoback, M.D. (Eds.), *Continental Scientific Drilling,* Heidelberg (Springer), 235–258, doi:10.1007/978-3-540-68778-8_6.

Saffer, D.M. and Marone, C., 2003. Comparison of smectite- and illite-rich gouge frictional properties: application to the up-dip limit of the seismogenic zone along subduction megathrusts. *Earth Planet. Sci. Lett.,* 215(1–2):219–235. doi:10.1016/S0012-821X(03)00424-2.

Tobin, H. and Kinoshita, M., 2006. NanTroSEIZE: the IODP Nankai Trough seismogenic zone experiment. *Sci. Drill.,* 2:23–27, doi:10.2204/iodp.sd.2.06.2006.

Authors

Harold Tobin, Department of Geology and Geophysics, University of Wisconsin, 1215 West Dayton Street, Madison, Wis., 53706 U.S.A., e-mail: htobin@wisc.edu.
Masataka Kinoshita, IFREE (Institute for Frontier Research on Earth Environment), JAMSTEC (Japan Agency for Marine-Earth Science and Technology), 2-15 Natsushima-cho, Yokosuka-city, Kanagawa 237-0061, Japan.

Scientific Drilling of the Terrestrial Cretaceous Songliao Basin

by Yongjian Huang, Chengshan Wang, and the Terrestrial Scientific Drilling of the Cretaceous Songliao Basin Science Team

Investigations of critical climate changes during the Cretaceous have the potential to enhance our understanding of modern global warming because the extreme variances are the best-known and most recent example of a greenhouse Earth (Bice et al., 2006). Marine Cretaceous climate archives are relatively well explored by scientific ocean drilling programs such as the Integrated Ocean Drilling Program (IODP) and its predecessors. However, Cretaceous terrestrial climate records are at best fragmentary (Heimhofer et al., 2005). The long-lived Cretaceous Songliao Basin of NE China is an excellent candidate to fill this gap and provide important ocean-continent linkages in relation to environmental change (Fig. 1). This basin, located within one of the largest Cretaceous landmasses (Scotese, 1988), acted for about 100 million years as an intra-continental sediment trap; the present-day area of the basin is about 260,000 km^2. It provides an almost complete terrestrial sedimentary record from the Upper Jurassic to the Paleocene (Chen and Chang, 1994). Large-scale geological and geophysical investigations of lacustrine sediments and basin structures demonstrate that a rich archive of Cretaceous paleoclimate proxies exists. For example, the basin includes the Jehol Biota, a terrestrial response to the Cretaceous oceanic anoxic events (OAEs), and a potential K/T boundary (Qiang et al., 1998). An ongoing drilling program is supported by the Ministry of Science and Technology of China and by the Daqing Oilfield. It allowed for recovering of nearly complete cores from Upper Albian to the Uppermost Cretaceous in two boreholes (SK-I, SK II; commenced in 2006, Fig. 1). However, the older Cretaceous sedimentary record of Songliao Basin has not yet been cored. For that reason, a scientific drilling program has been proposed to the International Continental Scientific Drilling Program (ICDP) to sample the deeper sedimentary record of the Songliao Basin through a new drill hole (Figs. 1 and 2).

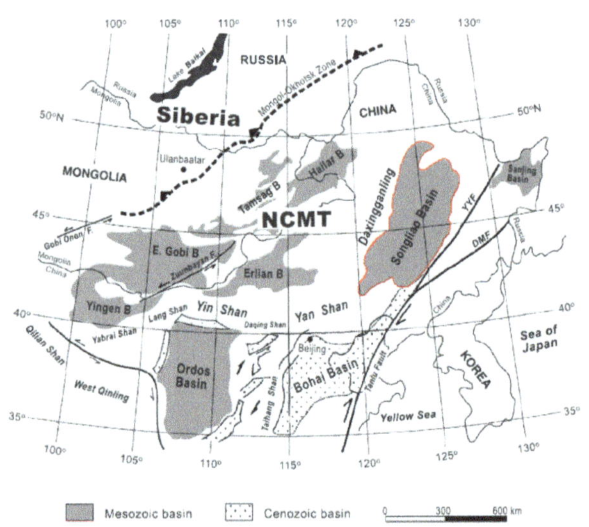

Figure 1. Location of the Songliao Basin in the tectonic framework of the North China–Mongolia Tract (NCMT). The NCMT is separated from the East Siberian Craton by the Mongolia-Okhotsk suture zone and is marked by the Yinshan-Yanshan Belt in the south and by the Daxing-Ganling Belt in the east. YYF=Yilan-Yitong fault; DMF= Dunhua-Mishan fault; HSZ = Honam Shear Zone; B= basin; F = fault.

In order to better constrain the scientific objectives, feasibility of deep drilling, and study of the core material, an ICDP workshop on "Deep Terrestrial Scientific Drilling Project of Cretaceous Songliao Basin" was held on 28–30 August 2007 in Daqing, China. The workshop was organized by the China University of Geosciences at Beijing and by the Daqing Oilfield and was jointly supported by the ICDP and sponsors from China: the Department of International Cooperation and Technology, Ministry of Land and Resources, China; Department of Basic Research, Ministry of Science and Technology, China (MOST); Department of Earth Science, National Natural Science Foundation of China (NSFC); and China Geological Survey.

About seventy participants (thirty-one from outside China) from eleven countries took part in the workshop. Three sessions were organized with more than fifty talks to meet the objectives of the workshop including "Evolution of Songliao Basin and East Asia, and Terrestrial Scientific Drilling Program", "Terrestrial Environment Change and Paleontological Response to Cretaceous Global Change" and "Cretaceous Paleoceanography".

During the workshop the following five key scientific goals for the project were defined:

- Improve understanding of the geodynamics of deep Earth, in particular the relation between the Cretaceous Super-Chron and Large Igneous Provinces;
- Quantify the biotic response to terrestrial environmental change and the deep biosphere (fossil DNA);
- Refine stratigraphic boundaries to improve the correlation between marine and terrestrial stratigraphy;
- Determine the terrestrial response to OAEs; and

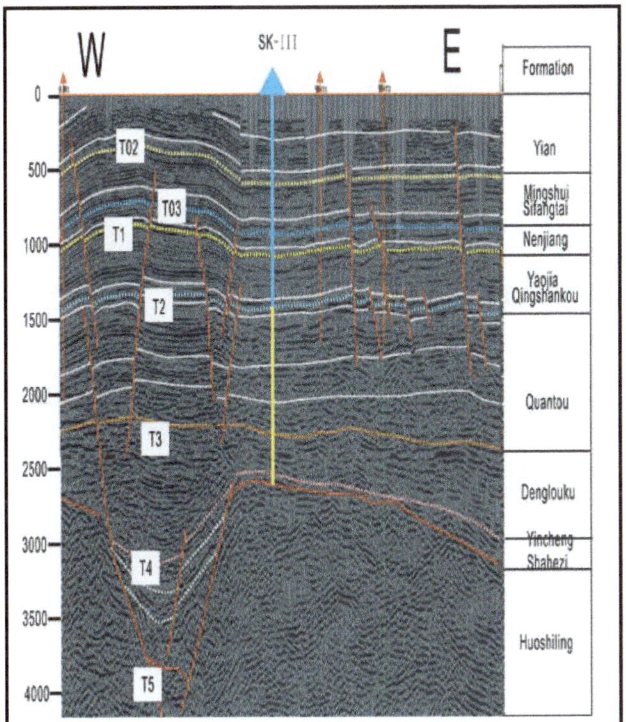

Figure 2. W-E Seismic profile showing the locale of the proposed Songke-III well with stratigraphic interpretation. The vertical blue line indicates the strata to be drilled in rotary mode. The yellow line indicates the planned wireline coring section down to the total depth of ~2700 m. This was done in order to avoid structural complications and to penetrate the most complete stratigraphic section.

- Investigate the formation of mass terrestrial hydrocarbon source rock.

In order to address these overarching research goals, workshop participants recommended that all boreholes (SK-I, SK-II, and the planned SK-III) should be incorporated into the planned ICDP project. The Upper Cretaceous core from SK-1 will especially serve to focus on the K/T boundary; SK-II cores will be analyzed to investigate mid-Cretaceous events; and samples from the proposed SK-III well will focus on the Lower Cretaceous including the J/K boundary, OAEs, and the evolution of the Jehol Biota. Preliminary results from existing boreholes and outcrop studies presented at the workshop showed great promise. During a field trip to the core curation facility of the Daqing Oilfield Company, the workshop participants were able to examine the cores of the SK-I and SK-II wells.

An important objective was to assess the feasibility of the deep drill hole (SK-III) in the Songliao Basin and its potential to recover strata older than Albian. Discussions of site selection concluded that the project should aim to core from the Albian Quantou to the Upper Jurassic Huoshiling Formations. Several site selection principles for drilling SK-III were defined as follows: to drill the most complete section, to find the least thickness of the overlying strata, and to preferentially drill fine-grained, clay-rich deposits. With these principles in mind, existing data were evaluated for identification of an optimum drilling location. The existing data base comprises, for example, over 250,000 km of seismic reflection profiles, some 3-D seismic reflection volumes, and about 50,000 wells (without coring). A preliminary site for SK-III has been selected in the center part of the basin (Fig. 2).

Finally, meeting participants agreed to form five strategic teams covering the objectives as outlined above. The goals of this science team are to compose a full drilling proposal and to submit it to ICDP and other potential funding agencies in early 2008.

References

Bice, K.L., Birgel, D., Meyers, P.A., Dahl, K.A., Hinrichs, K.-U., and Norris, R.D., 2006. A multiple proxy and model study of Cretaceous upper ocean temperatures and atmospheric CO2 concentrations. *Paleoceanogr.*, 21(2206):PA2002, doi:10.1029/2005PA001203.

Chen, P.-J., and Chang, Z.-L., 1994. Nonmarine Cretaceous stratigraphy of eastern China. *Cretaceous Research*, 5(3):245–257, doi:10.1006/cres.1994.1015.

Heimhofer, U., Hochuli, P.A., Burla, S., Dinis, J., and Weissert, H., 2005. Timing of Early Cretaceous angiosperm diversification and possible links to major paleoenvironmental change. *Geology*, 33:141–144, doi:10.1130/G21053.1.

Qiang, J., Currie, P.J., Norell, M.A., and Shu-An, J., 1998. Two feathered dinosaurs from northeastern China. *Nature*, 393:753–761, doi:10.1038/31635.

Scotese, C.R., Gahagan, L., and Larson, R.L., 1988. Plate tectonic reconstruction of the Cretaceous and Cenozoic ocean basins. *Tectonophysics*, 155:27–48, doi:10.1016/0040-1951(88)90259-4.

Authors

Yongjian Huang and Cheng-Shan Wang, Research Center for Tibetan Plateau Geology, China University of Geosciences, Xueyuan Road 29, Haidian District, 100083, Beijing, People's Republic of China, e-mail: huangyj@cugb.cn, **and the Terrestrial Scientific Drilling of the Cretaceous Songliao Basin Science Team**

Related Web Link

http://songliao.icdp-online.org

12

CPCP: Colorado Plateau Coring Project – 100 Million Years of Early Mesozoic Climatic, Tectonic, and Biotic Evolution of an Epicontinental Basin Complex

by Paul E. Olsen, Dennis V. Kent, and John W. Geissman

Introduction

Early Mesozoic epicontinental basins of western North America contain a spectacular record of the climatic and tectonic development of northwestern Pangea as well as what is arguably the world's richest and most-studied Triassic-Jurassic continental biota. The Colorado Plateau and its environs (Fig. 1) expose the textbook example of these layered sedimentary records (Fig. 2). Intensely studied since the mid-nineteenth century, the basins, their strata, and their fossils have stimulated hypotheses on the development of the Early Mesozoic world as reflected in the international literature. Despite this long history of research, the lack of numerical time calibration, the presence of major uncertainties in global correlations, and an absence of entire suites of environmental proxies still loom large and prevent integration of this immense environmental repository into a useful global picture. Practically insurmountable obstacles to outcrop sampling require a scientific drilling experiment to recover key sedimentary sections that will transform our understanding of the Early Mesozoic world.

To bring our insight into this critical time in Earth history to a new level, we developed the concept of the Colorado Plateau Coring Project (CPCP), an effort to recover continuous core spanning the early Mesozoic (Triassic-Jurassic) section of the Colorado Plateau and adjacent areas. The original basis for this was outlined at the 1999 International Continental Scientific Drilling Program (ICDP) and U.S. National Science Foundation (NSF) funded International Workshop for a Climatic, Biotic, and Tectonic, Pole-to-Pole Coring Transect of Triassic-Jurassic Pangea (http://www.ldeo.columbia.edu/~polsen/nbcp/pangeafinalreport.html, Olsen et al., 1999), section "Western Equatorial Pangea" (http://www.ldeo.columbia.edu/~polsen/nbcp/westpangea.html). Forty-five researchers from six countries attended the inaugural CPCP planning workshop held 13–16 November 2007, in St. George, Utah. The main goal was to develop a community consensus science plan for the CPCP (http://www.ldeo.columbia.edu/~polsen/cpcp/CPCP_home_page.html). The participants represented disciplines ranging from geochronology and physical stratigraphy through vertebrate paleontology and paleobotany. A plenary session with speakers highlighting the major science issues, precedents, and geoinformatics priorities was followed by thematic breakout groups. These included Stratigraphy, Geochronology and Magnetostratigraphy, Climate and Environments, Paleontology and Biotic Change, and Geoinformatics and Core-log Outcrop Integration. A half-day field trip was made to the St. George Dinosaur Discovery Site at Johnson Farm (http://www.sgcity.org/dinotrax/) and Warner Valley outcrops of Triassic-Jurassic strata (Fig. 3), and was followed by a plenary synthesis session. As developed by consensus at the workshop, the goal of the CPCP is to transform our understanding of the interplay between major biotic transitions, global climate change, plate position, and tectonics over 100 Ma of Earth histories (Fig. 2).

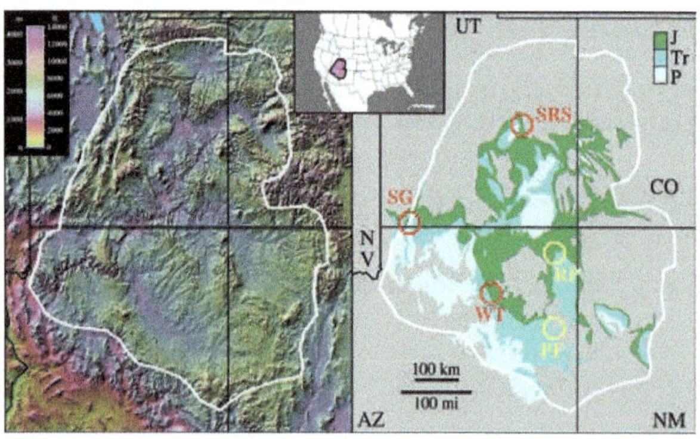

Figure 1. Map of the Colorado Plateau (white line) and adjacent areas: [Left] Shaded digital elevation map (courtesy of Andrew D. Birrell http://birrell.org/andrew/reliefMaps/ and http://birrell.org/andrew/copyright.html); [Right] Generalized geological map showing Permian and Triassic and Jurassic strata (modified from R. Blakely http://jan.ucc.nau.edu/~rcb7/Jurassic_erg_graphics.html). Core areas are: PF, Petrified Forest, Arizona; RP, Rock Point, Utah; SG, St. George, Utah; WT, Ward Terrace, Arizona; SRS, San Rafael Swell, Utah.

Overall Concept and Goals

Broad questions that could be profitably addressed in the Colorado Plateau venue include the following: what are the global or regional climate trends vs. plate position changes in "hot house" Pangea; how do largely fluvial systems respond to cyclical climate; what are the rates and magnitudes of the transition from the Paleozoic to essentially modern terrestrial ecosystems; and how does the stratigraphy reflect the interplay between growth in accommodation, uplift, and eustatic fluctuations? Based on these questions, the workshop identified eight goals attainable by coring of key Triassic/Jurassic sections:

1) Establishment of paleogeographic boundary conditions, particularly changes in paleolatitude.

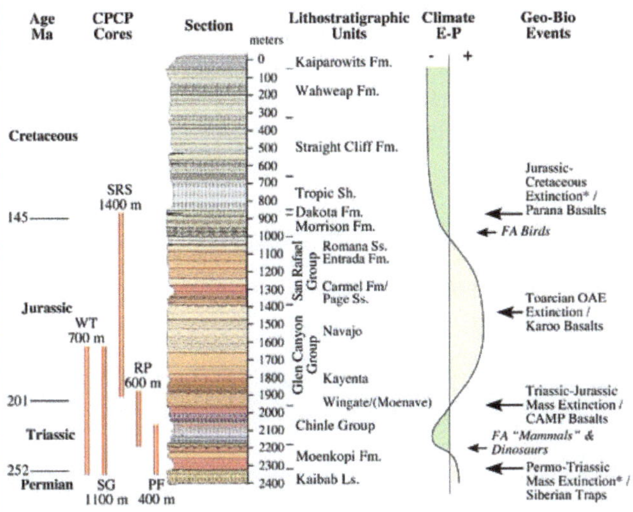

Figure 2. Details are given to the right.

Generalized Colorado Plateau section (Glen Canyon/Kaiparowits Plateau, based on http://jan.ucc.nau.edu/~rcb7/Glen_Can.jpg) with the cored sections recommended by the CPCP workshop participants, a generalized evaporation-precipitation (E-P) curve loosely based on climate sensitive facies, and some major geological and biological events (* actual boundary may or may not be present in rock section). See Fig. 1 for core area abbreviations. Note that the relative thicknesses of various stratigraphic units are generally different than what is shown in the color section and not the same between different coring areas.

The five major stratigraphic units shown here were deposited under dramatically changing climatic conditions that constitute the Triassic and Jurassic of the Colorado Plateau: the Early to Middle Triassic Moenkopi Formation (marginal marine and coastal sabka to semiarid fluvial and floodplain, minor eolian); the Late Triassic Chinle Group (humid to semiarid fluvial to lacustrine, some eolian near top); the latest (?) Triassic-Early Jurassic Glen Canyon Group (semiarid to arid, major eolian, fluvial, lacustrine); the Middle to Late Jurassic San Rafael Group (marine to coastal sabka, major eolian and fluvial arid to semiarid; from http://www.nps.gov/arch/); and the Late Jurassic Morrison Formation (semiarid to humid, fluvial and lacustrine). Determining the precise nature and origin of the large climatic transitions represented by these units is a major goal of the CPCP.

2) Development of the highest-resolution magnetic polarity stratigraphy for Early Triassic through Late Jurassic strata to facilitate detailed regional and global correlation.
3) Determination of how paleoclimates are expressed through time in the sedimentary record of western Pangea.
4) Identification of the precise stratigraphic position of major global biotic transitions (i.e., Permo-Triassic, Triassic-Jurassic, and Toarcian)
5) Refinement of lithostratigraphic and biostratigraphic correlations, considering regional unconformities and their possible relationship to eustacy and tectonics (c.f., Bachmann and Kozur, 2004).
6) Development of a chemostratigraphic ($\delta^{13}C$, cuticular CO_2 proxy, Nd, Sr, clays, etc.) reference sections.
7) Improvement of U-Pb zircon provenance stratigraphy and geochronology of ash beds.
8) Establishment of links among the temporal evolution of the Colorado Plateau sedimentary record, rifting of Pangea opening of the Atlantic Ocean, and emplacement of basalt provinces (Fig. 2).

Coring Plans

Given these overall goals, the workshop reached several conclusions that dictated the coring plan.

1) *Early Triassic through Late Jurassic Formations including the Morrison Formation should be cored* in order to cover the full range of climatic milieus represented by these rocks. Collectively, the group defined a three-tiered coring plan consisting of (1) three relatively thick (~1 km) synoptic intervals that together would yield an overlapping stratigraphic framework for the entire Jurassic and Triassic section, (2) two thinner (<500 m) cored sections that would tie to critical outcrop areas or to expanded critical intervals, and (3) a number of short sections to address more specific problems or provide regional coverage to the other five cores that are the nexus of the project.

2) *Superposition is paramount.* Evaluation of the critical Early Mesozoic transitions is required of all Early Mesozoic units in clear superposition. The five major stratigraphic units identified as major coring targets reflect, from oldest to youngest: arid (Moenkopi); humid to semiarid (Chinle); very arid (Glen Canyon and San Rafael); and return to semiarid and humid (Morrison) (Figs. 2 and 4). These climate transitions have been explained in several ways—translation of the North American plate from equatorial to mid-latitudes through zonal climate belts (Dickinson, 2005; Kent and Muttoni, 2003; Kent and Tauxe, 2004); large scale changes in the climate system involving changes in the non-zonal components of the climate system, particularly the monsoon (Kutzbach and Gallimore, 1989; Parrish, 1995; Rowe et al., 2007); or fluctuations in greenhouse gases like CO_2 (e.g., McElwain et al., 1999; Kürschner, 2001). These fundamentally different hypotheses remain untested because the temporal evolution of major boundary conditions, most notably latitude, has not been resolved to a useful level of precision (Tauxe and Kent, 2004; Tan et al., 2007).

3) *Internal time calibration needed.* Correlation of the Plateau sequence is presently based on low-resolution non-marine biostratigraphic approaches, and it does not provide clear biogeographic patterns or determination of the rates of biotic change in these very fossiliferous sequences. None of the major intervals of biotic change (Permo-Triassic; Triassic-Jurassic; or Toarcian) are located with precision in this succession. A combination of polarity stratigraphy along with geochronologic dates from ash deposits and dispersed grains will allow global correlations, including: 1) Triassic and Early Jurassic reference sections (Szurlies, 2007); 2) the astronomically calibrated polarity time scale (Kent and Olsen, 1999, 2008; Olsen and Kent, 1999; Hounslow et al., 2004; Kemp and Coe, 2007); 3) the Germanic basin (Bachmann and Kozur, 2004); 4) fully marine Tethyan sections (Muttoni et al., 2005; Channell et al., 2003; Gallet et al., 2007); 5) non-marine Jurassic to Early Cretaceous sequences of China (Feursich et al., 2002; Yao et al., 2003; Xu, 2005);

and 6) possibly, the marine magnetic anomaly M-sequence (Channell et al., 1995; Sager et al., 1998).

4) *Minimize hiatuses.* Acquiring as much stratigraphic scope as possible through the more continuous units and testing any paleomagnetic reversal sequences across geography by designing stratigraphic overlap between cores will help overcome significant unconformities and possible smaller and cryptic hiatuses. The thickest sections, least likely to be affected by rampant hiatuses, are not confined to a single area in the Colorado Plateau because of lateral shifts in the basin's depocenters. Several cores will be necessary to get the most favorable sections of each of the units. However, the thick sections proposed for coring are far from comparable surface outcrops, and thus, subsidiary sections more proximal to sources of the surface data must be cored as well. The goal is to provide a long enough section with unambiguous ties to the outcrop and sufficient stratigraphic scope to correlate with the main, long cores.

5) *Thick eolianite successions should be avoided* while key limestone and eolianite tongues must be intersected for age control by lateral correlation to outcrop.

Using these five principles above, three areas identified for long (~1 km) cores are, from the uppermost stratigraphic interval downward (Figs. 1 and 2): 1) *Dry Mesa* (east side of San Rafael Swell, UT), ~1400 m from basal Cretaceous Cedar Mountain Formation to the Permian Kaibab Formation; 2) *St. George* (west of the Hurricane fault), ~1100 m from locally basal Navajo Formation to Permian Kaibab Formation; and 3) *Tuba City* (north of Ward Terrace): ~700 m from basal Navajo Formation to Kaibab Formation. Although the St. George and Tuba City sections appear to span the same overall section, they are actually critical complements. The Moenkopi and latest (?) Triassic-Early JurassicTriassic-Early Jurassic Moenave formations are well-developed, cyclical, and probably relatively complete in the St. George area, yet the Chinle Group is very thin and erosionally truncated. In the Tuba City area, the Chinle is very well-developed, but the Moenkopi is very thin and erosionally truncated. Also, the Moenave lacks the well-developed, cyclical lacustrine strata present around St. George.

The two medium depth sections (Fig. 2) are targeted for 1) the *Rock Point* area (north of Round Rock, Utah), ~600 m from basal Wingate Formation to Permian Kiabab Formation,

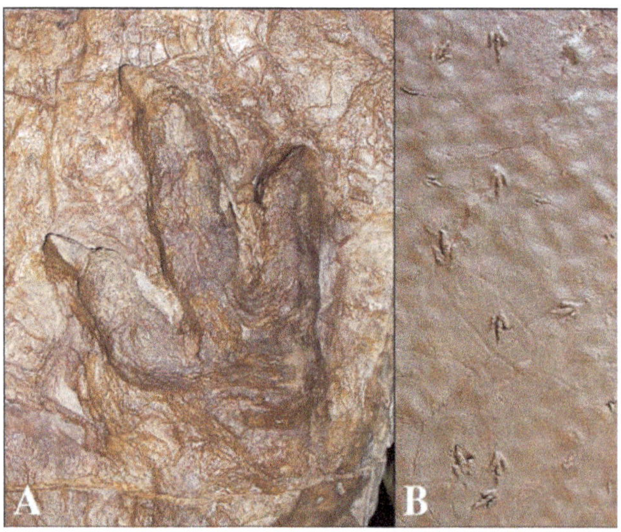

Figure 3. Examples of dinosaur tracks at the St. George Discovery Site at Johnson Farm in the lower Whitmore Point Member of the Moenave Formation: [A] Example of natural cast of *Eubrontes giganteus* (~35 cm long), plausibly made by a large theropod dinosaur similar to *Dilophosaurus* from the lower Kayenta Formation; [B] small theropod dinosaur footprints of the "Grallator" and "Anchisauripus" types (~10–15 cm long each). The importance of these fossil assemblages is that they document the terrestrial assemblages just after the Triassic-Jurassic extinction event that surprisingly is overwhelmingly dominated by carnivores.

Figure 4. Details are given below.

Although the American Southwest, including the Colorado Plateau, is famous for spectacular exposures in striking badlands, the thickest sections of key time intervals are often projected from the subsurface. The most continuous sections in outcrop are often exposed as inaccessible vertical cliffs or are heavily weathered and geochemically altered, precluding research at the appropriate level of detail. Because of low bedding dips, the assembly of sections spanning large stratigraphic thicknesses requires long-distance traverses that often compromise the essential assumption of superposition because of facies changes. The problems are amply revealed by repeated attempts in the literature at compiling long, high-resolution sections that describe the succession in any specific area. Long, continuously cored intervals tied to outcrop in key areas are required for further substantial progress.

Opportunities and problems of working only with outcrops present themselves in examples of superb near-100% outcrops, especially of muddy facies (Fig. 4). Mudstones are often bentonitic and with a partial volcanic source, which means there are datable ashes; however, it also means that the outcrop surfaces present horrendous sampling problems where freshness and physical integrity are at a premium as for geochemistry or paleomagnetics. Two examples include the Late Jurassic Morrison Formation (Fig. 4A and 4B) and the Late Triassic Chinle Group (Fig. 4C and 4D).

Fig. 4A, Brushy Basin Member of the Morrison Formation west of Green River, Utah (from http://en.wikipedia.org/wiki/Morrison_Formation); Fig. 4B, typical "popcorn" surface of bentonitic mudstone of the Brushy Basin Member of the Morrison Formation west of Green River, Utah (same source as 4A); Fig. 4C, "popcorn" surface of the Chinle Group near Bluewater Creek, NM (Stop 1 of Lucas et al., 2007), with fragmentary weathered metoposaur amphibian dermal bones; and Fig. 4D, outcrops of bentonitic Painted Desert Member of the Petrified Forest Formation of the Chinle Group in Petrified Forest National Park with the U-Pb dated Black Forest Bed (Riggs et al., 2003) outcropping as the white bed in foreground.

and 2) the *Petrified Forest* area (within Petrified Forest National Park, AZ), ~400 m from Sonsela (mid-Chinle) to Permian. The Rock Point area is a critical target because it exposes the Triassic-Jurassic boundary section in the largely eolian Wingate Formation, based on both vertebrate paleontology and magnetostratigraphy, and it is a natural complement to the lacustrine-dominated Moenave Formation near St. George. The Petrified Forest area has produced the bulk of the Chinle fauna and flora, as well as dated ash beds, and it is a natural complement to the Chinle section in the Tuba City area. The workshop concluded that the geophysical logs of the core holes and cores will be critical in tying cores to outcrop (Szurlies et al., 2003).

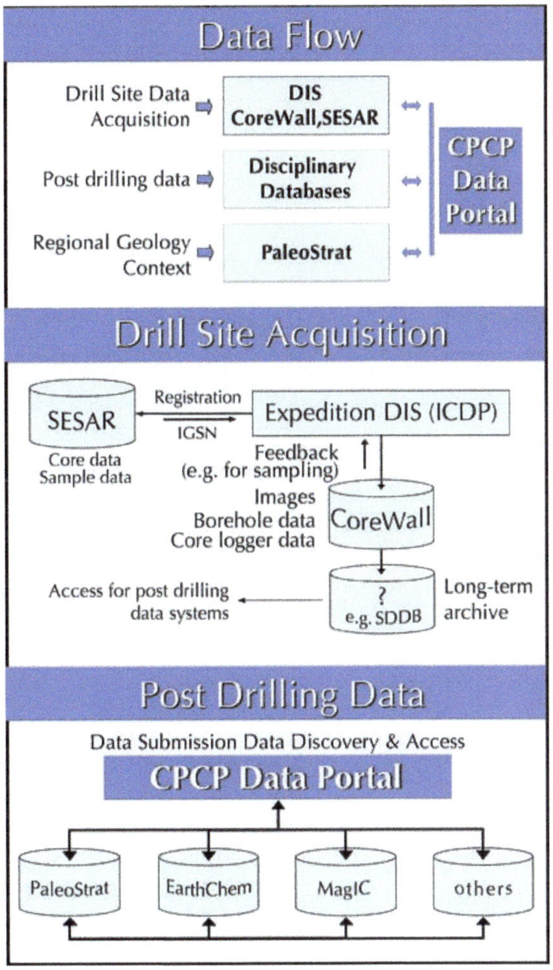

Figure 5. Details are given to the right.

Data management scenario for the CPCP showing the relationship between data gathering (i.e., from the cores and core holes) and the flow of information and samples derived from it to community access systems. Acronyms are: DIS, Drilling Information System; CoreWall, a real-time stratigraphic correlation, core description and data visualization system (http://www.evl.uic.edu/cavern/corewall/index.php); SESAR, System for Earth Sample Registration (http://www.geosamples.org/); PaleoStrat, a community digital information system for paleontology and sedimentary geology (https://www.paleostrat.com/); IGSN, International Geo Sample Number; SDDB, The Scientific Drilling Database is the repository for data from operations of ICDP (http://www.icdp-online.org/contenido/lakedb/front_content.php); EarthChem, system to facilitate the preservation, discovery, access, and visualization of geochemical datasets (http://www.earthchem.org/); MagIC, Magnetics Information Consortium (http://www.earthref.org/MAGIC/). A robust and effective data management/geoinformatics system, accessible through a CPCP Data Portal, will facilitate and support the science and provide the basis for education and outreach efforts. This will include coupling the Drilling Information System (DIS) of the ICDP with System for Earth Sample Registration (SESAR; www.geosamples.org), EarthChem (www.earthchem.org), CoreWall (www.evl.uic.edu/cavern/corewall), and PaleoStrat (www.paleostrat.org), a core-core hole-log integration system, and a novel digital framework of regional geology (Fig. 5).

Next Steps

The CPCP workshop endorsed aiming full drilling proposals to ICDP and U.S. National Science Foundation (NSF) Continental Dynamics. In preparation for this, there will be an international CPCP proposal planning workshop in Albuquerque, N.M., to be advertised in the near future.

References

Bachmann, G.H., and Kozur, H.W., 2004. The Germanic Triassic: correlations with the international chronostratigraphic scale, numerical ages and Milankovitch cyclicity. *Hallesches Jahrbuch Geowissenschaften B*, 26:17–62.

Channell, J.E.T., Erba, E., Nakanishi, M., and Tamaki, K., 1995. Late Jurassic-Early Cretaceous time scales and oceanic magnetic anomaly block models. *In* Berggren, W.A., Kent, D.V., Aubry, M.-P., and Hardenbol, J. (Eds.), *Geochronology, Time Scales and Global Stratigraphic Correlations*. Tulsa, Okla. (SEPM (Society for Sedimentary Geology)) SEPM Special Volume No. 54, p. 51–63.

Channell, J.E.T., Kozur, H.W., Sievers, T., Mock, R., Aubrecht, R., and Sykora, M., 2003. Carnian-Norian biomagnetostratigraphy at Silicka Brezova (Slovakia): correlation to other Tethyan sections and to the Newark Basin: *Palaeogeogr. Palaeoclimatol. Palaeoecol.*, 191:65–109.

Dickinson, W.R., 2005. Redbed paleomagnetism and Mesozoic paleolatitude of the Colorado Plateau. *Geol. Soc. Am. Abst. Prog.*, 37(7):362.

Feursich, F.T., Sha, J.G., Jiang, B.Y., and Pan, Y.H., 2002. High resolution palaeoecological and taphonomic analysis of Early Cretaceous lake biota, western Liaoning (NE-China). *Palaeogeogr. Palaeoclimatol. Palaeoecol.*, 253:434–457.

Gallet, Y., Krystyn, L., Marcoux, J., and Besse, J., 2007. New constraints on the End-Triassic (Upper Norian-Rhaetian) magnetostratigraphy. *Earth Planet. Sci. Lett.*, 255:458 –470.

Hounslow, M.W., Posen, P.E., and Warrington, G., 2004. Magnetostratigraphy and biostratigraphy of the Upper Triassic and lowermost Jurassic succession, St. Audrie's Bay, UK. *Palaeoceanogr. Palaeoclimatol. Palaeoecol.*, 213:331–358.

Kemp, D.B., and Coe, A.L., 2007. A nonmarine record of eccentricity forcing through the Upper Triassic of southwest England and its correlation with the Newark Basin astronomically calibrated geomagnetic polarity time scale from North America. *Geology*, 35(11):991–994.

Kent, D.V., and Muttoni, G., 2003. Mobility of Pangea: Implications for Late Paleozoic and Early Mesozoic paleoclimate. *In* LeTourneau, P.M., and Olsen, P.E. (Eds.), *The Great Rift Valleys of Pangea in Eastern North America, Volume 1, Tectonics, Structure, and Volcanism*. New York (Columbia University Press),11-20.

Kent, D.V., and Olsen, P.E., 1999. Astronomically tuned geomagnetic polarity time scale for the Late Triassic. *J. Geophys. Res.*,

104:12831–12841.

Kent, D.V., and Olsen, P.E., 2008. Early Jurassic magnetostratigraphy and paleolatitude from the Hartford continental rift basin (eastern North America): Testing for polarity bias and abrupt polar wander in association with the Central Atlantic Magmatic Province. *J. Geophys. Res.,* 113, B06105, doi:10.1029/2007JB005407.

Kent, D.V., and Tauxe, L., 2005. Corrected Late Triassic latitudes for continents adjacent to the North Atlantic. *Science,* 307:240–244.

Kürschner, W.M., 2001. Leaf sensor for CO_2 in deep time. *Nature,* 411:247–248.

Kutzbach, J.E., and Gallimore, R.G., 1989. Pangean climates: megamonsoons of the megacontinent. *J. Geophys. Res.,* 94:3341–3357.

Lucas, S.J., Heckert, A.B., Spielmann, J.A., Tanner, L.H., and Hunt, A.P., 2007. First Day: Middle and Upper Triassic stratigraphy, sedimentology, and paleontology of west-central New Mexico. *In* Lucas, S.J., and Spielmann, J.A. (Eds.), *Triassic of the American West, New Mexico Museum of Natural History & Science Bulletin 40,* 169–180.

McElwain, J.C., Beerling, D.J., and Woodward, F.I., 1999. Fossil plants and global warming at the Triassic-Jurassic boundary. *Science,* 285(5432):1386–1390.

Muttoni, G., Meco, S., and Gaetani, M., 2005. Magnetostratigraphy and biostratigraphy of the Late Triassic Guri Zi Section, Albania: Constraint on the age of the Carnian-Norian boundary. *Rivista Italiana di Paleontologia e Stratigrafia,* 111:233–245.

Olsen, P.E., and Kent, D.V., 1999. Long-period Milankovitch cycles from the Late Triassic and Early Jurassic of eastern North America and their implications for the calibration of the early Mesozoic time scale and the long-term behavior of the planets. *Phil. Trans. Roy. Soc. London A,* 357:1761–1787.

Olsen, P.E., Kent, D.V., and Raeside, R., 1999. International workshop for a climatic, biotic, and tectonic, pole-to-pole coring transect of Triassic-Jurassic Pangea. Newsletter, ICDP (Potsdam), 1:16–20.

Parrish, J.T., 1995. Geologic evidence of Permian climate. *In* Scholle, P.A., Peryt, T.M., and Ulmer-Scholle, D.S. (Eds.), *The Permian of Northern Pangea, Paleogeogr., Paleoclimatol. Stratigr. vol. 1,* Berlin (Springer Verlag), 53–61.

Riggs, M.R., Ash, S.R., Barth, A.P., Gehrels, G.E., and Wooden, J.L., 2003. Isotopic age of the Black Forest Bed, Petrified Forest Member, Chinle Formation, Arizona: An example of dating a continental sandstone. *Geol. Soc. Am. Bull.,* 115:1315–1323.

Rowe, C.M., Loope, D.B., Oglesby, R.J., Van der Voo, R., and Broadwater, C.E., 2007. Inconsistencies between Pangean reconstructions and basic climate controls. *Science,* 318:1284–1286.

Sager, W.W., Weiss, M.A., Tivey, M.A., and Johnson, H.P., 1998. Geomagnetic polarity reversal model of deep-tow profiles from the Pacific Jurassic "Quiet Zone" *J. Geophys. Res.,* 103:5269–5286.

Szurlies, M., Bachmann, G.H., Menning, M., Nowaczyk, N.R., and Käding, K.-C., 2003. Magnetostratigraphy and high-resolution lithostratigraphy of the Permian-Triassic boundary interval in Central Germany. *Earth Planet. Sci. Lett.,* 212(3-4):263–278.

Szurlies, M., 2007. Latest Permian to Middle Triassic cyclo- magnetostratigraphy from the Central European Basin, Germany: Implications for the geomagnetic polarity timescale: *Earth Planet. Sci. Lett.,* 261:602–619.

Tan, X., Kodama, K.P., Gilder, S., and Courtillot, V., 2007. Rock magnetic evidence for inclination shallowing in the Passaic Formation red beds from the Newark basin and a systematic bias of the Late Triassic apparent polar wander path for North America. *Earth Planet. Sci. Lett.,* 254:345–357.

Tauxe, L., and Kent, D.V., 2004. A simplified statistical model for the geomagnetic field and the detection of shallow bias in paleomagnetic inclinations: Was the ancient magnetic field dipolar? *In* Channell, J.E.T., Kent, D.V., Lowrie, W., and Meert, J. (Eds), *Timescales of the Paleomagnetic Field, AGU Geophysical Monograph* 145,101–115.

Xu, D.Y., 2005. Astro-geologic time scale and the advancements of cyclostratigraphy. *J. Stratigr.,* 29:635–640 (in Chinese with English abstract).

Yao, Y.-M., Fu, G.-B., Xu, D.-Y., Qin, J., and Yao, S., 2003. Preliminary study on the high-resolution cyclostratigraphy of the Jurassic system in Turpan-Hami basin, Xinjiang. *J. Stratigr.,* 27:122–128 (in Chinese with English abstract).

Authors

Paul E. Olsen, Lamont-Doherty Earth Observatory (LDEO) of Columbia University, 61 Route 9W, Palisades, N.Y., 10964-1000, U.S.A., e-mail: polsen@ldeo.columbia.edu.

Dennis V. Kent, Department of Earth & Planetary Sciences, Rutgers University, 610 Taylor Road, Piscataway, N.J. 08854, U.S.A. and Lamont-Doherty Earth Observatory, 61 Route 9W, Palisades, N.Y. 10964,, U.S.A.

John W. Geissman, Department of Earth and Planetary Sciences, University of New Mexico, Albuquerque, N.M., 87131, U.S.A.

Related Web Links

http://www.earthchem.org/
http://www.earthref.org/MAGIC/
http://www.evl.uic.edu/cavern/corewall/index.php
http://www.geosamples.org/
http://www.icdp-online.org/contenido/lakedb/front_content.php
http://www.ldeo.columbia.edu/~polsen/cpcp/CPCP_home_page.html
http://www.ldeo.columbia.edu/~polsen/cpcp/CPCP_dosecc_nsf.html
http://www.ldeo.columbia.edu/~polsen/nbcp/pangeafinal-report.html
http://www.ldeo.columbia.edu/~polsen/nbcp/westpangea.html
https://www.paleostrat.com/
http://www.sgcity.org/dinotrax/

Figure Credits

Fig. 1: courtesy of Andrew D. Birrell http://birrell.org/andrew/reliefMaps/ and http://birrell.org/andrew/copyright.html

13

ICDP Workshop on Borehole Monitoring at the Nankai Subduction Zone: Building a Land-Ocean Borehole Network to Study the Seismogenic Zone

by Takeshi Sagiya

Examples of subduction processes are currently investigated around the Kii Peninsula, Honshu Island by Japanese as well as international projects. The most well-known is the NanTroSEIZE project conducted by the Integrated Ocean Drilling Program (IODP). Drilling (riserless) with the drillship *Chikyu* started southeast off the Kii Peninsula in October 2007. In addition, the Japanese government launched a new project called Dense Ocean floor Network System for Earthquakes and Tsunamis (DONET) to install an integrated ocean bottom cable system for continuously monitoring earthquakes and tsunamis between the Kii Peninsula and the NanTroSEIZE drilling site. On land on the Kii Peninsula, an array of shallow boreholes for monitoring changes in groundwater level and strain is being constructed by Japan's National Institute of Advanced Industrial Science and Technology (AIST, Fig. 1). As part of the latter project, the International Continental Scientific Drilling Program (ICDP) organized and funded a workshop to prepare for land-based intermediate depth drilling to install monitoring equipment complementing the AIST and NantroSEIZE projects.

The "ICDP Workshop on Borehole Monitoring at the Nankai Subduction Zone: Building a Land-Ocean Borehole Network to Study the Seismogenic Zone" was conducted on 20–23 August 2007 at Nagoya University. Thirty-five scientists from four countries participated. Eighteen presentations addressed the seismic activity along the Nankai Trough subduction zone, geologic background of the Kii Peninsula, present status of various ongoing projects (NanTroSEIZE, DONET, etc.), and state-of-the-art observatory technologies.

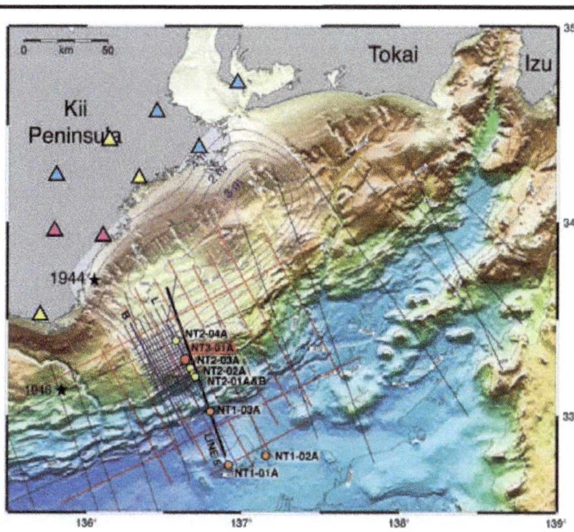

Figure 1. Location map of the Nankai Trough and boreholes in Kii Peninsula. Circles denote drilling sites in the NanTroSEIZE project. Triangles on land denote IST boreholes.

Other key topics included pore pressure measurements and technical aspects of the planned monitoring system and its installation. The discussion of the merits of pore pressure measurement led to the conclusion that such data must be one of the main parameters to be monitored in the Nankai Trough subduction complex. For strain measurements, two options were considered in detail. An integrated monitoring system of the Tono Research Institute of Earthquake Science in Mizunami, Japan, shows high sensitivity and good stability. Alternatively, optical fiber strain meters can be installed outside the borehole casing to allow using the boreholes for other purposes such as repeated downhole logs or permanent measurements inside casing. Another finding was that seismological monitoring and analysis should be included to study non-volcanic deep tremors.

A major part of the workshop was dedicated to preparation of a drilling proposal including discussion of prioritized scientific targets. Considering recent findings like non-volcanic deep low-frequency tremors (e.g. Obara, 2002), slow slip events (e.g. Ozawa et al., 2002), and ultra-slow earthquakes (e.g. Ito et al., 2007), participants agreed that the goal of drilling and monitoring is to resolve the complex behavior of the plate boundary megathrust at the deeper end of the locked zone.

A field trip was made to the Kii Peninsula to visit two geologic sites exhibiting Miocene igneous acidic rocks characterizing the southeastern part of the Kii Peninsula. The integrated borehole groundwater observatory constructed by AIST visited Kumano City (Fig. 2). The observatory was designed to detect precursory ground water level changes before a large megathrust earthquake (there were reports of ground water level change before the 1946 Nankai earthquake) and to investigate various

Figure 2. Field trip to the integrated borehole groundwater observatory of AIST.

plate boundary phenomena with precise observation with strainmeter, tiltmeter, and a seismograph. Finally, Kata village of Owase City, Japan was visited, where tsunami run-up heights were recorded and displayed on a telegraph pole (Fig. 3). The 1944 Tonankai and the 1854 Anse-Tokai earthquakes also shook this area. This display teaches us that the last earthquake in 1944 was a moderate one, and the region may have to prepare for a stronger one similar to the 1854 event.

The whole workshop was very successful in raising the understanding on various ongoing projects, introducing state-of-the-art monitoring techniques to potential project proponents as well as technical problems to be resolved. It also defined the goals and milestones towards a full drilling proposal planned for submission to ICDP by fall 2008.

Figure 3. Display of tsunami heights (in red on pole) in Kata village, Owase City.

Acknowledgements

The author would like to express sincere thanks to ICDP and the Japan Drilling Earth Science Consortium (J-DESC) for their financial support of the workshop.

References

Ito, Y., Obara, K., Shiomi, K., Sekine, S., and Hirose, H., 2007. Slow earthquake coincident with episodic tremors and slow slip events. *Science*, 315:503–506, doi:10.1126/science.1134454.

Obara, K., 2002. Nonvolcanic deep tremor associated with subduction in southwest Japan. *Science*, 296:1679–1681, doi:10.1126/science.1070378.

Ozawa, S., Murakami, M., Kaidzu, M., Tada, T., Sagiya, T., Yarai, H., and Nishimura, T., 2002. Detection and monitoring of ongoing aseismic slip in the Tokai region, central Japan. *Science*, 298: 1009–1012, doi:10.1126/science.1076780.

Author

Takeshi Sagiya, Research Center for Seismology, Volcanology, and Disaster Mitigation, Graduate School of Environmental Studies, Nagoya University, Furo-cho, Chikusa-ku, Nagoya 464-8601 Japan, e-mail: sagiya@seis.nagoya-u.ac.jp.

Upcoming Workshops

Marine Research Drilling in the Atlantic (Magellan WS Series)

10–12 September 2008, Montpellier, France

Participation on invitation only. Conveners: Marguerite Godard (Université Montpellier 2), Gretchen Früh-Green (ETH Zurich), and Christopher MacLeod (Cardiff University). **Contact: Dr. C. J. MacLeod**, e-mail: macleod@cardiff.ac.uk

Ultra-High Resolution Geological Records of Past Climate Change

29 Sept.–1 Oct. 2008, Potsdam, Germany

See details at: http://www.iodp.org/climate-ws-workshop/

Scientific Drilling of Sediment in Lake Ohrid (SCOPSCO)

13–17 October 2008, Ohrid, Republic of Macedonia

Scientific collaboration on past speciation conditions in Lake Ohrid (SCOPSCO), Republic of Macedonia, will be discussed to prepare drilling of sediments in the oldest lake in Europe. See details at: http://ohrid.icdp-online.org

Arctic Ocean History Workshop

3–5 November 2008, Bremerhaven, Germany
Application Deadline: 24 August 2008

To participate in this workshop, details are given at: www.oceanleadership.org/arctic. Partial travel support is available. Please contact **Bernard Coakley** (bernard.coakley@gi.alaska.edu) or **Ruediger Stein** (rstein@awi-bremerhaven.de) for further information.

Scientific Drilling for Human Origins

17–21 November 2008, Addis Ababa, Ethiopia

Following recent, successful lake drilling projects in Africa with implications for human origins, the time is right to consider new targets for drilling. A highly promising approach to this goal is to target lacustrine sedimentary sequences currently exposed on-land, in sedimentary basins of world-class importance to hominin evolutionary history. See details at: http://www.magadi.icdp-online.org

7th International Symposium for Subsurface Microbiology

Shizuoka, Japan, 16–21 November 2008, Regular Registration: 30 September 2008

The application of molecular approaches to the study of subsurface microbial ecology has been most encouraging in the past decade. As would naturally be expected, the deeper we delve, the more diverse the discoveries. For more detailed information, **contact: Prof. Kenji Kato** (Chair), **Sanami Nishida** (Secretary), e-mail: shizissm@ipc.shizuoka.ac.jp.

Distinguished Lecturer Program

Application Deadline 30 September 2008

ECORD is sponsoring an initiative for a lecture series to be given by leading scientists involved with the IODP. This program is designed to bring the exciting scientific discoveries of the IODP to the geosciences community. For more information, please contact: ESSAC Office (ECORD Science Support & Advisory Committee); e-mail: essac@cerege.fr, and apply via: http://www.doodle.ch/4nxs78whk6xmedh9.

IODP Conference on Future Scientific Ocean Drilling

The IODP is funded for the period 2003–2013, and is now starting to plan for its renewal beyond 2013. A community-wide, major conference on future directions of scientific ocean drilling is planned for 23–25 September 2009. The meeting location will be the University of Bremen, Germany. Based on this conference and other planning documents including past and future workshops, a science plan that defines new goals and effectively meets the challenges of future ocean drilling will be drafted during 2010. The conference will be open to all scientists with an interest in scientific ocean drilling. More information at http://www.iodp.org.

Chikyu and the boundless ocean view from IODP Exp. 316. Photo courtsey JAMSTEC/IODP.

Internships Available in Scientific Drilling

DOSECC Internships promote student involvement in projects where drilling has provided data and materials for study. Students can undertake research related to ongoing or past drilling efforts. The internships are open to undergraduate or graduate students, and primary and secondary school teachers worldwide. Internship funding will be available in the summer of 2009 and budgets of $2,000 to $5,000 are appropriate. See application details at: http://www.dosecc.org/html/internship.html. **Contact: David Zur**, e-mail: dzur@dosecc.org.

Distinguished Lecture Series

DOSECC's Distinguished Lecture Series makes available geoscientists to give talks or lectures on their areas of expertise. DOSECC covers travel costs for the lecturers, and the host institution is asked to provide food and lodging costs. Lecture themes include:

- Paleoclimate and Human Evolution;
- Global sea-level changes over the past 100 million years;
- Celestial Mechanics, Mass Extinctions, and Giant Volcanic Eruptions;
- 32 Million Years of Astronomical Forcing of Climate from Tropical, Triassic-Jurassic Pangea;
- New insights into African paleoclimate from the Lake Malawi Drilling Project;
- Sublacustrine Paleoseismology: Coring Earthquake Event Horizons in the Great Salt Lake; and
- Terrestrial Impact Cratering: The Earth's Record of Bombardment from Space.

Further details and speaker contact information is available at http://www.dosecc.org/html/distinguished_lectures.html. To apply, **contact: David Zur**, e-mail: dzur@dosecc.org.

ICDP Drilling Training

Geoscientists leading drilling missions need special management skills, but this know-how is often not part of their academc background. Industry helps by providing training, but scientists usually start learning on the job during a drilling project. The Operational Support Group of ICDP offers a comprehensive one-week training course covering key topics of scientific drilling to address this issue. The idea is not only to transfer specific knowledge and expertise but also to demonstrate the dependencies and linkages among different subjects involved, especially between engineering and science. Classes are taught by specialized instructors with practical industrial or academic experience in several basic modules. Content and training level can be customized (deadline for training proposals is 15 January each year), and costs for invited participants are paid by ICDP.

The 2007 ICDP Training was held in Germany at the drill site of the KTB from 5–9 November, attracting 33 scientists from 19 countries. A highlight of the training, in addition to the lectures at the fully equipped field lab and core repository, was an excursion to two drill sites, at a 4.5-km-deep geothermal well and a site where the newly developed scientific drilling facility InnovaRig was located.

ICDP training is usually held in conjunction with an active drilling project to bridge the gap between the classroom and practical applications in the field and to provide training on the job.

The next ICDP training is scheduled for 6–10 October 2008 in Germany. Further information: http://www.icdp-online.org

Contact: Thomas Wöhrl, ICDP, GFZ Potsdam, e-mail: woehrl@gfz-potsdam.de

MSPHDS Professional Development Program

Partnering with the MSPHDS (Minorities Striving and Pursuing Higher Degrees of Success in Earth System Science) Professional Development Program, the Consortium for Ocean Leadership supported four students to attend the IODP Science Steering and Evaluation Panel meeting in Busan, Korea. Mentored by panel members, the students gained an inside look at the proposal process for a large-scale science program and had the opportunity to meet and discuss science issues with international colleagues. One student commented, "The importance of writing succinct proposals with clear and concise objectives was a great lesson learned by observing the meeting." See details at http://oceanleadership.org/diversity/professionals

New Member of IODP:Australia

An agreement between the Australian Research Council and the Lead Agencies of IODP to make Australia an Associate Member of IODP effective 1 January 2008 was in the process of final signing by early July 2008. The Australian membership is intended to be amended by membership of New Zealand within a joint AUS-NZ consortium.

New Members of ICDP: Sweden and Switzerland

The Swedish Deep Drilling Program (SDDP) was initiated in early 2007 to address fundamental issues of the dynamic Earth system including (Palaeo)proterozoic orogens or ore genesis. The SDDP working group is planning a drilling program to attract funding for four world-class deep boreholes over a ten-year period.

Recently, the strongly anticipated financial support for a Swedish ICDP membership has been granted by the Swedish Research Council, greatly supporting the efforts of SDDP. Please visit the SDDP website at http://www.sddp.se for information.

Contact: Henning Lorenz, e-mail: henning.lorenz@geo.uu.se

ICDP Switzerland was initiated during the Swiss ICDP meeting on 1 February 2008 by about fifty scientists. Several are involved in ongoing or planned paleoclimate drilling projects and are discussing projects in Quaternary deep Alpine valleys and the Ivrea Zone. The Swiss National Science Foundation recently assured the Swiss participation for the next five years. **Contact: Flavio Anselmetti**, e-mail: flavio.anselmetti@eawag.ch, and web link: http://www.swiss-drilling.ethz.ch.

Joint EUROFORUM 2008

For the first time, IODP and ICDP held a joint EUROFORUM meeting during the EGU General Assembly in Vienna on 17 April 2008. This afternoon meeting provided eleven talks in a platform to exchange latest research highlights of various drilling projects covering new NanTroSEIZE results, paleoclimate science on various marine and terrestrial archives, as well as deep subsurface life and the planned volcanic risk research at Campi Flegrei.

The EUROFORUM was followed by a Joint Town Hall Meeting for latest program development updates and discussions.

First Joint IODP-ICDP Town Hall Meeting held in Japan

J-DESC sponsored the first joint IODP-ICDP town hall meeting on 27 May, while Japan Geoscience Union Meeting 2008 was also being held in Japan. The town hall event got ~100 scientists and graduate students together to exchange program news and other informaton. This will be an annual event.

The Thrill to Drill in Japanese

The ICDP brochure "The Thrill to Drill" was produced jointly by ICDP and J-DESC Japan in 2006 in English, and now its Japanese version is available from J-DESC or can be downloaded from www.j-desc.org.

Deep Earth Academy's Sea90E Poster and Web Site

Deep Earth Academy's poster and Web site won a 2008 Distinguished Achievement Award from the Association of Educational Publishers in the category of curriculum for teaching science for grades for six to eight, and more information at http://www.aepweb.org/awards/curric-win.htm. It features a bathymetry map, scientist profiles, student Q&As, and science challenges. See poster at http://www.ocean-leadership.org/learning/materials/posters/sea90poster.

Aurora Borealis: A European Project

The European Commission identified developing of the icebreaker *Aurora Borelis* for the highest scientific priority. A European Research Icebreaker Consortium (ERICON) of 15 institutions, funding agencies and companies from 10 European nations, including Russia, has been formed to develop management structures for this unique facility and to implement it into the European Research Area. The Alfred Wegener Institute for Polar and Marine Research coordinates the final engineering work for the development with funds from the German Federal Ministry for Science and Education. In December 2007, SCHIFFKO GmbH was awarded a design contract for the initial design concept (Fig.1), general arrangement planning, and full tender documentation. Currently the ice-tank and open water tests are being carried out. The new technical details include:

- 55 MW, twin hull icebreaker
- 201 m length, 49 m width
- 15 kn max. speed
- dynamic positioning, two moon pools
- riserless drilling, in max. 5000 m water for 1000 m core

Contact: Dr. Nicole Biebow, AWI, 27568 Bremerhaven, Germany, e-mail: nicole.biebow@awi.de, web link at : www.eri-aurora-borealis.eu.

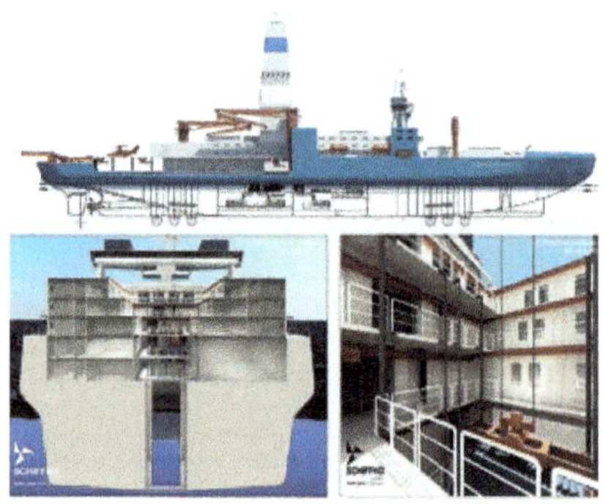

Figure 1. Side view of the *Aurora Borelis*, detailed cross sections.

D/V Chikyu Undergoes Repairs

Damage to the azimuth thruster gears on the *D/V Chikyu* was discovered during a dry-dock survey in the late spring of 2008. Fabrication and reassembly of six new gears are expected to be completed by January 2009, after which NanTroSEIZE drilling operations will resume (see back cover for schedule).

Comment: Magnetization of the Oceanic Crust

C. Harrison

In the March 2007 issue of *Scientific Drilling* there is a report of the "Mission Moho" meeting. Mission Moho is being proposed to the Integrated Ocean Drilling Program to investigate the formation and evolution of oceanic lithosphere. Deep drilling has been one of the main objectives of the drilling program ever since the first phase (Deep Sea Drilling Project) started in 1968. Shortly afterwards, a committee was formed to look at scientific objectives of deeper penetration into the oceanic crust. Part of the rationale for deeper penetration was to understand how the oceanic crust is magnetized and how this magnetization contributes to the formation of sea floor spreading magnetic anomalies. More recently the COMPLEX Meeting (May 1999, Vancouver, Canada http://www.odplegacy.org/PDF/Admin/Long_Range/COMPLEX.pdf) had a section on "How do marine magnetic anomalies relate to crustal architecture, cooling, and alteration". It is pointed out that "To fully understand the depth distribution of magnetization requires drilling of an intact crustal section". The CONCORD report (July 1997: Tokyo, Japan, http://www.odplegacy.org/PDF/Admin/Long_Range/CONCORD.pdf) on Ocean Riser Drilling posed the question "What role do the lower crust and upper mantle play in the origin of marine magnetic anomalies?" The Initial Science Plan (http://www.odplegacy.org/PDF/Admin/Long_Range/IODP_ISP.pdf) for IODP "Initiative: 21st Century Mohole" states that "...the source of marine magnetic anomalies will be much better understood when a complete section of the lower oceanic crust is available for analysis". These suggestions seem to have been forgotten by the "Mission Moho" group. The only place that crustal magnetization is discussed is in one phrase: "better understand the origin of magnetic anomalies". No mention was made of the desirability of drilling at higher latitudes in order to unscramble the magnetic signal. The main site for drilling lies at 6.5°N (but was probably formed closer to the equator). I have recently compiled a data set of paleomagnetic observations over the past 5 Ma in which Virtual Geomagnetic Poles from all latitudes were used. From this data set it is possible to determine, as a function of observation latitude,

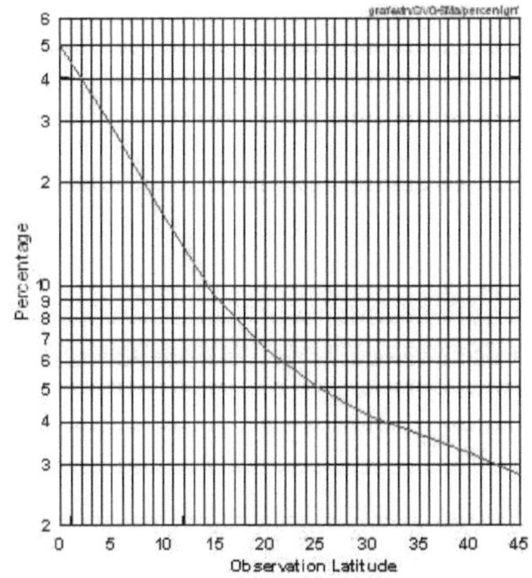

how frequently a field direction with positive inclination occurs when the core field is reversed (see figure). For a recording latitude of 6.5° fully 25% of the samples of positive inclination will have been produced by the secular variation during a time that the main field is of reversed polarity. Inclination is the only way of determining a sample's magnetic polarity in the absence of horizontal orientation, and an error rate of 25% renders the data relatively useless. It will be very difficult to unscramble the magnetization values gathered at such a place so as to be able to determine the external magnetic signature of the crustal rocks. Even at a recording latitude of 25° the probability is still 5% of confusing reversed with normal polarity. The IODP scientific advice structure should give serious consideration to choosing drilling sites at higher latitudes, preferably greater than 25°. Alternatively, they should ensure that a proven method of horizontal orientation of cores is available by the time "Mission Moho" starts.

Christopher Harrison, Rosenstiel School of Marine and Atmospheric Science (RSMAS), University of Miami, 4600 Rickenbacker Causeway, Miami, Fla. 33149, U.S.A.
e-mail: charrison@rsmas.maiami.edu

Response to: Magnetization of the Oceanic Crust

B. Ildefonse, et al.

We fully share Dr Harrison's concerns about the difficulty in deciphering the magnetic signal at low latitudes, such as the location of ODP/IODP Site 1256D , and we thank him for bringing everyone's attention to the critical need for a method to provide azimuthal orientation. Although this topic was not discussed in the workshop report published in the March 2007 issue of *Scientific Drilling*, it was extensively discussed at the Mission Moho Workshop. We would like to take this opportunity to emphasize the following points, which were also articulated in the Mission Moho proposal that was submitted to IODP in April 2007 :

- Although Site 1256D is the site for current deep drilling in the upper crust, it will not necessarily be the deep crustal, MoHole site. The latter has yet to be chosen. Site selection for deep drilling in the fast-spread ocean crust faces a few major trade-offs, dominated by water depth vs temperature at Moho, and also magnetic geometry vs weather window. Sites under consideration near San Diego and near Valparaiso would satisfy Dr Harrison's concerns.

- Site 1256D, the "6.5 °N main site" mentioned by Dr Harrison is favored for the near future for reasons including gaining engineering experience in deep gabbro holes and for better predicting the thermal structure, which will be useful even if we eventually choose a site at a higher latitude.

- We cannot emphasize enough the importance, for the future of scientific drilling in the ocean crust, to be able to reorient cores in a geographic reference frame. A Hard Rock Core Orientation (HRCO) system would be the ideal solution, and further development and implementation requires community support.

- Measuring the downhole magnetic field with logging tools does meet some of the objectives that Dr Harrison suggests in choosing a site far from the equator. Remanent magnetic data from samples collected over nearly 40 years of drilling still do not provide an unambiguous solution to the dominant source for lineated magnetic anomalies. This is partly related to the prominent effect of drilling induced remanence, and every effort to reduce this effect should be encouraged. While much can be learned from the magnetization of drillcore samples, detailed magnetic logging is arguably the best way to establish the relative contributions of the various parts of the crust to the magnetic anomalies. A reliable gyro-oriented tool should be designed, developed, and regularly used in IODP to achieve this goal.

Benoît Ildefonse, David M. Christie, Douglas S. Wilson, Jeffrey S. Gee, and the Mission Moho Workshop Steering Committee
(Natsue Abe, Shoji Arai, Wolfgan Bach, Donna Blackman, Bob Duncan, Emilie Hooft, Susan Humphris, and Jay Miller)
e-mail: benoit.ildefonse@gm.univ-montp2.fr

IODP - Expedition Schedule http://www.iodp.org/expeditions/

	ESO Operations *	Platform	Dates	Port of Origin
1	313 - New Jersey Shallow Shelf	MSP	May '09–Aug. '09	TBD
2	Great Barrier Reef	MSP	Sep. '09–Nov. '09	
	USIO Operations **	**Platform**	**Dates**	**Port of Origin**
3	317 - Canterbury Basin	*JOIDES Resolution*	12 Nov. '08–04 Jan. '09	Wellington, New Zealand
4	318 - Wilkes Land	*JOIDES Resolution*	04 Jan. '09–09 Mar. '09	Wellington, New Zealand
5	320 - Pacific Equatorial Age Transect (PEAT)	*JOIDES Resolution*	09 Mar. '09–09 May '09	Wellington, New Zealand
5	321 - PEAT/ Juan de Fuca Remedial Cementing	*JOIDES Resolution*	09 May '09–09 Jul. '09	Honolulu, Hawaii, U.S.A.
	CDEX Operations ***	**Platform**	**Dates**	**Port of Origin**
6	319 - NanTroSEIZE — Kumano Basin Riser	*Chikyu*	Mar '09 – mid Jun. '09	TBD
6	NanTroSEIZE — Sediment Inputs	*Chikyu*	Mid Jun. '09 – late Jul. '09	
6	NanTroSEIZE — Riserless Observatory Casing	*Chikyu*	Late Jul. '09 – Aug. '09	

MSP = Mission Specific Platform TBD = to be determined

* Dates subject to change pending final platform tenders.

** Dates of expeditions subject to change pending final delivery of vessel from shipyard.

*** CDEX schedule subject to final program approval.

ICDP - Project Schedule http://www.icdp-online.org/projects/

	ICDP Projects	Drilling Dates	Location
1	Iceland Deep Drilling Project	Jun. '08–'10	Krafla, Iceland
2	Lake El'gygytgyn Drilling Project	Apr. '08–May '10	Chukotka, Russia
3	PASADO (Lake Potrok Aike)	Sep. to Oct. '08	Patagonia, Argentina
4	New Jersey **	May to Jul. '09	Offshore New Jersey, U.S.A.
5	Lake Van	Summer 2009	Anatolia, Turkey

**IODP-ICDP joint project

All dates are approximate. Schedules are subject to approval.

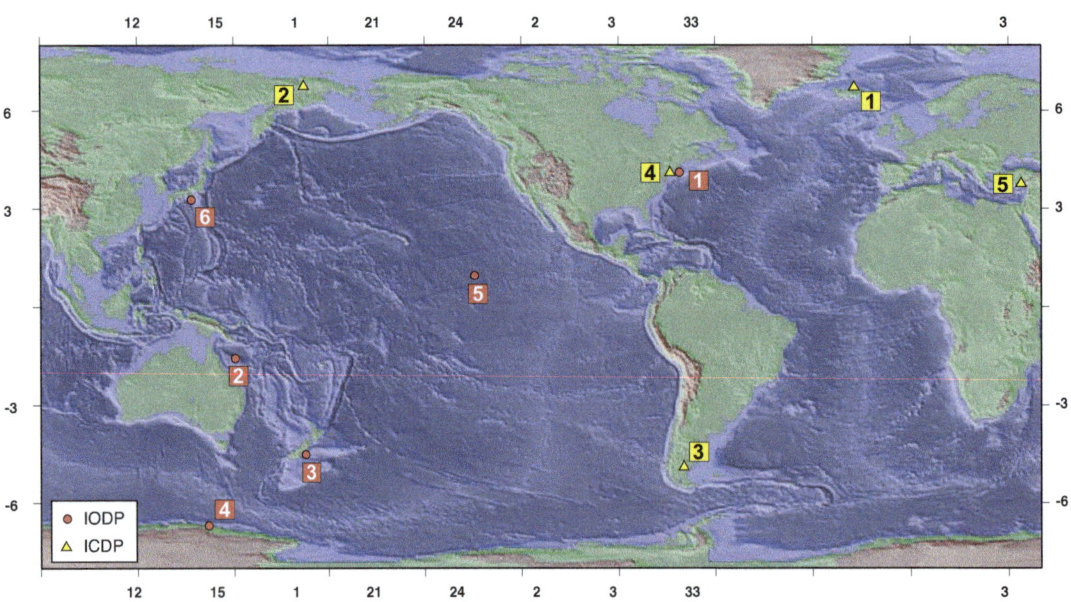

14

NanTroSEIZE Stage 1 Expeditions 314, 315, and 316: First Drilling Program of the Nankai Trough Seismogenic Zone Experiment

by Harold Tobin, Masataka Kinoshita, Juichiro Ashi, Siegfried Lallemant, Gaku Kimura, Elizabeth Screaton, Moe Kyaw Thu, Hideki Masago, Daniel Curewitz, and IODP Expeditions 314/315/316 Scientific Party

Introduction

Integrated Ocean Drilling Program (IODP) Expeditions 314, 315, and 316 were carried out as a unified program of drilling collectively known as Stage 1 of the Nankai Trough Seismogenic Zone Experiment, or NanTroSEIZE, the first multi-stage complex drilling project in IODP and the most ambitious effort in scientific ocean drilling to date (Tobin and Kinoshita, 2006b; Kinoshita et al., 2006). The first of four planned operational stages, the Stage 1 expeditions were conducted aboard the new riser-capable drilling vessel *Chikyu* from September 2007 to February 2008, and were *Chikyu*'s inaugural IODP expeditions. Seventy-one scientists from twelve countries took part in the three expeditions as shipboard participants, and thirty-three holes in total were drilled at eight sites, to as much as 1400 meters below the sea floor (Table 1, Fig. 1).

In the region offshore the Kii peninsula of Honshu, Japan, a transect of eight sites was drilled using riserless methods, targeting the accretionary complex in the frontal thrust region, the midslope mega-splay fault region, and the Kumano forearc basin region. Two of these sites were preparatory pilot holes for planned later deep-riser drilling operations, while the others targeted fault zone material at relatively shallow sub-bottom depth, where the faults were presumed to be aseismic. Expedition 314 was dedicated to *in situ* measurements of physical properties and borehole imaging through logging-while-drilling (LWD) in holes dedicated to that purpose. Expedition 315 was devoted to core sampling and downhole temperature measurements at one site in the mega-splay region and at another in the Kumano forearc basin. Expedition 316 targeted the frontal thrust fault and so-termed "mega-splay" (Tobin and Kinoshita, 2006a; Moore et al., 2007) fault region in the mid-slope environment. The initial scientific results on accretionary complex and fault zone structure, lithology and age, physical properties, and state of stress are summarized here and unified across the expeditions.

This article provides only an overview of the Stage 1 shipboard results; for the comprehensive results, please consult the IODP Expedition Report (Kinoshita et al., 2009) and the individual expedition summaries and site chapters therein.

Background and Objectives

The Nankai Trough subduction zone forearc is among the most intensively studied seismogenic and tsunamigenic plate boundary zones in the world, including previous DSDP and ODP drilling transects off Shikoku island southwest of the NanTroSEIZE study area. In the NanTroSEIZE transect, the offshore region exhibits a well-developed accretionary prism and forearc basin (Kumano basin), as well as a major thrust fault that splays from the deep plate interface at about 10 km depth beneath the Kumano basin and bisects the entire overriding prism, termed the "mega-splay" (Park et al., 2002; Moore et al., 2007). Similar mega-splay geometries are increasingly being identified in other subduction forearcs. The Nankai Trough in this region is considered a well-coupled megathrust (Wang and Hu, 2006) and exhibits

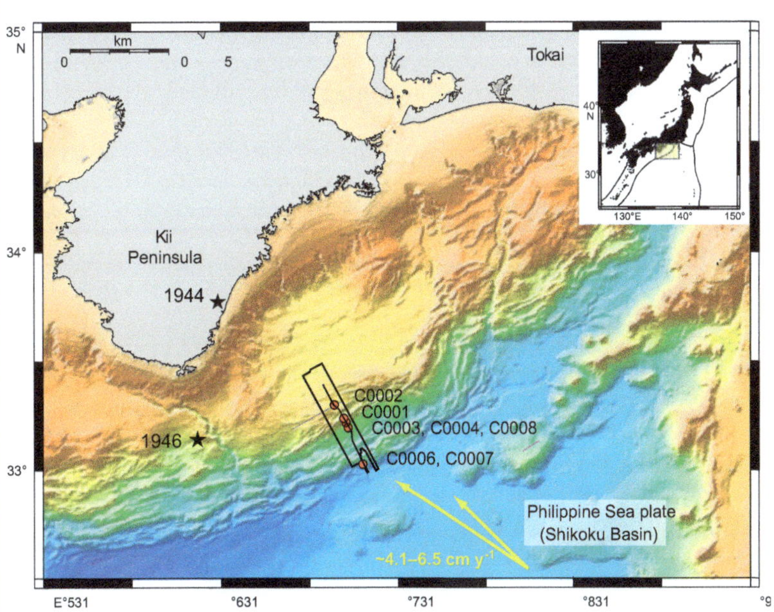

Figure 1. Location of sites drilled during the three expeditions of NanTroSEIZE Stage 1 (red circles) in the Nankai Trough off Kumano, Japan. Black box = location of the 2006 3-D seismic survey, black line = arbitrary line extracted from 3-D seismic volume (Fig. 3), yellow arrows = computed far-field convergence vectors between Philippine Sea plate and Japan (Seno et al., 1993; Heki, 2007). Stars = epicentral location of great earthquakes. Inset shows location of Nankai Trough.

no known interplate seismicity. The downgoing plate exhibited two ~M_W7 thrust and transpressional earthquakes in 2004 (Enescu et al., 2005), suggesting that, in contrast to many outer arc bending regions, this portion of the Nankai Trough's downgoing plate is in compression.

The NanTroSEIZE project is the first attempt to drill, sample, and instrument the seismogenic portion of a plate-boundary fault or megathrust within a subduction zone, identified as a very high priority in the IODP Initial Science Plan (2001). The fundamental goal of NanTroSEIZE science plan (Tobin and Kinoshita, 2006b) is to access, sample, and then instrument the faults in aseismic and seismogenic regions of the megathrust system. This involves drilling of key elements of the active plate boundary system at several locations off the Kii Peninsula, from the shallow onset of the plate interface to depths where earthquake slip and locking takes place (Figs. 1, 2). At this location, the plate interface and active mega-splay faults implicated in causing tsunami are accessible to drilling within the region of coseismic rupture in the 1944 Tonankai (magnitude 8.1) great earthquake (Baba et al., 2006). The final objective is to access and instrument the Nankai plate interface within the seismogenic zone through deep riser drilling, planned for a later stage in NanTroSEIZE.

Table 1. Drilling summary for Expeditions 314, 315, and 316 (pilot holes, including Site C0005 where no scientific drilling was done) and very short penetrations omitted).

Site & Hole	Expedition	Total Depth (mbsf)	Operations	Description
C0001D	314	976	LWD	Megasplay thrust sheet & riser pilot
C0001E	315	118	Core	
C0001F	315	249	Core	
C0001H	315	590	Core	
C0002A	314	1401	LWD	Kumano forearc basin and underlying prism
C0002B	315	1057	Core	
C0002D	315	204	Core	
C0003A	314	534	LWD	Megasplay thrust sheet
C0004B	314	400	Core	Megasplay fault zone
C0004C	316	135	Core	
C0004D	316	400	Core	
C0006B	314	885	LWD	Frontal thrust zone
C0006E	316	409	Core	
C0006F	316	603	Core	
C0007C	316	176	Core	Frontal thrust zone
C0007D	316	494	Core	
C0008A	316	358	Core	Slope deposits and prism
C0008C	316	176	Core	

The primary objectives of NanTroSEIZE Stage 1 were to complete the majority of the riserless coring and logging operations of the NanTroSEIZE transect as a whole—that is, all objectives besides the two planned deep riser-drilled sites into the plate boundary fault systems and the trench reference sites (see box; Tobin and Kinoshita, 2006a, 2006b).

NanTroSEIZE Stage 1 Riserless Drilling Expeditions: September 2007–February 2008

Scientific Targets

- The frontal thrust system at the toe of the accrtionary wedge
- The mid-wedge megasplay fault system
- 1000 m and 1400 m deep holes, respectively, at the two sites planned for later deep penetrations of the seismogenic zone faults

Expedition Breakdown of Drilling Activities

- Expedition 314: Logging-while-drilling (LWD) only transect of the all prism and forearc basin sites
- Expedition 315: Coring-focused drilling of the mega-splay fault thrust sheet and forearc basin
- Expedition 316: Coring-focused drilling of the shallow (<1000 mbsf) portions of the frontal thrust and mega-splay fault zone

Therefore *Chikyu* Expeditions 314–316 were planned to sample and characterize the sediments and relatively shallow levels of the major fault systems targeted for deep drilling in NanTroSEIZE to characterize fault properties outside the seismogenic zone, and to "pilot" the planned deep riser holes as a preparatory step towards borehole engineering for drilling to unprecedented depths (Tobin and Kinoshita, 2006a). As outlined below, we were successful in achieving a substantial majority of our objectives during Stage 1.

Stage 1 Expedition Strategy

NanTroSEIZE Stage 1 was divided into three individual expeditions, due to its five-month duration. The goals of Stage 1 were to obtain logs, cores, and downhole measurements through riserless drilling along the transect of sites (Tobin and Kinoshita, 2006a). Based on extensive past ODP experience of unsuccessful attempts to use wireline logging techniques in unstable accretionary prism environments, logging-while-drilling (LWD) was selected to obtain a comprehensive suite of logs; Expedition 314 was devoted entirely to LWD drilling of a first hole at each of the planned sites. Logging was performed using a state-of-the-art suite of LWD technology and included the measurement of natural gamma radiation, azimuthal gamma ray density, neutron porosity, full waveform sonic velocity, azimuthal resistivity imaging, zero-offset vertical seismic profile, ultrasonic caliper, and annular fluid pressure, though not all logs in this suite were collected at all sites (Kinoshita et al., 2009). The over-arching objective of the LWD program was to provide borehole data

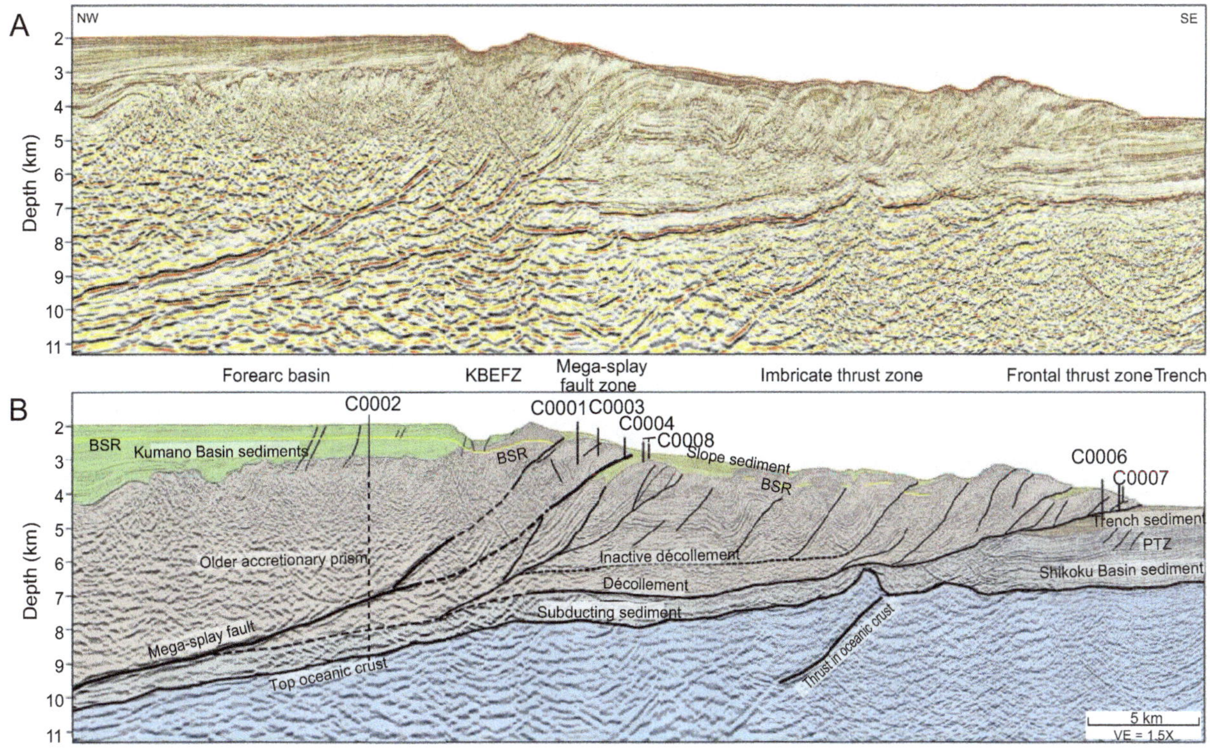

Figure 2. [A] Uninterpreted depth section, processed with pre-stack depth migration. [B] Interpreted version. Only a few representative faults are shown; dashed lines indicate less certain fault locations. Morpho-tectonic zones are shown between the two sections. Composite seismic line extracted from 3-D seismic volume, reproduced from Moore et al., (2009). The line crosses through (or near) all Stage 1 drill sites (solid vertical lines); location shown in Fig. 1. Location of planned Stage 3 ultra-deep drilling at Site C0002 is indicated by dashed vertical line. KBEFZ = Kumano Basin edge fault zone; PTZ = Protothrust zone; BSR=Bottom simulating reflection; VE = Vertical exaggeration.

to be used in conjunction with cores to document the geology, physical properties, mechanical state, fluid content, and stress conditions at the drilling site locations.

After Expedition 314 was completed, the same sites (for the most part) were revisited during Expeditions 315 and 316, in order to collect cores and other downhole measurements (e.g., temperature) in adjacent holes (typically 25–50 m away). Due to drilling challenges and other lost-time incidents causing contingency plans to be invoked a number of times, not all sites that were LWD logged were also cored, and vice versa. Multiple attempts to find sites to drill into the mega-splay and frontal thrust fault zones resulted in eight numbered sites drilled (note that at Site C0005, only pilot holes were drilled, and no scientific results were generated before this site was abandoned due to poor drilling conditions), with multiple holes at each one (Fig. 2, Table 1).

Site Descriptions and Results from Three Tectonic Domains

In this section, the principal results of drilling in each of the three tectonic regions are summarized, beginning with the two sites drilled in the frontal thrust region, moving to the four sites in the mid-slope, mega-splay fault zone region, and finally, the single Kumano basin site. This summary merges results from the Expedition 314 LWD logging drilling and the Exp. 315 or 316 coring and other measurements at each of these sites.

Frontal Thrust Region: Sites C0006 and C0007

Drilling at Sites C0006 (Expeditions 314 and 316) and C0007 (316 only) allowed examination of the frontal thrust region (Fig. 3). In Hole C0006B, LWD operations successfully drilled and logged to 885 meters below seafloor (mbsf), crossing the probable frontal thrust zone in the interval 657–711 mbsf (Expedition 314 Scientists, 2009a). The ambiguity exists because of the complex seismic reflectivity character, including apparent trench-fill channel deposit packages, and the lack of coring to this depth to elucidate the detailed structure. At 657 mbsf there is a zone of apparent high degree of fracturing/brecciation (based on LWD resistivity images) that is the most likely candidate for the frontal thrust. On Expedition 316, however, Holes C0006E and C0006F did not penetrate to this depth before drilling and coring was stopped because of poor borehole conditions and extremely poor core recovery. Several fault zones within the prism (above the frontal thrust) were penetrated and recovered, identified by concentrations of core-scale structural features, biostratigraphically-identified inversions of sediment age, and correlation with seismic reflection

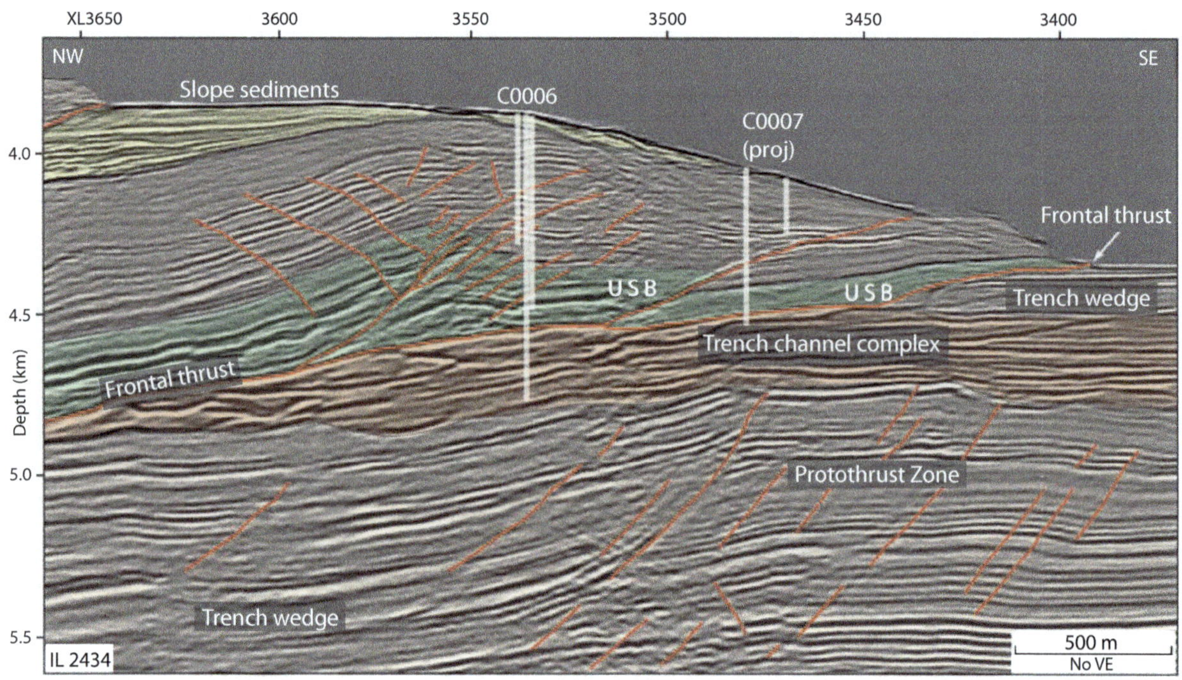

Figure 3. Inline seismic section for the frontal thrust region, Sites C0006 and C0007. Reproduced from Moore et al. (2009).

imaging (Expedition 316 Scientists, 2009a). LWD gamma and resistivity logs in Hole C0006B indicate that the interval beneath 711 mbsf is dominated by massive sand, consistent with a position below the frontal thrust in the trench fill deposits, correlative to the footwall cored at Site C0007 (see next paragraph).

The plate boundary frontal thrust was successfully cored at Site C0007 nearby (Figs. 4, 5), and thrust fault material ranging from breccia to fault gouge was recovered (Expedition 316 Scientists, 2009b). The lowermost part of Fault Zone 3 at 438 m core-based depth below seafloor (CSF) is intensely brecciated into fragments ~1–10 mm in size (Fig. 5). This 29-cm-thick breccia shows a foliated aspect with an anastomosing network of polished and striated surfaces. At the base of this zone, a dark 2-mm-thick layer sharply separates intensely brecciated hemipelagic mudstone above from unbroken hemipelagic mudstone and ash below. There is a biostratigraphic age reversal as well. These features indicate that the thin dark layer most likely represents extreme localization of slip associated with thrust faulting. This frontal thrust zone at 418–438 m CSF is located in a stratigraphic package of late Miocene mud, which appears to be similar to the Shikoku Basin facies that hosts the décollement in the Muroto and Ashizuri transects (Moore et al., 2001). The footwall proved to be remarkably coarse-grained, as it was dominated by sand and gravel trench-axial channel facies deposits, and core recovery was very poor in this interval.

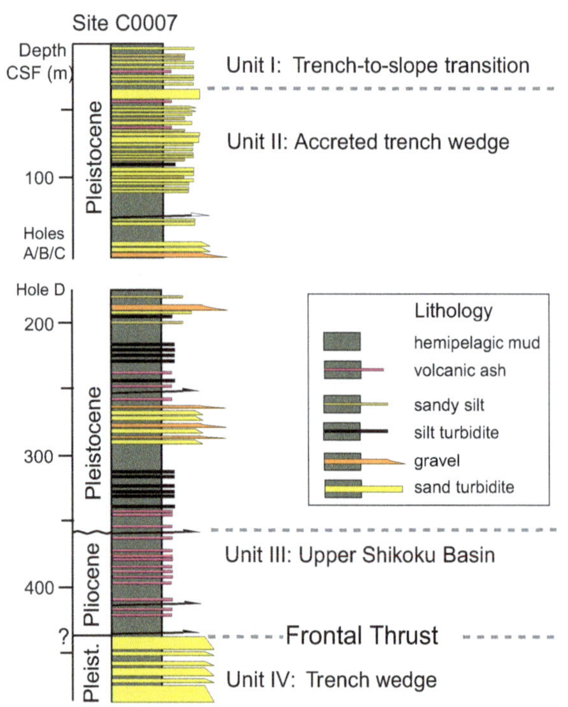

Figure 4. Hole C0007D summary of coring, lithology, and major faults.

Lithologic Unit I sediments at Site C0006 (Fig. 3) are interpreted as representing a transition from trench to slope deposition, while Unit II is composed of trench fill facies sediment; thus, the Unit I/II boundary records the uplift of trench material into the prism. Unit I sediments are younger than or the same age as sediments filling the basin behind the thrust. The age of this transition is 0.78–0.44 Ma. Seismic data show that about 6 km of overriding of trench channel facies by the frontal thrust has taken place since this fault initiation time (Fig. 3; Moore et al., 2009) without development of any further frontal accretion thrusts. How such a large amount of slip has concentrated within the fault zone and how the evolution of the fault zone has affected the characteristic features in this region are two important issues for post-expedition research. Furthermore, taking into account the relative plate motion velocity of ~4 cm yr^{-1} between the overriding Japanese islands and the Philippine Sea plate (Seno et al., 1993), we determine the relative total slip distance has to be ~40 km, not 6 km. Therefore, the majority of the total shortening must be taken up elsewhere within the accretionary prism.

The accretionary prism in the frontal thrust region is deformed by thrusting, as visible on the seismic profiles (Figs. 2, 3; Moore et al., 2009). Most of the thrusts inferred from the seismic profiles and LWD data were confirmed through age reversals, fault zones sampled in cores, and repetition of specific strata, but some additional faults were also defined during Expedition 316 drilling. Initial chemical analyses of interstitial fluid (porewater geochemistry) and microbial habitat around most of these "intraprism" thrusts do not indicate any signal of active fluid flow focused along these faults or in other permeable conduits.

In contrast to such intraprism thrusts, many normal faults are developed at the core scale and appear to be the youngest deformation feature. Clear slope-parallel mass sliding is observed from the submarine topography, seismic profiles, and shallow cores (Fig. 3). These facts suggest that the taper angle of the prism locally at the toe is presently above the critical wedge taper angle and is unstable. There is evidence that the system is currently in a period of wedge collapse, perhaps as a readjustment to the passage of subducted basement topography.

Porosity data suggest that a considerable thickness of material has been eroded or otherwise removed from the surface at Sites C0006 and C0007. Porosity is quite low at shallow depths below the surface, reaching 48% at 5 m and 34 m CSF at Sites C0006 and C0007, respectively. In contrast, porosity does not decrease to <50% until ~150 to ~200 mbsf, at other sites on the Stage 1 transect farther landward in the prism.

Mega-Splay Fault Region (Sites C0001, C0003, C0004, and C0008)

The out-of-sequence mega-splay fault zone branches into a number of individual splays as it enters the upper few kilometers of the subsurface (Fig. 2; Moore et al., 2009). A series of sites targeted the thrust sheet, fault zone, and footwall of the lower of these branches, which was judged to be the most active and most significant of the seismically imaged splays (Tobin and Kinoshita, 2006b; Moore et al., 2009).

Site C0001: Site C0001 was chronologically first in drilling, and targeted only the thrust sheet (not the underlying fault itself) in the most landward position (the fault zone was judged too deep at ~2000 mbsf to access with riserless drilling at this site). For NanTroSEIZE Stage 2, 3.5-km riser drilling had been planned at Site C0001, located at a small bench on the hanging wall of the main branch of the mega-splay fault (Fig. 6) where a small slope basin with coherent layered reflectors is developed, overlying a more seismically-chaotic thrust sheet above the splay fault. In Stage 1, pilot-scale drilling of the uppermost ~1000 mbsf at this site was executed to test conditions for riser drilling and to set the surface casing for this later effort. The results of this pilot, detailed below, indicate that this may be a difficult location for a riser site, and that plan is being reevaluated.

We first drilled an LWD-only hole at this site (Expedition 314 Scientists, 2009b), achieving 976 m of penetration (just 24 m short of the planned depth, but sufficient to achieve all logging objectives). Difficult hole conditions were encountered, including an interval of high drill string torque due to apparent borehole caving, enlargement, and partial collapse around 460–540 mbsf, an interval that became known as "the sticky zone". Logging results suggested that the sediments of the hanging wall thrust sheet are primarily muds and mudstones with some silty to sandy (or ash-bearing) sediments, overlain by ~200 m of hemipelagic slope deposits. Resistivity imaging suggested widespread fracturing and variable bedding orientation,

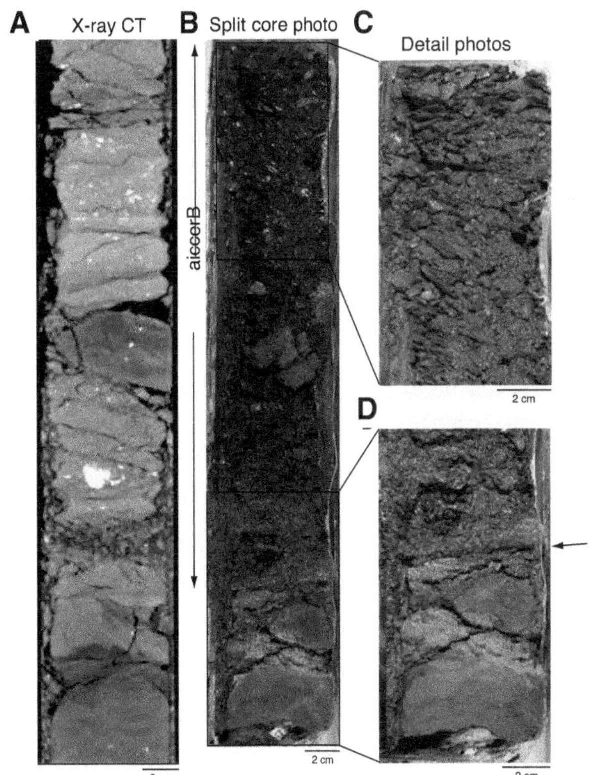

Figure 5. Fault core example from the frontal thrust region at Site C0007 (Expedition 316). Appearance of concentrated deformation in lower portion of frontal thrust fault Zone 3, at a depth of 438 mbsf. [A] CT image and [B] core photograph of finely brecciated hemipelagic mudstone (interval 316-C0007D-29R-2, 37–73 cm). [C] Close-up of finely brecciated interval. [D] Thin (2-mm-thick) dark layer (arrow) developed at base of finely brecciated interval..

indicative of strong deformation, especially below ~460 mbsf.

Subsequent coring at this site during Expedition 315 confirmed and greatly added to these observations (Fig. 6B). Results of the coring document the age and make-up of the deformed mega-splay thrust sheet and overlying slope apron deposits, as well as structural features and inferences on evolution of the stress state (Expedition 315 Scientists, 2009a). Despite numerous attempts in multiple holes at this site, the challenges posed by "the sticky zone" proved too great to advance the coring beyond 458 mbsf. The cored interval includes the slope basin (Unit I) and the top ~250 m of the underlying accretionary prism (Unit II). Unit I is composed mainly of Quaternary to late Pliocene silty clay and clayey silt with volcanic ash, while Unit II is made up of older (late-Pliocene to Miocene) mud-dominated sediments. The boundary between Units I and II, identified at 207 mbsf, is an unconformity located immediately below a thick sand layer. Unit II exhibits much more deformation at core and seismic scales, and a range of small-scale structures including normal, thrust, and strike-slip faults. Kinematic solutions computed from normal and thrust faults are consistent with northeast-southwest extension and northwest-southeast shortening, respectively, suggesting that tectonic shortening during emplacement of the thrust sheet and contemporaneous or later extension perpendicular to that are recorded. The extension may be evidence of near-surface collapse of the slope.

We found it interesting to note that at the frontal thrust area, interstitial water geochemistry and organic geochemistry data do not show clear evidence for advection of chemically-distinct fluids from depth.

Site C0003: This proposed site was intended to begin the downdip transect of the mega-splay fault system by sampling a relatively shallow, presumably aseismogenic, point on the fault zone at ~800 mbsf (Fig. 7). Unfortunately, we encountered very difficult drilling conditions at this site, such as especially pronounced caving and

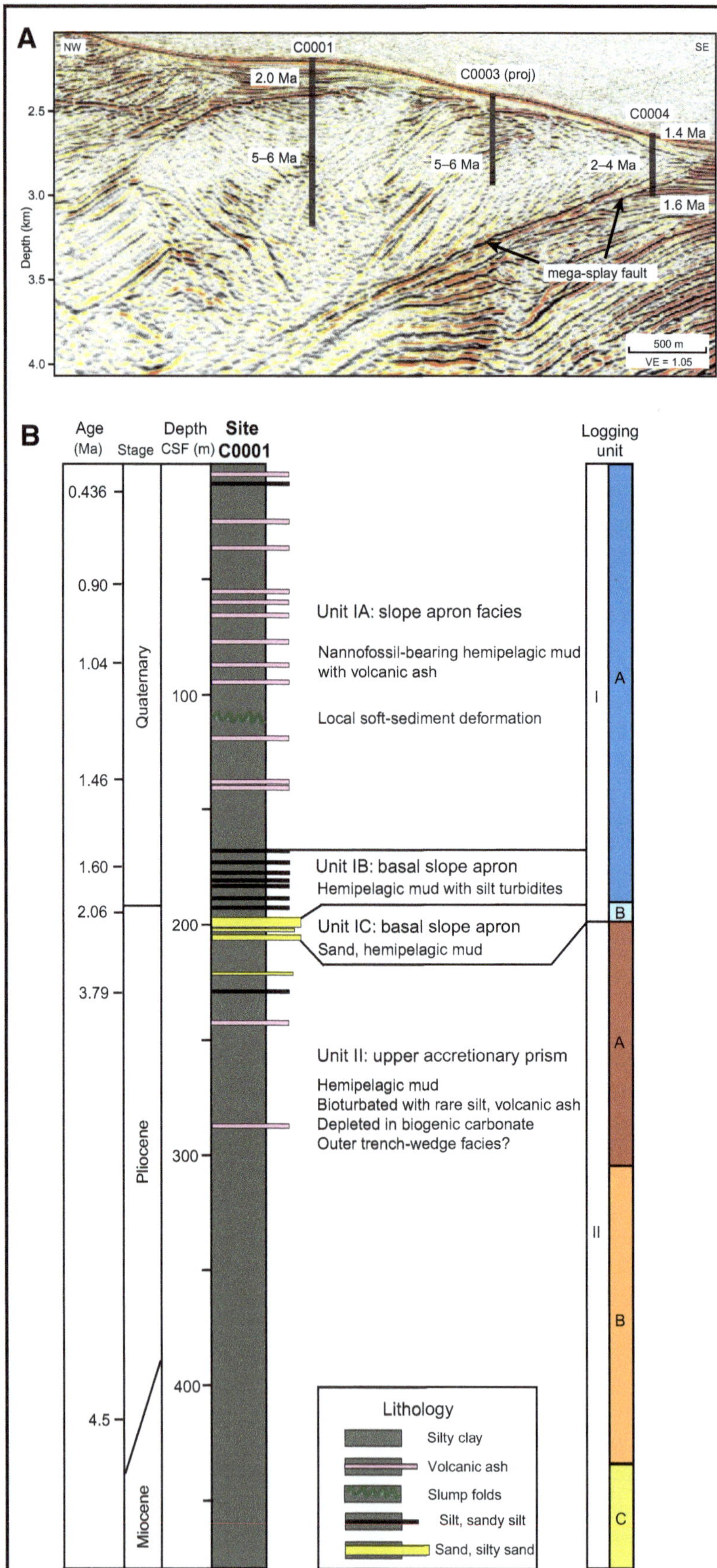

Figure 6. [A] Inline PSDM seismic section through the mega-splay fault thrust sheet sites, showing total drilled depth at each site. Age labels show maximum sediment depositional ages based on nannofossil biostratigraphy. [B] Site C0001 summary column of lithostratigraphy, logging units, and age of sediments determined through biostratigraphy and paleomagnetic data.

washout of likely fault zones and possibly sandy intervals, which caused the drill string to become irretrievably stuck before reaching the primary objective (Expedition 314 Scientists, 2009c). While drillers worked to free the stuck pipe, it parted and we were ultimately unable to recover the bottom hole assembly with the LWD tool string and ~200 m of drill collars at this site. Nevertheless, real-time data transmission provided substantial logging data from the seafloor to ~530 m LWD depth below seafloor (LSF). No further drilling or coring was attempted here; we instead moved to Site C0004 (see next section).

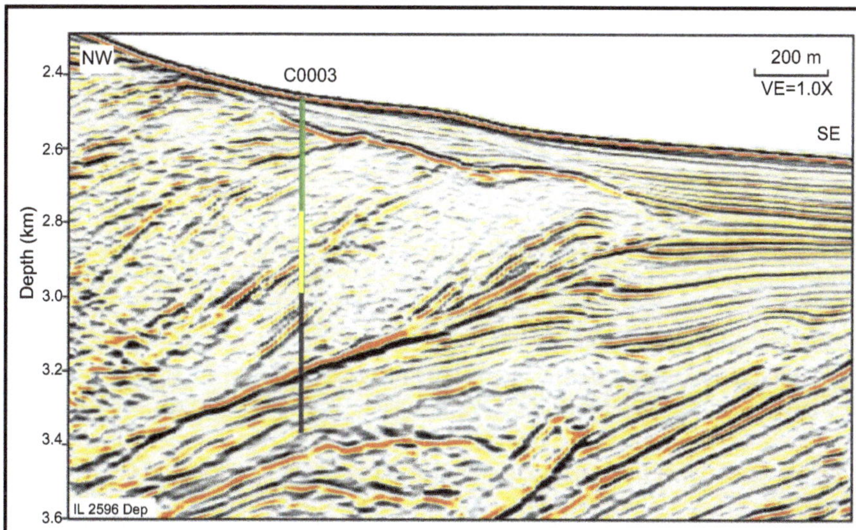

Figure 7. Site C0003 on Inline 2596. Black line is the originally planned total depth of 1000 mbsf; green line is actual depth achieved when the drill string became stuck. Yellow line is approximate length of stuck and abandoned bottom-hole assembly . The reflector that intersects the borehole at ~2.9 km depth is interpreted as a fault zone, and is the probable source of dated cavings in the hole.

Consistent with the other midslope sites, the lithology at C0003 is apparently composed of relatively coarse silt-sand and hemipelagic mud in the shallowest portion, underlain by more clay-rich, generally homogeneous muddy deposits interrupted by prominent zones of washouts interpreted as brecciated intervals, indicating likely faults (Expedition 314 Scientists, 2009c). The seismic reflection imaging shows that this site was drilled into a series of at least three individual thrust sheets within the hanging wall of the mega-splay fault system (Fig. 7). Therefore, major washout zones identified in logs at ~240 m and 420–450 m LSF are likely to be strongly brecciated damage zones from subsidiary thrusts, analogous to fault breccia recovered in core at other sites.

The seismic interpretation is consistent with substantial uplift of the thrust sheet bringing older and more deeply buried accretionary prism rocks to within 500 m of the seafloor. Log density data show relatively constant and high values for this depth; these suggest that anomalously well-indurated rocks make up the thrust sheet. When the broken section of drill pipe was recovered to the rig floor after loss of the LWD string, it was plugged with numerous large chunks (up to 5–8 cm in diameter) of cavings that had come from an unknown position in the hole. This material had a nannofossil age of late Miocene (5.5–7.2 Ma), bulk density of 2.1 g cm^{-3}, and P-wave velocity of ~2.1 km s^{-1}, also consistent with the thrust sheet rocks having been uplifted from greater depth within the accretionary complex.

Site C0004: Site C0004 is located seaward of Sites C0001 and C0003 (Fig. 6). It targeted the prominent splay fault reflector at a depth of approximately 300 mbsf. The overlying thrust sheet, the mega-splay fault zone, and ~100 m of the footwall sediments were successfully drilled with LWD logging (Expedition 314) and coring (Expedition 316). Drilling at this site also examined the youngest sediments on the slope overlying the accretionary prism, which consist of slowly deposited hemipelagic marine sediments and redeposited material from upslope. This redeposited material provides information about past slope failures, which may be related to past mega-splay movement, earthquakes, and tsunamigenesis. This will be the subject of future work on the core samples. The top of the prism corresponds to a prominent unconformity (age gap = ~1My) that displays pyrite and other mineralization. Structural observations of core material from the fault zone and two age reversals suggested by nannofossils indicate a complex history of deformation. Sediments under the fault zone were sampled to understand their deformation, consolidation, and fluid flow history. Further results and discussion of Site C0004 are summarized below in the "Key Results" section.

Site C0008: Drilling at Site C0008 targeted the slope basin seaward of the mega-splay fault (Fig. 2B) as a complement to Site C0004, which is about 1 km farther landward. This basin records the history of fault movement. Sediments are Pleistocene to late Pliocene hemipelagic silts and clays with ashes (Expedition 316 Scientists, 2009c). Several gravelly sequences were identified and interpreted as indicative of mass transport complexes. Sediments of the slope basin at Site C0008 provide a "reference site" for the sediments underthrust beneath the mega-splay fault. Comparison of the interval 190–200 m CSF in Hole C0008A (with an average porosity of 50%) and the correlated interval 320–330 m CSF in the mega-splay footwall section of Hole C0004D (average porosity of 43%) suggests the sediments are dewatering during underthrusting. Evidence for lateral flow from the C0004 area toward Site C0008 is provided by C1/C2 (methane/ethane; one indicator of biogenic vs. thermogenic gas origin) ratios that are slightly lower than expected for biogenic production at the estimated *in situ* temperature, but they are similar to those at Site C0004. Lateral flow along sand layers could transmit fluids from areas of higher temperature because of greater burial beneath the splay fault, driving fluids out along permeable sandy layers.

Kumano Forearc Basin Region: Site C0002

Site C0002 sampled the Kumano forearc basin and the underlying old accretionary prism material (Figs. 2, 8). LWD drilling during Expedition 314 was very successful here, with a single hole achieving 1401 m penetration—the deepest LWD hole in scientific ocean drilling to date—and we show a comprehensive log summary in Figure 8A (Expedition 314 Scientists, 2009d). Approximately 940 m of this drilling was through the sediments of the Kumano forearc basin, whereas the lower ~460 m accessed the underlying rocks interpreted to be the accretionary prism formed through earlier frontal accretion.

This site is slated to be the centerpiece of the NanTroSEIZE project, with six or more kilometers of drilling planned to access the plate boundary at seismogenic zone depths in a later stage (Fig. 2). Therefore, understanding the physical properties and documenting tectonic processes and lithology in the hanging wall of the plate boundary here were important goals of Stage 1 drilling, not the least because of the need for high-quality borehole engineering parameters to plan riser drilling. Scientifically, the hole records the timing and history of the mega-splay system in the formation and filling of the Kumano basin, and it was used to document the present-day stress regime in this portion of the margin (see "Summary of Key Results" below).

During Expedition 315, Site C0002 was drilled again (in two separate holes) to 1057 m CSF, with coring of the intervals 0–204 m CSF and 475–1057 m CSF (Fig. 8B, 8C). The Kumano Basin is a young feature (mostly Quaternary) with a high sedimentation rate (>800 m Ma^{-1}) overlying a late Miocene (5–6 Ma) accretionary prism. This is much younger than most of the Tertiary Shimanto belt outcropping onland. Biostratigraphic data show that the transition from Pliocene to late Miocene strata occurs as a marked age gap around 922 m CSF.

We cored across the basal unconformity of this Kumano forearc basin at ~922 m CSF and another 135 m into the accretionary prism (Expedition 315 Scientists, 2009b). The forearc basin sequence was divided into two units based on lithofacies. All units are dominated by mud and mudstone; however, Units I and II contain more sand and silt intercalation and have a much faster sedimentation rate than the basal deposits of Unit III. Underlying accretionary prism materials contain moderately more lithified—and much more deformed—sediments.

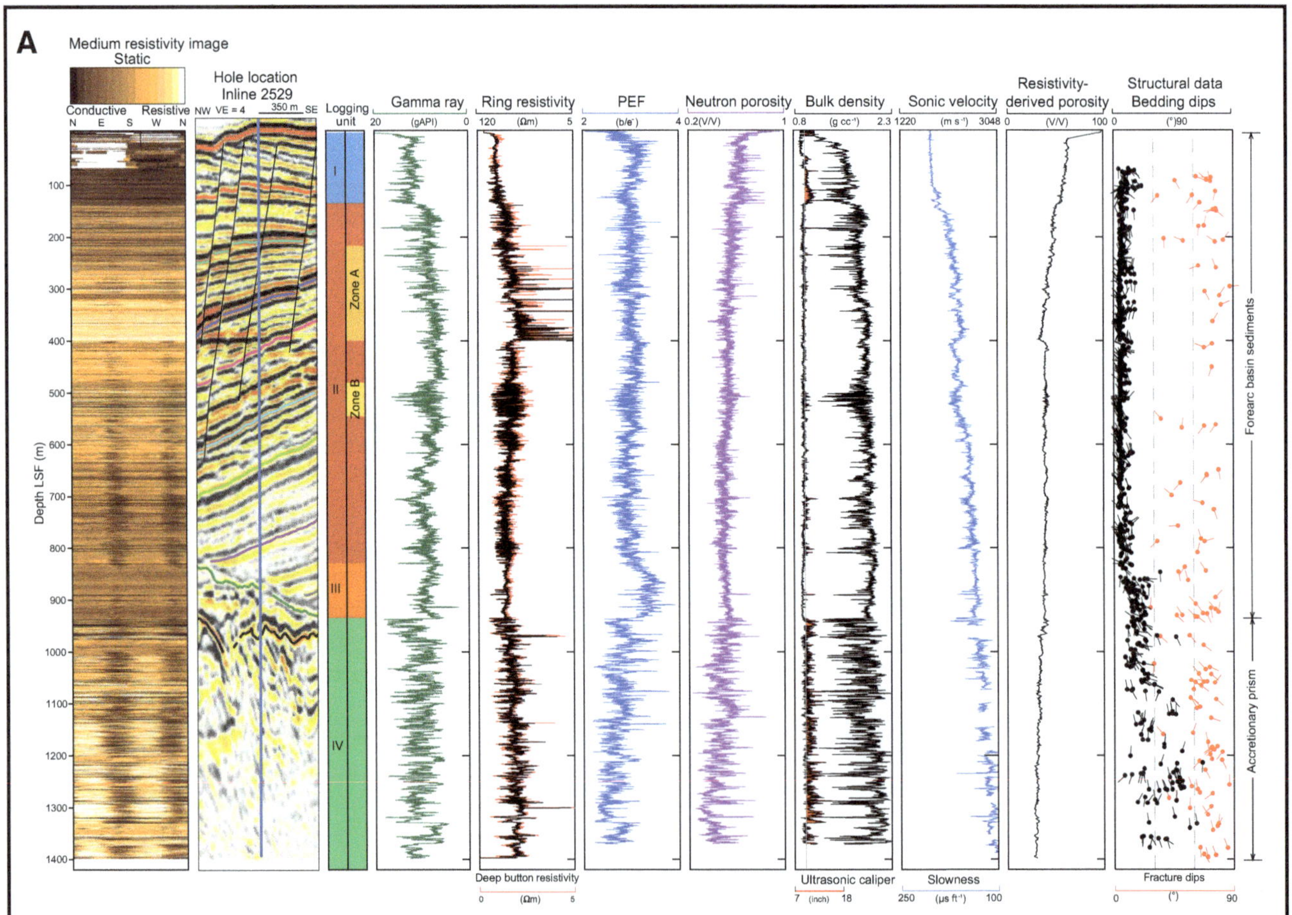

Figure 8 [A]. Site C0002 Results – Summary plots of major LWD logs for the full 1400-m interval drilled. (Figure 8. [B] shown on p.12 and [C] shown on p.13)

At Site C0002, a well-developed bottom simulating reflector (BSR) is imaged in the seismic data, and LWD logs recorded comprehensive *in situ* information about the nature of this gas hydrate BSR (Expedition 314 Scientists, 2009d). Resistivity logs showed a pattern of elevated resistivity background and spikes for ~200 m above the BSR depth at ~400 m LSF (Fig. 8). The gamma response indicated that the spikes are in especially sandy intervals as thick as 1–2 m, which were interpreted as the coarse basal beds in turbidite deposits. Sonic and resistivity responses are consistent with pore space in these sands being partially filled and "cemented" with gas hydrate. In contrast, similar sands below the BSR reflector depth show no elevated resistivity response. In a zone 80 m deeper and ~70 m thick, a low resistivity response in the sandy beds suggests a potential gas-charged interval beneath the gas hydrate stability field. The logging data indicate that the BSR reflectivity is a response to a small velocity high from hydrate cement in the hydrate stability zone and to a more significant velocity low caused by the presence of uncemented sediments and/or free gas below the stability field.

Ongoing log analysis integrated with 3-D seismic analysis will quantify the amount of pore space charged with hydrate in the zone above the BSR and the total amount of gas in this region of the Kumano Basin section. During Expedition 316, Site C0002 was cored late in the expedition as a contingency operation. The BSR interval and overlying apparent gas hydrate-rich zone was not cored in order to allow time for deeper objectives..

Summary of Key Results

Lithology, Structure, and Recent Activity of Mega-Splay Fault and Associated Thrust Sheet (Sites C0001, C0004, and C0008):

A key question for the NanTroSEIZE transect is whether or not the mega-splay style of fault development indicates that tsunamigenic coseismic slip comes near to the surface preferentially in the mid-slope region, as opposed to extending along the basal décollement out toward the trench. Based on seismic and tsunami inversion studies combined with seismic reflection research, various workers have suggested that the splay fault system in this area slipped during the 1944 Tonankai M_w 8.2 earthquake to generate the observed tsunami (Baba et al., 2006; Park et al., 2002). From a 3-D seismic reflection investigation, Moore et al. (2007) also suggest recent historical and geological accumulation of displacement along the mid-slope mega-splay fault. The age reversal from Pliocene to Pleistocene documented beneath the splay fault during Expedition 316 (Fig. 6) is consistent with geologically recent activity. However, age resolution is, of course, insufficient to document historical fault activity.

Integration of seismic imaging, coring, and logging results for Site C0004 provides evidence for recent activity of the splay fault. The splay fault clearly thrusts the hanging wall prism over younger slope sediments in the footwall; however, the youngest (less than about 1 Ma) slope sediments that cover the fault appear not to be cut by the fault. In addition, the lack of a slope break on the seafloor above the fault might also suggest that this splay fault is presently not active but ceased activity in the

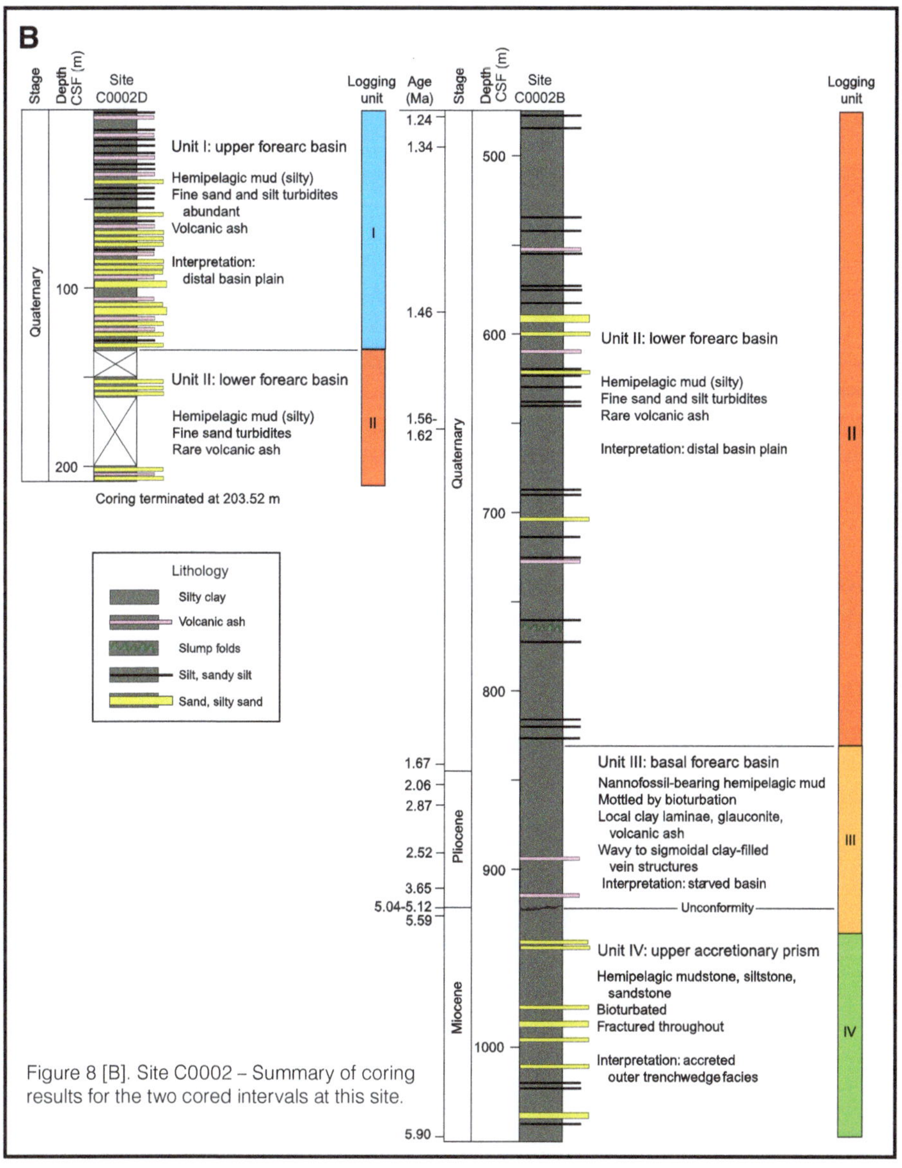

Figure 8 [B]. Site C0002 – Summary of coring results for the two cored intervals at this site.

recent past. However, the following shipboard results interpreted in the context of seismic reflection imaging provide support for the alternate interpretation that the splay fault is active as a blind thrust, in which the tip of the fault has not propagated to the surface but remains buried. Shallow cover sediments were found in cores to be composed of repeated mass transport complexes associated with repeated slope collapses and rip-up debris generation. Pleistocene cover sediments dip steeply approximately parallel to the slope and are cut by numerous normal faults, expressed at core and seismic scales (Expedition 316 scientists, 2009d). Slip on deeper levels of the fault zone could be manifested in some combination of folding and layer-parallel slip in the shallow slope sediments draping the uppermost portion of the mega-splay, as well as the possible triggering of numerous marine slump or slide deposits.

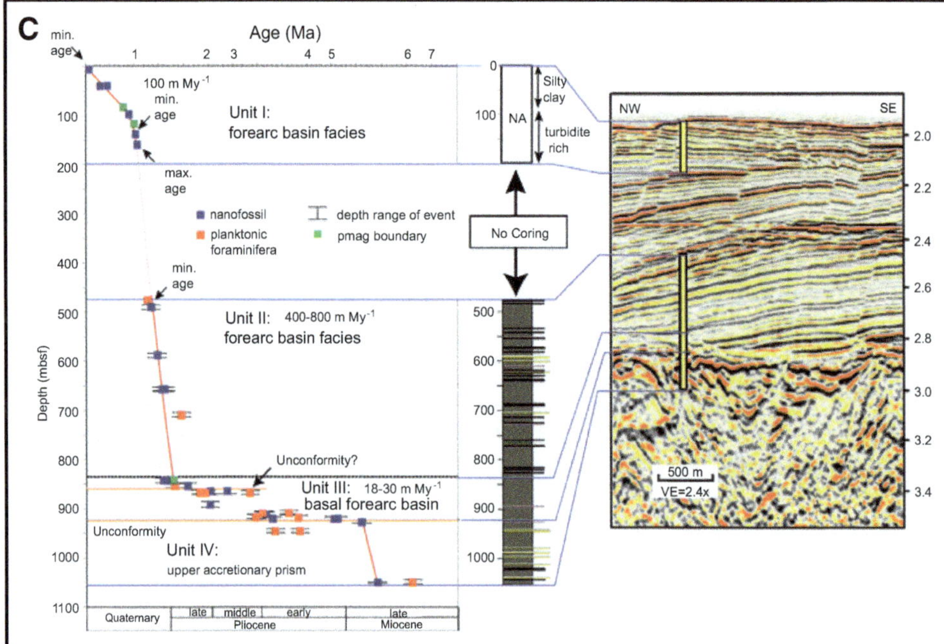

Figure 8 [C]. Site C0002 – Biostratigraphic age determinations for the cored intervals, illustrating contrast between prism, slope drape, and forearc basin sediment accumulation timing and rates.

Despite inferences of recent activity on the splay fault system, no porosity inversion is observed beneath the splay fault; this contrasts with previous results from the décollement of the Muroto transect (Screaton et al., 2002), in which a clear porosity inversion across the fault likely reflects fluid overpressure. Unlike the Muroto transect décollement, the splay fault system observed at Site C0004 has permeable pathways for dewatering provided by the observed sand and coarse ash layers.

Age of Cover Sequence and Uplifted Accretionary Prism Units

Rocks of the thrust sheet below 500 m LSF at Site C0001 and also of the cavings sampled at Site C0003 are anomalously dense relative to their present depth of burial, with high sonic/seismic velocity, indicating relatively advanced lithification. This suggests significant uplift and exhumation along the splay thrust. These inferences are consistent with the age determinations made on cores, showing that this thrust sheet contains rocks several millions of years old, in contrast to the immediately overlying Pleistocene and younger slope deposits (Fig. 6). Thrust sheet material drilled at C0004 is substantially younger than the analogous material at C0001 or C0003, suggesting that the mega-splay thrust sheet contains internal structural imbrication and has incorporated material progressively as it advanced.

At all of the sites landward of the frontal thrust area, units interpreted to be an uplifted section that was frontally accreted to the prism are covered by varying amounts of slope drape. In the case of Site C0002, this slope drape is in turn covered with the thick, dominantly Pleistocene Kumano forearc basin section (Fig. 2). Biostratigraphy (predominantly by nannofossil zones) and magnetostratigraphy define the ages of sediments above and below these boundaries, which also record time gaps of various durations between the prism section and cover (Figs. 6, 8C).

In the frontal thrust region at Site C0006, the transition from uplifted trench sediments into the overlying slope apron cover sediments is dated to 0.436–0.78 Ma, which presumably records the timing of initial frontal thrust activity. Only twenty-seven meters of younger slope cover overlies this boundary, and this implies potential removal of young deposits through slumping or mass wasting.

Moving to the mid-slope mega-splay fault region, the oldest sediment at the base of the slope cover at Site C0004 (dated at 1.46 Ma) rests on sediments that are 1.1. My older than that. Just a short distance landward at Site C0001, in the same apparent overall thrust complex (Figs. 2, 6), the age of the base of the slope apron is 2.0 Ma—substantially older than C0004—resting on a nearly 4 Ma accreted section. Ten kilometers further landward in the Kumano basin at Site C0002, the slope apron lies on top of the accreted complex at a depositional age of 3.79 Ma and rests on >5 Ma sediments in the accreted section. This landward progression of successively older dates marking the uplift and surface exposure of Shikoku basin sediments is consistent with progressive growth of the accretionary prism through late Miocene to Quaternary time.

Furthermore, the slope apron at Site C0002 accumulated very slowly until ~1.5 Ma, then more than 800 meters of forearc basin turbidites accumulated rapidly (Fig. 8C). The implication is that the onset of splay fault uplift of a pronounced outer arc high and/or capture of a significant turbidite sediment source for the basin was abrupt in the early Pleistocene.

Indicators of Stress Regime

Borehole breakouts and present-day stress orientations: Borehole breakouts were observed at all four sites for which we have imaging data. They also show very systematic orientations (Fig. 9). In a vertical borehole, the orientation of breakouts, or compressive borehole wall failures upon drilling, is a well-established indicator of the orientation of the horizontal maximum principal stress in the present-day stress field (Zoback et al., 2003). Drilling-induced tensile fractures (DITF) were observed less frequently, primarily at Site C0001 in the splay fault thrust sheet. Breakout orientations at Sites C0001, C0004, and C0006 all indicate northwest–southeast azimuths of the maximum horizontal principal stress (SH_{max}) (Fig. 10). In the thrust-dominated tectonic environment, this is consistent with trench-normal shortening, although strike-slip and/or normal faulting stress states are also permissible (see next section). By contrast, at Site C0002 in the Kumano Basin, the orientation of SH_{max} is northeast–southwest at 134°, or very nearly perpendicular to that in the more trenchward sites (Figs. 9, 10). This is consistent with a normal faulting stress state that extends through the basin section (which, in fact, exhibits numerous normal faults) and also in the upper 400 m (at least) of the underlying older accretionary prism section; however, given that stress magnitudes are unknown, the breakout orientation here also does not preclude strike-slip or reverse faulting stress state. This contrast between the outer accretionary prism and the forearc basin region (including the buried prism rocks beneath) suggests that the tectonic stress orientations differ markedly in the upper part of the prism at sites just a few kilometers apart along our transect. Implications of these data will be explored in post-expedition research to understand the mechanical state of the prism and basins.

Paleostress from core-based structural data: Core-scale structures can be used to infer paleostress regimes, while breakouts respond to present-day stress. Oriented structural data from cores drilled on Expedition 315 in Holes C0002C and C0002D (Kumano Basin site) show small-scale structures consistent with the interpretation that the present-day stress field is extensional and directed NE-SW. This agrees very well with the interpretation of breakout data for the present-day stress field (Expedition 315 Scientists, 2009b).

In the minor structures observed in the mega-splay thrust sheet at C0001 and at the frontal thrust at C0006, reverse and normal faults were documented (with minor strike-slip as well). In general, there is an overall indication at both of these sites that the youngest and/or most numerous small faults are normal and record extensional conditions striking NW-SE and dipping sub-equally to NE and SW (Expedition 315 Scientists, 2009a; Expedition 316 Scientists, 2009a). In the case of Site C0006, this was interpreted by the ship-board party as evidence of geologically recent collapse of an oversteepened frontal wedge (Expedition 316 Scientists, 2009a). By contrast, at Site C0004, normal faults were not prominent in the front of the mega-splay thrust sheet (except in a thin and shallow slope cover sequence of 0–78 mbsf), and reverse faults dominate, though they were not numerous and generally were not in oriented intervals of the core.

Site C0001, on the other hand, shows reverse faulting beneath the slope cover but overprinted by normal faults recording NE-SW directed extension, even in the thrust sheet, similar to C0006 at the frontal thrust area, and

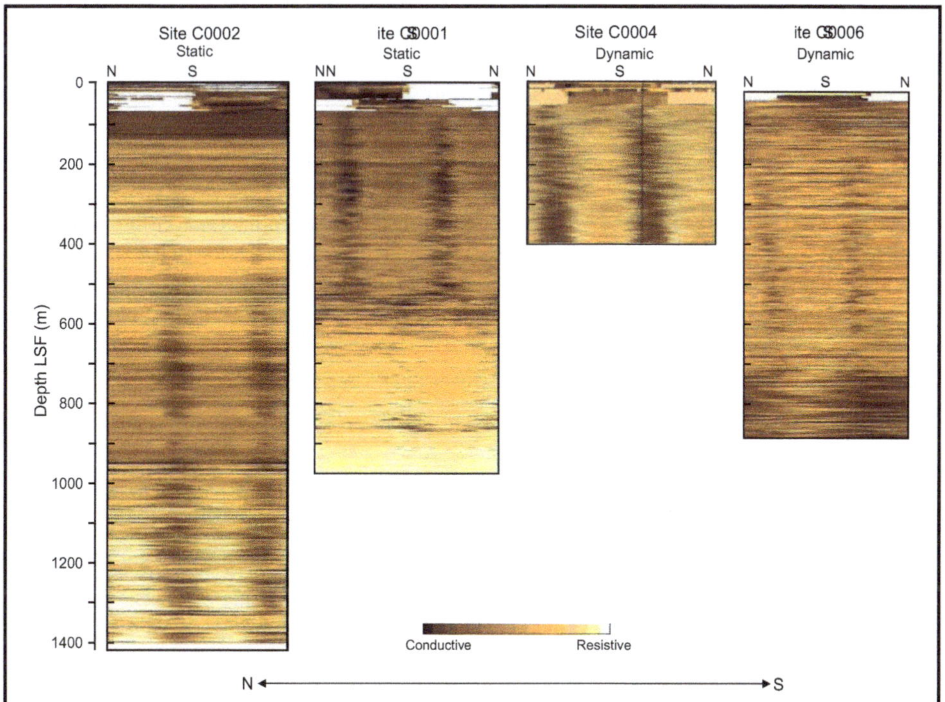

Figure 9. Resistivity at bit images synoptically showing full depth extent of LWD logged intervals at Sites C0001, C0002, C0004, and C0006. Vertical paired dark bands are images of borehole breakouts that have been interpreted in order to generate the average SH_{max} azimuths shown in Figure 10.

approximately orthogonal to the present-day SH_{max} direction. These apparently heterogeneous paleostress states at the various sites along the lower-slope portion of the transect (i.e., all except for C0002) may not be in conflict if the three principal stresses differ only modestly at these shallow depths, and small changes in their relative magnitude can effect a "flip" in stress/faulting regime. As the wedge develops, shortens, rides over basement topography, it might respond by successive and frequent flips in faulting regime, especially in the upper 1000 meters or so. The remarkably uniform orientation of the breakouts down each hole would typically be interpreted as evidence of a high-differential stress, homogeneous stress state environment. Further analysis of the stress conditions and this apparent paradox awaits post-cruise efforts.

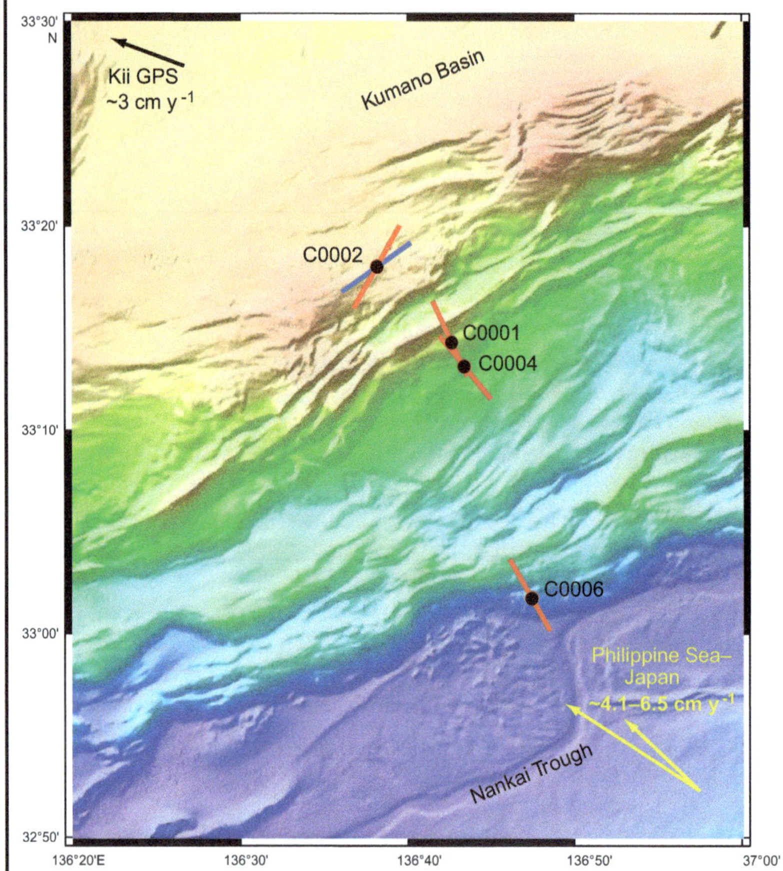

Figure 10. Azimuth of maximum horizontal stress estimated from borehole breakout images at Sites C0001, C0002 (in basin above 936 mbsf), C0004, and C0006 (red lines). Blue line = estimate for older accretionary prism formation below 936 mbsf at Site C0002.

Thermal Regime

Acquisition of a transect of good quality downhole temper-ature profiles and thermal conductivity data were an important part of Stage 1 drilling. Thermal regime may be closely tied to fault stability criteria (Hyndman et al., 1995), and the prediction of temperature at plate boundary fault zone depths is dependent on well-defined heat flow and thermal properties. Prediction of temperature at deeper levels is also crucial for future deep well planning and the mechanical specifications of long-term borehole instruments to be installed during NanTroSEIZE Stage 4, because operating temperature places severe constraints on available instruments.

Downhole temperature measurements using the APCT3 and Davis-Villinger Temperature Probe (DVTP) were successfully conducted across the Stage 1 transect. For details of the data from each site, see the respective site chapters in the Expedition Report (Kinoshita et al., 2009). In general, good linear gradients indicative of predominantly conductive heat flow were found at all sites, with the exception of some depths at Site C0006.

From SE to NW (i.e., seaward to landward along the transect), the results are as follows. In the frontal thrust region at Site C0007, the thermal gradient is 42°C km^{-1} (heat flow of 53 mW m^2). At the nearby Site C0006, a thermal gradient of 27°C km^{-1} (heat flow of 33 mW m^2) was computed. This is surprisingly low relative to regional data and the other sites. Moving to the mega-splay sites, we find the best fitting thermal gradients were 51°C km^{-1} and 57°C km^{-1} in Holes C0008A and C0008C, respectively. Very close by at Site C0004, the measured gradient was 52°C km^{-1}. At Site C0001, the measured gradient was 44°C km^{-1} (heat flow of 47 mW m^2). In the Kumano basin, the thermal gradient at Site C0002 was 43°C km^{-1}. All of these gradients are consistently lower—by 10% to >50%—than those estimated from surface measurements (Kinoshita et al., 2008). The reason for this is not clear. The extremely low heat flow observed at frontal thrust Sites C0006 and C0007 might be related to stratigraphic or structural fluid pathways developed in this region, perhaps facilitating circulation of seawater down into the thrust sheet, though geochemical evidence of seawater circulation was not identified in porewater analyses at those sites. Also, lower heat flow relative to that predicted by surface probe measurements is consistent across all the sites, even those where such circulation seems less likely. The measured gradients appear to be very nearly linear over the depth range of the available data, so discrepancies between borehole and surface measurements are probably not attributable to an advective component. The borehole vs. surface heat flow discrepancy remains enigmatic.

Conclusion

Major scientific accomplishments of NanTroSEIZE Stage 1 include the following:

Successful core sampling of the presumed aseismic portions of major faults and the near-fault environment at the mega-splay fault system and the frontal thrust area;

In situ measurements of physical properties and borehole imaging with LWD instruments, documenting present-day horizontal stress orientations through borehole breakout interpretation; and

Documenting of the lithology, structural features, and age of sediments in and below the thrust sheets, fault zones, slope cover, and forearc basin. In particular, age determination of the forearc basin fill, slope cover, and sections overridden by thrusts provides a framework for understanding the materials and processes in the seismogenic zone, by providing information on its development and evolution.

As these were the pioneering IODP expeditions for *Chikyu*, the tremendous amount of technically-challenging drilling that was accomplished is a testament to the skill of the drilling and operations teams on board and in support onshore. The scientific goals of the NanTroSEIZE transect have been framed from its earliest conception as a comparison of the properties and state of faults and wall rock between the shallow aseismic portion and the deeper, seismogenic system. Taken as a whole, the NanTroSEIZE Stage 1 transect of sites has documented the structure, sedimentology, tectonic history, present-day stress conditions, hydrologic and geochemical regime, and other aspects of the margin for the region shallower than ~1400 m below seafloor. Thus the crucial first part of the comparison has been achieved, setting the stage for deep riser drilling to the seismogenic zone.

References

Baba, T., Cummins, P.R., Hori, T., and Kaneda, Y., 2006. High precision slip distribution of the 1944 Tonankai earthquake inferred from tsunami waveforms: possible slip on a splay fault. *Tectonophysics*, 426(1–2):119–134. doi:10.1016/j.tecto.2006.02.015.

Enescu, B., Mori, J., and Ohmi, S., 2005. Double-difference relocations of the 2004 off the Kii peninsula earthquakes. *Earth, Planets Space*, 57(4):357–362.

Expedition 314 Scientists, 2009a. Expedition 314 Site C0006. In: Kinoshita, M., Tobin, H., Ashi, J., Kimura, G., Lallemant, S., Screaton, E.J., Curewitz, D., Masago, H., Moe, K.T., and the Expedition 314/315/316 Scientists, *Proc. IODP, 314/315/316*. Washington, DC (Integrated Ocean Drilling Program Management International, Inc.), doi:10.2204/iodp.proc.314315316.123.2009.

Expedition 314 Scientists, 2009b. Expedition 314 Site C0001. In: Kinoshita, M., Tobin, H., Ashi, J., Kimura, G., Lallemant, S., Screaton, E.J., Curewitz, D., Masago, H., Moe, K.T., and the Expedition 314/315/316 Scientists, *Proc. IODP, 314/315/316*. Washington, DC (Integrated Ocean Drilling Program Management International, Inc.), doi:10.2204/iodp.proc.314315316.123.2009.

Expedition 314 Scientists, 2009c. Expedition 314 Site C0003. In: Kinoshita, M., Tobin, H., Ashi, J., Kimura, G., Lallemant, S., Screaton, E.J., Curewitz, D., Masago, H., Moe, K.T., and the Expedition 314/315/316 Scientists, *Proc. IODP, 314/315/316*. Washington, DC (Integrated Ocean Drilling Program Management International, Inc.), doi:10.2204/iodp.proc.314315316.123.2009.

Expedition 314 Scientists, 2009d. Expedition 314 Site C0002. In Kinoshita, M., Tobin, H., Ashi, J., Kimura, G., Lallemant, S., Screaton, E.J., Curewitz, D., Masago, H., Moe, K.T., and the Expedition 314/315/316 Scientists, *Proc. IODP, 314/315/316*. Washington, DC (Integrated Ocean Drilling Program Management International, Inc.), doi:10.2204/iodp.proc.314315316.123.2009.

Expedition 315 Scientists, 2009a. Expedition 315 Site C0001. In: Kinoshita, M., Tobin, H., Ashi, J., Kimura, G., Lallemant, S., Screaton, E.J., Curewitz, D., Masago, H., Moe, K.T., and the Expedition 314/315/316 Scientists, *Proc. IODP, 314/315/316*. Washington, DC (Integrated Ocean Drilling Program Management International, Inc.), doi:10.2204/iodp.proc.314315316.123.2009.

Expedition 315 Scientists, 2009b. Expedition 315 Site C0002. In: Kinoshita, M., Tobin, H., Ashi, J., Kimura, G., Lallemant, S., Screaton, E.J., Curewitz, D., Masago, H., Moe, K.T., and the Expedition 314/315/316 Scientists, *Proc. IODP, 314/315/316*. Washington, DC (Integrated Ocean Drilling Program Management International, Inc.), doi:10.2204/iodp.proc.314315316.124.2009.

Expedition 316 Scientists, 2009a. Expedition 316 Site C0006. In: Kinoshita, M., Tobin, H., Ashi, J., Kimura, G., Lallemant, S., Screaton, E.J., Curewitz, D., Masago, H., Moe, K.T., and the Expedition 314/315/316 Scientists, *Proc. IODP, 314/315/316*. Washington, DC (Integrated Ocean Drilling Program Management International, Inc.), doi:10.2204/iodp.proc.314315316.134.2009.

Expedition 316 Scientists, 2009b. Expedition 316 Site C0007. In: Kinoshita, M., Tobin, H., Ashi, J., Kimura, G., Lallemant, S., Screaton, E.J., Curewitz, D., Masago, H., Moe, K.T., and the Expedition 314/315/316 Scientists, *Proc. IODP, 314/315/316*. Washington, DC (Integrated Ocean Drilling Program Management International, Inc.), doi:10.2204/iodp.proc.314315316.134.2009.

Expedition 316 Scientists, 2009c. Expedition 316 Site C0008. In: Kinoshita, M., Tobin, H., Ashi, J., Kimura, G., Lallemant, S., Screaton, E.J., Curewitz, D., Masago, H., Moe, K.T., and the Expedition 314/315/316 Scientists, *Proc. IODP, 314/315/316*. Washington, DC (Integrated Ocean Drilling Program Management International, Inc.), doi:10.2204/iodp.proc.314315316.134.2009.

Expedition 316 Scientists, 2009d. Expedition 316 Site C0008. In: Kinoshita, M., Tobin, H., Ashi, J., Kimura, G., Lallemant, S., Screaton, E.J., Curewitz, D., Masago, H., Moe, K.T., and the Expedition 314/315/316 Scientists, *Proc. IODP, 314/315/316*. Washington, DC (Integrated Ocean Drilling Program Management International, Inc.), doi:10.2204/iodp.proc.314315316.134.2009.

Heki, K., 2007. Secular, transient and seasonal crustal movements in Japan from a dense GPS array: implication for plate dynamics in convergent boundaries. In: Dixon, T., and Moore, C. (Eds.), *The Seismogenic Zone of Subduction Thrust Faults*. New York (Columbia University Press), 512–539.

Hyndman, R.D., Wang, K., and Yamano, M., 1995. Thermal constraints on the seismogenic portion of the southwestern Japan subduction thrust. *J. Geophys. Res.*, 100:15373–15392, doi:10.1029/95JB00153.

Kinoshita, M., Kanamatsu, T., Kawamura, K., Sibata, T., Hamamoto, H., and Fujino, K., 2008. Heat flow distribution on the floor of Nankai Trough off Kumano and implications for the geothermal regime of subducting sediments. *JAMSTEC Rep. Res. Dev.*, 8:13–28.

Kinoshita, M., Moore, G., von Huene, R., Tobin, H., and Ranero, C.R., 2006. The Seismogenic Zone Experiment. *Oceanography*, 19(4):28–38.

Kinoshita, M., Tobin, H., Ashi, J., Kimura, G., Lallemant, S., Screaton, E.J., Curewitz, D., Masago, H., Moe, K.T., and the Expedition 314/315/316 Scientists, 2009. *Proc. IODP, 314/315/316*, Washington, DC (Integrated Ocean Drilling Program Management International, Inc.), doi:10.2204/iodp.proc.314315316.134.2009.

Moore, G.F., Bangs, N.L., Taira, A., Kuramoto, S., Pangborn, E., and Tobin, H.J., 2007. Three-dimensional splay fault geometry and implications for tsunami generation. *Science*, 318(5853):1128–1131. doi:10.1126/science.1147195.

Moore, G., Park, J.-O., Bangs, N.L., Gulick, S.P., Tobin, H.J., Nakamura, Y., Saito, S., Tsuji, T., Yoro, T., Tanaka, H., Uraki, S., Kido, Y., Sanada, Y., Kuramoto, S., and Taira, A., 2009. Structural and seismic stratigraphic framework of the NanTroSEIZE Stage 1 transect. In: Kinoshita, M., Tobin, H., Ashi, J., Kimura, G., Lallemant, S., Screaton, E.J., Curewitz, D., Masago, H., Moe, K.T., and the Expedition 314/315/316 Scientists, *Proc. IODP*, 314/315/316. Washington, DC (Integrated Ocean Drilling Program Management International, Inc.), doi:10.2204/iodp.proc.314315316.134.2009.

Moore, G.F., Taira, A., Klaus, A., et al., 2001. *Proc. ODP, Init. Repts., 190*. College Station, Texas (Ocean Drilling Program), doi:10.2973/odp.proc.ir.190.2001.

Park, J.-O., Tsuru, T., Kodaira, S., Cummins, P.R., and Kaneda, Y., 2002. Splay fault branching along the Nankai subduction zone. *Science*, 297(5584):1157–1160. doi:10.1126/science.1074111.

Screaton, E., Saffer, D., Henry, P., and Hunze, S., 2002. Porosity loss within the underthrust sediments of the Nankai accretionary complex: implications for overpressures. *Geology*, 30(1):19–22. doi:10.1130/0091-7613.

Seno, T., Stein, S., and Gripp, A.E., 1993. A model for the motion of the Philippine Sea plate consistent with NUVEL-1 and geological data. *J. Geophys. Res.*, 98(B10):17941–17948. doi:10.1029/93JB00782.

Tobin, H.J., and Kinoshita, M., 2006a. Investigations of seismogenesis at the Nankai Trough, Japan. *IODP Sci. Prosp.*, NanTroSEIZE Stage 1. doi:10.2204/iodp.sp.nantroseize1.2006.

Tobin, H.J., and Kinoshita, M., 2006b. NanTroSEIZE: the IODP Nankai Trough Seismogenic Zone Experiment. *Sci. Drill.*, 2:23–27. doi:10.2204/iodp.sd.2.06.2006.

Wang, K., and Hu, Y., 2006, Accretionary prisms in subduction earthquake cycles: The theory of dynamic Coulomb wedge. *J. Geophys. Res.*, 111(B6):B06410 doi:10.1029/2005JB004094.

Zoback, M.D., Barton, C.A., Brudy, M., Castillo, D.A., Finkbeiner, T., Grollimund, B.R., Moos, D.B., Peska, P., Ward, C.D., and Wiprut, D.J., 2003. Determination of stress orientation and magnitude in deep wells. *Int. J. Rock Mech. Min. Sci.*, 40(7–8):1049–1076. doi:10.1016/j.ijrmms.2003.07.001.

Authors

Harold Tobin, Department of Geoscience, University of Wisconsin – Madison, 1215 West Dayton Street, Madison, Wis. 53706, U.S.A.

Masataka Kinoshita, Institute for Research on Earth Evolution/JAMSTEC, Kanazawa-ku, Yokohama, Kanagawa 236-0001, Japan.

Juichiro Ashi, Ocean Research Institute, The University of Tokyo, 1-15-1 Minamidai, Nakano-ku, Tokyo, 164-8639, Japan.

Siegfried Lallemant, Département des Sciences de la Terre et de l'Environnement, Université de Cergy-Pontoise, Cergy-Pontoise 95031 Cedex, France.

Gaku Kimura, Department of Earth and Planetary Science, Graduate School of Science, The University of Tokyo, Tokyo, Japan.

Elizabeth Screaton, Department of Geology, University of Florida, P.O. Box 112120, Gainesville, Fla. 32611, U.S.A.

Moe Kyaw Thu, Center for Deep Earth Exploration/JAMSTEC, 3173-25 Showa-machi, Kanazawa-ku, Yokohama, Kanagawa 236-0001, Japan.

Hideki Masago, Center for Deep Earth Exploration/JAMSTEC, 3173-25 Showa-machi, Kanazawa-ku, Yokohama, Kanagawa 236-0001, Japan.

Daniel Curewitz, Center for Deep Earth Exploration/JAMSTEC, 3173-25 Showa-machi, Kanazawa-ku, Yokohama, Kanagawa 236-0001, Japan, now at: Department of Earth Sciences, Syracuse University, Syracuse, N.Y. 13244, U.S.A.

and IODP Expeditions 314/315/316 Scientific Party

15

Magnetic Viscosity for Cyclostratigraphic Logging of Argillaceous Sediment

by Julien Thiesson, Emilia Huret, Alain Tabbagh, and Bruno Galbrun

Introduction

Magnetic susceptibility (MS) is currently used as a directly representative proxy for the study of climatic variations, and for cyclostratigraphic studies. It depends on the concentration of magnetic minerals in the rocks, but does not allow identifying the magnetic minerals. In the case of argillaceous sediments, the paramagnetism of clay particles often plays a major role in determining the magnitude of their magnetic susceptibility, while the presence of ferrimagnetic iron oxides or sulfides cannot be assessed using susceptibility measurements alone. Among the different methods that can be used to detect ferrimagnetic particles magnetic viscosity (MV) characterizing the delay corresponding to the acquisition or loss of induced magnetization, has the same advantages as MS. Its measurement is direct, rapid and has been proven to be very efficient in detecting the presence of secondary ferrimagnetic minerals in soils. A MV measurement technique was tested on cores taken from a borehole, in Callovian-Oxfordian formations in the eastern Paris Basin (France) (Fig. 1). Although the MV values are very small, they have cyclic variations of which strongest values are found at the base of the Lower Oxfordian. These values, when correlated to MS, Gamma Ray (GR) and a sequential interpretation of a borehole drilled close-by are found to be associated with the maximum clay fraction in the core samples, a MS maximum, and a major transgression event. Consequently, a significant increase in ferrimagnetic minerals can be associated with this event.

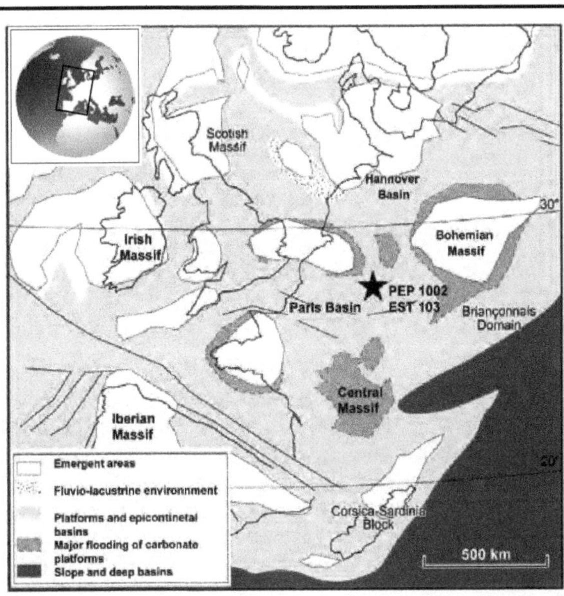

Figure 1. Paleogeographic map of the Paris Basin (indicated by stars) and surroundings (modified after Pellenard, 2003 and Dercourt et al., 1993).

Spectral analysis of the high resolution MV and MV/MS records reveals dominant patterns, identified as 95 ky eccentricity cycles, identical to those found using MS and GR analysis. Magnetic viscosity, which is sensitive to specific ferrimagnetic grains, thus appears to be a well suited proxy for cyclostratigraphic analysis, adapted to the Mesozoic homogeneous argillaceous series. Its sensitivity is complementary to, and extends the abilities of, magnetic susceptibility.

Cyclostratigraphy constitutes a remarkable tool to estimate the duration of sedimentary sequences. The method depends on the existence of climatic proxy records, whose high-resolution acquisition enables temporal resolutions to the scale of Milankovitch cycles (i.e., eccentricity, obliquity, and precession cycles with durations of 405 ky, 95 ky, 40 ky, 20 ky, and 18 ky). The quality of the records is often sufficient to allow the cyclostratigraphic series to be correlated with theoretical astronomical forcing over the past 40–60 My (Laskar et al., 2004). In such studies, magnetic properties—in particular the magnetic susceptibility (MS)—are powerful tools for cyclostratigraphic analysis (Crick et al., 2000; Latta et al., 2006). The properties, concentration, and grain size of the magnetic minerals are affected by erosion, transport, sedimentation, and/or diagenesis. MS fluctuations in a sedimentary sequence can thus reflect paleoclimatic variations and provide a record of orbital cycles. This technique has been frequently used in recent cyclostratigraphic studies; it is also a very promising diagnostic tool for the analysis of Mesozoic series (Weedon et al., 1999; Huret, 2006; Boulila et al., 2008a, 2008b). MS can be measured at low cost with a very high resolution on core samples, or by downhole logging with a high sampling rate, of the order of one centimeter. The measurements can be made in a few seconds on core, and they are simple to implement and non-destructive. The MS in argillaceous sediments is generally dominated by paramagnetic minerals, although a significant fraction of ferrimagnetic iron oxide or iron sulfide grains may also contribute. In question is however whether the ferrimagnetic fraction is more sensitive to climate changes, and thus more reliable in cyclostratigraphic studies.

It is possible to determine the presence of ferrimagnetic grains by computing the remanent magnetization, if the total field and the MS logging analysis are combined (Desvignes et al., 1992), either by measuring the magnetic viscosity (MV). Our aim was to assess the use of MV in mineralogical and sedimentological change interpretations. We tested the method on samples from the PEP 1002 borehole (Andra Underground Laboratory, France) and compared the observed MV variations with those determined using MS and sedimentology.

Magnetic Viscosity

When an external magnetic field is applied to a material, it acquires an induced magnetization with a delay which can be measured. In the case of an AC-applied field, this delayed reaction is described by a complex susceptibility: $\kappa = \kappa_{ph} - i\kappa_q$ in the frequency domain. In the time domain, for an applied step impulse function, this delay corresponds either to an increase in magnetization after the applied field has been switched on, or to a decrease in magnetization after it has been switched off. Magnetic viscosity has been observed experimentally for a wide variety of rocks, from soils (Mullins and Tite, 1973) to oceanic basalts (Trigui and Tabbagh, 1990). Theoretical approaches (Néel, 1949, 1950; Mullins and Tite, 1973) have established that for single domain grains and for wall displacement induced by thermal fluctuations in massive multi-domain grains, a quantitative relationship exists between the quadrature susceptibility (κ_{qu}, where ω is the angular frequency and κ_{td} is the time domain susceptibility) the decrease of in-phase susceptibility as a function of frequency, and time domain susceptibility variations:

$$\frac{2}{\pi}\kappa_{qu} = -\frac{\partial \kappa_{ph}}{\partial \ln \omega} = \frac{\partial \kappa_{td}}{\partial \ln t}$$

This relationship has been experimentally verified (Bloemendal et al., 1985; Dabas et al., 1992) in the presence of hyperfine grains. The role of hyperfine grains, at the transition between the superparamagnetic (SP) and stable single domain (SSD) states, has been confirmed (Dearing et al., 1996; Worm, 1998) and has been demonstrated to be very significant in soils with grain sizes below 30 nm as an indicator of the presence of secondary ferrimagnetic minerals (SFM).

In sediments, the MS corresponds to the sum of the contributions from paramagnetic and ferrimagnetic minerals, while the MV is affected only by single domain hyperfine grains. Therefore, the ratio MV/MS can be used to normalize with respect to the total magnetic particle content.

Geological Context at the Studied Boreholes

The studied boreholes truncate the Callovian-Oxfordian formation of the eastern Paris Basin (Fig. 1). They comprised homogeneous argillaceous series (around 130 m thick) deposited in a marine shelf environment. This formation is included in a second-order transgressive/regressive cycle (Hardenbol et al., 1998), with a maximum flooding surface situated near the base of the Oxfordian stage in the Mariae zone (Pellenard et al., 1999; Pellenard, 2003; Ferry et al., 2007). MS studies of many boreholes at the Bure site show that this evolution reflects the major transgressive/regressive changes of second and third orders, in which the transgressive phase is represented by an increase in MS and the regressive phase corresponds to a decrease in MS (Huret, 2006).

The rocks consist of approximately 50% clay minerals, 25% silt (quartz, feldspar, muscovite), and 25% carbonate (calcite and dolomite) whereas micas and pyrite are minor constituents (<5%). The MS is influenced mainly by the paramagnetic clay minerals and partly by ferrimagnetic minerals; significant concentrations are situated at the Callovian-Oxfordian boundary and in a clay maximum at the base of the lower Oxfordian (Esteban et al., 2006).

Two data sets from boreholes in this stratigraphy were used. The PEP 1002 borehole in the Callovian-Oxfordian argillaceous formation was accessed via the floor of the Bure Underground Laboratory at the depth of 490-m. This 19-m-long borehole is vertical. The studied interval covers approximately the top of the upper Callovian and the base of the lower Oxfordian (Fig. 2). The EST 103 hole was drilled on the Bure site, at a distance of approximately 200 m from the PEP 1002 borehole. The time interval of the EST 103 borehole could be defined as ranging from the top of the upper Callovian to the base of the lower Oxfordian (Thierry et al., 2006), by evaluating ammonite deposits. The clay content, GR value, and MS value maxima occur at depths of approximately 500 m in the EST 103 borehole and 490+3 m in PEP 1002. This corresponds to the maximum flooding surface (MFS).

Measurements and Data Processing

When the MS2B (Bartington Ltd) dual-frequency instrument was used, no difference was observed between the in-phase magnetic susceptibilities at 0.465 kHz and 4.65 kHz. This means that the quadrature susceptibility is smaller than 6.8 x 10^{-7} S.I. (for frequency dependent susceptibility of 1%, and a low frequency susceptibility of 10^{-4} S.I.). Thus, we chose the more sensitive time-domain electromagnetic technique (TDEM). A 0.084-m-diameter coincident loop sensor was built in order to be able to surround the core. Measurements were performed with TS6 electronics (Protovale Oxford Ltd), with a 2-cm sampling interval. The data were digitally recorded with thirty values stacked at each measurement point. To correct for thermal and electronic drifts, we took zeroing calibrations at regular intervals.

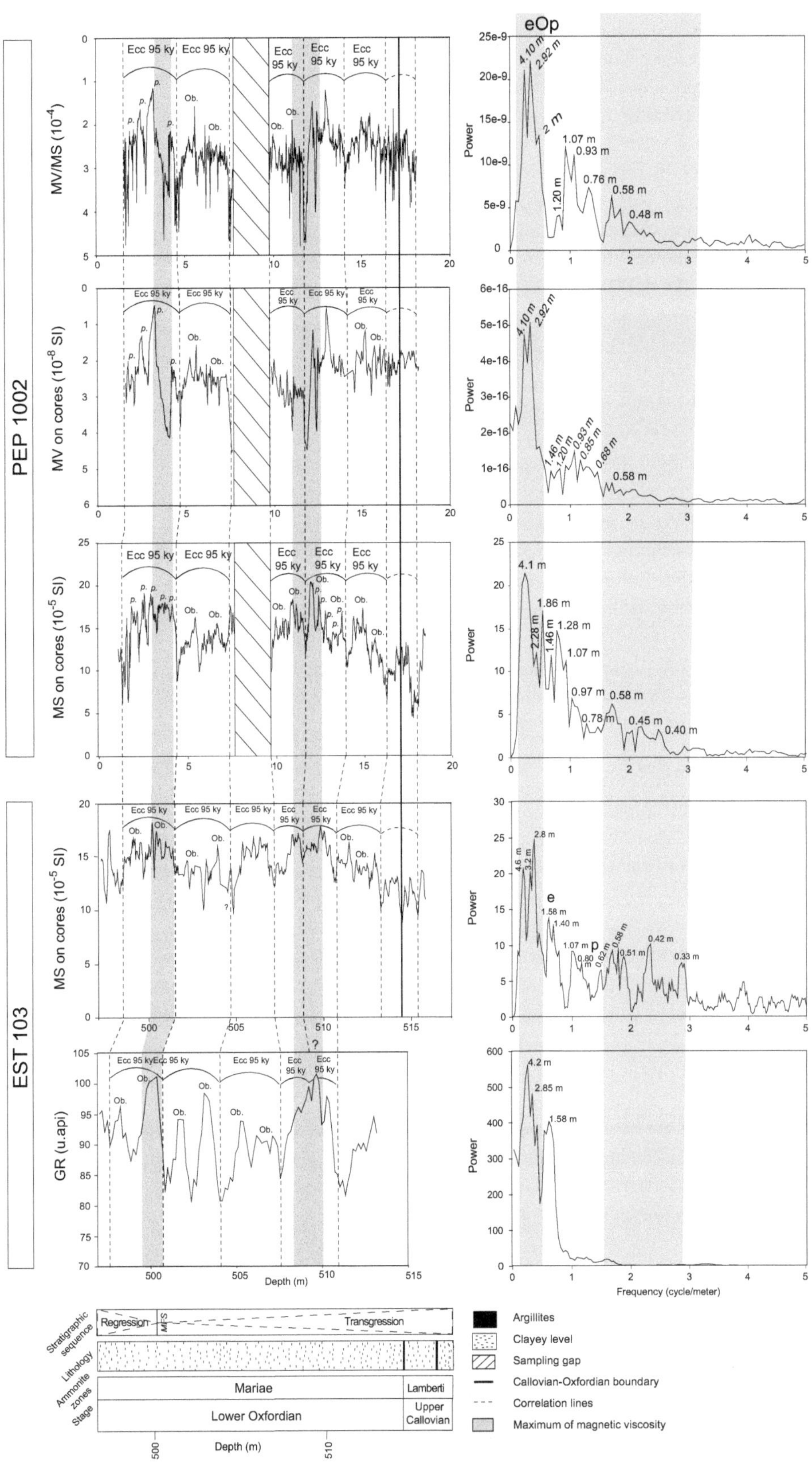

Figure 2. Biostratigraphy (Thierry et al., 2006), lithology (Huret, 2006), sequence stratigraphy interpretation and analysis of Gamma Ray (GR) well logging, and Magnetic Susceptibility (MS), Magnetic Viscosity (MV) and MV/MS ratio of the EST 103 and PEP 1002 boreholes. The light grey bands indicate zones with the strongest MV values. Presentation of 2π –MTM power spectrum after trend rejection. Identification of orbital cycles in the recordings (3 point smoothing of the MS and MV curves), symbols: eccentricity 95 ky (e), obliquity (O) and precession (p).

As the core is not continuous and presents some fractures and gaps, the data had to be corrected for these effects. The signal decay was computed as a function of core length, using an abacus, for several gap sizes. The final data reduction step involved transforming the measured voltage into apparent quadrature susceptibility, after calibrating the global loop and instrument response using a small 5-mm-diameter metallic sphere (Tabbagh and Dabas, 1996). In order to provide a good fit for the experimental values, the total instrument sensitivity was increased so that its total dynamic range corresponded to 10^{-7} S.I., in good agreement with the absence of a measurable frequency dependence of the MS2B instrument.

Magnetic Viscosity Results

The MV log is presented in Fig. 2, together with the MS (2-cm sampling interval), the MV/MS ratio recordings, and the GR (10-cm sampling interval) and the MS (4-cm sampling interval) recordings from the EST 103 borehole. These recordings are correlated with the biostratigraphic data. The values of MV we obtained are very low ($<5 \times 10^{-8}$ S.I.), which is roughly 100 times lower than in soils (Thiesson et al., 2007). This result can be explained by the very small secondary ferrimagnetic mineral content (possibly due to Fe^{++} ions dissolved in water during deposition and compaction).

The long-term evolution and high-frequency variations of the MS are similar between the EST 103 and PEP1002 boreholes. The MS maxima in the PEP 1002 hole are located at depths ranging from 3 m to 5 m, and those between 10 m and 13 m are correlated with strong GR values characterizing clay-rich zones. Moreover, the GR log presents the same high frequency variations as the MS recording (Fig. 2). These similarities confirm that the presence of paramagnetic clay minerals is the dominant contributor to MS. Correlations between MV and MS are less evident, although they present the same cyclic variations (of the order of 3 m in thickness). They appear to be anti-correlated, rather than correlated; nevertheless, the strongest MV values, which occur at 3.5 m, 7.5 m, and 11–13 m, correspond to maxima of both MS (values varying, 15–20 x 10^{-5} SI) and GR (Fig. 2). It should be noted that the relative variations in MV (variations divided by the mean value) are clearly higher than those in MS.

The main discrepancy, together with the main high and medium frequency variations observed in the MV, lies between 3 m and 4 m. This is associated with a strong relative increase in MV, followed by a strong relative decrease. By comparison, the sedimentology and stratigraphical information corresponds to a MS maximum and, after comparison with the EST103 borehole, is found to be correlated with the position of the second-order maximum flooding surface (MFS) known to exist at the base of the Lower Oxfordian. This level also corresponds to a clay-rich maximum in the EST 103 borehole, with strong GR values. A second interval with strong MV, MS, and GR values is observed between the depths of 11 m and 13 m, but is not associated with a particular lithological or stratigraphical change. However, the base of the Mariae zone can be recognized as the most argillaceous level of the series (Pellenard et al., 1999; Esteban et al., 2006).

The highest MS values, found at 3 m depth in the PEP 1002 borehole, indicate that the MFS may correspond to (i) a greater quantity of paramagnetic clay particles, or (ii) a greater quantity of magnetite grains of terrigenic origin (Esteban et al., 2006), or (iii) greater *in situ* generation of biogenic minerals (greigite or magnetite). The strong changes in MV suggest that the processes involved are even more complex than suggested by MS alone; inside this period of high susceptibility, rapid and strong changes occurred in the sediment transfer modes, and possibly in the oxygen content of water. In the absence of a detailed mineral analysis, it is not possible to determine exactly what happened during the flood event.

We would like to emphasize that, for the MFS event recorded at 3–4 m, the MV/MS ratio provides a large quantity of useful information. It not only highlights changes that have already been observed using GR or MS, as for the case just above 11 m, but also reveals more complex, detailed variations which have yet to be understood. As the relative variations in MS are of limited extent (20% maximum) and are correlated with clay content, the MV/MS ratio can be initially interpreted as expressing the small ferrimagnetic particle content of the clay. Variations in this ratio could thus provide an indication of the sediment's origin.

Frequency Content

MV data exhibited a higher relative variability than that recorded using MS (or GR). However, in order to assess its relevance for cyclostratigraphic studies, the frequency content of the relevant recordings must also be analyzed.

Time series analysis of the PEP 1002 and EST 103 well logs and core data were carried out using the "Multi-Taper Method" (Thomson, 1982, 1990). This method, which we implemented with Matlab® software, estimates the power spectral density of a time series through the application of a set of orthogonal tapers, while maintaining suitable confidence levels. In spite of the presence of some gaps in the PEP 1002 data and of the limited length of the recordings, our analyses were performed over the full series. The data was re-sampled every 2 cm, using linear interpolations (except for the GR, which has a constant sampling step), and the linear trend was removed. The results are shown in Fig. 2, for which the data was smoothed (3-point average) to improve the contrast of the high frequency variations.

Identification of the orbital cycles was achieved using the frequency ratio method (Mayer and Appel, 1999), which consists of comparing the frequency ratio of pairs of dominant cycles observed in the series, with those known to characterize the Earth's orbital cycles (i.e., 18.2 ky and 21.9 ky for precession, 37.7 ky for obliquity, and 95 ky for eccentricity) during the Jurassic period (Berger and Loutre, 1994). GR analysis of the EST 103 reveals a power spectrum with only two major peaks, with periods of 2.85 m and 4.2 m, and a weaker peak at 1.58 m, visible on the curve (Fig. 2). This power spectrum does not have small high-frequency variations similar to those associated with the MS data, because a longer (10 cm) sampling step was used. The MS spectrum of the EST 103 hole presents several peaks with strong amplitudes, the first identical in frequency to those of the GR spectrum, with periods between 4.6 m and 2.8 m clearly visible (Fig. 2). Others periods, close to 1.4 m and 1.58 m, 1.07 m and 0.8 m, and 0.62 m and 0.33 m, are also characterized by comparatively strong amplitudes. By comparing the terrestrial orbital frequency ratios with those derived from the GR and MS recordings at the EST 103 hole (Table 1), we can associate the core cycles (with periods of 2.8–4.1 m) with 95-ky eccentricity cycles. The core cycles with periods of 1.58–0.62 m can be associated with obliquity cycles, and the core cycles with periods 0.33–0.80 m can be associated with precession cycles. These periodicity allocations agree with the cyclostratigraphic analysis performed on cores of another hole (EST 322) drilled 10 km from EST103 and PEP 1002 (Boulila et al., 2008b). The fact that the observed cycles are spread over a large range of frequencies can be explained by variations in the rate of sedimentation. The sedimentation rate, in the MS recordings associated with 95-ky eccentricity cycles, increases from that corresponding to the 2.8–3.2 m thicknesses at the bottom of the recording, to that resulting in a thickness of approxi-mately 4 m at the top (Fig. 2).

Table 1. Durations and duration ratios of orbital cycles, defined for the Late Jurassic (Berger and Loutre, 1994), frequency ratios of the principal cycles observed in MS and GR recording spectra from the EST 103 hole, and of the principal cycles observed in MS, MV and MS/MV recording spectra from the PEP 1002 hole.

		Precession	Obliquity	Eccentricity	P/O	P/E	O/E
Orbital Cycles (Later Jurassic)		18.2 ka 21.9 ka	37.7 ka	95 ka	0.483 0.581	0.191 0.230	0.397
		Dominant Periods			Ratio		
Intervals		a	b	c	a/b	a/c	b/c
EST 103	GR		1.58	4.10			0.385
	MS	0.33 0.58 0.80	0.62 1.07 1.40	1.58 2.80 3.20	0.532 0.542 0.571	0.209 0.207 0.250	0.419 0.382 0.437
PEP 1002	MS	0.40 0.58	0.78 0.97	1.86 2.28	0.513 0.597	0.215 0.254	0.419 0.425
	MV	0.58	1.20	2.92	0.483	0.199	0.411
	MS/MV	0.48 0.58	0.93 1.20	2.00 2.92	0.516 0.483	0.240 0.199	0.465 0.411

The spectral analysis of MS in PEP 1002 is more complex. Eccentricity cycles are expressed, as for the EST 103 borehole, by a 4.1-m thickness. However, other cycles such as obliquity and precession are more difficult to identify because the spectrum contains many close frequency peaks with strong amplitudes. This effect is certainly amplified by the presence of a core sample gap (7.76–9.26 m). Similar spectral characteristics have already been observed by Weedon (1989); however, the frequency ratio of the major peaks allowed Milankovitch cycles to be identified with frequencies in agreement with those found at the EST 103 borehole (Table 1). MS stratigraphic correlations between the two boreholes show similar trends and clearly present the same high-frequency cycles. In this correlation, some thickness variations are also expressed by small differences in MS measurements.

The MV spectrum for PEP 1002 in Fig. 2 is clearer than that produced from the MS recordings. It has the same two strong peaks corresponding to periods of 2.92 m and 4.1 m, associated with the 95-ky eccentricity cycles. Although the higher frequency cycles are not very strong, the use of frequency ratios enabled obliquity and precession cycles to be identified. The spectrum of the MV/MS ratio is the most representative of all the spectra, with all Milankovitch cycles expressed with strong, distinct amplitude peaks. The MV and MV/MS curves (inversed in Fig. 2) clearly show the eccentricity cycles corresponding to thickness periods in the range of 2 m to 4 m.

Conclusions

The results presented in this paper are the first on direct magnetic viscosity measurements of argillite samples. Measurements of PEP1002 cores are shown to be informative, and they demonstrate that MV measured by the TDEM technique appears to be a potentially promising climate proxy for high-resolution cyclostratigraphic studies. The resolution and lateral sensitivity of MV measurements should be similar to that of MS and could be adjusted by appropriate definition of the coil geometry. The potential of MV as a proxy for stratigraphic climate analysis is its ability to constrain the ferromagnetic iron oxide or sulfide grain fractions. This capability potentially can be used to improve the identification of sedimentary sources, changes in mineralogy which occurs during transport, and sedimentation process. It is now necessary to determine precisely the advantages and drawbacks of this type of measurement and the extent to which it can be complementary to MS and other petrophysical analyses. The construction of a MV wire-logging tool could take advantage of the experience acquired in surface prospecting instruments.

Acknowledgements

We wish to thank Patrick Landais, Alain Trouiller, and Hervé Rebours of the ANDRA (French *Agence Nationale pour la gestion des Déchets Radioactifs*) for authorizing and assisting with the measurement of the PEP1002 cores at the Bure site. We are grateful to Linda Hinnov for fruitful discussions and to Slah Boulila for the preparation of the illustrations provided in this paper.

References

Berger, A., and Loutre, M.F., 1994. Astronomical forcing through geological time. *In* de Boer, P.L, and Smith, D.G. (Eds.), *Orbital Forcing and Cyclic Sequences,. Int. Ass. Sed. Sp. Pub.*, 19:15–24.

Bloemendal, J., Barton, C.E., and Radhakrishnamurty, C., 1985. Correlation between Rayleigh loops and frequency-dependent and quadrature susceptibility: application to magnetic granulometry of rocks. *J. Geophys. Res.*, 90:8759–8792, doi:10.1029/JB090iB10p08789.

Boulila, S., Galbrun, B., Hinnov, L.A., and Collin, P.Y., 2008a. High-resolution cyclostratigraphic analysis from magnetic susceptibility in a Lower Kimmeridgian (Upper Jurassic) marl-limestone succession (La Méouge, Vocontian Basin, France). *Sed. Geol.*, 203:54–63, doi:10.1016/j.sedgeo.2007.10.004.

Boulila, S., Hinnov, L.A., Huret, E., Collin, P.Y., Galbrun B., Fortwengler, D., Marchand, D., and Thierry, J., 2008b. Astronomical calibration of the Early Oxforiden (Vocontian and Paris basins, France): consequences of revising the Late Jurassic time scale. *Earth Planet. Sci. Lett.*, 276:40–51, doi:10.1016/j.epsl.2008.09.006.

Crick, R.E., Ellwood, B.B., El Hassani, A., and Feist, R., 2000. Proposed magnetostratigraphy susceptibility magnetostratotype for the Eifelian-Givetian GSSP (Anti-Atlas, Morocco). *Episodes*, 23:93–101.

Dabas, M., Jolivet, A., and Tabbagh, A., 1992. Magnetic susceptibility and viscosity of soils in a weak time varying field. *Geophys J. Int.*, 108:101–109, doi:10.1111/j.1365-246X.1992.tb00841.x.

Dearing, J.A., Dann, R.J.L., Hay, K., Lees, J.A., Loveland, P.J., Maher, B.A., and O'Grady, K., 1996. Frequency-dependent susceptibility measurements of environmental materials. *Geophys. J. Int.*, 124:228–240, doi:10.1111/j.1365-246X.1996.tb06366.x.

Dercourt, J., Ricou, L.E., and Vrielynck, B., 1993. *Atlas Tethys palaeoenvironmental maps*. Paris, (Gauthier-Villars), 307pp.

Desvignes, G., Barthès, V., and Tabbagh, A., 1992. Direct determination of natural remanent magnetization effect in a hole drilled in layered ground from magnetic field and susceptibility logs. *Geophysics*, 57:872–884, doi:10.1190/1.1443300.

Esteban, L., Bouchez, J.L., and Trouiller, A., 2006. The Callovo-Oxfordian argillites from Eastern Paris basin: magnetic data and petrofabrics. *C. R. Geoscience*, 338:867–881.

Ferry, S., Pellenard, P., Collin, P.Y., Thierry, J., Marchand, D., Deconinck, J.F., Robin, C., Carpentier, C., Durlet, C., and Curial, A., 2007. Synthèse des données stratigraphiques récentes sur les formations bathoniennes à oxfordiennes de l'Est du Bassin de Paris. *Mém. Soc. Géol. Fr.*, 178:37–52.

Hardenbol, J., Thierry, J., Farley, M.B., Jacquin, T., de Graciansky, P.C., and Vail, P.R., 1998. Mesozoic and Cenozoic sequence chronostratigraphic framework of European basins. *In* de Graciansky, P.-C., Hardenbol, J., Jacquin, T., and Vail, P.R. (Eds.), *Mesozoic and Cenozoic Sequence Stratigraphy of European Basins, Spec. Publ. SEPM 60*, Tulsa, Okla., U.S.A. (SEPM Society for Sedimentary Geology), 3–13.

Huret, E., 2006. Analyse cyclostratigraphique des variations de la susceptibilité magnétique des argilites callovo-oxfordiennes de l'Est du Bassin de Paris: application à la recherche de hiatus sédimentaires. PhD thesis, Université Pierre et Marie Curie, Paris, France.

Laskar, J., Robutel, P., Joutel, F., Gastineau, M., Correia, A.C.M., and Levrard, B., 2004. A long-term numerical solution for the insolution quantities of the Earth. *Astron. Astrophys.*, 428:261–285, doi:10.1051/0004-6361:20041335.

Latta, D.K., Anastasio, D.J., Hinnov, L.A., Elrick, M., and Kodama, K.P., 2006. Magnetic record of Milankovitch rhythms in lithologically noncyclic marine carbonates. *Geology*, 34:29–32, doi:10.1130/G21918.1.

Mayer, H., and Appel, E., 1999. Milankovitch cyclicity and rock-magnetic signatures of paleoclimatic changes in the early Cretaceous Biancone Formation of the Southern Alps, Italy. *Cret. Res.*, 20:189–214, doi:10.1006/cres.1999.0145.

Mullins, C.E., and Tite, M.S., 1973. Magnetic viscosity, quadrature susceptibility and frequency dependence of susceptibility in single domain assemblies of magnetite and maghemite. *J. Geophys. Res.*, 78:804–809, doi:10.1029/JB078i005p00804.

Néel, L., 1949. Théorie du traînage magnétique des ferromagnétiques en grains fins avec application aux terres cuites. *Annales Géophysique*, 5:99–136.

Néel, L., 1950. Théorie du traînage magnétique des substances massives dans le domaine de Rayleigh. *Le Journal de Physique et le Radium*, 11:49–61.

Pellenard, P., 2003. Message terrigène et influences volcaniques au Callovien-Oxfordien dans les bassins de Paris et du Sud-Est de la France. *Soc. Géol. Nord, Publ. 31*, 362pp.

Pellenard, P., Deconinck, J.F., Marchand, D., Thierry, J., Fortwengler, D., and Vigneron, G., 1999. Contrôle géodynamique de la sédimentation argileuse du Callovien-Oxfordien moyen dans l'Est du bassin de Paris : influence eustatique et volcanique. *C. R. Acad. Sci. Paris*, 328:807–813.

Tabbagh, A., and Dabas, M., 1996. Absolute magnetic viscosity determination using time domain electromagnetic devices. *Archaeol. Prospect*, 3:92–100.

Thierry, J., Marchand, D., Fortwengler, D., Bonnot, A., and Jardat, R., 2006. Les ammonites du Callovien-Oxfordien des sondages Andra dans l'Est du Bassin de Paris: bio-chronostratigraphie, intérêts paléoécologique et paléogéographique. *C. R. Geoscience*, 338:834–853, doi:10.1016/j.crte.2006.05.005.

Thiesson, J., Tabbagh, A., and Flageul, S., 2007. TDEM magnetic viscosity prospecting using a Slingram coil configuration. *Near Surface Geophysics*, 5(6):363–374.

Thomson, D.J., 1982. Spectrum estimation and harmonic analysis. *Proc. IEEE*, 70:1055–1096, doi:10.1109/PROC.1982.12433.

Thomson, D.J., 1990. Time series analysis of Holocene climate data. *Phil. Trans. R. Soc. Lond.*, 330:601–616.

Trigui, M., and Tabbagh, A., 1990. Magnetic susceptibilities of oceanic basalts in alternative fields. *J. Geomag. Geoelect.*, 42:621–636.

Weedon, G P., 1989. The detection and illustration of regular sedimentary cycles using Walsh power spectra and filtering, with examples from the Lias of Switzerland. *J. Geol. Soc. Lond.*, 146:133–144, doi:10.1144/gsjgs.146.1.0133.

Weedon, G.P., Jenkyns, H.C., Coe, A.L., and Hesselbo, S.P., 1999. Astronomical calibration of the Jurassic time scale from cyclostratigraphy in British mudrock formations. *Phil. Trans. Royal Soc. Lond.*, 357:1787–1813, doi:10.1098/rsta.1999.0401.

Worm, H.U., 1998. On the superparamagnetic stable single domain transition for magnetite and frequency dependent susceptibility. *Geophys. J. Int.*, 133:201–206, doi:10.1046/j.1365-246X.1998.1331468.x.

Authors

Julien Thiesson, Université Pierre et Marie Curie, UMR 7619, Sisyphe, case 105, 4 place Jussieu, 75252 Paris cedex 05, France and INRA, Centre de Recherche d'Orléans, UR 0272 Science du Sol, 2163 Avenue de la Pomme de Pin, CS 40001 ARDON, 45075 ORLEANS Cedex 2, France, e-mail : julien.thiesson@upmc.fr.

Emilia Huret, Andra, 1/7 rue Jean Monnet, Parc de la croix blanche, 92298 Châtenay Malabry cedex, France, e-mail: emilia.huret@andra.fr.

Alain Tabbagh, Université Pierre et Marie Curie, UMR 7619, Sisyphe, case 105, 4 place Jussieu, 75252 Paris cedex 05, France, e-mail: alain.tabbagh@upmc.fr.

Bruno Galbrun, Université Pierre et Marie Curie, ISTeP, case 117, 4 place Jussieu, 75252 Paris cedex 05, France, e-mail: bruno.galbrun@upmc.fr.

16

Drilling for the Archean Roots of Life and Tectonic Earth in the Barberton Mountains

by Eugene G. Grosch, Nicola McLoughlin, Maarten de Wit, and Harald Furnes

Introduction

In the Barberton Scientific Drilling Program (BSDP) we successfully completed three drill holes in 2008 across strategically selected rock formations in the early Archean Barberton Greenstone Belt, South Africa. This collaborative project's goal is to advance understanding of geodynamic and biogeochemical processes of the young Earth. The program aims to better define and characterize Earth's earliest preserved ocean crust shear zones and microbial borings in Archean basaltic glass, and to identify biogeochemical fingerprints of ancient ecological niches recorded in rocks. The state-of-the-art analytical and imaging work will address the question of earliest plate tectonics in the Archean, the $\delta^{18}O$ composition, the redox state and temperature of Archean seawater, and the origin of life question.

The BSDP project serves as a first phase towards a large-scale deep drilling project which will also encompass the development of a local Earth science research and education center. During the period July–August 2008, a first phase of drilling yielded three boreholes with a total of 800 m of core with 99% core recovery in the early Archean Barberton Greenstone Belt (BGB), South Africa. Five decades of scientific research on the 3.5–3.1-Ba-old BGB has established the mountain ranges of Barberton as a world heritage site and a focus for international scientists interested in early Earth processes. The rocks around Barberton are unique in that they represent relatively intact and undisturbed remnants of preserved ancient seafloor and continental crust that have largely escaped tectonometamorphic reworking since the time they formed (Schoene et al., 2008).

Outstanding questions about early Earth—as well as Earth systems and linkages between physical, chemical, and biological processes operating within and on the early Earth—include the following areas of interest:

- plume-driven versus plate-tectonic-driven mechanisms for the early Archean
- Earth's early magnetic field and exogenous influx from space
- volcanology, origin of komatiites, and thermal gradients on early Earth

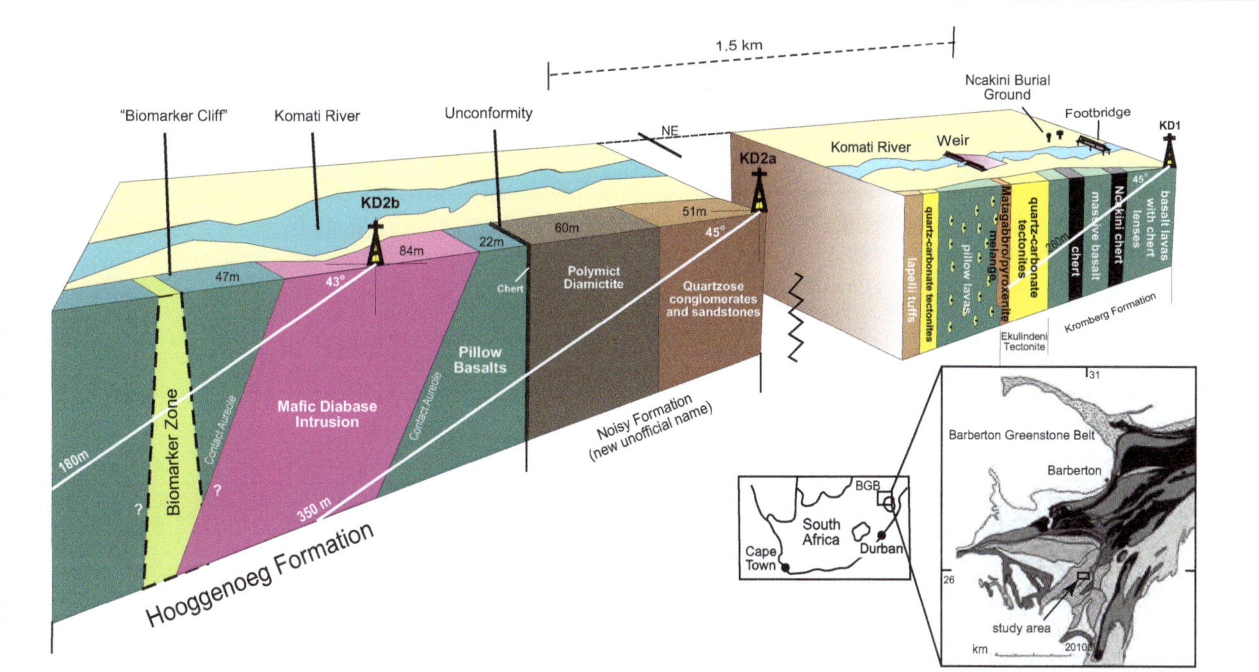

Figure 1. Three-dimensional block diagram with simplified geology depicting the positions and shallow drilling angles at drill sites KD1, KD2a, and KD2b and location maps of the study area with South Africa and simplified geology of Barberton Greenstone Belt (BGB map modified from Kareem, 2005).

- chemical, structural, and tectonic make-up of the oldest greenstone belts
- crustal and mantle fluid processes, chemistry of ancient seawater composition
- climate and atmospheric conditions at 3.4–3.5 Ga
- origin of life and early life-sustaining environments

Addressing these topics using surface samples has, however, been plagued by a number of complications related to limited exposure and secondary processes like weathering.

Diamond core drilling, which retrieves fresh, representative samples from depth at strategic locations, has the advantage of providing continuous information with depth, and avoiding a number of complexities related to surface processes.

Geological Background and Drill Site Locations

The BGB forms part of the easternmost margin of the Kaapvaal Craton as a NE-SW trending tectono-metamorphic belt. Traditionally, rocks in the southern part of the belt have been ascribed to the 3.2–3.5 Ga Swaziland Supergroup, which in turn has been further subdivided into the Onverwacht, Fig Tree, and Moodies Groups (Viljoen and Viljoen, 1969; Brandl et al., 2006). Drilling was aimed at focusing on the mostly tholeiitic basalts of the 3.47–3.44-Ga mid-upper Onverwacht Group of the Swaziland Supergroup in the southwesternmost BGB (Fig. 1). Drill sites are located in the Songimvelo Nature Reserve, near Ekulindeni (Kromdraai). The exposed rocks include the Mendon, Kromberg, and Hooggenoeg Formations of the Onverwacht Group.

The first hole (KD1, S26°02'15.3", E31°00'00.0", total depth of 261 m) allowed sampling into the Upper Kromberg Formation, a section of pillow lavas and sheet flow with interlayered cherts and chert xenoliths that overlies an enigmatic early carbonated and silicified oceanic shear zone known as the Ekulindeni Shear Zone. Drilling stopped in pyroxenitic units just beyond a carbonated shear zone. Studies on core samples will allow definition of the chemical alteration and microbial activities from an Archean marine bottom-surface into its underlying sub-surface fault zones. Results will also help to reconstruct the chemical composition and temperature of the associated seawater.

The second and third holes (KD2a, S26°01'50.0", E30°59'32.6" and KD2b, S26°01'25.3", E30°59'16.6", total depths of 350 m and 180 m, respectively) are located about 1.5 km farther NE along the Komati river on a flat open river terrace. The main objective of these holes was to drill across a basal unconformity in the Hooggenoeg Formation between a set of subaerial siliciclastics consisting of diamictites, conglomerates, and sandstones and underlying basaltic pillow lavas with interbedded cherts. Although these conglomerates were formerly placed in the Hooggenoeg Formation by Viljoen and Viljoen (1969), ongoing geological mapping in the area has lead to a revised stratigraphy (de Wit et al., in prep.). The pillow lava basalts in the Hooggenoeg Formation contain early Archean biomarkers or ichnofossils described by Furnes et al. (2004), in the type area known as the "Biomarker Cliff". Field and preliminary core observations suggest that in this site the top of the pillow lava pile was exposed to the Archean atmosphere at the unconformity. To avoid an 84-m-thick mafic/diabase intrusion about 22 m away from the unconformity, we relocated the drill rig for a third hole (KD2b) 172 m further north (Fig. 1). Drilling at KD2b intercepted the type section for the biogenic structures that occur in pillow lava margins and typify Earth's 3.4-Ga oceanic environments, including an early hydrothermal ocean-floor-type metamorphic overprint (Furnes et al., 2004). This part of the section will allow the physio-chemical conditions of the biological communities in the pillow margins and their preservation window to be better defined.

Figure 2. Drill rig aligned and set to drill at the shallowest possible angle (45°) relative to the dip and dip direction of the steep near-vertical orientation of the geological units along the Komati River. The >3-m-long rod is composed of basalt or diamictites.

Drilling Methods

The drill rig was oriented perpendicular to the bedding or structural orientation of the lithological units, using the average strike and dip readings of rock outcrops at the site.

Given the steep, generally vertical orientation of the geologic units, drilling for all three boreholes was carried out at the shallow angle of about 45° and a drilling direction towards 320° to 340° (Fig. 2). Drilling at shallow angles raised a number of difficulties, including borehole wall collapses in the alluvial overburden. Due to frictional drag of the inner tube on the inside of the drill stem rods, the elapsed time between runs increased significantly as a consequence of the flat angle. The overshot instrument attached to the wireline for lowering the innertube down the hole was modified with a heavy metal rod weight to overcome frictional energy downhole to shorten down time and to facilitate efficient latching of the innertube onto the core barrel. The potentially delicate nature of bio-textures and chemical biomarker (carbon) signatures to be encountered made it vital not to alter the original Archean chemistry of the drilled rock. Furthermore, since data on the redox state of the Archean atmosphere and seawater composition is one of the analytical goals, reducing oxidation was imperative. Therefore, a number of steps were taken during the drilling project to ensure that the original rock chemistry remained as intact as possible.

- Only clean water was used as drilling mud.
- A clean drill rig was used and repeatly washed, and the site was kept clean of potential contaminants including smoke at all times.
- Only one person handled, washed, and dried all core immediately after retrieval.
- Cores were packed in wooden boxes with wood chips for support during shipping.

Core recovery for all three holes was typically above 98% providing up to 3-m rods of core (Fig. 2). Only in the case of intercepting the first chert horizon, namely the "Footbridge Chert" in KD1, did core recovery temporarily drop to 92% due to partial fracturing of the chert. Initial stages of each hole involved emplacement of HQ (63.5-mm diameter) size metal casing to stabilize the hole.

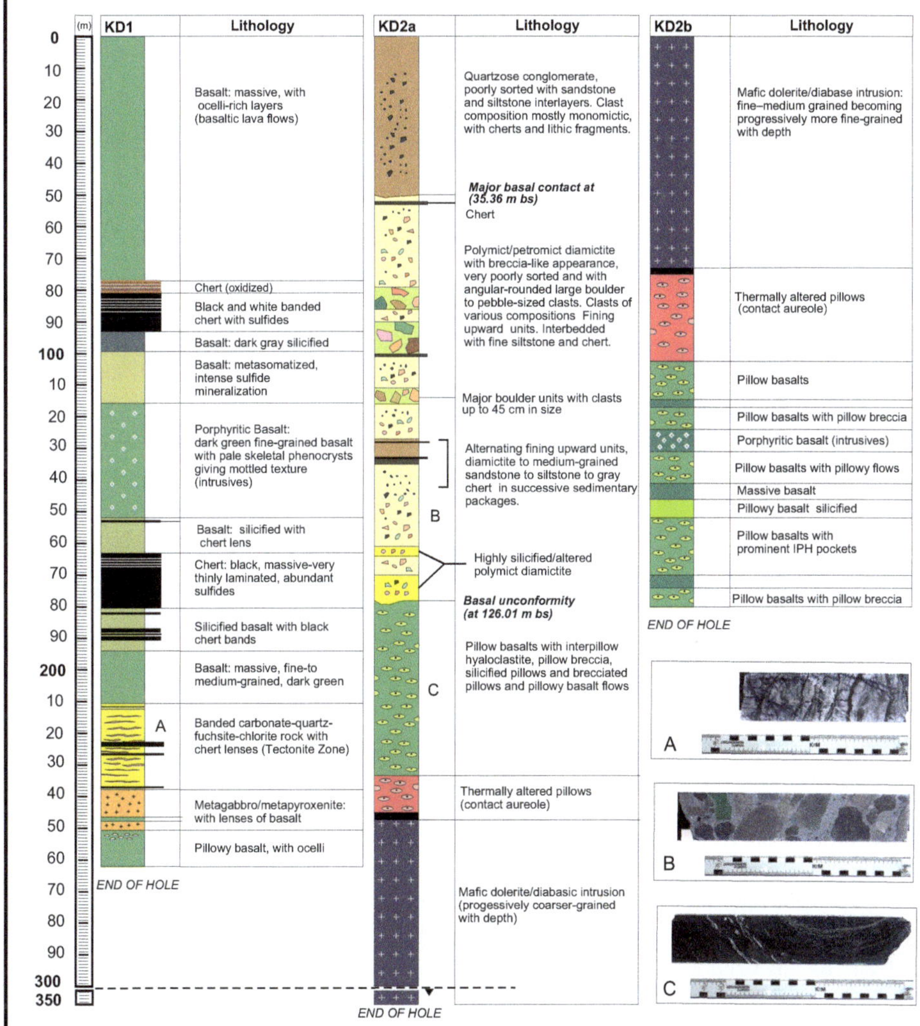

Figure 3. Simplified diagrams of geological logs for KD1, KD2a, and KD2b. The labelled photographs of selected core sections in the inset [A–C] correspond to the labels indicated at different depths in the three logs. [A] Carbonate-quartz and fuchsite-chlorite bands in the tectonite zone of KD1. [B] Example of shape, size, and variety of clasts occurring in the diamictite unit that directly overlies pillow basalts at the unconformity in KD2a. [C] Drill core section through a pillow basalt consisting of an ocelli-rich pillow core, outer ocelli-poor glassy pillow rim, and veined and sulfide-bearing interpillow hyaloclastite.

Water flow rate downhole generally varied at 30–33 L min^{-1}. Optimum revolutions per minute were around 580, typical for basalts at a torque of around 3 kN. Given the flat 45° angle, the weight on the bit for the drill rig was adjusted to around 2–2.5 kN using a soft matrix NQ corebit (47.6-mm diameter). The very hard chert horizons were cored at 450 rpm. A core orientation tool was used to fix the orientation for paleomagnetic studies on basalts.

Preliminary Results

Core log data from KD1 (total length 261 m) records about 54.5 m of basalts from the Mendon Formation before the top of the Kromberg Formation was intercepted (Fig. 3). The section of drill core representing this formation includes two major chert units of about 12 m each in an 83-m-thick section of variably metasomatized and silicified massive intrusives and lava flows. The first of these chert units appears to record an

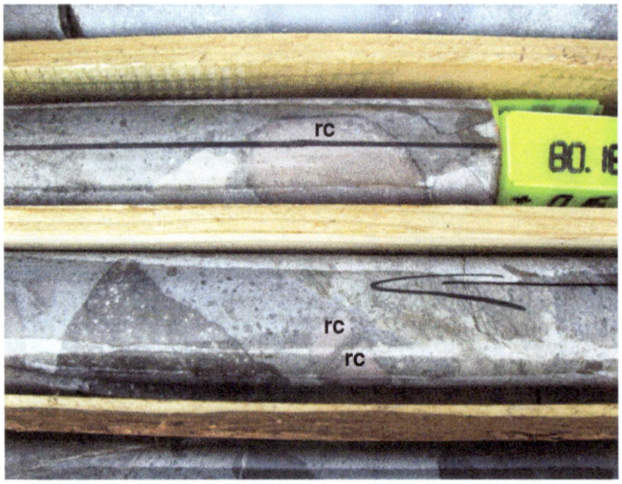

Figure 4. Distinct chemical weathering rim around light pink rhyolitic clasts (rc) possibly recording weathering conditions and/or redox conditions on the early Earth's surface.

upper 2.5-m-thick oxidized horizon, followed by an 8.5-m lower part consisting of black massive occasional white banded chert containing sulfides (implying a more reduced environment). The second chert unit was intercepted at a depth of about 115 m below surface; it occurs at the center of a 30-m-wide zone of highly silicified basalt that is generally associated with a number of discontinuous black chert lenses and veins. Near the end of KD1, a 20-m-thick zone occurs with thin black chert veins and consists of carbonate and quartz bands and veins in a matrix of dark green fuchsite (Cr-muscovite), chlorite, and magnetite. In surface outcrop this zone is as thick as 67 m, and it is adjacent to a 72-m-thick unit of highly foliated metapyroxenite. It is currently interpreted to represent a remnant component of an early oceanic shear zone. The carbonate-quartz tectonite zone, as seen in the drill core, is less than one-third of its thickness on surface, indicating discontinuous and variable thickness. Drilling stopped in a metapyroxenitic unit consisting of a melange of metagabbros and pillow basalt screens for about 17 m beyond the end of the carbonate-quartz zone. Micro- to nano-scale analysis of the rock units in KD1 will allow comparison of various potential ecological niches across an Archean seafloor environment into deeper underlying sub-seafloor tectonite zones. Such a comparison has not been previously carried out. Carbonaceous cherts in the upper and lower parts of KD1—representing marine seafloor and hydrothermal cherts, respectively—are the primary analytical target in the search for early life microtextures and chemical fingerprints in KD1. These rocks will be investigated using laser Raman microspectroscopy and bulk rock geochemical analysis. Calcite-quartz extensional veins/bands in the lower tectonite zone will be analyzed for their $\delta^{18}O$ composition to determine hydrothermal seawater fluid temperatures and water-rock ratios. These results will be integrated with *in situ* stable isotope and synchrotron X-ray analyses of individual fluid inclusions in the lower tectonite zone, combined with a micro-structural fabric analyses of deformed rocks. High-resolution Secondary Ion Microprobe Analyses of various sulphide generations present throughout KD1 aims to investigate mass-dependent and possible mass-independent sulphur isotope fractionation as a geochemical tool to identify microbial interactions and improve our understanding of the early Earth's sulfur cycle.

Drill hole KD2a penetrated a basal unconformity at around 126 m below surface, between a lower marine section of pillow lavas and interbedded cherts that are abruptly overlain by subaerial diamictites, conglomerates, and sandstones. In the drill core, clasts in diamictite units directly above the unconformity are poorly sorted, polymictic in composition, and generally angular giving the rocks a breccia-like appearance suggestive of glacial diamictites. Many of the clasts, particularly those of the rhyolitic and dacitic composition, display a prominent weathering halo or chemical alteration rim (Fig. 4). Preliminary measurement of clast size in the core indicates that the entire unit coarsens upward. Overlying these 'diamictites' are more typical fluvial facies, conglomerates, and sandstones. However, pronounced soft sediment deformation features are also present and may suggest an offshore marine turbidite facies origin. A sedimentological provenance study involving a detailed documentation of the petrological characteristics of the clasts, U-Pb dating of detrital zircons from selected depths in the core, and reconstructing a chemical weathering profile will aim at understanding of the depositional environment(s) of these enigmatic sediments and possibly constraining the atmospheric conditions on the Earth's surface at ~3.45 Ga. The lower part of drill hole KD2a penetrated a section of thermally altered pillow basalts in a metamorphic contact aureole and ended in a medium-grained mafic intrusive at a depth of 350 m.

Figure 5. Photomicrograph of microbial biotextures in a surface sample of inter-pillow breccia from the Hooggenoeg Formation of the Barberton Greenstone Belt. The meta-volcanic glass comprises a greenschist facies assemblage of chlorite and quartz with titanite-filled tubular textures. These are curvilinear and unbranched, and they radiate from a central band of microcrystalline titanite. Some taper towards their ends and are segmented by cross-cutting chlorite. These textures are interpreted to be mineralized tunnels created by microbes that etched volcanic glass on the Archean sub-seafloor. This is a composite z-stacked high-resolution image showing tubes at different focal depths.

The third hole, KD2b, intercepted at depth the "Biomarker Zone" originally described in surface outcrop by Furnes et al. (2004). Detailed petrographic work on surface samples from this zone has shown a range of morphological types of ichnofossils (McLoughlin et al., 2007; see also Fig. 5). The recent *in situ* U-Pb dating of titanite infilling the microbial ichnofossils has revealed an Archean age (Fliegel et al., 2008). Pillow lavas in the drill core of KD2b will be investigated for early Archean ichnofossils from a depth of around 130 m onwards to further investigate morphological diversity and antiquity of the bioalteration textures by applying state-of-the-art analytical and imaging techniques (e.g., Raman spectroscopy, X-ray synchrotron imaging). A detailed study of the continuous drill core section recovered from depth in KD2b aims to reconstruct the spatial distribution of these microbial ichnofossils with depth through an ancient oceanic lithosphere. In addition, we aim to determine the preservation conditions of the biosignatures and the prevailing life-sustaining hydrothermal conditions that existed in the early oceanic crust.

KD2b also penetrated thermally altered pillow basalts in a metamorphic contact aureole around the northern margin of the mafic intrusion that will be used as a thermal and reversal test for paleomagnetic studies. To date, only volcanic rocks of late Archean age (~2.7 Ga) have yielded detailed paleomagnetic signatures, and we anticipate that this Barberton drill core will provide new insights into the strength of the much earlier ~3.45 Ga Archean magnetic field and the size of the young Earth's inner core.

Future Plans for Drilling in Barberton

The present drilling project serves as a preliminary stage for a more ambitious super-deep drilling project to penetrate one of two ancient crustal-scale suture zones in the BGB. A magneto-telluric transect across the BGB has been conducted over the period March–May 2009 to start identifying the target zone. This will be followed by a seismic survey and an on-site ICDP workshop to plan a proposal for deep drilling. In addition, we plan to establish an advance drilling technology school and develop a rural science education center linked to a world class 'Early Earth' museum near this drill site.

Acknowledgments

We thank Annelize Steyn, Johan Eksteen, Shane Plumkel, and Property Mokoene from the Mpumalanga Tourism and Parks Agency (MTPA), Songimvelo Nature Reserve, South Africa for their support and valuable environmental advice prior to and during the drilling operation. Thanks are also due to Fred Daniel of Nkomazi Wilderness for his continued support of our Barberton research. This joint project was funded by the Center for Geobiology (Norwegian Centre of Excellence) at the University of Bergen, Norway and the Africa Earth Observatory Network (AEON) at the University of Cape Town, South Africa. Drillers in Training (South Africa) and Atlas Copco Exploration Products Africa are thanked for their technical support. This is an AEON contribution number 58.

References

Brandl, G., Cloete, M., and Anhaeusser, C.R., 2006. Archaean greenstone belts. *In* Johnson, M.R., Anhaeusser, C.R., and Thomas, R.J. (Eds.), *The Geology of South Africa*, Johannesburg and Pretoria (Geological Society of South Africa, and Council for Geosciences), 9–56.

Fliegel, D., McLoughlin, N., Simonetti, A., de Wit, M., and Furnes, H., 2008. *In situ* dating of the oldest morphological traces of life on Earth. *Eos Trans. AGU*, 89(53) Fall Meet. Supplement, Abstract P51A-1395.

Furnes, H., Banerjee, N.R., Muehlenbachs, K., Staudigel, H., and de Wit, M., 2004 Early life recorded in Archean pillow lavas. *Science*, 304:578–581, doi:10.1126/science.1095858.

Kareem, K., 2005. Komatiites of the Weltevreden Formation, Barberton Greenstone Belt, South Africa: implications for the chemistry and temperature of the Archean mantle. PhD dissertation, The Department of Geology and Geophysics of the Louisiana State University, 227 pp.

McLoughlin, N., Brasier, M.D., Wacey, D., Green, O.R., and Perry, R.S., 2007. On biogenicity criteria for endolithic microborings on early Earth and beyond. *Astrobiology*, 7(1):10–11, doi: 10-26. doi:10.1089/ast.2006.0122.

Schoene, B., Bowring, S.A., and de Wit, M.J., 2008. Assembly and stabilization of the eastern kaapvaal craton: a structural-thermochronological perspective. *Tectonics*, 27:TC5010, doi:10.1029/2008TC002267, 2008 (27 pgs).

Viljoen, M.J., and Viljoen, R.P., 1969. The geology and geochemistry of the lower ultramafic unit of the Onverwacht Group and a proposed new class of igneous rocks. *Geol. Soc. S. Afr., Sp. Pub.*, 2:55–86.

Authors

Eugene G. Grosch and Maarten de Wit, Africa Earth Observatory Network (AEON) and the Department of Geological Sciences, University of Cape Town, Rondebosch 7700, South Africa, e-mail: geogene@gmail.com.
Nicola McLoughlin and Harald Furnes, Department of Earth Science and Centre for Geobiology, University of Bergen, Allegaten, 41, N-5007 Bergen, Norway.

Related Web Links

http://www.aeon.uct.ac.za/
http://www.uib.no/geobio/en/
http://www.uib.no/geobio/en/public-outreach/field-activities

Figure and Photo Credits

Fig. 1, 3: Eugene G. Grosch (AEON, Cape Town)
Fig. 2: Photo by Harald Furnes (CGB, Bergen)
Fig. 4: Photo by Eugene G. Grosch (AEON, Cape Town)
Fig. 5. Photomicrograph by Nicola McLoughlin (CGB, Bergen)

17

The Laguna Potrok Aike Scientific Drilling Project PASADO (ICDP Expedition 5022)

by Bernd Zolitschka, Flavio Anselmetti, Daniel Ariztegui, Hugo Corbella, Pierre Francus, Christian Ohlendorf, Frank Schäbitz and the PASADO Scientific Drilling Team

Introduction

In the Southern Hemisphere long, continuous, and high-resolution series of terrestrial paleoclimatic data are scarce, and they are only slowly emerging. Globally speaking, the most extreme oceanic character is encountered between 40°S and 60°S (Fig. 1). In this latitudinal belt ninety-eight percent of water is juxtaposed to only two percent of land—Patagonia and a few sub-Antarctic islands. Therefore, records from Patagonia are a key to a better evaluation of inter-hemispheric linkages and differences in the climate system, especially as the Southern Ocean plays a key role for a proper understanding of the global climate system (Kaiser et al., 2007). Moreover, this region—close to the Andean volcanic chain—is one of the source regions for southern hemispheric dust. It is also subject to shifts in polar to mid-latitude pressure fields and precipitation regimes related to the Southern Hemispheric Westerlies and the Antarctic Oscillation. Patagonia thus potentially provides unique terrestrial records of variations in (1) climate, (2) hydrology, (3) erosion and deposition of atmospheric dust, and (4) volcanic activity. Additionally, links can be established to ice cores from Antarctica and to marine records from the South Atlantic where dust and tephra of Patagonian provenance have been deposited (Ackert, 2009; Narcisi et al., 2005; Sugden et al., 2009). For southernmost South America most lake sediments extend in time not beyond the Late-Glacial. However, Laguna Potrok Aike (Fig. 1) is older than the numerous Patagonian glacial lakes and offers the opportunity for volcanological studies.

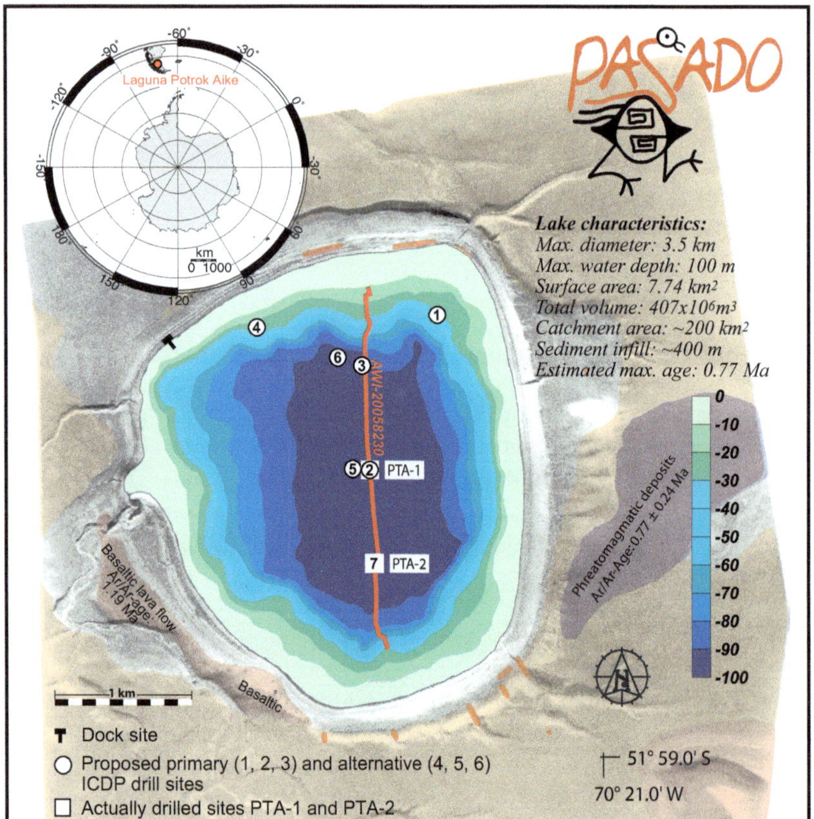

Figure 1. Bathymetric map of Laguna Potrok Aike merged with an aerial photograph and geological data (Corbella, 2002; Zolitschka et al., 2006). Inset map shows the site location in sub-Antarctic South America. Color-coded young fluvial and lacustrine deposits (gray), a mid-Pleistocene basalt lava flow (red), phreatomagmatic tephra deposits (pink), moraine till (yellow), Tertiary Santa Cruz Formation (orange), multi-channel seismic line of Fig. 2 (red line), and planned (numbered white circles) and drilled (numbered and labeled white squares) coring sites.

Laguna Potrok Aike developed in the Pliocene to late Quaternary Pali Aike Volcanic Field (Zolitschka et al., 2006). Located 100 km east of the Andean volcanic arc, this northwest-southeast oriented back arc volcanic belt is about 50 km wide and more than 150 km long. The oldest outcropping geological strata in the catchment area of Laguna Potrok Aike are Lower Miocene fine-grained fluvial sediments of the Tertiary Santa Cruz Formation (Blisniuk et al., 2005). During the Plio-Pleistocene (3.5–1.0 Ma) glaciers advanced northward from the Magellan Strait and covered the area now occupied by Laguna Potrok Aike. Thus, the surface of its catchment area today is dominated by fluvioglacial sediments and basal moraine tills of the most extensive glaciation (Coronato et al., 2004) with occasional exposures of basalts and volcanic spatter cones. The youngest glaciation that influenced the catchment area ended around 0.76 Ma (Singer et al., 2004) and left behind terminal moraines in Chile located ~20 km south of Laguna Potrok Aike. Around 0.77 Ma, a phreatomagmatic explosion was triggered, creating a maar which developed into the lake of Laguna Potrok Aike (113 m a.s.l.). The lake is 100 m deep with a maximum diameter of 3.5 km (Fig. 1), and it has a surface area of 7.74 km^2 (Zolitschka et al., 2006). The bathymetry is simple and pot-shaped, typical for a maar lake (Figs. 1, 2). The crater depression itself, only partly occupied by the lake, has a larger diameter of 5 km. Because this basin has not been scoured by young glaciations, the crater lake of Laguna Potrok Aike is an ideal site for the recovery of long paleoenvironmental and paleoclimatic records. The lacustrine sedimentary infill potentially covers several glacial to interglacial cycles up to present times; it is underlain by coarse volcaniclastic rocks deposited immediately after the phreatomagmatic maar eruptions ceased.

The maar lake, with currently episodic tributaries, must be regarded as a subsaline terminal lake with a pH of 8.8 and a salinity of 2.2‰ due to the prevailing semiarid climate in the dry steppe environment of southernmost South America (Zolitschka et al., 2006). However, geomorphologic evidence suggests that overflow occurred northward in the past. Despite its great depth and large water volume, the lake has not developed a persistent seasonal stratification today. In response to the strong westerly winds, the lake is currently polymictic.

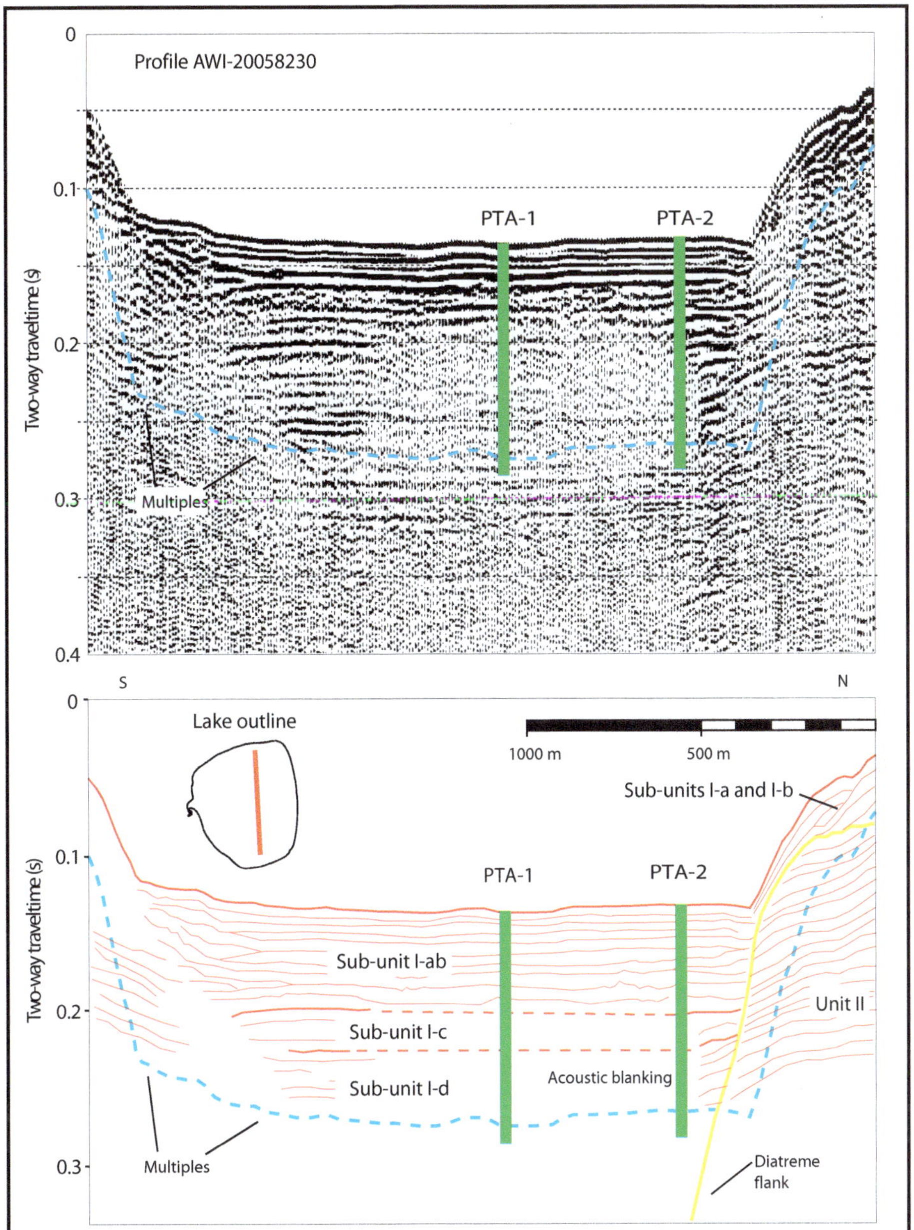

Figure 2. S-N multi-channel seismic reflection profile from Laguna Potrok Aike with uninterpreted seismic section (top) and interpreted line drawing (bottom). Drill sites PTA-1 and PTA-2 are marked. Inset (top left) shows outline of Laguna Potrok Aike with multi-channel seismic line (red) from Fig. 1.

Rainfall within the catchment area of Laguna Potrok Aike is highly variable and is mainly related to northward and southward shifts of the Southern Hemispheric Westerlies. A west wind belt over Southern Patagonia causes drier conditions (lee effect of the Andes), while a northward migration of the Westerlies allows rain-bringing easterly winds to introduce precipitation from the Atlantic Ocean (Haberzettl et al., 2007b; Mayr et al., 2007). Therefore, a paleorecord from this site has a large potential to act as a cornerstone for paleodata—climate model intercomparison (Meyer and Wagner, 2008; Wagner et al., 2007). It makes the water budget extremely susceptible to the regional precipitation-evaporation balance. In addition, a multitude of surficial and subaqueous lake level terraces can be found. A continuous 16,000-year record with high temporal resolution and a low-resolution 53,000-year record exist for Laguna Potrok Aike (Anselmetti et al., 2009; Haberzettl et al., 2007a, 2008; Mayr et al., 2007, 2009; Wille et al., 2007).

Objectives, Project Planning and Drilling Operations

The recent discovery of an exceptionally thick (~400 m) sediment sequence for Laguna Potrok Aike (Anselmetti et al., 2009; Gebhardt et al., 2009) has sparked an international research effort directed towards an interlinked approach that addresses two key objectives: (1) the evolution of maar craters and (2) quantitative climatic and environmental reconstruction.

Recovery of more than 400 m of sediments from a water depth of 100 m requires rotary drilling. The goal of the PASADO (Potrok Aike maar lake Sediment Archive Drilling prOject) deep drilling project was to recover triplicate cores from two sites and one core from a third site with inclined drilling (sites 1–6 in Fig. 1). The anticipated target depth ranged from 50 m to 600 m blf (below lake floor). Drill sites 2 and 5 (Fig. 1) had highest priority where seismic data indicated the potential for a penetration down to at least 400 m blf within the center of the basin, where hiatuses related to lake level fluctuations are less likely to exist (Fig. 2). Penetration was also planned into the volcaniclastic deposits below the 400 m or more of lake sediments. The second drilling target is located on the circum-lacustrine lake level terrace at 35 m water depth (sites 1 and 4 in Fig. 1). An erosional hiatus was detected there (Haberzettl et al., 2008). It is anticipated that this discontinuous record and the record from the lake center would jointly provide information about hydrological variations. The inclined drilling site is located near the northern end of the lake at 90 m water depth (sites 3 and 6 in Fig. 1) where a deviated hole was intended to be drilled through the crater wall and into basement rocks to investigate the transition zone formed during maar eruptions.

Drilling operations took place from September 2008 to November 2008. Prior to drilling, many logistical and technical challenges had to be overcome. The main obstacle was the lack of harbor facilities as well as the inaccessibility of the lake shore for a crane. Thus, the drilling platform had to be launched in a unique manner with a specially designed carriage called the "Launcher". This device made it possible to assemble the platform, including the drill rig and other heavy parts of equipment, on land and then lower the assembled barge downhill across the lake level terraces into a water depth sufficient to reach flotation. From the floating platform, drilling was carried out with the GLAD800 coring system operated by the Consortium for Drilling, Observation and Sampling of the Earth's Continental Crust (DOSECC) (Fig. 3). However, due to weather-related and technical downtime, only one of the primary sites (site 2, PTA-1) was drilled in quadruplicate and one additional site in the deep center of the lake (site 7, PTA-2) was drilled in triplicate. These sustained an average core recovery of 92.1% (PTA-1) and 98.8% (PTA-2) (Table 1; Fig. 1). A total of 533 m of cores were obtained reaching a maximum depth of 101.5 m blf at the deepest hole (PTA-2C). Almost all coring was done with the Hydraulic Piston Corer (HPC), and one core run was drilled with the Extended Nose Corer (EXN). Rotary drilling was not applied because sediments at 100 m blf were still rather soft. The planned comprehensive downhole logging program combined with seismic experiments could not be conducted, due to the fact that all holes had to be abandoned in emergency because of unseasonably strong westerly winds.

Figure 3. Platform *R/V Kerry Kelts* with the GLAD800 drilling system (center).

In the field laboratory all cores were logged with the multi-sensor core logger (MSCL) provided by the ICDP. Magnetic susceptibility was obtained in 2-cm intervals for all cores. GRAPE (gamma ray attenuation porosity evaluation) density was not operational because we were not allowed to import the gamma ray source into Argentina. P-wave velocity measurements did not provide any signal, probably because the coupling between inner liner wall and sediment was imperfect. All core catcher samples were studied in the field laboratory for their physical and chemical properties including pH, Cl^-, Ca^{2+}, electrical conductivity, water content, and dry density. Additionally, an initial lithologic description was carried out consisting of digital photography and macroscopic as well as microscopic description,(using smear slides for the latter). In addition, biotic activities related to early diagenetic processes were investigated as follows. This "deep biosphere" research tracks microbial activity through depth (Amend and Teske, 2005). On-site analyses included the detection of living organisms using ATP (adenosine triphosphate) measurements accompanied by sampling under the most sterile conditions possible. Immediate chemical fixation of the samples secured the best preservation of natural conditions for further laboratory studies. Sediment cores were then stored in a refrigerated container shipped to the GEOPOLAR Lake Core Repository at the University of Bremen where core splitting, scanning, and subsampling takes place.

Table 1. Location of drill sites, water and penetration depths as well as core recovery rate

Site	Latitude (Deg.S)	Longtitude (Deg.W)	Water Depth (m)	Penetration Depth (m blf)				Average Recovery
				Hole A	Hole B	Hole C	Hole D	
PTA-1	51.96403	70.37595	98	88.1	71	62.3	100.4	92.1
PTA-2	51.97053	70.37551	95	88.5	21.1	101.5		98.8

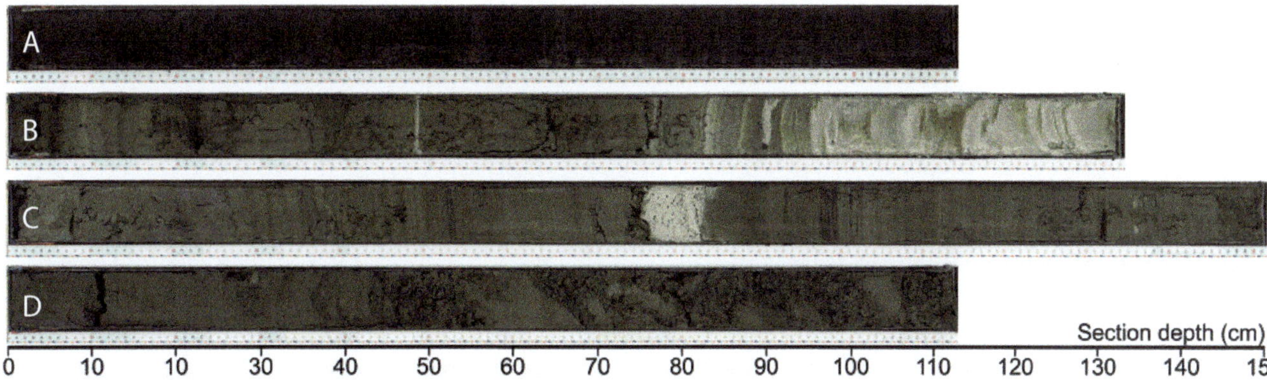

Figure 4. Selected split core sections from site PTA-2 (5022-2A). From top to bottom these represent [A] Holocene dark gray, laminated and carbonaceous silts starting from 6.39 m blf; [B] clastic and volcaniclastic, greenish-gray laminated and carbonaceous silts from 13.52 m blf; [C] clastic and volcaniclastic, greenish-gray laminated silts from 28.54 m blf; [D] clastic, greenish-gray sand and gravel from 82.52 m blf.

Preliminary Results

During the ICDP field work we recovered the longest environmental and climatic record (so far) for South America south of the tropics. However, lacustrine deposits were not penetrated completely, and volcaniclastic sediments were not reached. The total catch of cores was only 29% of what was proposed; thus, one of the two objectives of PASADO—the study of the basal volcaniclastic sediments—was not addressed.

All cored holes were correlated by applying MSCL (magnetic susceptibility core logger) data. The longest (101.5 m) lacustrine record consists mainly of lacustrine mud. Despite this dominance, grain sizes vary considerably from layers of almost pure clay and sandy sections (especially below 50 m blf) to gravel beds (Fig. 4). Some of the latter are related to volcanic ash layers consisting of sands with coarser scoria components. Surprisingly enough, rounded gravels were also recovered from the center of Laguna Potrok Aike. This can only be explained by either drastic lake level lowering until fluvial activity reached the coring site or seasonal lake freezing during the last glacial with snowmelt-related slush flows deposited onto the ice cover and coarse particles relocated as drop stones. We favor the latter interpretation, because lake level lowering should cause precipitation of evaporates like they presently occur at Laguna Maar Bismarck, a modern analogue for such conditions in the Pali Aike Volcanic Field.

There is a clear Holocene signal in the core catcher data (Fig. 5). However, the presence of older interglacial (Eemian; Oxygen Isotope Stage = OIS 5.5) is missing within the recovered sections. This finding is consistent with linear extrapolation of the age model developed for piston cores of the last 16 ka (Haberzettl et al., 2007b), and it suggests an estimated age of ~85 ka for the lowermost sediment recovered. Preliminary microfossil analyses of core catcher samples show a clear difference between Holocene and glacial conditions in the upper part of the profile, and reveal increases in diatom frequency and pollen of shrub taxa for the lowermost part of the profile. Based on these data, the obtained sediment record may extend back to OIS 5.1.

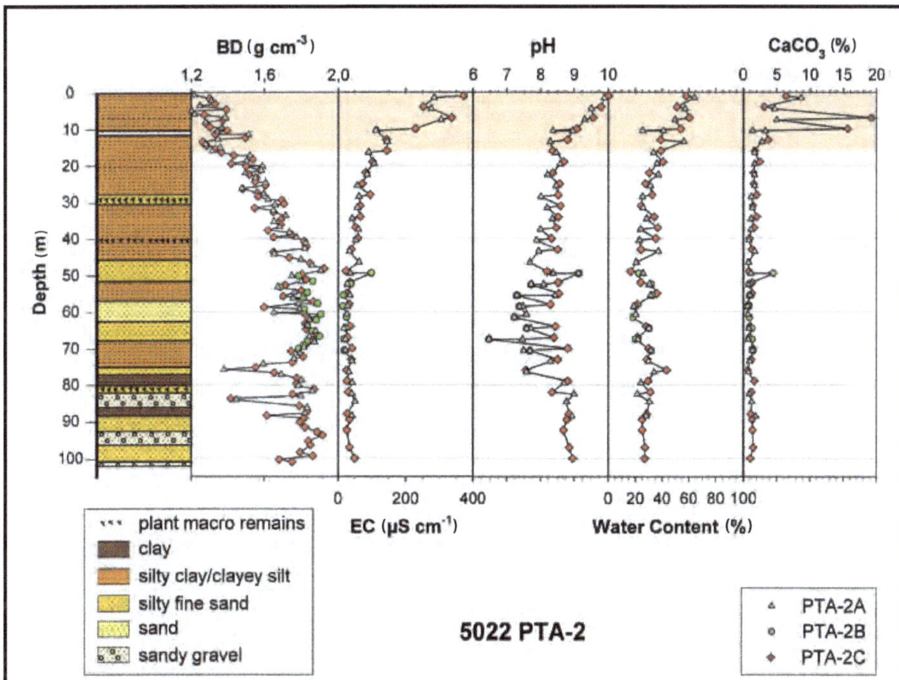

Figure 5. Lithostratigraphy and records of bulk density (BD), electrical conductivity (EC), pH, water content, and carbonate content ($CaCO_3$) based on core catcher samples from site PTA-2 (combined holes A, B, and C) as obtained in the field laboratory. The Holocene section of the record is shaded in an orange hue (top).

Summary

The 770-ka-old maar lake of Laguna Potrok Aike was the target of the international deep drilling project PASADO, carried out in the framework of the ICDP from September to November 2008. Although the project could not recover the entire lacustrine sediment infill and volcaniclastics due to unusually windy conditions and technical problems, PASADO drilled seven holes at two sites in the central deep lake basin to a maximum depth of 101.5 m blf. A total of 533 m of sediment cores was gathered with a mean recovery rate of 95.5%. Currently, sediments of this longest Patagonian lake record to date are subsampled and analyzed. First detailed scientific results are expected for the year 2010.

Acknowledgements

We are grateful to Beau Marshall and the DOSECC drilling crew as well as to Capitan Jorge Daniel Moreteau and his son Javier for their excellent technical skills. Without them the PASADO field work would not have been possible. We also thank the members of Instituto Nacional de Tecnología Agropecuaria (INTA) Santa Cruz, in particular Guillermo Clifton and Gabriel Oliva, for providing logistical support and the ground for the field camp at their experimental field station Potrok Aike. Many thanks go as well to Cristobal Zimmermann of the German Embassy in Buenos Aires. Without his steady support many customs affairs would have been unsolvable for us.

Preparatory studies were supported by the German Federal Ministry of Education and Research (BMBF) in the framework of the German Climate Research Program (DEKLIM and SALSA) and through the projects POTROK and ASADO within the Priority Program "ICDP" of the German Science Foundation (DFG).

This research is supported and partially funded by the ICDP. Co-funding for drilling as well as for currently ongoing scientific investigations is provided by the German Science Foundation (DFG), the Swiss National Science Foundation (SNF), the Natural Sciences and Engineering Research Council of Canada (NSERC), the Swedish Vetenskapsradet (VR), and the University of Bremen.

The PASADO (ICDP Expedition 5022) Scientific Drilling Party

B. Zolitschka (Co-Chief Scientist), C. Ohlendorf (Co-Chief Scientist), F. Anselmetti, D. Ariztegui, H. Corbella, S. Delpit, R. Farah, V. Förster, D. Fortin, P. Francus, C. Gebhardt, T. Grelle, T. Haberzettl, A. Hahn, S. Kastner, P. Kliem, O. Langkamp, A. Lisé-Pronovost, C. Mayr, F. Müller, D. Penshorn, C. Recasens, P.-S. Ross, F. Schäbitz, G. St-Onge, A. Stopp, D. Veres, and A. Vuillemin.

References

Ackert, R.P., Jr., 2009. Patagonian dust machine. *Nature Geosci.*, 2:244–245, doi:10.1038/ngeo485.

Amend J.P., and Teske A., 2005. Expanding frontiers in deep subsurface microbiology. *Palaeogeogr., Palaeoclimatol., Palaeoecol.*, 219:131-155, doi: 10.1016/j.palaeo.2004.10.018

Anselmetti, F.S., Ariztegui, D., De Batist, M., Gebhardt, A.C., Haberzettl, T., Niessen, F., Ohlendorf, C., and Zolitschka, B., 2009. Environmental history of southern Patagonia unravelled by the seismic stratigraphy of Laguna Potrok Aike. *Sedimentology*, 56:873–892, doi:10.1111/j.1365-3091.2008.01002.x.

Blisniuk, P., Stern, L., Chamberlain, C., Uldleman, B., and Zeitler, P., 2005. Climatic and ecological changes during Miocene surface uplift in the Southern Patagonian Andes. *Earth Planet. Sci. Lett.*, 230:125–142, doi:10.1016/j.epsl.2004.11.015.

Corbella, H., 2002. El campo volcano-tectónico de Pali Aike. *In* Haller, M. (Ed.), *Geología y Recursos Naturales de Santa Cruz*, Buenos Aires (Asociación Geológica Argentina), 285–302.

Coronato, A., Martinez, O., and Rabassa, J., 2004. Glaciations in Argentine Patagonia, southern South America. *In* Ehlers, J., and Gibbard, P.L. (Eds.), *Quaternary Glaciations - Extent and Chronology, Part III: South America, Asia, Africa, Australasia, Antarctica. Developments in Quaternary Science, Vol. 2c,* Amsterdam, The Netherlands (Elsevier), 49–67.

Gebhardt, A.C., De Batist, M., Niessen, F., Anselmetti, F.S., Ariztegui, D., Kopsch, C., Ohlendorf, C., and Zolitschka, B., 2009. Origin and evolution of the Laguna Potrok Aike maar, Southern Patagonia. *Basin Research*, in review.

Haberzettl, T., Corbella, H., Fey, M., Janssen, S., Lücke, A., Mayr, C., Ohlendorf, C., Schäbitz, F., Schleser, G.-H., Wessel, E., Wille, M., Wulf, S., and Zolitschka, B., 2007a. A continuous 16,000 year sediment record from Laguna Potrok Aike, southern Patagonia (Argentina): sedimentology, chronology, geochemistry. *The Holocene*, 17:297–310, doi:10.1177/0959683607076437.

Haberzettl, T., Kück, B., Wulf, S., Anselmetti, F., Ariztegui, D., Corbella, H., Fey, M., Janssen, S., Lücke, A., Mayr, C., Ohlendorf, C., Schäbitz, F., Schleser, G., Wille, M., and Zolitschka, B., 2008. Hydrological variability and explosive volcanic activity in southeastern Patagonia during Oxygen Isotope Stage 3 and the Holocene inferred from lake sediments of Laguna Potrok Aike, Argentina. *Palaeogeogr., Palaeoclim., Palaeoecol.*, 259:213–229, doi:10.1016/j.palaeo.2007.10.008.

Haberzettl, T., Mayr, C., Wille, M., and Zolitschka, B., 2007b. Linkages between southern hemisphere westerlies and hydrological changes in semi-arid Patagonia during the last 16,000 years. PAGES News, 15(2):22–23.

Kaiser, J., Lamy, F., Arz, H., and Hebbeln, D., 2007. Variability of sea-surface temperatures off Chile and the dynamics of the Patagonian Ice Sheet during the last glacial period based on ODP Site 1233. *Quaternary Intl.*, 161:77–89, doi:10.1016/j.quaint.2006.10.024.

Mayr, C., Lücke, A., Maidana, N.I., Wille, M., Haberzettl, T., Corbella, H., Ohlendorf, C., , Schäbitz, F., Fey, M., Janssen, S., and Zolitschka, B., 2009. Isotopic and geochemical fingerprints on lacustrine organic matter from Laguna Potrok Aike (southern Patagonia, Argentina) reflect environmental changes during the last 16,000 years. *J. Paleolimnol.*, 42:81–102, doi: 10.1007/s10933-008-9249-8.

Mayr, C., Wille, M., Haberzettl, T., Fey, M., Janssen, S., Lücke, A.,

Ohlendorf, C., Oliva, G., Schäbitz, F., Schleser, G.H., and Zolitschka, B., 2007. Holocene variability of the Southern Hemisphere westerlies in Argentinean Patagonia (52°S). Quat. Sci. Rev., 26:579–584, doi:10.1016/j.quascirev.2006.11.013.

Meyer, I., and Wagner, S., 2008. The Little Ice Age in southern Patagonia: comparison between paleoecological reconstructions and downscaled model output of a GCM simulation. *PAGES News*, 16:12–13.

Narcisi, B., Petit, J.R., Delmonte, B., Basile-Doelsch, I., and Maggi, V., 2005. Characteristics and sources of tephra layers in the EPICA-Dome C ice record (East Antarctica): implications for past atmospheric circulation and ice core stratigraphic correlations. *Earth Planet. Sci. Lett.*, 239:253–265, doi:10.1016/j.epsl.2005.09.005.

Singer, B.S., Ackert, R.P., and Guillou, H., 2004. $^{40}Ar/^{39}Ar$ and K-Ar chronology of Pleistocene glaciations in Patagonia. *Geol. Soc. Am. Bull.*, 116:434–450, doi:10.1130/B25177.1.

Sugden, D.E., McCulloch, R.D., Bory, A.J.-M., and Hein, A.S., 2009. Influence of Patagonian glaciers on Antarctic dust deposition during the last glacial period. *Nature Geosci.*, 2:281–285, doi:10.1038/ngeo474.

Wagner, S., Widmann, M., Jones, J., Haberzettl, T., Lücke, A., Mayr, C., Ohlendorf, C., Schäbitz, F., and Zolitschka, B., 2007. Transient simulations, empirical reconstructions and forcing mechanisms for the Mid-Holocene hydrological climate in Southern Patagonia. *Clim. Dyn.*, 29:333–355, doi:10.1007/s00382-007-0229-x.

Wille, M., Maidana, N.I., Schäbitz, F., Fey, M., Haberzettl, T., Janssen, S., Lücke, A., Mayr, C., Ohlendorf, C., Schleser, G.H., and Zolitschka, B., 2007. Vegetation and climate dynamics in southern South America: the microfossil record of Laguna Potrok Aike, Santa Cruz, Argentina. *Rev. Palaeobot. Palynol.*, 146:234–246, doi:10.1016/j.revpalbo.2007.05.001.

Zolitschka, B., Schäbitz, F., Lücke, A., Clifton, G., Corbella, H., Ercolano, B., Haberzettl, T., Maidana, N., Mayr, C., Ohlendorf, C., Oliva, G., Paez, M.M., Schleser, G.H., Soto, J., Tiberi, P., and Wille, M., 2006. Crater lakes of the Pali Aike Volcanic Field as key sites of paleoclimatic and paleoecological reconstructions in southern Patagonia, Argentina. *J. South Am. Earth Sci.*, 21:294–309, doi:10.1016/j.jsames.2006.04.001.

Authors

Bernd Zolitschka, Geomorphology and Polar Research (GEOPOLAR), Institute of Geography, University of Bremen, Celsiusstraße FVG-M, D-28359 Bremen, Germany, e-mail: zoli@uni-bremen.de.
Flavio Anselmetti, Swiss Federal Institute of Aquatic Science and Technology (Eawag), Department of Surface Waters, Überlandstrasse 133, CH-8600 Dübendorf, Switzerland.
Daniel Ariztegui, Section of Earth & Environmental Sciences, University of Geneva, Rue de Maraîchers 13, CH-1205 Geneva, Switzerland.
Hugo Corbella, Argentine Museum of Natural History, Av. Angel Gallardo 470, Buenos Aires, Argentina.
Pierre Francus, Institut National de la Recherche Scientifique, Centre Eau, Terre et Environnement, 490 rue de la couronne, Québec, QC G1K 9A9, Canada.
Christian Ohlendorf, Geomorphology and Polar Research (GEOPOLAR), Institute of Geography, University of Bremen, Celsiusstraße FVG-M, D-28359 Bremen, Germany.
Frank Schäbitz, Seminar for Geography and Education, University of Cologne, Gronewaldstr. 2, D-50931 Cologne, Germany.
and the PASADO (ICDP Expedition 5022) Science Team.

Related Web Links

http://www.pasado.uni-bremen.de
http://www.icdp-online.org/contenido/icdp/front_content.php?idart=2185

Figure and Photo Credits

Fig. 1: Corbella, 2002 and Zolitschka et al., 2006 (both with modifications); inserted map source: Online Map Creation (http://www.aquarius.geomar.de)
Fig. 2: Data by Gebhardt et al. 2009
Fig. 3: Photo by B. Zolitschka, University of Bremen
Fig. 4: Photo by C. Ohlendorf, University of Bremen

18

Bio-Archive Core Storage and Subsampling Procedure for Subseafloor Molecular Biological Research

by Noriaki Masui, Yuki Morono, and Fumio Inagaki

Introduction

The Integrated Ocean Drilling Program (IODP) provides unprecedented opportunities to study the deep subseafloor biosphere. Subseafloor microbes play important roles in biogeochemical cycling of carbon, nitrogen, sulfur, metals, and other elements on geologic timescales; however, their growth and metabolic characteristics remain largely unknown because most subseafloor microbes are phylogenetically distinct from known isolates and resistant to culturing in laboratories. Given the significance of this region as potential habitats of subseafloor life, cored materials (or portions thereof) should be frozen in long-term storage to provide opportunities for future molecular analyses arising from rapid biotechnological developments (D'Hondt et al., 2007). Here we report a semi-aseptic subsampling technique for frozen core samples using an electric saw system in a clean booth. This subsampling technique enables us to share and distribute samples from the same depth for multiple analyses, such as DNA and lipid biomarkers, without thawing the frozen core.

Problems with Current Storage Conditions

The activity of subseafloor microbial life is generally very slow (D'Hondt et al., 2002, 2004); however, it impacts biogeochemical cycling on Earth on geological timescales, and moreover it provides clues for understanding the origin and evolution of life. Microscopic observations as well as molecular (e.g., DNA) surveys demonstrate widespread microbial presence in the marine subsurface environment (Parkes et al., 1994, 2000; Inagaki et al., 2006; Lipp et al., 2008; Morono et al., 2009). Given the significance of cored materials for studying deep subseafloor life and the biosphere, legacy samples called "Bio-archives" are extremely useful for studying distributions of biomass, diversity, community composition, habitability, and metabolic activity. To achieve bio-archive storage and sample distribution, new sampling techniques are in high demand to minimize damage and contamination.

During the long history of scientific drilling, most cores have been stored at room temperature or around 4°C in open storage rooms. However, contaminating microbes from the air or scientists themselves easily grew under these storage conditions (Fig. 1) and thus caused critical damage to core quality. Oxygen penetration stimulates growth of contaminating aerobic bacteria as well as germination of buried fungal and bacterial spores. Since aerobic bacteria grow on organic matter in cored sediments, the dramatic change in conditions during storage easily kills subseafloor microbial life. After cell death, fragile bio-molecules such as RNA, enzymes, sugar chains, and intact polar lipids rapidly degrade via abiotic hydrolysis and enzymatic reactions, subsequently promoting the activity of contaminating microbes. In addition to preventing such degradation, freezing cores also prevents dissolution of carbonates during long-term storage and hence maintains the fine textures of carbonate fossils. Therefore, preserving frozen core materials under appropriate conditions is necessary for microbiological and biogeochemical studies.

Electric Band Saw System

During previous ODP/IODP expeditions, most microbiological samples were obtained as whole round cores (WRC) and stored at -80°C. One of the difficulties of frozen WRC handling is subsampling for molecular biological or biogeochemical analyses in shore-based laboratories. In the past, to collect subsamples from a frozen WRC, cores were thawed on a clean bench, then an innermost sample was collected using a heat-sterilized spatula or an autoclaved tip-cut syringe, and the rest of the sample was returned to the deep freezer for subsequent analyses. This freeze-thaw cycle enhanced damage of endolithic and sedimentary biomolecules and increased the probability of experimental contamination. Alternatively, frozen cores could be manually broken with a heat-sterilized chisel without the risk of melting; however, collecting the innermost part of the core that had avoided

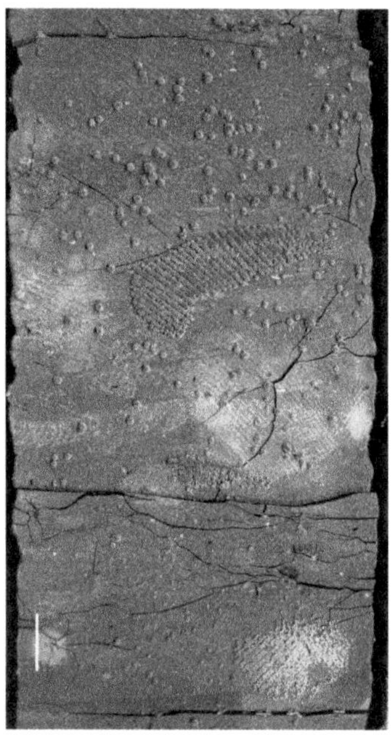

Figure 1. An example of contaminating microbial colonies growing on the core surface after long-term storage (ID: 190-1174B-63R-3W). Bar indicates 1 cm.

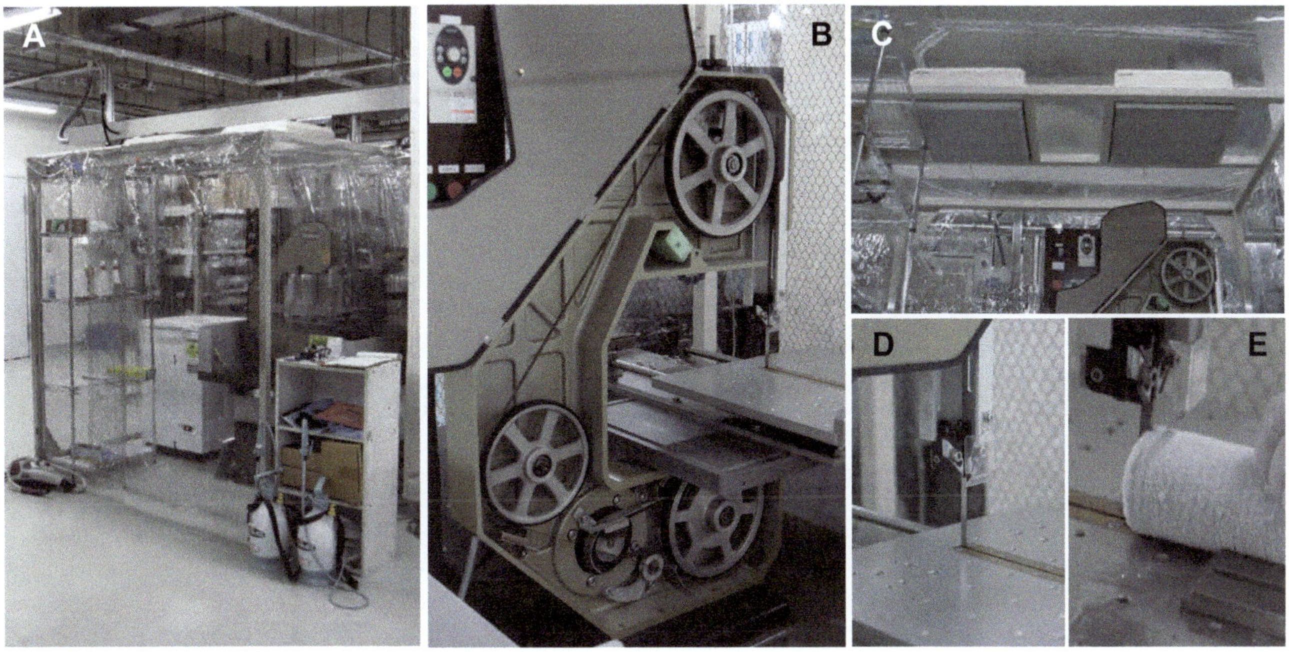

Figure 2. [A]Overview of the electric band saw system for microbiological frozen core. Motor-driven electric band saw machine [B] and -50°C freezer is deployed in a clean booth equipped with two HEPA-filter units [C]. Diamond-tipped band saw cuts frozen core samples under clean conditions [D, E].

potential drilling fluid contamination was difficult.

For better analytical use of frozen WRC and other types of frozen geological materials such as rocks and sediments, we constructed a semi-aseptic electric band saw system. A diamond powder-etched band saw machine (RYOWA Co. Ltd., Japan) and a -50°C freezer for temporary core storage were set up in a clean booth equipped with two HEPA-filter units (Model ATML-2-E252, NIPPON MUKI Co. Ltd., Japan) (Fig. 2). The electric saw smoothly cuts frozen WRCs under clean air in a short time without melting the core. In standard applications, the electric band saw is cooled with water to prevent frictional heat buildup; however, this step is not necessary for frozen cores, which are even better for maintaining aseptic conditions. Cutting the frozen core also eases multiple rounds of subsampling without disturbing the stratigraphic conditions. To prevent cross-contamination, it is necessary to clean the saw's parts, such as the saw blade and its wheels, using a steam-heat cleaner (Fig. 3) followed by an ethanol spray and a Milli-Q water rinse.

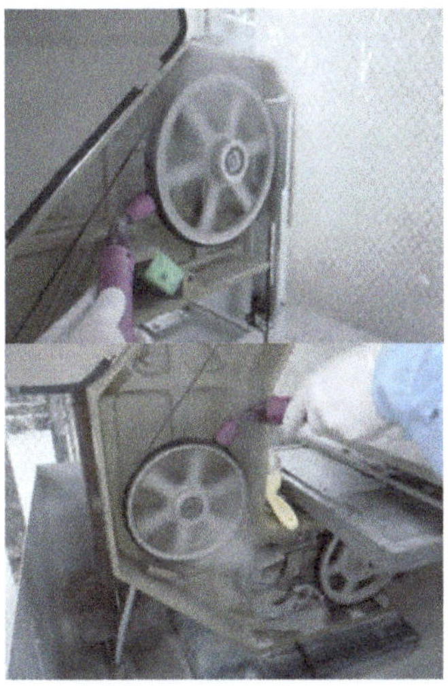

Figure 3. Cleaning run. Band saw and three wheels must first be cleaned with heat-steam followed by an ethanol spray-wash and a Milli-Q rinse.

Example of Frozen Whole Round Core Subsampling Possibilities

The electric saw system described herein enables quick subsampling of frozen cores into various shapes. For examples, the frozen WRC is first cut in half vertically, and one of the halves is cut into a plate shape (Fig. 4). Then the plate can be cut into bars (5 mm x 10 mm). A 6.5-cm-diameter core, which is the usual ODP/IODP core size, yields forty-six of these subsampled bars. As many of the frozen bar samples as needed can be used for molecular analyses. For experiments requiring larger volumes, half of the WRC may be used. Overall, this protocol is flexible with respect to subsample volume, and it eases sharing the exact same horizon among multiple analyses.

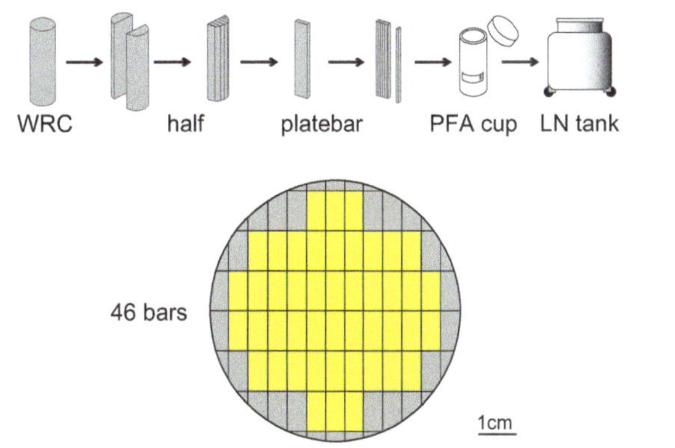

Figure 4. An example of frozen whole round core (WRC) subsampling. The WRC is first cut in half lengthwise, then one half is sectioned into plates, and each plate is cut into bars. Cut core samples are put in PFA cups and stored in a liquid nitrogen tank at -160°C (or deep freezer at -80°C).

Storage of Frozen Bioarchives

For optimal preservation, the initial freezing step is important. Flash freezing with liquid nitrogen (LN) is better than normal (slow) freezing. Cores should then ideally be subsampled with the electric saw system and stored either in a LN tank or -80°C deep freezer until examined. For effective preservation of RNA, the sample should be put into a commercial RNA-stabilizing solution (e.g., RNAlater®, Ambion), the supernatant removed after several hours' incubation at 4°C, and then the sample put into the deep freezer. RNAlater®-ICE (Ambion) may be useful for recovering RNA from frozen samples during thawing. Sample preservation in acetone may be an alternative method to preserve RNA (Fukatsu, 1999).

For long-term storage we are currently performing trials using a LN tank (Fig. 3), whose (gas phase) temperature is approximately -160°C (T liquid phase = -196°C) . It is well recognized that lower temperatures are better for long-term storage of fragile biomolecules, and an oxygen-free nitrogen atmosphere is suitable for preserving oxygen-sensitive enzymes and anaerobic microbes. For samples stored in a LN tank, a cup made of PFA (perfluoroalkoxylalkane) is the best choice despite its price, because PFA withstands temperatures ranging from -196°C to 200°C and hence is also autoclavable. PFA cups are also useful for storing methane hydrate samples; liquid nitrogen can be poured directly on the samples in a cup. These options should be practiced onboard in order to preserve biological samples most effectively.

Future Developments

Several possibilities are conceivable for future uses of bioarchived cores. First, if this electric saw system can be deployed on drilling platforms, frozen samples at the same horizons can be distributed among researchers, and the rest can be returned for sedimentological and paleontological studies or standard archiving. This process will minimize sample volume and avoid sample location differences among multiple analyses; however, we need to modify the system for ease of use, smaller sample sizes, and reduced costs. The cleaning step must be modified to save time and to simplify its execution. In addition, we are currently testing an improved mini-core sampling system using an autoclavable metal tube with a saw tip, which allows onboard subsampling not only of frozen samples but also of hard rocks.

Acknowledgements

The authors thank K. Aoike (JAMSTEC-CDES) and RYOWA Co. Ltd. for useful suggestions on the electric saw system, and T. Terada (Marine Works Japan, Ltd.) for technical assistance. The photographic data was provided by the ODP/IODP.

References

D'Hondt, S., Inagaki, F., Ferdelman, T., Jørgensen, B.B., Kato, K., Kemp, P., Sobecky, P., Sogin, M., and Takai, K., 2007. Workshop Report: Exploring subseafloor life with the Integrated Ocean Drilling Program. *Sci. Drilling*, 5:26–37.

D'Hondt S., Jørgensen, B.B., Miller, D.J., Batzke, A., Blake, R., Cragg, B.A., Cypionka, H., Dickens, G.R., Ferdelman, T., Hinrichs, K.U., Holm, N.G., Mitterer, R., Spivack, A., Wang, G.Z., Bekins, B., Engelen, B., Ford, K., Gettemy, G., Rutherford, S.D., Sass, H., Skilbeck, C.G., Aiello, I.W., Guerin, G., House, C.H., Inagaki, F., Meister, P., Naehr, T., Niitsuma, S., Parkes, R.J., Schippers, A., Smith, D.C., Teske, A., Wiegel, J., Padilla, C.N., and Acosta, J.L.S., 2004. Distributions of microbial activities in deep subseafloor sediments. *Science*, 306:2216–2221, doi:10.1126/science.1101155.

D'Hondt, S., Rutherford, S., and Spivack, A.J., 2000. Metabolic activity of subsurface life in deep-sea sediments. *Science*, 295:2067–2070, doi:10.1126/science.1064878.

Fukatsu, T., 1999. Acetone preservation: a practical technique for molecular analysis. *Mol. Ecol.*, 8:1935–1945, doi:10.1046/j.1365-294x.1999.00795.x.

Inagaki, F., Nunoura, T., Nakagawa, S., Teske, A., Lever, M., Lauer, A., Suzuki, M., Takai, K., Delwiche, M., Colwell, F.S., Nealson, K.H., Horikoshi, K., D'Hondt, S., Jørgensen, B.B., 2006. Biogeographical distribution and diversity of microbes in methane hydrate-bearing deep marine sediments on the Pacific Ocean Margin. *Proc. Natl. Acad. Sci. U.S.A.*, 103:2815–2820, doi:10.1073/pnas.0511033103.

Lipp, J.S., Morono, Y., Inagaki, F., and Hinrichs, K.-U., 2008. Significant contribution of Archaea to extant biomass in marine subsurface sediments. *Nature*, 454:991-994, doi:10.1038/nature07174.

Morono, Y., Terada, T., Masui, N., and Inagaki, F., 2009. Discriminative detection and enumeration of microbial life in marine sediments. *ISME J.*, 3 :503–511, doi:10.1038/ismej.2009.1.

Parkes, R.J., Bragg, B.A., and Wellsbury, P., 2000. Recent studies on bacterial populations and processes in subseafloor sediments: A review. *Hydrogeol. J.*, 8:11–28, doi:10.1007/PL00010971.

Parkes, R.J., Cragg, B.A., Bale, S.J., Getliff, J.M., Goodman, K., Rochelle, P.A., Fry, J.C., Weightman, A.J., and Harvey, S.M., 1994. Deep bacterial biosphere in Pacific Ocean sediments. *Nature* (London, U. K.), 371:410–413, doi:10.1038/371410a0.

Authors

Noriaki Masui, Yuki Morono, and Fumio Inagaki, Geomicrobiology Group, Kochi Institute for Core Sample Research, Japan Agency for Marine-Earth Science and Technology (JAMSTEC), Monobe B200, Nankoku, Kochi 783-8502, Japan, e-mail: inagaki@jamstec.go.jp.

19

"Core on deck!" The End of SODV and the Return of the *JOIDES Resolution* as the IODP Riserless Vessel

by Peggy Delaney and Sean Higgins

In February 2009, after a 3-year operational hiatus in order to undergo a major retrofit, the *JOIDES Resolution* (or simply the *JR*, as it has been called since 1985) reverberated with the trademark phrase, "Core on deck!". For the crew and technical staff aboard during the *JR*'s sea trials cruise in January–February 2009, it brought shouts of joy and tears of relief from years of rebuilding the *JR* in a Singapore shipyard, and a realization that we were "really" going back to carrying out international expeditions. But what does the phrase mean to marine geoscientists around the world? For most scientists, it means that one of the only tools to explore sediments and rocks in the deep sub-seafloor throughout the world's oceans is finally available again. This unique research facility is our community's equivalent to the Hubble telescope and provides the access we need to the deep ocean. Accumulating an impressive array of accomplishments, the *JR* (Fig. 1) has served as the sole scientific drilling vessel for the Ocean Drilling Program (ODP, 1985–2003) and for the initial phase of riserless drilling for the Integrated Ocean Drilling Program (IODP, 2003–2006). Over 120 expeditions since 1985 have retrieved samples encompassing a latitudinal range from 80.5°N to 70.8°S, a water depth range from 85 m to nearly 6000 m, sub-bottom depths as deep as 2.1 km, and a total depth range (water depth plus penetration) over 6.9 km.

As part of the United States' contribution to IODP, the National Science Foundation—through Geosciences Directorate and through its Major Research Equipment and Facilities Construction (MREFC) Account—provided more than $100 million in funding for the conversion and upgrade of the *JR* after it was selected in a public competition. This project, known by the acronym "SODV" (Scientific Ocean Drilling Vessel), has been an extended journey for those involved in bringing the *JR* back to life, unlike any project any of us could have imagined when it started out in the spring of 2006. Following on the heels of Hurricane Katrina and a worldwide boom in oil prices that led to the biggest shipbuilding boom in oil industry history, rebuilding the *JR* could not have faced a more complex set of obstacles. Delivery of SODV came in January 2009, more than a year later than we had planned and only after near-heroic efforts from both the Transocean and Overseas Drilling Limited ship's crew and technical staff from the U.S. Implementing Organization (USIO) (Texas A & M, Lamont Doherty Earth Observatory's Borehole Research Group and the Consortium for Ocean Leadership). The management of all these organizations also played a critical role in pushing for and accomplishing the SODV project.

Figure 1. The pre-modernized *JOIDES Resolution*, an ~143-m-long, dynamically positioned drill ship, with a derrick extending >61 m above the water line, began operations in 1978 as the *SEDCO/BP 471* and was converted in 1985 for use by the Ocean Drilling Program (ODP). The derrick height and overall size allow this vessel to be able to access a wide variety of ports, including passing under the Bridge of the Americas (Panama) and the Golden Gate Bridge (San Francisco, California). The name came from the JOIDES (Joint Oceanographic Institutions for Deep Earth Sampling) advisory structure and after the *HMS Resolution*, the vessel of exploration commanded by Captain James Cook. By the end of ODP and during initial IODP operations, the *JOIDES Resolution* (*JR*) included a fully equipped lab stack, the scientific heart of expeditions, equivalent in its fitting and capabilities to shore-based research laboratories.

The SODV project has its roots in a decade-long process of imagining the next many years of scientific ocean drilling

Figure 2. The Core Deck labs are but a portion of the all-new and larger lab facilities. The scientific equipment and systems in the labs were newly acquired, upgraded, or refurbished, based on experience gained in past expeditions and consultation with the international scientific advisory structure of IODP.

and translating this vision into a reality that guided the teams responsible for the conversion of a riserless drilling vessel for use by IODP. Recollecting the events, this all started in the Conceptual Design Committee (CDC) Report (2000) or the international Conference on Multi-Platform Exploration of the Ocean (COMPLEX) report in 1999 and, finally, the summarized requirements for a new riserless vessel in the USIO's Briefing Book (2005). This conversion project was guided by several overarching priorities (e.g. to provide more reliable riserless drilling vessel; to meet or exceed modern health and safety requirements; and science, sampling, and logging capabilities along with operational efficiency) and by a structure built to ensure science community input at multiple levels. One of the areas of greatest importance is preserving and enhancing the drilling, coring, sampling, and logging capabilities within the financial and physical limitations of the program. The project was focused on reliable, cost- and time-efficient drilling rig operations. Central goals were core quality and core recovery, including the ability to make holes with good progress in difficult lithologies where great depths must be achieved, and the ability to deploy the widest variety of downhole logging instruments and sampling tools.

In March 2009, the refurbished *JR* arrived in Honolulu harbor on an unusually crisp morning after six weeks of sea trials that took the ship on a >6000-mile odyssey from an emotional departure in Singapore to Guam to the Ontong Java Plateau and finally to Hawaii. After the *JR* just gained access to the science laboratory spaces in late November 2008, installation of over sixty instruments, an entire IT infrastructure, and more than 100 new software applications was completed, thus readying the *JR* (or nearly so) for its first science party since the end of 2005 (Fig. 2). This expedition (320T) was an intense testing period by both USIO staff and community participants, the Readiness Assessment Team (RAT) who produced an important baseline readiness report that documented current lab status and priority items that still needed attention. But, perhaps more importantly, the ship itself, the life supporting vessel systems, its critical drilling systems, newly refurbished derrick and passive heave systems, and new downhole logging system were also tested. After the ship had been guttee from the derrick forward, the *JR* was indeed alive and kicking. All conversion and capabilities of the *JR* are listed at http://www.oceanleadership.org/programs-and-partnerships/ship-conversion/ and some examples are given below:

Examples of ship conversion:

- 34% increase in laboratory spaces (includes offices and conference space)
- Redesigned laboratory layout for better core handling and workflow
- Streamlined accessibility between laboratories and accommodations areas
- Enclosed, sound controlled core splitting and sampling room
- Improved core flow and integration of science activities on the Core Deck
- Refrigerated core storage for >8000 m of core
- New ergonomic core description stations and software
- New and enhanced laboratories
- Refurbished sub-sea camera system
- Improved core handling capability
- New and expanded data servers and storage systems
- Refined and expanded berthing
- Updated safety and environmental systems

As of the start of Expedition 320 in March 2009 (the first of two expeditions focused on studying Cenozoic climate history of the equatorial Pacific), we no longer have to imagine what it might be like on the riserless scientific ocean drilling ship for the IODP in 2009 and beyond. Based on the many tours given in the first two port calls, scientists wondering if the *JR* would really be different have seen the remarkable transformation of the ship, from the ship's new core (Fig.2), chemistry/biology, and downhole labs, to the light-filled galley on the main dec, to the all new integrated living quarters and the impressive new bridge (http://iodp.tamu.edu/labs/ship.html).

Ultimately, the true test of the *JR* is its ability to obtain quality cores and geophysical logs in a variety of environments. Expeditions 320 and 321 immediately put the drilling systems, pipe handling, and coring systems to work in over 5 km of water in the central Pacific Ocean. The refurbished passive heave system, new wireline heave compensator, and a new rig instrumentation system to capture details of ship's drilling performance will all be put to the test in our upcoming expeditions (see back cover). The *JR* has been retrofitted with modern technology and laboratories. In the coming years as it sails the world's oceans, it will again try to live up to its heritage (named after Captain Cook's *HMS Resolution*). The scientific progress that IODP is able to achieve with the *JR* is now up to a new generation of marine scientists.

Hawaiian blessing of ship in May 2009 port call in Honolulu. From left to right: Kahu Kordel Kekoa (Chaplain for Kamehameha Schools Kapâlama), Arden Bement (Director of NSF), Bob Gagosian (President, Consortium for Ocean Leadership), Brian Taylor (Dean of SOEST, University of Hawaii), Bjorn Kjerfve (Dean of Geosciences, Texas A&M Unviersity.

Authors

Peggy Delaney, Professor of Ocean Sciences, Ocean Sciences Department/Institute of Marine Sciences, 1156 High Street, University of California, Santa Cruz, Calif., U.S.A., e-mail: Delaney@ucsc.edu.

Sean Higgins, Associate Director, Ocean Drilling Programs, Consortium for Ocean Leadership, U.S. Implementing Organization, Washington, DC, U.S.A., e-mail: shiggins@oceanleadership.org

Related Web Links

http://www.iodp-usio.org/About/default.html
http://iodp.tamu.edu/labs/ship.html
Ship Conversion and Capabilities: http://www.oceanleadership.org/programs-and-partnerships/ship-conversion/ship-conversion-and-capabilities/

20

The U-tube: A New Paradigm for Borehole Fluid Sampling

by Barry Freifeld

Introduction

Fluid samples from deep boreholes can provide insights into subsurface physical, chemical, and biological conditions. Recovery of intact, minimally altered aliquots of subsurface fluids is required for analysis of aqueous chemistry, isotopic composition, and dissolved gases, and for microbial community characterization. Unfortunately, for many reasons, collecting geofluids poses a number of challenges, from formation contamination by drilling to maintaining integrity during recovery from depths. Not only are there substantial engineering issues in retrieval of a representative sample, but there is often the practical reality that fluid sampling is just one of many activities planned for deep boreholes. The U-tube geochemical sampling system presents a new paradigm for deep borehole fluid sampling. Because the system is small, its ability to integrate with other measurement systems and technologies opens up numerous possibilities for multifunctional integrated wellbore completions. To date, the U-tube has been successfully deployed at four different field sites, each with a different deployment modality, at depths from 260 m to 2 km. While the U-tube has proven to be highly versatile, these installations have resulted in data that provide additional insights for improving future U-tube deployments.

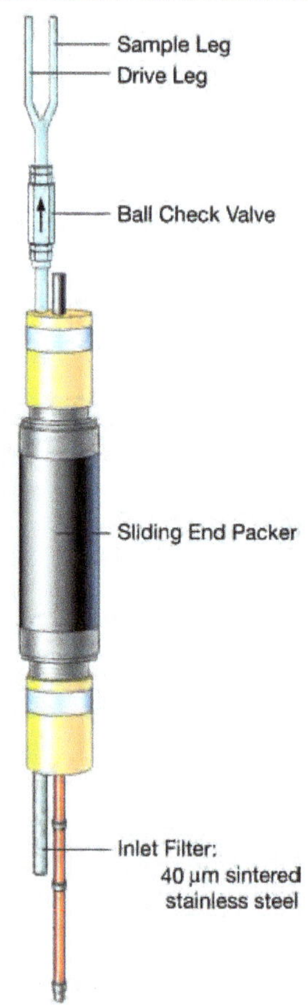

Figure 1. Schematic layout of the U-tube sampling system as deployed at the Frio Site, Dayton, Texas.

The U-Tube

The U-tube sampler is a simple positive fluid displacement pump which uses a high pressure gas drive. Wood (1973) demonstrated this methodology for shallow vadoze zone sampling using a porous cup inlet. Figure 1 shows a schematic of the U-tube sampler as configured for installation beneath a wellbore packer. The heart of the U-tube is the ball check-valve, which contains the only moving part of the sampling system in the borehole. A loop of tubing that terminates at the surface forms the "U". The check-valve is located beneath a "tee" at the base of the U and it permits fluid to enter the loop but closes (by application of gas from the surface) when pressure in the U is increased above hydrostatic. A sintered stainless steel filter terminates the inlet beneath the check-valve to prevent the check-valve from plugging.

To collect a sample, the U is first filled by venting the sample and drive legs to the atmosphere, thus allowing fluid to rise to the formation hydrostatic level. The sample is recovered by supplying high-pressure N_2 (or another inert gas) to the drive leg, closing the check-valve, and forcing fluid out of the sample leg. While conceptually the U-tube is straightforward, the ability to independently control pressures on both the drive and sample legs presents numerous options for collecting and processing of fluids at the surface. By repeating sample collection cycles, the borehole fluid can be flushed and fresh fluid pulled in from the formation. The volume of the sample depends on the formation pressure as well as the diameter and wall thickness of the tubing used, but for an installation at a depth of 1 km, a 6 mm or 9 mm diameter tube with a 1.2 mm wall thickness has a U-tube volume of 20 L or 68 L, respectively.

The Frio Brine Pilot—CO_2 Storage and Multiphase Fluids

The Frio Brine Pilot was the first deep borehole application in which I deployed a positive displacement pump incorporating small diameter tubing and a check-valve. The Frio Brine Pilot, Dayton, Texas, was a demonstration program for the geosequestration of CO_2 in saline aquifers led by the Texas Bureau of Economic Geology (Hovorka et al., 2006). In September 2004, 1600 t of CO_2 were injected into the Frio Sandstone at a depth of 1.5 km over a span of ten days. A second experiment conducted in 2006 at the same site consisted of an injection of 350 t of CO_2 in sand at a depth of 1.6 km. The objectives of the Frio Brine Pilot tests included understanding the hydrological and geochemical impact of the CO_2 on the subsurface and evaluation of various monitoring tools that would inform us as to the shape and distribution of the CO_2 plume. In addition to surface and cross-hole seismic monitoring, we used fluid sampling from an observation well placed 30 m up-dip from an injector well to determine changes in the geochemistry as the plume swept past, and interpret the arrival of added tracers.

The constraints on the fluid sampling program were many. We had to do the following:

- Minimize impact on the flow-field
- Permit wireline logging operations
- Coexist with other downhole sensors, which included a tubing deployed seismic source, a string of hydrophones, and a pressure/temperature sensor
- Conduct frequent (hourly) sampling
- Allow for on-site real-time analysis of pH, alkalinity, and gas composition
- Enable functioning in a water-only, gas/water, or gas-only system
- Accurately determine the ratio of supercritical CO_2 to formation brine in the sampled fluid

The standard sampling technologies that were considered and rejected included electrical submersible pumps (unable to operate in two-phase systems), wire-line deployed flow-through samplers (unable to operate at high frequency during a 24-hour-a-day, multi-week experiment), and gas lifting (it would be difficult to control the flow rate, and the large volumes produced would disturb the flow-field. It would also be difficult to interpret downhole gas chemistry because of the large volume of lifting gas.).

The Frio Brine Pilot U-tube design consisted of a loop of 3/8" × 0.049" wall 316 L stainless steel tubing, which at a depth of 1.5 km contains a volume of 118 L. Four 13 L high pressure stainless steel cylinders were used to store the sampled fluid on the surface for analysis. The U-tube control lines and other instruments were strapped onto production tubing using cable protectors (Cannon Services Ltd., Stafford, Texas, U.S.A.). To isolate the perforated interval in the $5\frac{1}{2}$" casing, a pneumatic packer with feedthroughs to accommodate various lines was located at the base of $2\frac{7}{8}$" production tubing. From a geochemical sampling standpoint, it would have been best to locate the bottom of the packer just above the top perforation, minimizing the volume below the packer. However, to accommodate logging operations, the packer was set 14.3 m above the top perforation. The inlet port to the U-tube—a sintered stainless cylinder (60 cm long × 19 mm diameter, 40 µm pore size; Mott Corporation, Farmington, Conn., U.S.A.)—was located 30 cm below the bottom of the packer. The volume of the cased interval between the bottom of the packer and the top of the perforations was 177 L, with the 6.1 m perforated interval having an additional 75.4 L volume. Every two cycles of the U-tube would purge a volume equal to the dead-space between the top of the perforated interval and the U-tube inlet.

In contrast to the simplicity of the downhole U-tube system, the surface processing of the recovered fluids was quite complex. The U-tube surface completion consisted of six different manifold assemblies, allowing for careful control of sampling conditions and complete system purging between samplings. To facilitate reproducible sampling and minimize the potential for operator error, the sampling operation was automated with pneumatically operated valves. Control software was developed using the graphical programming language Labview (National Instruments, Austin, Texas, U.S.A.). Figure 2A shows the four 13 L sample vessels along with associated valving and gauges, with strain gauges mounted at the base of each cylinder to measure the bulk

Figure 2. The Frio Brine Pilot surface sampling systems: [A] Four 13 L high pressure sample vessels mounted in sliding sleeves and resting on strain gauges to measure fluid density. [B] Computer-operated valve manifolds used to operate the U-tube samplers.

density of the sampled fluid. Figure 2B shows the bank of automated valves needed to operate the various manifolds. Lifting the water from the sampling interval to the ground surface via the U-tube and collecting and analyzing the samples involved an eight-step process described in detail in Freifeld et al. (2005). In short, the process consists of (1) filling the U-tube with borehole fluid by venting the drive and sample legs to the atmosphere; (2) purging the gas sample manifold, sample vessels, and liquid sample manifold with N_2 gas to remove any residual fluid from these systems prior to sampling; (3) bleeding down the N_2 to atmospheric pressure inside all of the process tubing and sample vessels; (4) evacuating the N_2 from the sample cylinders; and (5) charging the U-tube drive leg with N_2 to close the downhole check-valve and force fluid to the surface (steps 6–8 described below). The first 20–25 L of fluid produced from the sample leg was discarded because it was considered potentially phase-segregated. At the end of Step 5, bulk samples were collected from the sample leg in small, transportable high-pressure cylinders for later laboratory analysis, as well as in open containers for immediate on-site measurement of pH and alkalinity. Step (6) consisted of directing the sample fluid from the sample line into the evacuated high-pressure cylinders until a pressure gauge indicated that the cylinders were at formation pressure. The final two steps consisted of (7) directing a gas sample from the top of the sample cylinders to a quadrupole mass spectrometer (Omnistar, Pfeiffer Vacuum Technology, Nashua, N.H., U.S.A.) and (8) N_2 purging the remaining fluid from the U-tube, so that it could repeat the cycle for the next sample.

Some problems arose during the operation.

1. Hydrate formation and freezing occurred at some of the vent values because of Joule-Thompson cooling.
2. The large volumes of gases and high moisture and salt content bypassed various collection systems and disabled rotary vane and oil-based vacuum pumps.
3. The carefully developed automated sampling routines had to be modified as gas contents increased and system behavior changed.

In the end, it was learned that leaving the sample vessels filled with N_2 at one atmosphere both eliminated the complex evacuation cycle and facilitated the accurate estimation of dissolved gas concentration, by acting as a known reference standard (Freifeld and Trautz, 2006). The combination of limiting the pressure drop across vent valves and using heat lamps and heat trace tape stopped the formation of blockages in the system.

Despite considerable difficulty in keeping the complex system operating, forty-two aqueous samples were acquired over the first fifty-two hours of sampling, and rapid changes in geochemical conditions were observed as the CO_2 plume traversed the monitoring well (Kharaka et al., 2006). Figure 3 shows the gas composition for CO_2, O_2, Ar, and CH_4 measured

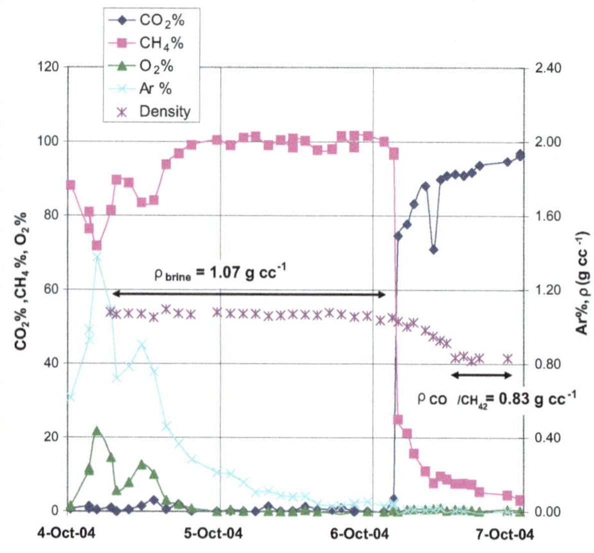

Figure 3. Frio Brine Pilot geofluid sample gas composition and density. Note that the presence of oxygen is indicative of atmospherically contaminated well-workover fluids.

on the quadrupole spectrometer, as well as sample fluid density. The presence of O_2 is an indicator of contamination with workover fluids that had equilibrated with air, and is not naturally found in the highly reducing environment of the Frio brine. Eventually, the sample streams changed from liquid to predominantly supercritical CO_2, and the self-lifting U-tube no longer needed compressed N_2 to drive the sample to the surface. While the system was self-lifting, the sample cylinders (based upon their known volume and measured pressure) were used to quantify the gas withdrawn from the ground and to estimate the time at which the gas entered the U-tube from the formation. The tracer travel times from the injection well to the U-tube sampler were used to estimate the formation CO_2 saturation. A geochemical interpretation of the rock-brine-CO_2 interaction has been presented by Kharaka et al. (2006), and a discussion on the quadrupole mass spectrometer methodology can be found in Freifeld and Trautz (2006).

Further Borehole Tests

A borehole in the Amargosa Valley (Yucca Mountain, Nevada) was drilled to a total depth of 400 m using a dual-wall reverse air rotary process to minimize subsurface contamination. The well was fitted with four U-tubes, two installed for permanent (redundant) use at two separate locations (264 mbgs and 348 mbgs), along with a distributed thermal perturbation sensor (DTPS) (Freifeld et al., 2008b). Sand and bentonite backfill materials were installed after lowering the instrument string in the open borehole strapped onto a 2" stainless steel piezometer (Fig. 4). Compressed gas cylinders of N_2 were used to operate the U-tubes. The initial produced fluid showed significant contamination with suspended bentonite, but after repeated operation, the fluid turbidity dropped. A single outlet valve on the N_2 cylinder

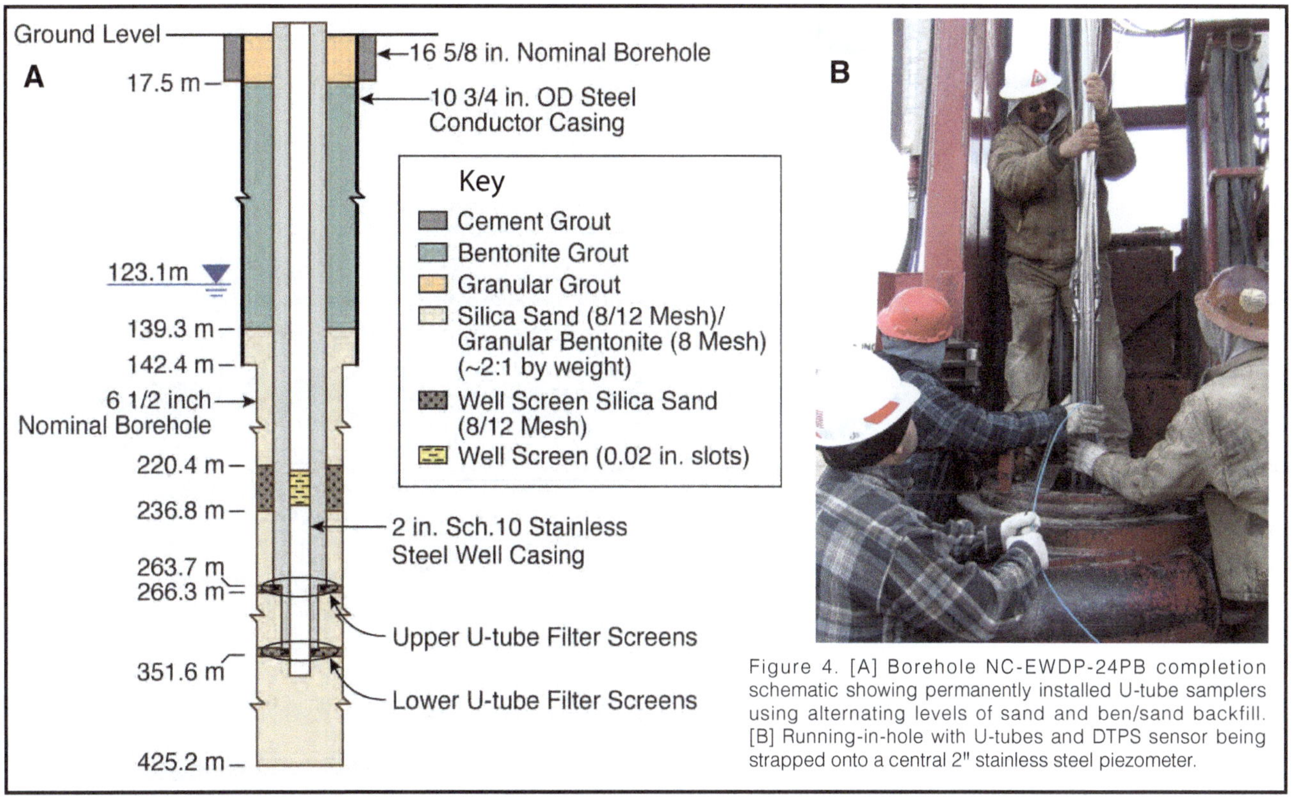

Figure 4. [A] Borehole NC-EWDP-24PB completion schematic showing permanently installed U-tube samplers using alternating levels of sand and ben/sand backfill. [B] Running-in-hole with U-tubes and DTPS sensor being strapped onto a central 2" stainless steel piezometer.

Figure 5. Multi-crane lift deployment of 32 m-long bottomhole assembly consisting of U-tubes, seismic sensors, and pressure temperature sensors at the CO2CRC Otway Project Site, Victoria, Australia.

pressure regulator controlled the flow of gas to the U-tube drive leg, and a separate valve was on the sample outlet.

The CO2CRC Otway Project (SW of Melbourne, Australia) is a program to demonstrate the storage of CO_2 in depleted natural gas formations (Sharma et al., 2007). The project drilled a well 300 m downdip of a depleted gas well to inject up to 100,000 t of CO_2 into the gas-bearing formation and monitor the previous gas well to assess geophysical and geochemical changes induced by injected gas.

The small-diameter casing (3½") posed the largest challenge to deploying a network of multilevel fluid sampling instruments. To instrument the well, a bottomhole-assembly (BHA) consisting of three U-tubes, two pressure-temperature sensors, and a string of seismic sensors was designed and fabricated. The entire BHA was lifted and installed as one 32 m-long assembly (Fig. 5) and lowered 2 km using conventional oil field sucker rod. The uppermost U-tube was installed near the top of the gas cap, while U-tube 2 and U-tube 3 were installed 1.5 m and 6 m below the gas-water contact, respectively. At the top of the BHA was a custom dual-element, double fixed-end packer. All three U-tubes have been functioning since their initial deployment in October 2007, with weekly sampling operations ongoing. The CO_2 injection caused the gas-water contact to be pushed down, and U-tube 2 and U-tube 3 have transitioned from sampling liquids to self-lifting and sampling predominantly gas. One significant hurdle overcome in the operation of the U-tubes was the unexpected presence of natural waxy alkanes in the fluids being produced (Boreham et al., 2008). As the wax traveled up the U-tube sample lines and cooled, it coated the U-tube lines, as evidenced by reduced flows, and eventually completely occluded the U-tube 1 sample line. The initial plug was dislodged using a high-pressure piston pump, and a solvent composed primarily of dialkyl- and trialkylbenzenes was used to eliminate future wax build up.

At the High Lake massive sulfide deposit in Nunuvut Territory, Canada, an existing boring was deepened for the purpose of characterizing sub-permafrost microbial communities. A borehole observatory consisting of a U-tube and pressure/temperature sensor was installed beneath a pneumatic packer. The borehole observatory was lowered by hand to a depth of 350 m, using the stainless steel U-tube and packer inflation lines as the strength member. By monitoring the pressure as the sample reservoir chamber filled, we estimated the formation hydraulic conductivity to be 2×10^{-11} m s^{-1} (Freifeld et al., 2008b). Despite heating of the borehole at 20 W m^{-1}, the presence of heavily mineralized zones (pyrite > 50%) along with temperatures of about -6°C led to the eventual freezing of fluid in the sample tubes. Seven samples were collected prior to the blockage of the tubes. This past summer a U-tube deployment below permafrost near Kangerlussuaq, Greenland used an insulating jacket around the sampling and heater lines to sucessfully prevent freezing while sampling.

Summary and Future Plans

The U-tube was successful in recovering fluid samples under diverse conditions from depths between 260 m and 2 km. In all four field deployments presented, the same downhole sampling equipment was used for fluid recovery, despite wide variations in deployment modality. Future plans for U-tube systems include deployments at CO_2 storage demonstration sites near Cranfield, Mississippi, at a depth of 3.1 km, along with an installation near Cholla, Arizona. Given the small, unobtrusive size of U-tubes, many other applications are possible, including in extreme environments such as in geothermal fields or subglacial lakes. The ability for U-tubes to provide minimally altered fluids and to be integrated with other measuring devices makes the U-tube an ideal choice for recovery of geofluids.

Acknowledgments

The field programs were supported by numerous government, industrial, and academic institutions, along with large diverse scientific teams. I would like to thank Susan Hovorka, Robert Trautz, and Paul Cook for making the initial deployment of the U-tube system possible. Funding for Berkeley Lab was provided by the U.S. Department of Energy under Contract No. DE-AC02-05CH11231.

References

Boreham, C.J., Chen, J., and Hong, Z., 2008. Baseline study on sub-surface petroleum occurrences at the CO2CRC, Otway Project, western Victoria. Poster presented at PESA Eastern Australasian Basins Symposium III, Sydney, 14–17 September, 2008: 489–499.

Freifeld, B.M., and Trautz, R.C., 2006. Real-time quadrupole mass spectrometer analysis of gas in borehole fluid samples acquired using the U-tube sampling methodology. *Geofluids* 6(3):217–224. doi:10.1111/j.1468-8123.2006.00138.x.

Freifeld, B.M., Chan, E., Onstott, T.C., Pratt, L.M., Johnson, A., Stotler, R., Holden, B., Frape, S., Pfiffner, S.M., DiFurio, S., Ruskeeniemi, T., and Neill, I., 2008b. Deployment of a deep borehole observatory at the High Lake Project Site, Nunavut, Canada. *Proc. Ninth Intl. Conf. on Permafrost*, LBNL-78E, Fairbanks, Alaska, U.S.A., 29 June–3 July 2008.

Freifeld, B.M., Finsterle, S., Onstott, T.C., Toole P., and Pratt, L.M., 2008a. Ground surface temperature reconstructions: using *in situ* estimates for thermal conductivity acquired with a fiber-optic distributed thermal perturbation sensor, *Geophys. Res. Lett.*, 35:L14309, doi:10.1029/2008GL034762.

Freifeld, B.M., Trautz, R.C., Yousif, K.K., Phelps, T.J., Myer, L.R., Hovorka, S.D., and Collins, D., 2005. The U-Tube: A novel system for acquiring borehole fluid samples from a deep geologic CO_2 sequestration experiment. *J. Geophys. Res.*, 110:B10203, doi:10.1029/2005JB003735.

Hovorka, S.D., Benson, S.M., Doughty, C., Freifeld, B.M., Sakurai, S., Daley, T.M., Kharaka, Y.K., Holtz, M.H., Trautz, R.C., Nance, H.S., Myer, L.R., and Knauss, K.G., 2006. Measuring permanence of CO_2 storage in saline formations: the Frio experiment: *Environ. Geosci.*, 13:105–121 doi:10.1306/eg.11210505011.

Kharaka, Y.K., Cole D.R., Hovorka S.D., Gunter W.D., Knauss K.G., and Freifeld, B.M., 2006. Gas-water-rock interactions in Frio Formation following CO_2 injection: Implications for the storage of greenhouse gases in sedimentary basins. *Geology*: 34(7):577–580 doi:10.1130/G22357.1.

Sharma, S., Cook, P., Berly, T. and Anderson, C., 2007. Australia's first geosequestration demonstration project - the CO2CRC Otway Basin Pilot Project. *APPEA J.*, 47(1): 257–268.

Wood, W., 1973. A technique using porous cups for water sampling at any depth in the unsaturated zone. *Water Resour. Res.*, 9(2): 486–488, doi:10.1029/WR009i002p00486.

Author

Barry M. Freifeld, Earth Sciences Division, Lawrence Berkeley National Laboratory, MS 90-1116, One Cyclotron Road, Berkeley, Calif. 94720, U.S.A., e-mail: BMFreifeld@lbl.gov.

Acquiring High to Ultra-High Resolution Geological Records of Past Climate Change by Scientific Drilling

by Juergen Thurow, Larry C. Peterson, Ulrich Harms, David A. Hodell, Heather Cheshire, Hans-Juergen Brumsack, Tomohisa Irino, Michael Schulz, Valerie Masson-Delmotte, and Ryuji Tada

Introduction and Workshop Goals

Scientific drilling on land and sea has played a key role in advancing our knowledge of climate change. It has helped to demonstrate the effects of orbital variations on climate, revealed evidence for extreme warm events in the past and for the timing of Antarctic ice growth, and provided insights into the hydrologic balance of lake systems around the world. Now, with attention increasingly focused on the likely manifestation of future climate change, the challenge to understand past climates at societally relevant, high-resolution timescales has become ever more critical. Sediments and other archives that preserve climate information on timescales approaching those of instrumental records have much to offer to our understanding of how the climate system works (Fig. 1). These records, ideally with a sub-annual to centennial resolution, provide a unique opportunity to evaluate the global operation of the ocean-continent-atmosphere system on human timescales and to appraise the relative importance of each part of the system.

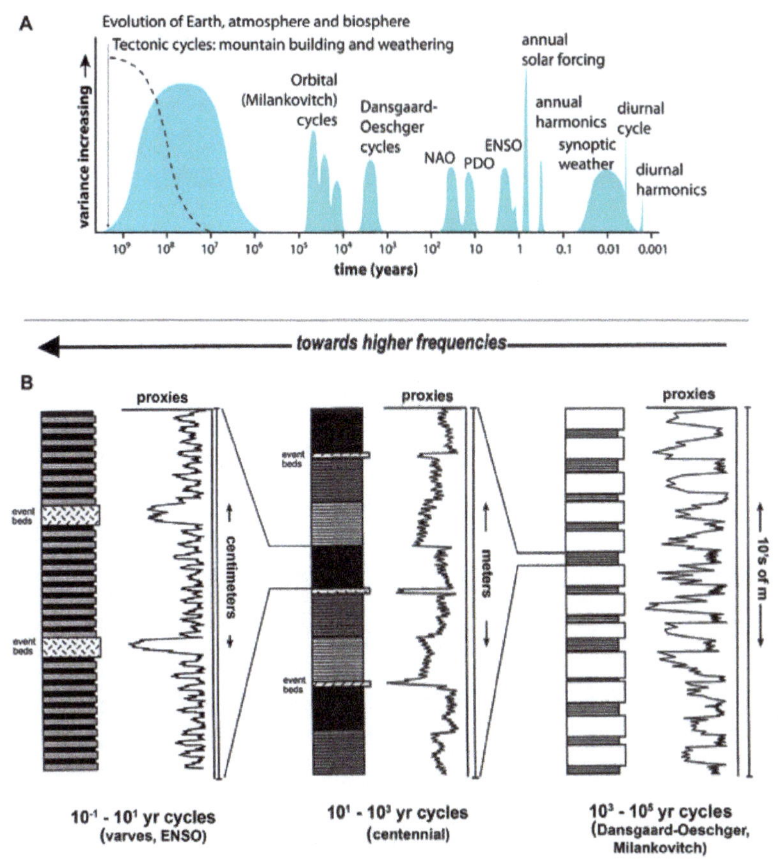

Figure 1. [A] The National Oceanic and Atmospheric Administration (NOAA) climate timeline http://www.ncdc.noaa.gov/paleo/ctl/ (the North Atlantic Oscillation (NAO), the Pacific Decadal Oscillation (PDO), El Niño-Southern Oscillation (ENSO)). The resolution required of sedimentary records depends on the timescale of climate variability. [B] Different orders of climate cycles and their expression in sedimentary archives paralleled by changes in proxies. Higher frequency cycles are superimposed on the lower frequency cycles. The ability to resolve high frequency cycles (e.g., annual to centennial) in the geologic record is generally limited by the nature of the archive (i.e., whether they are sediments, corals, ice cores, or speleothems), the recording rate of that archive (dictated by sedimentation, growth, or accumulation rate), and the potential for post-formation disturbance (e.g., bioturbation in sediments).

Long high-resolution records of excellent quality acquired through ocean drilling have contributed significantly to our current understanding of climate (Ocean Drilling Program (ODP) Site 893, Santa Barbara Basin, California [Behl and Kennett, 1996; Hendy and Kennett, 2000]; ODP Site 1002, Cariaco Basin [Peterson et al., 2000; Hughen et al., 2004]). Comparable-resolution records recovered from lake drilling (Hodell et al., 2008; Scholz et al., 2007) and analyses of speleothems (Wang et al., 2001) have significantly enhanced the climate "portfolio" (Fig. 2). However, the existing inventory of high-resolution records generated by scientific coring and drilling in the oceans and on land (including ice) is currently limited in both number and global coverage (Voelker, 2002; Clement and Peterson, 2008). The acquisition and analysis of many more detailed multi-proxy records of past environmental conditions from all areas of Earth are needed to allow researchers to better understand why and how the climate system responds rapidly to external and internal forcing and how the various high-frequency oscillations of the climate system interact over longer time intervals.

A group of sixty-four international scientists—experts in the fields of marine, terrestrial, ice-cores and Earth-system modeling—met at the German Research Centre for Geosciences (GFZ), Potsdam, from 29 September through 1 October 2008, with the objective of identifying key climate questions best addressable with high- to ultra-high resolution records and designing scientific drilling strategies to recover them. The workshop was initiated and funded by the Integrated Ocean Drilling Program (IODP) and the International Continental Drilling Program (ICDP). Its primary aim was to find and/or explore key areas with potentially high-resolution records, in order to coordinate and integrate existing proposals from within IODP/ICDP and from other scientific programs and to develop new drilling proposals incorporating the objectives outlined in this workshop summary. The ultimate goal was to identify a global array of potential coring sites spanning different time intervals to fully understand the causes and consequences of rapid environmental/climate change.

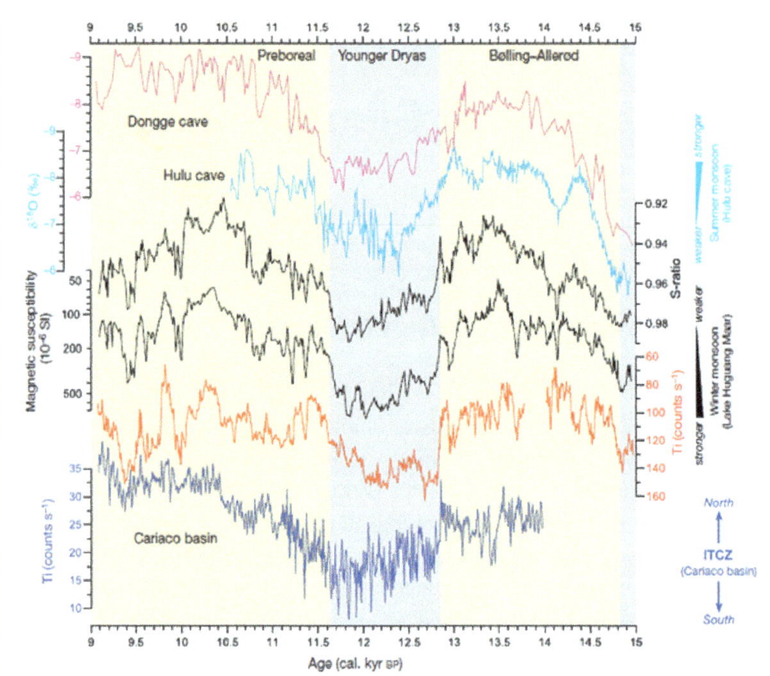

Figure 2. Comparison of selected high-resolution speleothem, lake, and marine records for the last deglaciation, with the well-known Younger Dryas cold event delineated by the gray bar (from Yancheva et al., 2007). Oxygen isotope records from Dongge and Hulu Cave stalagmites (China) are a proxy for rainfall and monsoon strength, as are the sediment proxy records (magnetic susceptibility, S-ratio, and titanium [Ti] counts) for Lake Huguang Maar in coastal China. Titanium abundance in sediments from Cariaco Basin, off northern Venezuela, is a proxy for regional rainfall and riverine delivery related to the changing position of the Intertropical Convergence Zone (ITCZ). These widely separated and hydrologically-sensitive records illustrate the potential for recovering high-resolution paleoclimate information from a variety of geologic archives.

The workshop opened with an invited series of keynote presentations designed to summarize our current understanding of climate change on sub-millennial timescales and to high-light recent advances and outstanding scientific issues. This was followed by talks that focused on the scientific missions, program structure, and proposal submission and review process for ICDP and IODP, and by short statements from all of the workshop participants about their research interests. A complete list of speakers and topics can be found on the workshop website (http://www.iodp.org/climate-ws-workshop/ Based on break-out group discussions, five overlapping climate themes emerged that are graphically represented by the petals of the flower diagram in Fig. 3. The interrelated petals are the main drivers of climate variability and, as such, form and inform the questions that can be addressed by high-resolution records. Within each of these themes, the nature and origin of abrupt climate change form a common thread.

Understanding the Causes and Consequences of Climate Variability

Climate Modes

High-resolution sedimentary archives from ocean and terrestrial systems provide a means of tracking climate variability throughout much of the globe at interannual to decadal to centennial/millennial timescales. They can be used to study past variability in the most important elements of the modern climate system, such as the El Niño-Southern Oscillation (ENSO), monsoons, the North Atlantic Oscillation (NAO), the Pacific Decadal Oscillation (PDO), the Arctic Oscillation (AO), and the Southern Annular Mode (SAM) (Fig. 1). Each of these modes or elements has been defined using instrumental climate and weather data, yet little is known about centers of action or centers of impact and about how these modes may have varied in the past under different boundary conditions.

Sedimentary archives can provide us not only with high fidelity records of past climate mode behavior—such as the millennial record of ENSO from its source region in the western Pacific (Tudhope et al., 2001)—but also with records of their teleconnected impacts around the globe (Wang et al., 2006). In this way, questions can be answered regarding the stationarity of ENSO-associated droughts or PDO-associated changes in North American temperature or snowfall (Mantua and Hare, 2002). Accordingly, the Potsdam Workshop identified as high priority for both IODP and ICDP the development of a series of climate mode reference sections

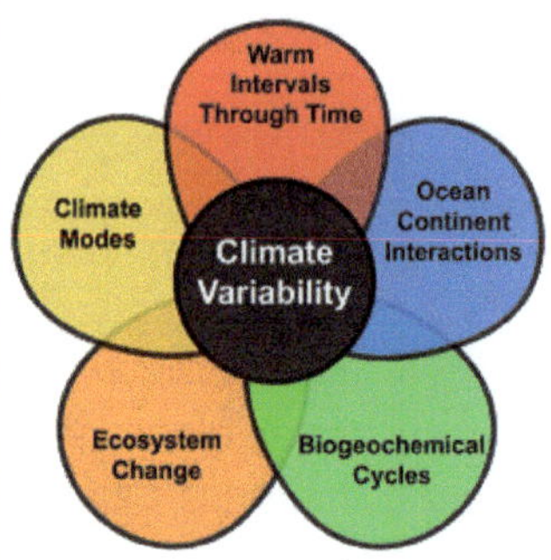

Figure 3. The five research themes shown here as "flower petals" emerged from workshop discussions as being central to the overarching goal of understanding the causes and consequences of climate change variability.

as well as response records from key parts of the globe. One suggestion was to focus on establishing the record of past variability in the global Walker Circulation (the zonal atmospheric overturning circulation that drives ENSO) as well as African and Asian monsoons. In the case of the Walker Circulation and ENSO, the acquisition of records that span the tropical Pacific from west (e.g., Papua New Guinea) to east (potentially Galapagos) should be a goal, as well as in regions that today show evidence of strong teleconnections, such as Ecuador, Peru, northeast Brazil, and South Africa. Combined with a similar latitudinal transect of high-resolution lake records from the western Americas, it should be possible to begin sorting out the relative forcing of ENSO as compared to such phenomena as latitudinal shifts in zonal circulation and changes in tropical convection.

During the last glacial period, the rapid temperature excursions over Greenland that have come to be known as Dansgaard-Oeschger events (Dansgaard et al., 1984, 1993) define a mode of climate variability that is critical to understand. The excursions appear to have a global imprint, though many regions are greatly undersampled (e.g., the Southern Hemisphere), and improved knowledge is needed of the spatial and temporal expression of these events. Their signature often seems to be manifested in variables other than temperature—for example, shifts in tropical rainfall patterns (Clement and Peterson, 2008). Abrupt, rapid climate change of this type, were it to occur in the future, could potentially put modern society under extreme stress. An urgent need exists to identify the 'symptoms'—the concatenation of events—that lead to the threshold of Earth's climate becoming markedly hotter or colder over the course of a decade or two. Paleoclimate records have revealed evidence of stressed ecosystems that are in disequilibrium during times of rapid climate change, with commonplace local extinctions. Future climate events have the potential to affect millions of people, most probably by disruption of the food chain or water supplies.

Geologic archives that preserve and can be sampled at high temporal resolution will help us answer questions about climate modes and abrupt change such as the following: What is the climate sensitivity to natural forcing, and how will anthropogenic actions influence this sensitivity? What changes might we expect in the Walker Circulation in the (warmer) years ahead? How variable can ENSO be? Is millennial variability present that shows ENSO patterns? What happened to ENSO during times of Pliocene warmth? Is there a real decadal mode in the Pacific? Is the PDO a distinct and separate element of the climate system? What is the role of the meridional overturning circulation (MOC) in the climate system on timescales ranging from decadal to millennial (Keenlyside et al., 2008; Stanford et al., 2006)? Is there a relation between the MOC and AO? Can sediment records be used to establish a definitive link between high latitude sea ice and the mean position of the Intertropical Convergence Zone (ITCZ) (Broecker, 2006)?

A number of strategies and data needs were identified to quantify conditions contributory to abrupt climate change in the late Quaternary. These include the development of proxies and drilling targets for reconstruction of sea-ice variability on various timescales; evaluating the sensitivity of ice sheets and sea level to past climate change to answer such questions as whether the West Antarctic Ice Sheet collapsed during previous interglacials; and examining whether inter-hemispheric phase-relationships known in the 100-kyr world were the same as earlier in the Pleistocene in the 41-kyr world.

Warm Intervals through Time

The most recent Intergovernmental Panel on Climate Change Report (IPCC, 2007) predicted that present and near-future global warming will produce boundary conditions very different from those of the last two million years; however, there is currently no clearly identifiable analogue for either the present interglacial or for anthropogenic climate warming. Models suggest that the Greenland Ice Sheet and perhaps portions of the West Antarctic Ice Sheet will begin to disappear as global climate warms (Gregory et al., 2004; Hu et al., 2009), but we don't really know if this happened previously within the warmest intervals of the late Quaternary. How sensitive were ice sheets (and sea level) to past climate change, especially during periods warmer than today (e.g., Marine Isotope Stages 5e, 11, 31)? What were the rates of change? Did the West Antarctic Ice Sheet collapse during previous interglacials?

Interglacial periods can be regarded as anomalies within the climate norm of the 100-kyr glacial cycles, which display the classic sawtooth pattern of abrupt beginnings that gradually subside towards the background norm (Fig. 4). Each interglacial is unique, and as such none can be used as a perfect analogue for the continued evolution of the present interglacial or for future climate change. One possible strategy is to compare all late Quaternary interglacials to generate 'an ideal interglacial' and a model for end-member states (Ruddiman, 2005). Re-examination of extant material and drilling a dedicated array of widely spaced high-resolution records are useful means of achieving this. Since oceans are buffers of climate change and lakes are amplifiers, transects of both marine and terrestrial records are prerequisites for achieving a complete picture of the patterns and magnitude of interglacial warmth.

Though it is natural to focus on recent interglacials because of the similarities in non-anthropogenic forcing, much can be learned from the study of older but more extreme warm intervals in Earth's history. Can extreme warm events like those of the mid-Cretaceous (Wilson and Norris, 2001) and the Paleocene-Eocene Thermal Maximum

(PETM; Zachos et al., 2005; Weijers et al., 2007) be used to make predictive models that are of value today? Were other forcing mechanisms that are not known today operative in those events? These extreme departures from 'background' climate should be manifested in high-resolution records of the hydrologic cycle (Kraus and Riggins, 2007), atmos-pheric and oceanic circulation, and vegetation dynamics on land. A concerted and coordinated effort to examine time-equivalent studies over the widest possible geographical range is one approach to globally quantifying the precursor conditions and effects of these extreme warm events themselves.

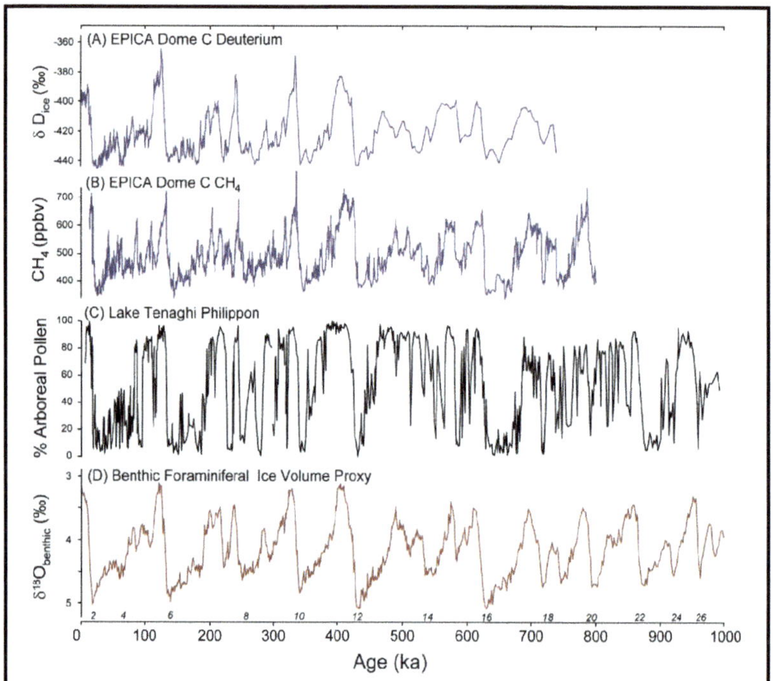

Figure 4. A one million year comparison of ice core and marine core data to a terrestrial record of pollen variability from Lake Tenaghi Philippon, Greece. (A) δD temperature proxy from Epica Dome C ice core, Antarctica (EPICA Community Members, 2004); (B); EPICA Dome C record of atmospheric methane concentrations (Loulergue et al., 2008); (C) percentages of arboreal pollen in sediments from Lake Tenaghi Philippon, Greece (Tzedakis et al., 2006); (D) composite benthic foraminiferal $\delta^{18}O$ proxy for global ice volume variations, with glacial marine isotope stages labelled as 2 to 26 (Liesecki and Raymo, 2005). Note that the pollen record shows clear evidence of orbital and sub-orbital scale vegetation responses to Quaternary climate change and illustrates the potential for recovering long, continuous sediment sequences in lacustrine settings.

Ocean-Continent Interactions

IODP and ICDP can make significant potential contributions to understanding past climate variability connected to the Ocean-Continent Interactions theme. This theme is as broad as it is central to human societies. For example, regional sea-surface temperature variability fundamentally affects heat and moisture supplies to nearby continents ("upstream" effects) (Shanahan et al., 2009), such as in the major low-latitude monsoon systems. Understanding monsoon dynamics under different climate regimes (Wang et al., 2005) is of critical importance because a significant fraction of the human population depends on the monsoons and their life-giving rains for survival. Furthermore, investigating the interrelationships between climate, biotic evolution, and human cultural development will continue to define the effects of abrupt climate change (including droughts and floods) on life (Bartov et al., 2003; Haug et al., 2003; deMenocal, 2004). On the continents, "downstream" effects such as changes in continental weathering, hydrologic balance, and vegetation response can be reconstructed based on terrestrial archives such as lakes (Baker et al., 2001; Cheddadi et al., 2005). Oceanic archives recovered adjacent to the continents also record the direct influence of continental inputs, such as sediment from rivers and winds, and nutrients that play an important role in near-coastal bioproductivity and biogeochemical cycling (Mulitza et al., 2008).

Ocean-continent interactions in the high latitudes have been shown to have far-reaching effects on climate variability on all timescales. Investigating the extent of continental ice sheets, timing of ice sheet ablation, and relationships between Northern Hemisphere and Southern Hemisphere high-latitude climate (European Project for Ice Coring in Antarctica (EPICA) Community Members, 2006) are fundamental to understanding the role of the oceans and ice sheets in abrupt climate change. These studies depend on improved proxies of ice sheet dynamics, sea-ice, meltwater input, and sediment provenance, among others. Furthermore, scientific drilling has just begun to elucidate the connections between high-latitude climate and low-latitude hydrology—for instance, the teleconnections between sea ice and monsoons or sea ice and the position of the ITCZ (Broecker, 2006).

One challenge the community will face in these efforts is the difficulty of finding paired ocean and continent archives in appropriate locations and with adequate resolution to investigate upstream and downstream processes. Nevertheless, new opportunities for enhancing the study of ocean-continent linkages are emerging from the recent integration of novel proxies for continental hydrology and vegetation (e.g., elemental ratio data in bulk sediments, molecular and isotopic composition of leaf waxes; Schefuss et al., 2005; Huang et al., 2007) with classic proxies such as dust, pollen, and clay mineralogy, which can be found in both oceanic and continental archives. Accordingly, workshop participants specifically recommended that scientific drilling proposals on ocean-continent interactions be solicited that focus on three broad topics of critical interest: (1) the major low-latitude monsoons, (2) the hydrologic balance in continental regions critical for human societies, and (3) ice sheet dynamics (Clarke, 2005), all of which will enhance understanding of abrupt climate change.

Biogeochemical Cycles

The cycling of biogenic and geochemical fluxes has important effects on all Earth systems. However, the large number of unknowns associated with these cycles must be addressed at the highest possible resolution if past climate changes and all their complexities are to be correctly evaluated and future changes realistically predicted. We should know what role changes in the carbon cycle played as forcing or feedback during past climate changes (Cao and Woodward, 1998). In the endeavor to find analogues and to produce predictive models for ongoing climate warming, questions have to be addressed about the biotic response to ocean acidification (Fabry et al., 2008), the range of glacial-interglacial CO_2 variability, and the controls on atmospheric greenhouse gas composition as recorded in ice cores.

The employment of high-resolution sedimentary records in these contexts has great strengths and advantages; not only does high-resolution sampling more accurately reveal the details and complexities of the system (e.g., better definition of the climate norm and anomalies), but it is also more likely to yield un-aliased climate signals. In order to understand the specific workings of the climate system under different boundary conditions, we must undertake investigations of the carbon cycle using well-dated, well-calibrated, high-resolution records. Fundamental questions that remain to be answered include the following: What is the role of the carbon cycle in the transition from 41-kyr to 100-kyr cycles in the mid-Pleistocene (Fig. 4; Bintanja and van de Wal, 2008)? What is the role of ocean circulation in the carbon cycle, and how does carbon storage affect the cycle? What is the impact of carbon storage on metal cycles, and how useful are non-traditional metal isotope systems for tracing global paleoenvironmental change? What affects the CO_2 exchange between the surface ocean and the atmosphere on different timescales (Bender et al., 1994)? Is there a threshold level of CO_2 when the oceans switch to a different mode? In addition, the higher frequency behavior of the carbon cycle during the extreme warm climates of the mid-Cretaceous and PETM should be investigated at the highest possible resolution in order to gain insights into the possible extremes of our future world.

Ecosystem Change and Biodiversity

The resilience of ecosystems to climate change is of critical importance in the face of projected warming and altered precipitation regimes (Cramer et al., 2001; Parmesan, 2006). Determining whether systems bend or break under climate loading and where tipping points occur from one state to another can be profitably investigated through high-resolution paleoclimatic records. Indeed, a critical insight into paleoclimates is that they behave non-linearly, with sudden changes of direction. Determining the ecosystem impact of such events will require improved understanding of taphonomy, higher taxonomic resolution (possibly DNA-based identifications), and finer temporal resolution sampling than has commonly been applied, especially in terrestrial sequences. Long, high-resolution terrestrial records of past climate variability comparable to those of marine and ice cores must be studied (Fig. 4; Tzedakis et al., 2006). The 1-Myr-long record from tropical Lake Bosumtwi (Koeberl et al., 2007) and the 3-Myr record from the arctic Lake El'gygytgyn (Brigham-Grette et al., 2007) are outstanding examples that such archives can be found.

The impact on ecosystems of the arrival of new agents of change—such as humans or fire, or the cascading consequences of extinctions and immigrations—are additional issues that need to be resolved. Because terrestrial ecosystems are where people live, work, and generate the majority of global income, it is essential that their responses to climate change are understood and, if possible, anticipated.

Towards The Development of Synergies and Strategies

Other important discussions at the workshop focused on the practicalities of adopting a synergistic approach to climate research and of how to improve present techniques and develop new ones. There were strong feelings in common among the participants that a more tightly coordinated collaboration between paleoceanographers, terrestrial paleoclimatologists, paleoglaciologists, and modelers is required for climate change science to move beyond incremental advances.

There was lively debate about the nature and quality of the various proxies used to infer past climatic and oceanographic conditions. People acknowledged that there needs to be a more universal understanding of the limitations of proxies specific to each segment of the paleoclimate community in order to improve the correlation and synthesis of existing and future records. In particular, proxy development should focus on providing more quantitative data for climate modeling. Existing proxies need to be re-examined; we have the capacity to measure/analyze an ever-widening suite of sediment parameters, but do we always know what they mean/represent?

It is clear that there is a different culture of sampling between the ice core and marine/terrestrial communities. The latter tend to divide a core in time, while ice core groups tend to take each sample and divide it among multiple investigators. In order to maximize the information gained, highly coordinated research efforts are needed that focus on multiple proxies. In the future, cooperating groups will need to acquire more sediment than is presently considered necessary to allow for the many proxies now possible to analyze. This could be achieved by drilling wider diameter cores and/or additional parallel holes. Close coordination between separate investigating groups (e.g., shipboard versus non-shipboard) needs to be established early in the process of planning for drilling projects. In addition, the widest possible range of questions that could be addressed by analyzing the sediment should be considered right from the start, and thus the planning requirements of each potential investigating group could be met. This type of forward planning will be necessary to achieve coordinated temporal and spatial arrays.

To the extent that existing material from previously drilled sites is available, these archives should be used when practical to address new problems and/or apply new techniques whose development postdates the drilling and first wave of science to emerge. With regards to the acquisition of new sediment, core quality is critical. Scientific drilling has gone on long enough that many individuals with longstanding experience have moved on or retired. To avoid costly mistakes and wasted time, records of how best to core particular types of sediments should be kept so that this experience is not lost.

The continuing improvement of standard chronological techniques and the development of new methods are necessary prerequisites to producing high quality high-resolution climate records. Ideally, chronologies should be absolute, as can be achieved by the counting of annual layers in ice cores or distinct varves in lacustrine or marine sediments. In the absence of absolute age control, dating is usually based on a combination of techniques including radiocarbon, foraminiferal oxygen isotope stratigraphies (in marine sequences), paleomagnetism, and tephrochronology. In addition, new techniques that are under development include exposure dates on tufa (Rich et al., 2003); optically stimulated luminescence (OSL) dating at millennial resolution (Stevens et al., 2007); U/Th for tephra chronologies (Farley et al., 2002); development of microtephra stratigraphy (Freibel, 2008); measurements of iridium flux (Gabrielli et al., 2004); and stable isotope correlations with speleothems (Lorrey et al., 2008). A new method of ice core dating being explored involves N_2/O_2 variations in the ice at Dome Fuji tuned to local insolation at 70°S (a method that assumes no lag between insolation forcing and the N_2/O_2 response; Kawamura et al., 2007). Such an N_2/O_2 timescale is potentially very useful to the combined marine and ice core communities and will likely be adopted if it can be shown that paleointensity can be correlated to cosmonuclides. To be able to apply this to high-resolution sediment studies means that the magnetic paleointensity record has to be systematically developed from marine sections with millennial-scale records. Paleointensity records tend to be better in high latitude/low oxygen environments, and sediment drifts are an ideal archive (Channell et al., 2000). Locations need to be systematically sought out in order to build up paleointensity databases needed for calibration of the Quaternary.

There was considerable discussion over the concept of developing marine 'reference' cores akin to the long ice core records for better global correlation of climate records. The Greenland ice core records have become, almost by default, the de facto "type-section" for late Pleistocene climate change in that they serve as the measuring stick against which other climate records are inevitably compared. One drawback, however, is that the ice cores represent conditions at essentially one location in the high northern latitudes and are inappropriate for study of much of the climate system (Shackleton et al., 2000, 2004). The development of a counterpart series of marine reference sections would require the generation of a few internationally coordinated, multi-replicated, multi-parameter, high-resolution records (Fig. 5; Alley, 2003). However, it was acknowledged that the great size and horizontal and vertical variability in the oceans make identification of a "few" reference sections a difficult task. In selecting the locations for such reference sites, it was felt that drilling targets should be chosen so as to address specific paleoclimatic questions rather than locating them simply to achieve some degree of geographic coverage. For example, a reference section for understanding the MOC might actually consist of a depth

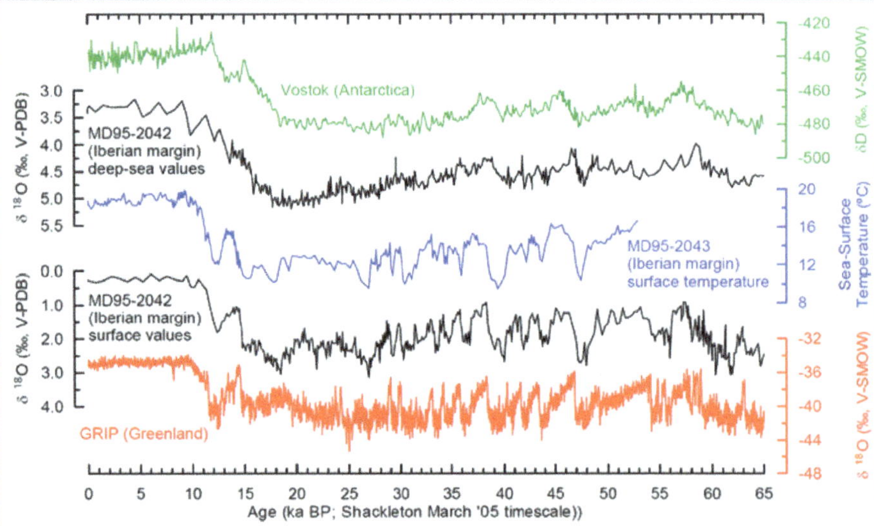

Figure 5. The development of a limited number of well-dated marine 'reference sections' for the late Quaternary would serve as a valuable complement to Greenland and Antarctic ice core records, which by their nature only record high-latitude processes in the two hemispheres. One recommended location for such a reference section is off the southwest coast of Iberia (see no. 35 with 'star' in Fig. 6) near the site of long Calypso cores MD95-2042 and MD95-2043, for which pre-existing data (Shackleton et al., 2000, 2004) have already demonstrated high and uninterrupted sedimentation rates and a clear record of abrupt climate events in the mid-latitudes that can be correlated to those in Greenland. Extending the record at this site to encompass multiple glacial-interglacial cycles can only be accomplished by scientific ocean drilling.

transect, while reference sections for ENSO could consist of drilled sequences from El Niño-sensitive areas such as Indonesia or Tahiti. If such a suite of marine reference sections can be developed over time, they will provide marine sediment analogues to the polar ice cores, but will be temporally and spatially more extensive.

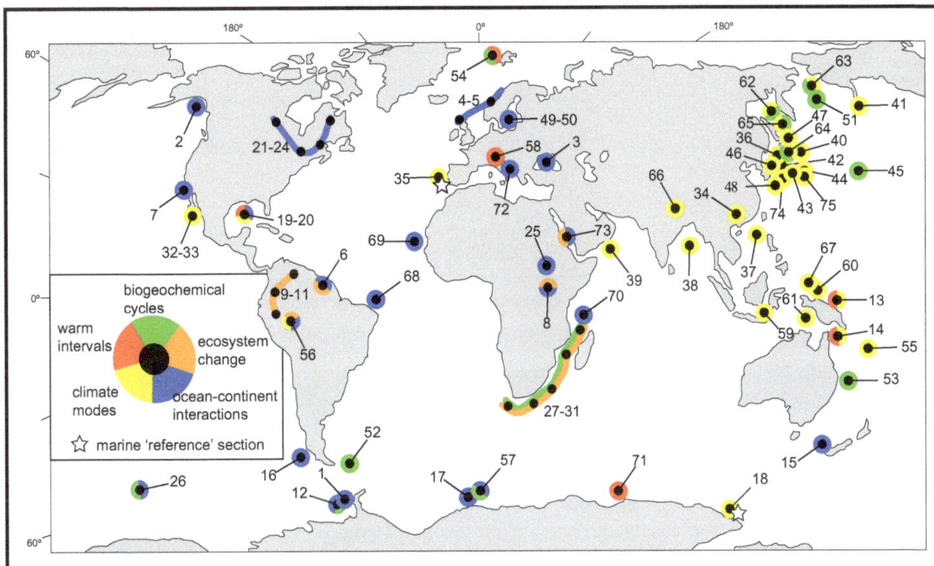

Figure 6. Compilation of potential drilling sites proposed by workshop participants to address the five "petal" themes of the flower diagram shown in Fig. 3. Site numbers are linked to information that can be found in Appendix 1 (http://high-resolutoin.icdp-online.org).

As part of the workshop, participants were asked to submit information on potential targets for high-resolution continental and oceanic drilling that satisfied specific requirements of record length and resolution (annual to millennial) and which address questions related to the key research themes highlighted in Fig. 3 and this report. The resulting array of drilling targets, identified in Fig. 6 and keyed there to theme and to the site-specific information listed in Appendix 1, shows that an impressive number of locations are available globally where the potential exists to recover sedimentary sequences suitable for high-resolution study. The themes/targets incorporate many sites already in the review system as part of existing ICDP/IODP drilling proposals (Appendix 1, http://high-resolution.icdp-online.org).

A series of four hypothetical transects emerged from discussions as potential examples of how combined IODP and ICDP drilling could maximize scientific progress in paleoclimate research. These are presented here to illustrate how strategically placed sites can be combined to address broad climate questions and themes.

1) The Core ENSO Signal and Teleconnections. Much can be learned about ENSO and its behavior through time, based on a combined land/sea transect from SE Asia/Papua New Guinea via Pacific Islands to the west coast of South and Central America. This might incorporate both high deposition rate marine records and drilled coral sequences spanning the tropical Pacific, as well as lakes along the western edge of the Americas. In addition, records from regions such as northeast Brazil, South Africa, and New Zealand that are strongly teleconnected to ENSO today would provide important information as to whether these linkages persist through time under different global boundary conditions.

2.) Baikal to Bengal. A roughly north-south transect was envisioned that would focus on the internal dynamics of the Asian monsoonal system. This would include sites spanning the latitude range from the Siberian arctic Lake El'gygytgyn to Lake Baikal to the Bengal Fan system, with potential intermediate sites located in the Amur River, the Tibetan Plateau and vicinity, India, and the Ganges Plain. Adding sites in and around the Arabian Sea would broaden the set of addressable questions by tying in regions affected instead by the Indian monsoon.

3.) Southern Hemisphere Tropics (South America) to Southern High Latitude Interactions. Southern hemisphere climate records are currently limited in number, but potential exists to drill latitudinal transects of sites ranging from the southern tropics of South America into the high southern latitudes, both on land and in coastal waters where high sedimentation rate sequences are most likely to be found. It was generally felt that South America offered an important opportunity in the southern hemisphere to develop a comprehensive regional database of sites.

4.) The Bipolar Seesaw, North-South Gradients, and Interhemispheric Linkages. Records from the last glacial indicate that the abrupt warmings and coolings recorded in Greenland ice appear to be out of phase with their counterpart temperature excursions in the high-latitude Southern Hemisphere, an observation consistent with changes in the Atlantic's meridional overturning circulation and its effects on ocean heat transport (EPICA Community Members, 2006). We know little about whether this "bipolar seesaw" is operative at other times in the past and of how changing temperature gradients affect such climatic variables as the distribution of tropical precipitation and ITCZ position. Transects of drill sites on land and in the ocean spanning the high to low latitudes in both hemispheres are the only way to examine how gradients and interhemispheric linkages have varied through time.

As already noted, workshop participants endorsed the concept of developing a limited number of globally distributed marine reference sections for the purpose of calibrating Quaternary climate at high resolution in key regions. It was recommended that one such early target area should be offshore of the southwest coast of Iberia (Figs. 5, 6) where high sedimentation rates, record continuity, and tight linkages to the rapid climate excursions in Greenland have already been demonstrated. In addition, a large data set of proxy records already exists from this location (Cayre et al., 1999; Sanchez-Goñi et al., 1999; Shackleton et al., 2004; Skinner et al., 2003), so we know that sediments are amenable to analysis using a wide variety of methods.

Proxy data generated from high-resolution geologic archives will seldom lead to an improved understanding of how the climate system works if those data are not linked into modern thinking about climate dynamics and integrated into an Earth-system modeling approach. The presence of modelers at the workshop ensured that data-model integration was a prominent component of the discussions.

Integration of Paleoclimatic Data and Earth-System Modelling

Over the past years Earth-system modeling has become increasingly important in the area of paleoclimatic research. Specifically, physically based models provide a valuable tool to assess conceptual models of past climate changes derived from paleoclimatic data. In addition, data-based reconstructions of past climate variations provide critical test grounds for Earth-system models that are usually tuned to capture present-day climate. Accordingly, combining data-based reconstructions and paleoclimate modeling offers a promising track to fully comprehend past climate dynamics and to test models used for assessing future climate changes (Ramstein et al., 2007; Mulitza et al., 2008).

Specific benefits that can result from the joint utilization of paleoclimatic reconstructions and paleoclimate modeling include a means for formulating and testing hypotheses (e.g., by quantifying the response of the climate system to known or potential forcings). Moreover, climate models provide a comprehensive framework for exploring couplings and feedbacks between the various components of the climate system. This type of analysis is of special relevance for detecting thresholds in the climate system. Finally, Earth-system models offer a link between past climate changes and projections of future climate, which is assessed by the same type of models.

The combination of paleoclimatic reconstructions and Earth-system modeling is thus very forward-looking with respect to disentangling climate dynamics on timescales of societal relevance. Both IODP and ICDP have the obligation to underpin the efforts of the IPCC in reducing the uncertainties associated with the projections of future climate changes. A pivotal contribution of paleoclimatic reconstructions arises through the possibility of testing IPCC-type Earth-system models beyond the range of climate variability documented by the instrumental record. A research strategy following this path would involve modelers from the early planning stage of an IODP or ICDP project. Such collaboration would not only allow for setting up joint hypotheses to be tested but also to choose drilling locations most sensitive to specific climate changes.

Summary and Recommendations

The Potsdam workshop attracted an overwhelming response from the paleoclimate community, and the enthusiastic discussions and debates that took place clearly demonstrated that the topic of scientific drilling of high-resolution archives is hot, timely, and widely interesting to the scientific community. Great synergy already exists between IODP and ICDP in the field of paleoclimate research. Workshop participants clearly felt that having these two programs work together to jointly address the key research themes identified in Potsdam (the "petals" of Fig. 3) is the most promising way forward.

We recommend that further strengthening of the links between IODP and ICDP be a goal of the leadership of both programs. Each program brings state-of-the-art drilling technology to bear on scientific questions that only these programs—and the research communities they support—can answer. Cooperation should also be sought with other major programs devoted to paleoclimate sampling, including IPICS (International Partnership for Ice Core Sciences) and IMAGES (International Marine Global Change Study). Climate, by its very nature, is a complex system, and neither land-based nor marine-based records are sufficient by themselves to provide a complete picture of the system's workings through time.

Closer cooperation between IODP and ICDP ideally should be extended to have standard procedures for core processing and treatment, agreed-upon suites of standard measurements and protocols, policies for data integration, and perhaps even common curation of cores. One obvious indication of a closer working relationship between the programs—and that would seem simple to achieve—would be to have site locations for ICDP drilling and links to the data included on the searchable online IODP digital site map. This would facilitate planning efforts for some of the large-scale transect opportunities envisioned at this workshop.

Continued development of drilling capabilities should be a high priority. Within ICDP, there seems to be a gap in drilling technology for the recovery of sediments between about 20 m and 100 m penetration (i.e., what can be achieved with standard piston coring and by deep drilling using the GLAD rig). We should also further explore the possibility of increasing core diameters in both programs to allow for larger sediment volumes for multi-proxy studies. Finally, increased flexibility in the duration of drilling campaigns, especially within IODP, would greatly facilitate the planning and proposal process for many high-resolution sites where sediments can often be recovered relatively quickly and efficiently.

The Potsdam workshop left no doubt that scientific drilling is absolutely critical to the future of paleoclimate research. The large array of potential drilling targets summarized here, together with the concept of transects and climate reference sections put forth, should ensure that high-resolution records appropriate for investigating the climate research themes identified in this report are available to the community at large.

Acknowledgements

The workshop Steering Committee thanks IODP and ICDP for their generous financial support of this workshop and our friendly hosts at GFZ-Potsdam for their care and attention. Thanks also go to Benjamin Flower, Mark Bush, Paul Baker, and Robert Dunbar for their contributions to the thematic summaries, and to the keynote speakers, discussion leaders, and volunteer reporters who kept diligent notes. Finally, we are grateful to all of the workshop participants for sharing their time, expertise and great enthusiasm for paleoclimate science.

References

Alley, R.B., 2003. Raising paleoceanography. *Paleoceanography*, 18:1085, doi:10.1029/2003PA000942.

Baker, P.A., Rigsby, C.A., Seltzer, G.O., Fritz, S.C., Lowenstein, T.K., Bacher, N.P., and Veliz, C., 2001. Tropical climate changes at millennial and orbital timescales on the Bolivian Altiplano. *Nature*, 409:698–701, doi:10.1038/35055524.

Bartov, Y., Goldstein, S.L., Stein, M., and Enzel, Y., 2003. Catastrophic arid perturbations of the East-Mediterranean climate linked to the North Atlantic Heinrich Events. *Geology*, 31:439–442, doi:10.1130/0091-7613(2003)031<0439:CAEITE>2.0.CO;2.

Behl, R.J., and Kennett, J.P., 1996. Brief interstadial events in the Santa Barbara Basin, NE Pacific, during the past 60 kyr. *Nature*, 379:243–246, doi:10.1038/379243a0.

Bender, M., Sowers, T., and Labeyrie, L., 1994. The Dole effect and its variations during the last 130,000 years as measured in the Vostok ice core. *Global Biogeochem. Cycles*, 8:363–376, doi:10.1029/94GB00724.

Bintanja, R., and van de Wal, R.S.W., 2008. North American ice-sheet dynamics and the onset of 100,000-year glacial cycles. *Nature*, 454:869–872, doi:10.1038/nature07158.

Brigham-Grette, J., Melles, M., Minyok, P., and Scientific Party, 2007. Overview and significance of a 250 ka paleoclimate record from El'gygytgyn Crater Lake, NE Russia. *J. Paleolimnol.*, 37:1–16, doi:10.1007/s10933-006-9017-6.

Broecker, W.S., 2006. Abrupt climate change revisited. *Global Planet. Change*, 54:211–215, doi:10.1016/j.gloplacha.2006.06.019.

Cao, M., and Woodward, F.I., 1998. Dynamic responses of terrestrial ecosystem carbon cycling to global climate change. *Nature*, 393:249–252, doi:10.1038/30460.

Cayre, O., Lancelot, Y., and Vincent, E., 1999. Paleoceanographic reconstructions from planktonic foraminifera off the Iberian Margin: temperature, salinity and Heinrich events. *Paleoceanogr.*, 14:384–396, doi:10.1029/1998PA900027.

Channell, J.E.T., Stoner, J.S., Hodell, D.A., and Charles, C.D., 2000. Geomagnetic paleointensity for the last 100 kyr from the sub-Antarctic South Atlantic: a tool for inter-hemispheric correlation. *Earth Planet. Sci. Lett.*, 175:145–160, doi:10.1016/S0012-821X(99)00285-X.

Cheddadi, R., de Beaulieu, J.L., Jouzel, J., Andrieu-Ponel, V., Laurent, J.M., Reille, M., Raynaud, D., and Bar-Hen, A., 2005. Similarity of vegetation dynamics during interglacial periods. *Proc. Natl. Acad. Sci. U.S.A.*, 102(39):13939–13943, doi:10.1073/pnas.0501752102.

Clarke, G.K.C., 2005. Subglacial processes. *Ann. Rev. Earth Planet. Sci.*, 33:247–276, doi:10.1146/annurev.earth.33.092203.122621.

Clement, A.C., and Peterson, L.C., 2008. Mechanisms of abrupt climate change of the last glacial period. *Rev. Geophys.*, 46:RG4002, doi:10.1029/2006RG000204.

Cramer, W., Bondeau, A., Woodward, F.I., Prentice, I.C., Betts, R.A., Brovkin, V., Cox, P.M., Fisher, V., Foley, J.A., Friend, A.D., Kucharik, C., Lomas, M.R., Ramankutty, N., Sitch, S., Smith, B., White, A., and Young-Molling, C., 2001. Global response of terrestrial ecosystem structure and function to CO_2 and climate change: results from six dynamic global vegetation models. *Global Change Biol.*, 7:357–373, doi:10.1046/j.1365-2486.2001.00383.x.

Dansgaard, W., Johnsen, S.J., Clausen, H.B., Dahl-Jensen, D., Gundestrup, N., and Hammer, C.U., 1984. North Atlantic climatic oscillations revealed by deep Greenland ice cores. *In* Hansen, J.E., and Takahashi, T. (Eds.), *Climate Processes and Climate Sensitivity, Geophys. Monogr. Ser., 29*, Washington, DC (American Geophysical Union), 288-298.

Dansgaard, W., Johnsen, S., Clausen, H.B., Dahl-Jensen, D., Gundestrup, N.S., Hammer, C.U., Hvidberg, C.S., Steffensen, J., Sveinbjörnsdottir, A.E., Jouzel, J., and Bond, G., 1993. Evidence for general instability of past climate from a 250-kyr ice core record. *Nature*, 364:218–220, doi:10.1038/364218a0.

deMenocal, P., 2004. African climate change and faunal evolution during the Pliocene/Pleistocene. *Earth Planet. Sci. Lett.*, 220:3–24, doi:10.1016/S0012-821X(04)00003-2.

EPICA Community Members, 2004. Eight glacial cycles from an Antarctic ice core. *Nature*, 429:623-628, doi:10.1038/nature02599.

EPICA Community Members, 2006. One-to-one coupling of glacial climate variability in Greenland and Antarctica. *Nature*, 444:195–198, doi:10.1038/nature05301.

Fabry, V.J., Seibel, B.A., Feely, R.A., and Orr, J.C., 2008. Impacts of ocean acidification on marine fauna and ecosystem processes. *ICES J. Mar. Sci.*, 65:414–432, doi:10.1093/icesjms/fsn048.

Farley, K.A., Kohn, B.P., and Pillans, B., 2002. The effects of secular disequilibrium on (U-Th)/He systematics and dating of Quaternary volcanic zircon and apatite. *Earth Planet. Sci.*

Lett., 201:117–125, doi:10.1016/S0012-821X(02)00659-3.

Freibel, C.G.S., 2008.Microstratigraphy of the Kibish hominin sites KHS and PHS, lower Omo Valley, Ethiopia. *J. Hum. Evol.*, 55:404–408.

Gabrielli, P., Barbante, C., Plane, J.M.C., Varga, A., Hong, S., Cozzi, G., Gaspari, V., Planchon, F.A.M., Cairns, W., Ferrari, C., Crutzen, P., Cescon, P., and Boutron, C.F., 2004. Meteoric smoke fallout over the Holocene epoch revealed by iridium and platinum in Greenland ice. *Nature*, 432:1011–1014, doi:10.1038/nature03137.

Gregory, J.M., Huybrechts, P., and Raper, S.C.B., 2004. Climatology: threatened loss of the Greenland ice sheet. *Nature*, 428:616, doi:10.1038/428616a.

Haug, G.H., Günther, D., Peterson, L.C., Sigman, D.M., Hughen, K.A., and Aeschliman, B., 2003. Climate and the collapse of Maya civilization. *Science*, 299:1731–1735, doi:10.1126/science.1080444.

Hendy, I.L., and Kennett, J.P., 2000. Dansgaard-Oeschger cycles and the California Current system: planktonic foraminiferal response to rapid climate change in Santa Barbara Basin, Ocean Drilling Program Hole 893A. *Paleoceanogr.*, 15:30-42, doi:10.1029/1999PA000413.

Hodell, D.A., Anselmetti, F.S., Ariztegui, D., Brenner, M., Curtis, J.H., Gilli, A., Grzesik, D.A., Guilderson, T.J., Müller, A.D., Bush, M.B., Correa-Metrio, A., Escobar, J., and Kutterolf, S., 2008. An 85-ka record of climate change in lowland Central America. *Quat. Sci. Rev.*, 27:1152–1165, doi:10.1016/j.quascirev.2008.02.008.

Hu, A., Meehl, G., Han, W., and Yin, J., 2009. Transient response of the MOC and climate to potential melting of the Greenland Ice Sheet in the 21st century. *Geophys. Res. Lett.*, 36:L10707, doi: 10.1029/2009GL037998.

Huang, Y., Clemens, S.C., Liu, W., Wang, Y., and Prell, W.L., 2007. Large-scale hydrological change drove the late Miocene C4 plant expansion in the Himalayan foreland and Arabian Peninsula. *Geology*, 35:531–534, doi:10.1130/G23666A.1.

Hughen, K., Lehman, S., Southon, J., Overpeck, J., Marchal, O., Herring, C., and Turnbull, J., 2004. ^{14}C activity and global carbon cycle changes over the past 50,000 years. *Science*, 303:202–207, doi:10.1126/science.1090300.

IPCC, 2007. Climate Change 2007: Synthesis Report. Contribution of Working Groups I, II and III to the Fourth Assessment Report of the Intergovernmental Panel on Climate Change [Pachauri, R.K., and Reisinger, A. (Eds.)], Geneva, Switzerland (Intergovernmental Panel on Climate Change), 104 pp.

Kawamura, K., Parrenin, F., Lisiecki, L., Uemura, R., Vimeux, F., Severinghaus, J.P., Hutterli, M.A., Nakazawa, T., Aoki, S., Jouzel, J., Raymo, M.E., Matsumoto, K., Nakata, H., Motoyama, H., Fujita, S., Goto-Azuma, K., Fujii, Y., and Watanabe, O., 2007. Northern Hemisphere forcing of climatic cycles in Antarctica over the past 360,000 years. *Nature*, 448:912–917, doi:10.1038/nature06015.

Keenlyside, N.S., Latif, M., Jungclaus, J., Kornblueh, L., and Roeckner, E., 2008. Advancing decadal-scale climate prediction in the North Atlantic sector. *Nature*, 453:84–88, doi:10.1038/nature06921.

Koeberl, C., Milkereit, B., Overpeck, J.T., Scholz, C.A., Amoako, P.Y.O., Boamah, D., Danuor, S., Karp, T., Kueck, J., Hecky, R.E., King, J.W., and Peck, J.A., 2007. An international and multidisciplinary drilling project into a young complex impact structure: the 2004 ICDP Bosumtwi Crater Drilling Project - An overview. *Meteor.Planet. Sci.*, 42:483–511.

Kraus, M.J., and Riggins, S., 2007. Transient drying during the Paleocene–Eocene Thermal Maximum (PETM): analysis of paleosols in the Bighorn Basin, Wyoming. *Palaeogeogr., Palaeoclimatol., Palaeoecol.*, 245:444–461, doi:10.1016/j.palaeo.2006.09.011.

Lisiecki, L.E., and Raymo, M.E., 2005. A Pliocene–Pleistocene stack of 57 globally distributed benthic $\delta^{18}O$ records. *Paleoceanogr.*, 20, PA1003, doi:10.1029/2004PA001071.

Lorrey, A., Williams, P., Salinger, J., Martin, T., Palmer, J., Fowler, A., Zhao, J-X., and Neil, H., 2008. Speleothem stable isotope records interpreted within a multi-proxy framework and implications for New Zealand palaeoclimate reconstruction. *Quat. Int.*, 187:52–75, doi:10.1016/j.quaint.2007.09.039.

Loulergue, L., Schilt, A., Spahni, R., Masson-Delmotte, V., Blunier, T., Lemieux, B., Barnola, J.M., Raynaud, D., Stocker, T.F., and Chappellaz, J., 2008. Orbital and millennial-scale features of atmospheric CH4 over the past 800,000 years. *Nature*, 435:383–386, doi:10.1038/nature06950.

Mantua, N.J., and Hare, S.R., 2002. The Pacific Decadal Oscillation. *J. Oceanogr.*, 58:35–44, doi:10.1023/A:1015820616384.

Mulitza, S., Prange, M., Stuut, J.-B., Zabel, M., von Doboneck, T., Itambi, A.C., Nizou, J., Schulz, M., and Wefer, G., 2008. Sahel megadroughts triggered by glacial slowdowns of Atlantic meridional overturning. *Paleoceanogr.*, 23:PA4206, doi: 10.1029/2008PA001637.

Parmesan, C., 2006. Ecological and evolutionary responses to recent climate change. *Ann. Rev. Ecol. Evol. Syst.*, 37:637–669, doi:10.1146/annurev.ecolsys.37.091305.110100

Peterson, L.C., Haug, G.H., Hughen, K.A., and Rohl, U., 2000. Rapid changes in the hydrologic cycle of the tropical Atlantic during the last glacial. *Science*, 290:1947–1951, doi:10.1126/science.290.5498.1947.

Ramstein, G., Kageyama, M., Guiot, J., Wu, H., Hély, C., Krinner, G., and Brewer, S., 2007. How cold was Europe at the Last Glacial Maximum? A synthesis of the progress achieved since the first PMIP model-data comparison. *Clim. Past*, 3:331–339.

Rich, J., Stokes, S., Wood, W., and Bailey, R., 2003. Optical dating of tufa via in situ aeolian sand grains: A case example from the Southern High Plains, USA. *Quat. Sci. Rev.*, 22:1145–1152, doi:10.1016/S0277-3791(03)00021-0.

Ruddiman, W.F., 2005. Cold climate during the closest Stage 11 analog to recent millennia. *Quat. Sci. Rev.*, 24:1111–1121, doi:10.1016/j.quascirev.2004.10.012.

Sanchez-Goñi, M.F., Eynaud, F., Turon, J.L., and Shackleton, N.J., 1999. High resolution palynological record off the Iberian margin: direct land-sea correlation for the Last Interglacial complex. *Earth Planet. Sci. Lett.*, 171:123–137, doi:10.1016/S0012-821X(99)00141-7.

Schefuss, E., Schouten, S., and Schneider, R.R., 2005. Controls on central African hydrology during the past 20,000 years. *Nature*, 437:1003–1006, doi:10.1038/nature03945.

Scholz, C.A., Johnson, T.C., Cohen, A.S., King, J.W., Peck, J.A., Overpeck, J.T., Talobot, M.R., Brown, E.T., Kalindekafe, L., Amoako, P., Lyons, R.P., Shanahan, T.M., Castañeda, I.S.,

Heil, C.W., Forman, S.L., McHargue, L.R., Beuning, K.R., Gomez, J., and Pierson, J., 2007. East African megadroughts between 135 and 75 thousand years ago and bearing on early-modern human origins. *Proc. Natl. Acad. Sci. U.S.A.*, 104(42):16416–164211, doi: 10.1073/pnas.0703874104.

Shackleton, N.J., Hall, M.A., and Vincent, E., 2000. Phase relationships between millennial-scale events 64,000-24,000 years ago. *Paleoceanogr.*, 15:565–569, doi:10.1029/2000PA000513.

Shackleton, N.J., Fairbanks, R.G., Chiu, T.-C., and Parrenin, F., 2004. Absolute calibration of the Greenland time scale: implications for Antarctic time scales and for $\Delta^{14}C$. *Quat. Sci. Rev.*, 23:1513–1522, doi:10.1016/j.quascirev.2004.03.006.

Shanahan, T.M., Overpeck, J.T., Anchukaitis, K.J., Beck, J.W., Cole, J.E., Dettman, D.L., Peck, J.A., Scholz, C.A., and King, J.W., 2009. Atlantic forcing of persistent drought in West Africa. *Science*, 324:377–380. doi: 10.1126/science.1166352.

Skinner, L.C., Shackleton, N.J., and Elderfield, H., 2003. Millennial-scale variability of deep-water temperature and $\delta^{18}O_{dw}$ indicating source variations in the northeast Atlantic, 0-34 cal. Ka BP. *Geochem. Geophys. Geosyst.*, 4:1098, doi:10.1029/2003GC000585.

Stanford, J.D., Rohling, E.J., Hunter, S.E., Roberts, A.P., Rasmussen, S.O., Bard, E., McManus, J., and Fairbanks, R.G., 2006. Timing of meltwater pulse 1a and climate responses to meltwater injections. *Paleoceanogr.*, 21:PA4103, doi: 1029/2006PA001340.

Stevens, T., Armitage, S.J., Lu, H., and Thomas, D.S.G., 2007. Examining the potential of high sampling resolution OSL dating of Chinese loess. *Quat. Geochronol.*, 2:15–22, doi:10.1016/j.quageo.2006.03.004.

Tudhope, A.W., Chilcott, C.P., McCulloch, M.T., Coo, E.R., Chappell, J., Ellam, R.M., Lea, D.W., Lough, J.M., and Shimmield, G.B., 2001. Variability in the El Niño-Southern Oscillation through a glacial-interglacial cycle. *Science*, 291:1511–1517, doi:10.1126/science.1057969.

Tzedakis, P.C., Hoogheimstra, H., and Pälike, H., 2006. The last 1.35 million years at Tenaghi Philippon: revised chronostratigraphy and long-term vegetation trends. *Quat. Sci. Rev.*, 25:3416–3430, doi:10.1016/j.quascirev.2006.09.002.

Voelker, A.H.L., 2002. Global distribution of centennial-scale records for marine isotope stage (MIS) 3: a database. *Quat. Sci. Rev.*, 21:1185–1212, doi:10.1016/S0277-3791(01)00139-1.

Wang, P., Clemens, S., Beaufort, L., Braconnot, P., Ganssene, G., Jian, Z., Kershaw, P., and Sarnthein, M., 2005. Evolution and variability of the Asian monsoon system: state of the art and outstanding issues. *Quat. Sci. Rev.*, 24:595–629, doi:10.1016/j.quascirev.2004.10.002.

Wang, X., Auler, A.S., Edwards, R.L., Cheng, H., Ito, E., and Solheid, M., 2006. Interhemispheric anti-phasing of rainfall during the last glacial period. *Quat. Sci. Rev.*, 25:3391–3403, doi:10.1016/j.quascirev.2006.02.009.

Wang, Y.J., Cheng, H., Edwards, R.L., An, Z.S., Wu, J.Y., Shen, C.-C., and Dorale, J.A., 2001. A high-resolution absolute-dated late Pleistocene monsoon record from Hulu Cave, China. *Science*, 294:2345–2348, doi:10.1126/science.1064618.

Weijers, J.W.H., Schouten, S., Sluijs, A., Brinkhaus, H., and Sinninghe Damsté, J.S., 2007. Warm Arctic continents during the Palaeocene-Eocene thermal maximum. *Earth Planet. Sci. Lett.*, 261:230–238, doi:10.1016/j.epsl.2007.06.033.

Wilson, P.A., and Norris, R.D., 2001. Warm tropical ocean surface and global anoxia during the mid-Cretaceous period. *Nature*, 412:425–429, doi:10.1038/35086553.

Yancheva, G., Nowaczyk, N.R., Mingram, J., Dulski, P., Negendank, J., Liu, J., Sigman, D.M., Peterson, L.C., and Haug, G.H., 2007. Influence of the intertropical convergence zone on the East Asian monsoon. *Nature*, 445:74–77, doi:10.1038/nature05431.

Zachos, J.C., Röhl, U., Schellenberg, S.A., Sluijs, A., Hodell, D.A., Kelly, D.C., Thomas, E., Nicolo, M., Raffi, I., Lourens, L.J., McCarren, H., and Kroon, D., 2005. Rapid acidification of the ocean during the Paleocene-Eocene Thermal Maximum. *Science*, 308:1611–1615, doi:10.1126/science.1109004.

Authors (Workshop Steering Committee)

Juergen Thurow, Department of Earth Sciences, University College London, Gower Street, London, WC1E 6BT, U.K., e-mail: j.thurow@ucl.ac.uk.

Larry C. Peterson, Rosenstiel School of Marine and Atmospheric Science, University of Miami, 4600 Rickenbacker Causeway, Miami, Fla., 33149, U.S.A., e-mail: lpeterson@rsmas.miami.edu.

Ulrich Harms, International Continental Scientific Drilling Program, GFZ German Research Centre for Geosciences, Telegrafenberg, 14473, Potsdam, Germany.

David A. Hodell, Department of Earth Sciences, University of Cambridge, Cambridge, CB2 3EQ, U.K.

Heather Cheshire, Department of Earth Sciences, University College London, Gower Street, London, WC1E 6BT, U.K.

Michael Schulz, Faculty of Geosciences/Center for Marine Environmental Research (MARUM), University of Bremen, D-28334, Bremen, Germany.

Hans-Juergen Brumsack, Institute for Chemistry and Biology of the Marine Environment (ICBM), University of Oldenburg, P.O. Box 2503, D-26111, Oldenburg, Germany.

Tomohisa Irino, Faculty of Environmental Earth Science, Hokkaido University, Kita 10, Nishi 5, Kita-ku, Sapporo, 060-0810, Japan.

Valerie Masson-Delmotte, Laboratoire des Sciences du Climat et de l'Environnement, IPSL/CEA-CNRS UMR 1571, Gif-sur-Yvette, France.

Ryuji Tada, Department of Earth and Planetary Science, Graduate School of Science, University of Tokyo, Science Bldg. No.1-734, 7-3-1 Hongo, Bunkyo-ku, Tokyo, 113-0033, Japan.

Related Web Links

http://www.iodp.org/climate-ws-workshop/
http://www.iodp.org
http://www.icdp-online.org
http://high-resolution.icdp-online.org

Figure Credits

Fig. 3 – Ben Flower (College of Marine Science, University of South Florida, St. Petersburg, U.S.A)

Testing the Extensional Detachment Paradigm: A Borehole Observatory in the Sevier Desert Basin

by Nicholas Christie-Blick, Mark H. Anders, Gianreto Manatschal, and Brian P. Wernicke

Low-angle normal faults or detachments are widely regarded as playing an important role in crustal extension and the development of rifted continental margins (Manatschal et al., 2007). However, no consensus exists on how to resolve the mechanical paradox implied by the gentle dips of these faults and by the general absence of evidence for associated seismicity (Sibson, 1985; Wernicke, 1995; Axen, 2004). As part of a new initiative to rationalize geological and geophysical evidence and our theoretical understanding of how rocks deform, a group of forty-seven scientists and drilling experts from five countries met for four days on 15–18 July 2008 to discuss the present status of the paradox and a borehole-based strategy for resolving it. The workshop was held at two venues in Utah (the Utah Department of Natural Resources in Salt Lake City, and Solitude Mountain Resort in the adjacent Wasatch Range), with a one-day field trip to the Sevier Desert basin of west-central Utah (Figs. 1, 2) to examine the general setting of potential drill sites and the footwall geology of the Sevier Desert detachment (Canyon Range).

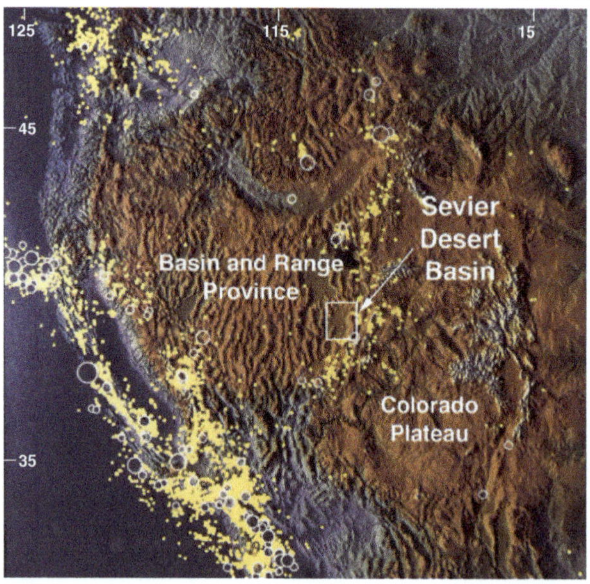

Figure 1. Regional physiography, seismicity, and location of the Sevier Desert basin in the western United States (from Simpson and Anders, 1992). White open circles are earthquakes greater than magnitude 6.0 from 1900 to 1975. Yellow dots are earthquakes greater than magnitude 2.5, occurring between 1970 and 1985. The north-trending band of seismicity adjacent to the Sevier Desert marks the eastern edge of the Basin and Range Province. Intense seismicity to the west and southwest, and including the western Basin and Range Province, corresponds with the diffuse transform plate boundary of which the San Andreas fault system is the most important element.

Interest in the Sevier Desert detachment (Fig. 3) relates to its large scale, geometric simplicity, severe misorientation with respect to σ_1, comparatively shallow depth, and compelling evidence for contemporary slip, as well as its accessibility from more or less flat public land (Von Tish et al., 1985; Niemi et al., 2004; Wills et al., 2005). The fault was first recognized in the mid-1970s, on the basis of seismic reflection data and commercial wells, as the subsurface contact between Paleozoic carbonate rocks and Cenozoic basin fill (McDonald, 1976). It is thought by most workers to root into the crust to the west of the Sevier Desert, to have large offset (as much as 47 km; DeCelles and Coogan, 2006), and to have been active since the late Oligocene at or near its present dip of 11° (GPS data and prominent Holocene scarps on steeply inclined hanging-wall faults that appear to sole downward into the detachment; Von Tish et al., 1985; Oviatt, 1989; Wernicke, 1995; Niemi et al., 2004). Whether the detachment fault crops out today at the eastern margin of the Sevier Desert basin is unresolved (Otton, 1995; Wills and Anders, 1999). No modern scarps have been observed there. Although no historic seismicity has been documented on the detachment, its scale is consistent with earthquake magnitudes at least as large as Mw = 7.0 (Wernicke, 1995). It is also possible that slip is currently taking place by aseismic creep. While dozens of low-angle normal faults have been recognized, at numerous locations in both extensional and orogenic settings—and by low angle we refer to the dip of a fault today, not necessarily its dip when it was active—the Sevier Desert detachment is one of very few that is sufficiently well-documented, active, and accessible from the surface that it might reasonably yield new insights about the conditions under which such faults slip.

A two-step drilling strategy emerged during workshop discussions. The first step (a pilot hole) is to re-enter one of several wells drilled by petroleum companies in the Sevier Desert basin (Wills et al., 2005), to deviate a few tens of meters above the base of the Cenozoic section, and to core through the detachment level to at least several tens of meters below the top of the Paleozoic section. Before embarking on a dedicated main hole, it is imperative to demonstrate that a fault is present (i.e., that fault rocks are present). The detachment interpretation, though generally accepted, currently depends entirely on geophysical data, not direct observation. It may be necessary to deviate and core through the detachment more than once to obtain definitive samples. The well provisionally selected for the pilot hole, and for technical as well as geological reasons, is ARCO Hole-in-Rock in the southern Sevier Desert (AHR in Figs. 2, 3B). The detachment is sufficiently deep at the Hole-in-Rock well (2774 m), and its hanging-wall offset is sufficiently large that fault rocks ought to be well-developed in both Cenozoic strata above and Paleozoic strata below. At the same time, the existence of late Pleistocene to Holocene fault scarps to the east of the well is consistent with recent displacement on the detachment at this location.

The second step (main hole) is to core, log, and make *in situ* measurements at a location between a few tens of meters and 4 km west (downdip) of ARCO Hole-in-Rock, and intersecting the interpreted detachment at a depth of 2800–3500 m (Fig. 3B). All surface scarps appear to be east of the Hole-in-Rock well at this latitude. The selection of a site 4 km or more to the west of ARCO Hole-in-Rock would permit the detachment fault to be penetrated within Paleozoic or Neoproterozoic strata west of the intersection between the basin's western bounding fault and the detachment.

The principal objective of the second hole is to establish an observatory at a depth and location most likely to allow monitoring of the full rate of extension across the Sevier Desert (~0.35 mm yr^{-1}; Niemi et al., 2004), and yet not so deep as to be prohibitively expensive. Among *in situ* measurements to be made in the vicinity of the detachment are the following: pore pressure, fracture permeability, fluid chemistry (including He), temperature, the orientation of stress axes, and the magnitude of differential stress. A borehole seismometer will be installed as part of a local array. A second objective of this main hole is to investigate the history of sediment accumulation and how the timing of basin development relates to exhumation of the detachment's footwall (based upon already published fission-track data for the Canyon Range; Stockli et al., 2001). A full suite of downhole logs (especially acoustic logging) will allow confident correlation with seismic reflection data. A byproduct of stratigraphic and geochronological analyses will be an extended lacustrine record of continental climate change since the late Oligocene.

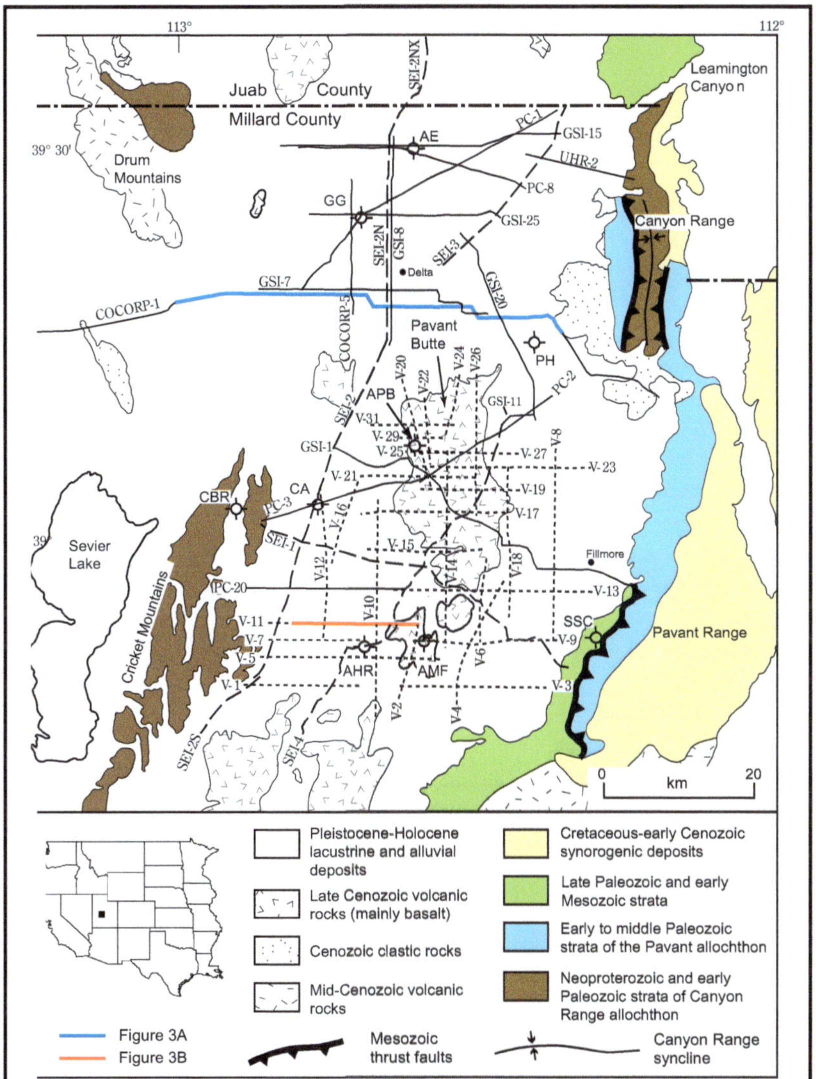

Figure 2. Generalized map of the Sevier Desert basin, west-central Utah, showing locations of wells and seismic reflection profiles (from Wills et al., 2005; see Fig. 1 for location). Wells: AE, Argonaut Energy Federal; AHR, ARCO Hole-in-Rock; AMF, ARCO Meadow Federal; APB, ARCO Pavant Butte; CA, Cominco American Federal; CBR, Chevron Black Rock; GG, Gulf Gronning; PH, Placid Henley; SSC, Shell Sunset Canyon. Solid, dashed, and dotted lines correspond with different seismic datasets. Part of Consortium for Continental Reflection Profiling (COCORP) Utah Line 1 (blue) is shown in Fig. 3A. A portion of Vastar Resources, Inc. Line V-11 (red) is shown in Fig. 3B. Thrust faults are the tightly folded Canyon Range thrust in the Canyon Range and the structurally lower Pavant thrust in the Pavant Range (teeth on upper plate). Inset map shows location relative to states in the western U.S.

A priority before any drilling is undertaken is to acquire new seismic reflection data in a grid encompassing both ARCO Hole-in-Rock and candidate locations for the proposed main hole. These data will be essential in establishing confidence in three-dimensional stratigraphic and structural geometry. Other pre-drill data that may be particularly useful—among many excellent ideas raised at the workshop—are high-resolution seismic and GPR (ground-penetrating radar) data combined with trenching across prominent fault scarps, and the establishment of closely spaced GPS stations aimed at determining more precisely how contemporary extension is distributed across the Sevier Desert.

Acknowledgments

We thank ICDP for sponsoring the workshop, the Utah Geological Survey for logistical assistance, and all of the participants for stimulating discussions.

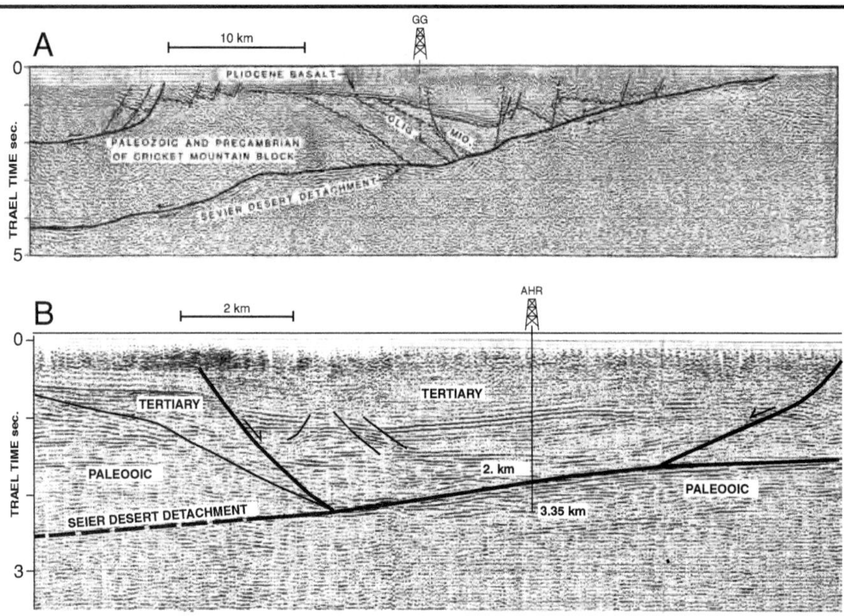

Figure 3. Examples of seismic reflection profiles from the Sevier Desert basin (see Fig. 2 for location). [A] Part of COCORP Utah Line 1, with interpretation of the Sevier Desert detachment from Von Tish et al. (1985). [B] A portion of Vastar Resources, Inc. Line V-11, with interpretation modified from Planke (1987). Projected wells: GG, Gulf Gronning; AHR, ARCO Hole-in-Rock. Line V-11, located ~3 km north of the Hole-in-Rock well, is the profile that best illustrates the geological structure in the vicinity of the well.

References

Axen, G.J., 2004. Mechanics of low-angle normal faults. *In* Karner, G.D., Taylor, B., and Driscoll, N.W. (Eds.), *Rheology and Deformation of the Lithosphere at Continental Margins.* New York (Columbia University Press), 46–91.

DeCelles, P.G., and Coogan, J.C., 2006. Regional structure and kinematic history of the Sevier fold and thrust belt, central Utah. *Geol. Soc. Amer. Bull.*, 118:841–864, doi:10.1130/B25759.1.

Manatschal, G., Müntener, O., Lavier, L.L., Minshull, T.A., and Péron-Pinvidic, G., 2007. Observations from the Alpine Tethys and Iberia-Newfoundland margins pertinent to the interpretation of continental breakup. *In* Karner, G.D., Manatschal, G., and Pinheiro, L.M. (Eds.), *Imaging, Mapping and Modelling Continental Lithosphere Extension and Breakup.* London (Geological Society, Spec. Publ. 282), 291–324.

McDonald, R.E., 1976. Tertiary tectonics and sedimentary rocks along the transition: Basin and Range province to plateau and thrust belt province, Utah. *In* Hill, J.G. (Ed.), *Symposium on Geology of the Cordilleran Hingeline.* Denver, Colorado (Rocky Mountain Association of Geologists), 281–317.

Niemi, N.A., Wernicke, B.P., Friedrich, A.M., Simons, M., Bennett, R.A., and Davis, J.L., 2004. BARGEN continuous GPS data across the eastern Basin and Range province, and implications for fault system dynamics. *Geophys. J. Int.*, 159:842–862, doi:10.1111/j.1365-246X.2004.02454.x.

Otton, J.K., 1995. Western frontal fault of the Canyon Range: Is it the breakaway zone of the Sevier Desert detachment? *Geology*, 23:547–550, doi:10.1130/0091-7613(1995)023<0547:WFFOTC>2.3.CO;2.

Oviatt, C.G., 1989. *Quaternary Geology of Part of the Sevier Desert, Millard County, Utah. Utah Geological and Mineral Survey Special Studies 70*, Salt Lake City, Utah (Utah Department of Natural Resources), 41 pp.

Planke, S., 1987. Cenozoic structures and evolution of the Sevier Desert basin, west-central Utah, from seismic reflection data. Master's thesis, University of Utah, Salt Lake City, Utah, 163 pp.

Sibson, R.H., 1985. A note on fault reactivation. *J. Struct. Geol.*, 7:751–754, doi:10.1016/0191-8141(85)90150-6.

Simpson, D.W., and Anders, M.H., 1992. Tectonics and topography of the western U.S. – an example of digital map making. *GSA Today*, 2:118–121.

Stockli, D.F., Linn, J.K., Walker, J.D., and Dumitru, T.A., 2001. Miocene unroofing of the Canyon Range during extension along the Sevier Desert Detachment, west central Utah. *Tectonics*, 20:289–307, doi:10.1029/2000TC001237.

Von Tish, D.B., Allmendinger, R.W., and Sharp, J.W., 1985. History of Cenozoic extension in central Sevier Desert, west-central Utah, from COCORP seismic reflection data. *AAPG Bull.*, 69:1077–1087.

Wernicke, B., 1995. Low-angle normal faults and seismicity: A review. *J. Geophys. Res.*, 100:20159–20174, doi:10.1029/95JB01911.

Wills, S., and Anders, M.H., 1999. Tertiary normal faulting in the Canyon Range, eastern Sevier Desert. *J. Geol.*, 107:659–682, doi:10.1086/314375.

Wills, S., Anders, M.H., and Christie-Blick, N., 2005. Pattern of Mesozoic thrust surfaces and Tertiary normal faults in the Sevier Desert subsurface, west-central Utah. *Am. J. Sci.*, 305:42–100, doi:10.2475/ajs.305.1.42.

Authors

Nicholas Christie-Blick and Mark H. Anders, Department of Earth and Environmental Sciences and Lamont-Doherty Earth Observatory of Columbia University, Palisades, N.Y. 10964-8000, U.S.A., e-mail: ncb@ldeo.columbia.edu.

Gianreto Manatschal, Université de Strasbourg, IPGS-EOST, 1 rue Blessig F-67 084, France.

Brian P. Wernicke, Division of Geological and Planetary Sciences, California Institute of Technology, Pasadena, Calif. 91125, U.S.A.

Web Link

http://sevier.icdp-online.org/

23

Understanding Paleoclimate and Human Evolution Through the Hominin Sites and Paleolakes Drilling Project

by Andrew Cohen, Ramon Arrowsmith, Anna K. Behrensmeyer, Christopher Campisano, Craig Feibel, Shimeles Fisseha, Roy Johnson, Zelalem Kubsa Bedaso, Charles Lockwood†, Emma Mbua, Daniel Olago, Richard Potts, Kaye Reed, Robin Renaut, Jean-Jacques Tiercelin, and Mohammed Umer

Understanding the evolution of humans and our close relatives is one of the enduring scientific issues of modern times. Since the time of Charles Darwin, scientists have speculated on how and when we evolved and what conditions drove this evolutionary story. The detective work required to address these questions is necessarily interdisciplinary, involving research in anthropology, archaeology, human genetics and genomics, and the earth sciences. In addition to the difficult tasks of finding, describing, and interpreting hominin fossils (the taxonomic tribe which includes *Homo sapiens* and our close fossil relatives from the last 6 Ma), much of modern geological research associated with paleoanthropology involves understanding the geochronologic and paleoenvironmental context of those fossils. When were they entombed in the sediments? What were the local and regional climatic conditions that early hominins experienced? How did local (watershed scale) and regional climate processes combine with regional tectonic boundary conditions to influence hominin food resources, foraging patterns, and demography? How and when did these conditions vary from humid to dry, or cool to warm? Can the history of those conditions (Vrba, 1988; Potts, 1996) be related to the evolution, diversification, stasis, or extinction of hominin species?

Most of the efforts to address these questions to date have centered on evidence from outcrops where the hominin fossils have been collected. Earth scientists have made great strides in understanding these contextual questions using fluvial, paleosol, and marginal lacustrine sediments associated with hominin fossils; however, this approach has its limitations. Outcrops, for example, cannot normally provide us with continuous, unweathered stratigraphic sections needed to address many questions relating events in hominin evolution and environmental change. The places where hominins actually lived (literally, above the water table) tend to have only discontinuous and relatively low resolution lithostratigraphic records of climate and other aspects of environmental change.

For these reasons the paleoanthropology community has turned to drill cores as a potential source of more highly resolved paleoenvironmental information. This concept is not new. Almost thirty years ago, the U.S. National Science Foundation (NSF) sponsored a workshop to examine the potential of recovering long sediment cores from the deepest and oldest of the modern African Rift Valley lakes, with a particular emphasis placed on how these records might inform our understanding of the environmental context of early hominin evolution (Lewin, 1981). In an influential paper, deMenocal (1995) demonstrated how northeastern African paleoclimate could be inferred from dust records encased in Deep Sea Drilling Project (DSDP) drill cores collected in the Gulf of Aden. This paper, as well as subsequent ones (deMenocal,

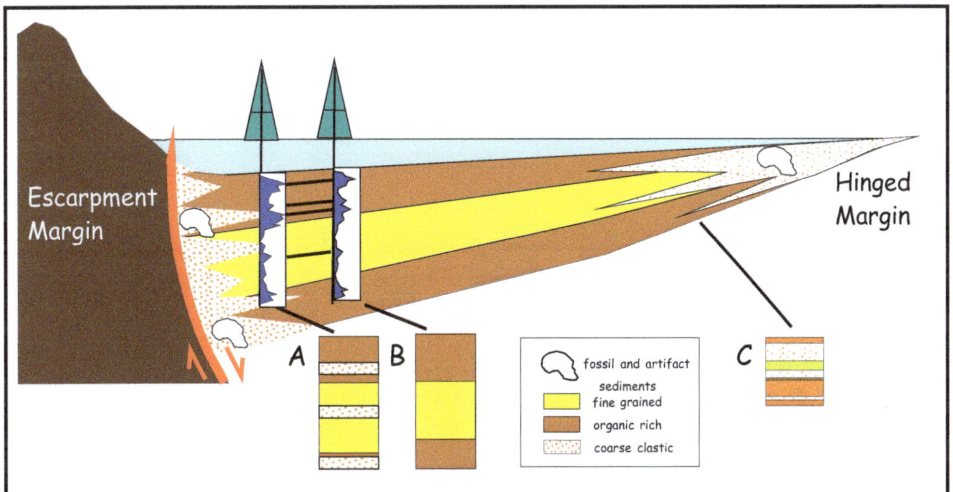

Figure 1. A conceptual model for drill site targeting in HSPDP (Hominin Sites and Paleolakes Drilling Project), using a half-graben rift basin depositional model. Note that a lake is shown in the diagrammatic cross section, although for the current phase of drilling we are targeting paleolake deposits where drilling from a barge can be avoided. The primary target (Drill Site B) would be far enough from the paleoshoreline to avoid most coarse clastic facies (pattern), with a relatively thick and highly resolved stratigraphic record. These types of offshore (fine-grained, organic rich) sediments (shown as yellow and brown solid colors) are most likely to yield the continuous paleoclimatic records useful for addressing current questions about hominin evolution/climate interactions. In contrast, a hinged margin site (C), even though potentially in close proximity to fossil and artifact localities, will yield a thinner and more discontinuous record, and one in which high stand lake deposits are over-represented. A more proximal target on the escarpment margin (A) will yield a dominantly coarse-grained stratigraphic record that is also less desirable for paleoclimate studies, although it may provide a useful secondary target for correlation into the hominin-bearing outcrops and/or records of tectonic uplift affecting the watershed.

2004; Feakins et al., 2005) by the same research group, provided the first more-or-less continuous offshore record of Neogene climate for hominin evolution occurring nearby terrestrially, and also set out a research agenda for investigating the possible role of orbital forcing mechanisms on the timing of hominin evolution. The idea of drilling the African Great Lakes ultimately came to fruition in 2005 with the successful completion of the Lake Malawi Drilling Project (Scholz et al., 2006), which had important paleoclimate implications for hominin evolution and the expansion of anatomically modern humans out of Africa (Cohen et al., 2007).

Despite these successes, the role of scientific drilling in answering questions of importance to paleoanthropology is still hampered by the location of the drill core records currently available relative to where hominin fossils have actually been found. Both the Gulf of Aden and Lake Malawi lie at considerable distances from the important fossil hominin and artifact sites in East Africa. Additionally, in the case of the DSDP dust records, the signal itself integrates climate history over a vast area of northeastern Africa and Arabia. To address questions about local environmental conditions, a drilling effort was needed to specifically target these regions. The Hominin Sites and Paleolakes Drilling Project (HSPDP) grew out of this need.

During 17–21 November 2008, over sixty scientists from thirteen countries met in Addis Ababa, Ethiopia for a workshop entitled "Scientific Drilling for Human Origins: Exploring the Application of Drill Core Records to Understanding Hominin Evolution" to chart a new path forward in the application of drill core studies to understanding human evolution. This workshop followed up on an earlier (2005) meeting on the broader theme of "Paleoclimates and Human Evolution" (Cohen et al., 2006), as well as two years of exploratory research and site surveys at potential drill sites.

Major themes of the 2008 workshop were laid out in introductory talks by Andy Cohen and Rick Potts; they concerned general criteria guiding the HSPDP in its attempt to identify promising drilling sites and what we might learn from drilling that would be of interest to paleoanthropology. Ideal candidate sites for drilling would individually provide cores with very high temporal resolution and continuous records of basin scale environmental history and regional scale paleoclimate, would have the potential for excellent geochronologic control, and would be amenable to a wide range of analytical techniques (Fig. 1). To allow us to address questions of environment/hominin interactions on a local scale, ideal drilling locations would also be in close proximity and be geologically and chronologically related to important hominin fossil and artifact sites. In particular, lacustrine deposits can provide an extraordinary range of paleoecological, geochemical, and sedimentological records of past climate, and typically yield much higher resolution records than coeval fluvial deposits or paleosols. In the East African Rift Valley, we have focused on relatively deep, paleolake deposits within the same depositional basins as important hominin fossil finds. Within rift half-graben lake basins, our search for ideal drilling targets has centered on what would have been the offshore portions of escarpment margins of half-grabens, where water depths would have been deepest and sediment accumulation rates highest over extended periods. Secondary objectives might include collecting cores from more proximal locations to the paleoshorelines, to tie in the paleoclimate records directly to fossil/archaeological sites via correlation and to collect records of tectonic activity and other processes within the watersheds. HSPDP has focused efforts on identifying several basins, spanning a range of "critical intervals" in hominin evolution. By comparing multiple lacustrine records, we have a realistic opportunity to address questions, for example, about the relationship between accurate local climatic records, possible orbital forcing mechanisms, or shorter (millennial scale) events with hominin evolutionary patterns such as adaptive radiations, species appearances, and species extinctions. With a highly detailed environmental record from each targeted basin, we can also compare the evolutionary responses to climate change in other biota to contrast possibly unique hominin responses to environmental variation (Potts, 1996). Interbasinal comparisons will also be critical for evaluating evolutionary hypotheses that invoke interactions between climate forcing and tectonic history in the rift (Sepulchre et al., 2006).

Precise measurements of climate change and rift dynamics may be associated with such things as the origin of *Homo*, major transitions in stone technology, rapid brain size increase, or the dispersal of *H. sapiens*; consequently, the goal is to substantially improve our ability to characterize the environmental factors that can be considered central to an understanding of human evolution.

Interregional and inter-temporal investigations will generate paleoenvironmental data for a series of repeated historical experiments, whose outcomes can be compared. As Potts suggested at the workshop, with drill core records in the basins where hominins have been recovered, we could evaluate the hypothesis that episodes of increased magnitude in environmental variability (e.g., recorded in paleo-precipitation proxies) correlate in time and space with hominin species diversification events and/or technological innovations. Through paleoclimate modeling approaches grounded in better data from core records, we will be better able to understand the dynamics that link climatic and evolutionary histories. Vegetation simulations using fully coupled General Circulation Models (GCMs) have already proven useful for understanding Quaternary climate changes relevant to the evolution of *H. sapiens* (Cowling et al., 2008).

To keep total project costs realistic given a goal of drilling multiple hominin basins, the HSPDP decided early on to pursue on-land drilling targets, accessible by a truck-mounted drill rig, rather than trying to return to extant lakes, where the logistics complexity and costs would be much greater. In 2007, HSPDP received funding from NSF to conduct site logistics and subsurface geophysical surveys at several promising localities (Fig. 2), which were the focus of much of the 2008 workshop discussion. All of these represented areas of long-standing geological and paleoanthropological research, where thick sequences of paleolake deposits occur in close proximity to important hominin or archaeological sites.

Chris Campisano described the scientific opportunities provided by the Northern Awash River Valley in the Afar region (Fig. 3), targeting the Middle Pliocene (~3.8–2.9 Ma) Hadar Formation lake deposits. Nearby fluvio-lacustrine and paleosol deposits of this formation and the overlying Busidima Formation have yielded spectacular fossil and artifact discoveries, including the iconic Lucy, First Family, and Selaam (Dikika baby) fossils, all representing the early hominin species Australopithecus afarensis (Johanson and White, 1979; Johanson et al., 1982; Alemseged et al., 2005; Behrensmeyer, 2008; Campisano and Feibel, 2008; Wynn et al., 2008), as well as the oldest known stone tools (Semaw et al., 1997). All of the *A. afarensis* fossils (representing ~90% of the fossils of this taxon discovered to date) accumulated around the margins of a relatively deep rift lake, the deposits of which would be the target of this drilling project (Fig. 2).

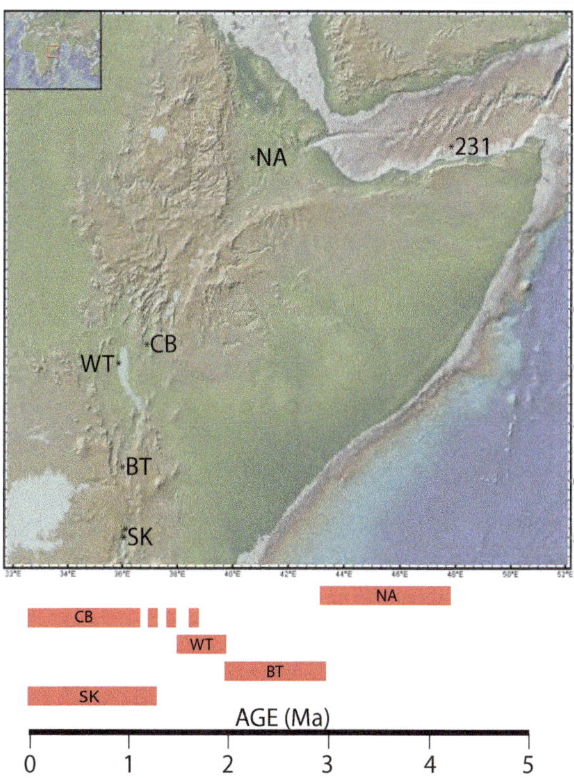

Figure 2. Map showing areas under active consideration for HSPDP drilling and stratigraphic ranges covered by each area. NA=Northern Awash River Valley, Northern Afar; CB=Chew Bahir; WT=West Turkana; BT=Baringo-Tugen Hills; SK=Southern Kenya. Gulf of Aden DSDP Drill Site 231, discussed in the text, is also shown. Base map developed with GeoMapApp©.

Figure 3. Outcrop photo of Middle Pliocene laminated diatomites from near the depocenter of a paleolake in the Northern Afar, one of the likely target locations and intervals for HSPDP drilling. Hill is about 8 m high.

Prior and ongoing geological research have focused drilling exploration efforts on the paleo-depocenter of a large Hadar basin lake, whose depositional cycles appear linked to global climate forcing (Campisano and Feibel, 2007; Dupont-Nivet, et al., 2008). A HSPDP team led by Roy Johnson conducted reflection seismic site surveys of potential drilling targets in this depocenter in 2008 with promising results.

Craig Feibel discussed potential onshore drilling targets on the west side of modern Lake Turkana in northern Kenya (Fig. 2). Plio-Pleistocene deposits rich in artifacts and fossils of hominins and other vertebrates (Harris et al., 1988; Feibel et al., 1989; Roche et al., 2004) accumulated in this region on the margins of large paleolakes (proto-Lake Turkana) (Lepre et al., 2007). A thick sequence of 2.0–1.5-Ma lake beds from one of these lacustrine phases would provide a continuous record of climate and environmental history during a particularly important interval of hominin evolution. Seminal events—such as evolution of several species of our own genus *Homo*, the first long-distance transport of stone tools, and the appearance of Acheulean (e.g., hand axe) types of tools, and the first expansion of hominins out of Africa—all occur during this critical time interval. A HSPDP seismic reflection profile and logistics survey conducted in 2008, coupled with decades of prior stratigraphic research on local outcrops, provide clearly defined drilling targets for sampling the offshore paleolake deposits spanning this interval.

Two areas in the southern Kenya Rift Valley (Fig. 2) that have been evaluated for drilling logistics were also considered in detail. Kay Behrensmeyer explained the potential for drilling around the Pleistocene (~1.2–0.5 Ma) Olorgesailie lake basin. For many years, Olorgesailie has been recognized as one of the most important archaeological sites in the world for documenting Acheulean (and more recently, Middle Stone Age) technology, and the deposits also contain a rich vertebrate fossil record, including hominins (Isaac and Isaac, 1977; Potts and Deino, 1995; Potts et al., 2004).

The fossils and stone tools occur on the margins of a Pleistocene lake, whose diatomaceous sediments would be one drilling target (Behrensmeyer et al., 2002; Owen et al., 2008). A terminal sump for the basin, the Koora Graben, forms another attractive target for obtaining an even higher resolution and more continuous paleoclimatic and paleohydrologic record for the basin. Robin Renaut discussed the potential for drilling nearby Lake Magadi, the second site of interest in the Rift Valley. This soda lake is well-known among sedimentologists and sedimentary geochemists (Eugster, 1980) and is also surrounded by a number of important Early and Middle Pleistocene archaeological and fossil hominin sites. The lake is currently the site of active trona production, making the playa surface readily accessible. It was drilled for mineral exploration purposes in the 1960s (unfortunately, the cores have been lost), and several studies have demonstrated the great potential of this basin for producing high resolution paleoclimate records (Taieb et al., 1991; Damnati and Taieb, 1995). Because Lake Magadi and Olorgesailie are in close proximity (~20 km apart) but occur in completely separate rift sub-basins (see Fig. 2) with no hydrologic connection, drill core records from both could allow paleoanthropologists to separate regional paleoclimatic effects on early hominin habitats from local watershed processes for the first time.

The workshop also provided opportunities to discuss other potential drilling targets. Alan Deino explained the potential for drilling Late Pliocene lake beds from the Tugen Hills region of the Lake Baringo Basin (central Kenya rift). The 3–2-Ma interval encompassed by these lake beds is also a period of important events in this time period, such as the first documented stone tools, the first fossils of *Homo*, and the evolution of several other hominin species. A sequence of well-dated, diatomaceous lake beds in the Tugen Hills area displays strong cyclicity interpreted as correlating to orbital (Milankovitch) time scales, making this an optimal site for testing ideas about linkages between hominin evolutionary events and climatic variability (Deino et al., 2006; Kingston et al., 2007). Giday WoldeGabriel discussed the potential of drilling in the Chew Bahir Basin in southern Ethiopia. This area sits on the interface between the Kenyan and Ethiopian rifts, regions whose biogeographic histories, despite their proximity, appear to be quite distinct. The area is also in close proximity to important Early Pleistocene fossil and archaeological sites (Suwa et al., 2007). The potential drilling target, a playa surface similar to Lake Magadi, will be the subject of preliminary field studies in late 2009. At the conclusion of the workshop, a call to the paleoanthropology and paleoclimate communities was issued by HSPDP for additional suggestions for sites that might be incorporated into future drilling proposals.

Much of the workshop was spent discussing analytical methods that might be used to study the cores collected. The wide range of techniques common to paleolimnological investigations (diatoms, ostracodes, pollen, stable isotopes, scanning XRF, sedimentology) were well-represented by researchers at the meeting. Additionally, leaders in exciting new methods in organic geochemistry, such as the TEX_{86}, MBT, and leaf wax studies, raised the possibility of obtaining much improved quantitative reconstructions of temperature and precipitation for the basins under study. Making the most of all of these methods will require the continuous, unweathered samples afforded only by drilling. Geochronology was a major focus of discussion, with many participants considering how best to transfer our existing (outcrop-based) understanding of the stratigraphy and age models for each basin to core records. Because of the wide range of time scales under consideration, many different techniques will ultimately be used. It was evident from the lively discussion that HSPDP could be a proving ground for new dating and correlation methods and the extension in time of existing ones. Breakout groups considered how coordination between research groups might be achieved between the different drilling areas to obtain intersite comparability. This ultimately will be critical for synthesizing results and examining those patterns in the environmental history-hominin evolution relationship that transcend individual locations or time intervals. The application of appropriate statistical techniques to infer trends and periodicities (Trauth et al., 2009) will be essential for the success of HSPDP, as will a vigorous modeling component to provide explanatory theory for understanding any potential climate-hominin evolution linkages that we uncover.

A drilling project focusing on human evolution in Africa offers unprecedented educational opportunities for communities near the drill sites, for educators, and for museums. The number of sites that might be drilled will provide welcome opportunities for research training for African geoscience students. Nontechnical outreach programs for local communities have proven successful for many drilling projects in the past (including at Lake Malawi in East Africa), and there was great enthusiasm at the meeting for developing a vigorous program along these lines for many different types of audiences. Because human evolution is a high profile target for scientific endeavors, many opportunities also exist for major museums to become involved in disseminating the results of HSPDP worldwide.

HSPDP opens up a new direction for scientific drilling. Although the focus of the immediate project is East Africa, the concept of developing synergies between the Earth science and paleoanthropology communities can (and almost certainly will) be applied worldwide, wherever the questions of relationships between human evolution and Earth system history arise. Future phases in such research might also involve collaborative ICDP/IODP efforts to collect new marine cores in critical locations and/or marine/terrestrial coring transects, as well as new drilling efforts on the extant and ancient deep rift lakes. "Drilling for human origins" promises to become an exciting new research direction in the years ahead.

Acknowledgements

We thank the U.S. National Science Foundation (Grant #0725553) and the International Continental Scientific Drilling Program for support of the site survey (NSF) and drilling workshop (ICDP and NSF) work discussed in this report. We also thank the governments of Kenya (Ministry of Science and Technology) and Ethiopia (ARCCH) for providing research permits and support for this project.

References

Alemseged, Z., Wynn, J.G., Kimbel, W.H., Reed, D., Geraads, D., and Bobe, R., 2005. First hominin from the Basal Member of the Hadar Formation, Dikika, Ethiopia and its geological context. *J. Hum. Evol.*, 49:499–514, doi:10.1016/j.jhevol.2005.06.001.

Behrensmeyer, A.K., 2008. Paleoenvironmental context of the Pliocene A.L. 333 "First Family" hominin locality, Hadar Formation, Ethiopia. In: Quade, J., and Wynn, J.G. (Eds.), *The Geological Context of Human Evolution in the Horn of Africa. Geological Society of America Special Paper,* 446: 203–214.

Behrensmeyer, A.K., Potts, R., Deino, A., and Ditchfield, P., 2002. Olorgesailie, Kenya: a million years in the life of a rift basin. In: Renaut, R.W., and Ashley, G.M. (Eds.), *Sedimentation in Continental Rifts, SEPM Spec. Publ.*, 73:97–106.

Campisano, C.J., and Feibel, C.S., 2007. Connecting local environmental sequences to global climate patterns: evidence from the hominin-bearing Hadar Formation, Ethiopia. *J. Hum. Evol.*, 53:515–527, doi:10.1016/j.jhevol.2007.05.015.

Campisano, C.J., and Feibel, C.S., 2008. Depositional environments and stratigraphic summary of the Hadar Formation at Hadar, Afar Depression, Ethiopia. In: Quade, J., and Wynn, J.G., eds. *The Geology of Early Humans in the Horn of Africa*, Boulder, Colo. (Geological Society of America), 179–201.

Cohen, A.S., Ashley, G.M., Potts, R., Behrensmeyer, A.K., Feibel, C., and Quade, J., 2006. Paleoclimate and Human Evolution Workshop. *Eos Trans. Am. Geophys. Union*, 87(16):161, doi:10.1029/2006EO160008.

Cohen, A.S., Stone, J.R., Beuning, K.R., Park, L.E., Reinthal, P.N., Dettman, D., Scholz, C.A., Johnson, T.C., King, J.W., Talbot, M.R., Brown, E.T., and Ivory, S.J., 2007. Ecological consequences of early Late-Pleistocene megadroughts in tropical Africa. *Proc. Nat. Acad. Sci.*, 104:16422–16427, doi:10.1073/pnas.0703873104.

Cowling, S.A., Cox, P.M., Jones, C.D., Maslin, M.A., Peros, M., and Spall, S.A., 2008. Simulated glacial and interglacial vegetation across Africa: implications for species phylogenies and trans-African migration of plants and animals. *Global Change Biol.*, 14:827–840, doi:10.1111/j.1365-2486.2007.01524.x.

Damnati, B., and Taieb, M., 1995. Solar and ENSO signatures in laminated deposits from Lake Magadi (Kenya) during the Pleistocene/Holocene transition. *J. Afr. Earth Sci.*, 21:373–382, doi:10.1016/0899-5362(95)00094-A.

Deino, A.L., Kingston, J.D., Glen, J.M., Edgar, R.K., and Hill, A., 2006. Precessional forcing of lacustrine sedimentation in the late Cenozoic Chemeron Basin, Central Kenya Rift. *Earth Planet. Sci. Lett.*, 247:41–60, doi:10.1016/j.epsl.2006.04.009.

deMenocal, P.B., 1995. Plio-Pleistocene African climate. *Science*, 270:53–59, doi:10.1126/science.270.5233.53.

deMenocal, P.B., 2004. African climate change and faunal evolution during the Plio-Pleistocene. *Earth Planet. Sci. Lett.*, 220:3–24, doi:10.1016/S0012-821X(04)00003-2.

Dupont-Nivet, G., Sier, M., Campisano, C.J., Arrowsmith, R., DiMaggio, E., Reed, K., Lockwood, C., Franke, C., and Häusing, S., 2008. Magnetostratigraphy of the hominin-bearing Hadar Formation (Ledi-Geraru, Ethiopia), and regional evidence for environmental change ca. 3.2 Ma. In: Quade, J., and Wynn, J.G. (Eds.), *The Geology of Early Humans in the Horn of Africa*. Boulder, Colo. (Geological Society of America), 67–86.

Eugster, H.P., 1980. Lake Magadi, Kenya and its precursors. In: Nissenbaum, A., (Ed.), *Hypersaline Brines and Evaporites*. Amsterdam (Elsevier), 195–232.

Feakins, S.J., deMenocal, P.B., and Eglinton, T.I., 2005. Biomarker records of late Neogene changes in northeast African vegetation. *Geology*, 33:977–980, doi:10.1130/G21814.1.

Feibel, C.S., Brown, F.H., and McDougall, I., 1989. Stratigraphic context of fossil hominids from the Omo Group deposits: Northern Turkana Basin, Kenya and Ethiopia. *Am. J. Phys. Anthro.*, 78:595–622, doi:10.1002/ajpa.1330780412.

Harris, J.M., Brown, F.H., and Leakey, M. G., 1988. Geology and paleontology of Plio-Pleistocene localities west of Lake Turkana, Kenya. *Contributions in Science*, 399:1–128.

Isaac, B., and Isaac, G.L., 1977. *Olorgesailie: Archeological Studies of a Middle Pleistocene Lake Basin in Kenya*. Chicago (University of Chicago Press), 272pp.

Johanson, D.C., and White, T.D., 1979. A systematic assessment of early African hominids. *Science*, 203:321–330, doi:10.1126/science.104384.

Johanson, D.C., Taieb, M., and Coppens, Y., 1982. Pliocene hominids from the Hadar Formation, Ethiopia (1973-1977): stratigraphic, chronological and paleoenvironmental contexts, with notes on hominid morphology and systematics. *Am. J. Phys. Anthro.*, 57:373–402, doi:10.1002/ajpa.1330570402.

Kingston, J.D., Deino, A., Hill, A., and Edgar, R., 2007. Astronomically forced climate change in the Kenyan Rift Valley 2.7-2.55 Ma: implications for the evolution of early hominin ecosystems. *J. Hum.Evol.*, 53:487–503, doi:10.1016/j.jhevol.2006.12.007.

Lepre, C.J., Quinn, R.L., Joordens, J.C.A., Swisher, C.C., and Feibel, C.S., 2007. Plio-Pleistocene facies environments from the KBS Member, Koobi Fora Formation: implications for climate controls on the development of lake-margin hominin habitats in the northeast Turkana Basin (northwest Kenya). *J. Hum. Evol.*, 53:504–514, doi:10.1016/j.jhevol.2007.01.015.

Lewin, R., 1981. Lake bottoms linked with human origins. *Science*, 211:564–566, doi:10.1126/science.211.4482.564.

Owen, R.B., Potts, R., Behrensmeyer, A.K., and Ditchfield, P., 2008. Diatomaceous sediments and environmental change in the Pleistocene Olorgesailie Formation, southern Kenya Rift Valley. *Palaeogeogr. Palaeoclimatol. Palaeoecol.*, 269:17–37, doi:10.1016/j.palaeo.2008.06.021.

Potts R., 1996. Evolution and climatic variability. *Science*, 273:922–923, doi:10.1126/science.273.5277.922.

Potts, R., and Deino, A., 1995. Mid-Pleistocene change in large mammal faunas of the southern Kenya rift. *Quat. Res.*, 43:106–113, doi:10.1006/qres.1995.1010.

Potts, R., Behrensmeyer, A.K., Deino, A., Ditchfield, P., and Clark, J., 2004. Small mid-Pleistocene hominin associated with East African Acheulean technology. *Science*, 305:75–78, doi:10.1126/science.1097661.

Roche, H., Brugal, J.-P., Delagnes, A., Feibel, C., Harmand, S., Kibunjia, M., Prat, S., and Texier, P.-J., 2004. Plio-Pleistocene archaeological sites in the Nachukui Formation, West Turkana, Kenya: synthetic results 1997-2001. *Comptes Rendus Palevol*, 2:663–673.

Scholz, C.A., Cohen, A.S., Johnson, T.C., and King, J.W., 2006. The 2005 Lake Malawi Scientific Drilling Project. *Sci. Drill.*, 2:17–19, doi:10.2204/iodp.sd.1.04.2006.

Sepulchre, P., Ramstein, G., Fluteau, R., Schuster, M., Tiercelin, J.J., and Brunet, M., 2006. Tectonic uplift and Eastern Africa aridification. *Science*, 313:1419–1423, doi:10.1126/science.1129158.

Semaw, S., Renne, P., Harris, J.W.K., Feibel, C.S., Bernor, R.L., Fesseha, N., and Mowbray, K., 1997. 2.5-million-year-old stone tools from Gona, Ethiopia. *Nature*, 385:333–336, doi:10.1038/385333a0.

Suwa, G., Asfaw, B., Haile-Selassie, Y., White, T., Katoh, S., Woldegabriel, G., Hart, W.K., Nakaya, H., and Beyene, Y., 2007. Early Pleistocene *Homo erectus* fossils from Konso, southern Ethiopia. *Anthro.Sci.*, 115:133–151, doi:10.1537/ase.061203.

Taieb, M., Barker, P., Bonnefille, R., Damnati, B., Gasse, F., Goetz, C., Hillaire, C., Icole, M., Masault, M., Roberts, N., Vincens, A., and Willamson, D., 1991. Histoire paléohydrologique du lac Magadi (Kenya) au Pleistocène superieur. *C. R. Acad. Sci., Sér. 2*, 313:339–346.

Trauth, M.H., Larrasoaña, J.C., and Mudelsee, M., 2009. Trends, rhythms and events in Plio-Pleistocene African climate. *Quat. Sci. Rev.*, 28:399–411, doi:10.1016/j.quascirev.2008.11.003.

Vrba, E.S., 1988. Late Pliocene climatic events and hominid evolution. In: Grine, F.E. (Ed.), *Evolutionary History of the "Robust" Australopithecines*. New York (Aldine de Gruyter), 405–426.

Wynn, J.G., Roman, D.C., Alemseged, Z., Reed, D., Geraads, D., and Munro, S., 2008. Stratigraphy, depositional environments, and basin structure of the Hadar and Busidima Formations at Dikika, Ethiopia. In: Quade, J., and Wynn, J.G., (Eds.), *The Geological Context of Human Evolution in the Horn of Africa. Geological Society of America Special Paper*, 446:87–118.

Authors

Andrew Cohen, Department of Geosciences, University of Arizona, Tucson, Ariz., U.S.A., e-mail: cohen@email.arizona.edu.

Ramon Arrowsmith, School of Earth and Space Exploration, Arizona State University, Tempe, Ariz., U.S.A., e-mail: ramon.arrowsmith@asu.edu.

Anna K. Behrensmeyer, Department of Paleobiology, National Museum of Natural History, Washington, DC, U.S.A., e-mail: behrensa@si.edu.

Christopher Campisano, Institute of Human Origins, Arizona State University, Tempe, Ariz., U.S.A., e-mail: campisano@asu.edu.

Craig Feibel, Departments of Geological Sciences and Anthropology, Rutgers University, New Brunswick, N.J., U.S.A., e-mail: feibel@rci.rutgers.edu.

Shimeles Fisseha, Department of Earth Sciences, Addis Ababa University, Addis Ababa,Ethiopia, e-mail:wmikel99@gmail.com.

Roy Johnson, Department of Geosciences, University of Arizona, Tucson, Ariz., U.S.A., e-mail:johnson6@email.arizona.edu.

Zelalem Kubsa Bedaso, Department of Geology, University of South Florida, Tampa, Fla., U.S.A., e-mail:zelex24@yahoo.com.

Charles Lockwood (†deceased), Department of Anthropology, University of Wittwatersrand, Johannesburg, South Africa.

Emma Mbua, Department of Paleontology, National Museums of Kenya, Nairobi, Kenya, e-mail:enmbua@yahoo.com.

Daniel Olago, Department of Geology, University of Nairobi, Nairobi, Kenya, e-mail: dolago@uonbi.ac.ke.

Richard Potts, Human Origins Program, National Museum of Natural History, Washington, DC, U.S.A., e-mail:pottsr@si.edu.

Kaye Reed, Institute of Human Origins, Arizona State University, Tempe, Ariz., U.S.A., e-mail:kreed.iho@asu.edu.

Robin Renaut, Department of Geological Sciences, University of Saskatchewan, Saskatoon, Saskatchewan, Canada, e-mail:robin.renaut@usask.ca.

Jean-Jacques Tiercelin, CNRS, Université de Rennes, Rennes, France, e-mail:jean-jacques.tiercelin@univ-rennes1.fr.

Mohammed Umer, Department of Earth Sciences, Addis Ababa University, Addis Ababa, Ethiopia, e-mail:moha_umero@yahoo.com.

Web Links

http://magadi.icdp-online.org
http://malawidrilling.syr.edu/photos/Outreach%20Program/index.html

Photo Credits

Fig. 3: Photo by R. Johnson.

Rapid Response Fault Drilling Past, Present, and Future

by Emily E. Brodsky, Kuo-Fong Ma, Jim Mori, Demian M. Saffer, and the participants of the ICDP/SCEC International Workshop

Introduction

New information about large earthquakes can be acquired by drilling into the fault zone quickly following a large seismic event. Specifically, we can learn about the levels of friction and strength of the fault which determine the dynamic rupture, monitor the healing process of the fault, record the stress changes that trigger aftershocks and capture important physical and chemical properties of the fault that control the rupture process. These scientific and associated technical issues were the focus of a three-day workshop on Rapid Response Fault Drilling: Past, Present, and Future, sponsored by the International Continental Scientific Drilling Program (ICDP) and the Southern California Earthquake Center (SCEC). The meeting drew together forty-four scientists representing ten countries in Tokyo, Japan during November 2008. The group discussed the scientific problems and how they could be addressed through rapid response drilling. Focused talks presented previous work on drilling after large earthquakes and in fault zones in general, as well as the state of the art of experimental techniques and measurement strategies. Detailed discussion weighed the tradeoffs between rapid drilling and the ability to satisfy a diverse range of scientific objectives. Plausible drilling sites and scenarios were evaluated. This is a shortened summary of the workshop report that discusses key scientific questions, measurement strategies, and recommendations. This report can provide a starting point for quickly mobilizing a drilling program following future large earthquakes. The full report can be seen at http://www.pmc.ucsc.edu/~rapid/.

Scientific Motivation

Scientists have long been asking questions about the physics of earthquakes in the hope that their catastrophic effects can somehow be mitigated. We can now generally delineate the regions where earthquakes are likely to occur. We can use seismic waves to measure the size and determine the location of an earthquake. We can model idealized ruptures and predict the resulting ground motions. We can design buildings that are resistant to the shaking near an earthquake. These advances have had direct benefits for the hundreds of million of people globally who are exposed to earthquake hazards.

However, further progress is largely hampered by a lack of fundamental data about the forces and processes on a fault deep beneath the ground during an earthquake. Knowing how much stress is exerted on the fault during slip is a prerequisite for predicting when the stress will break the fault and how far the fault will slip. We do not know how the fault evolves after an earthquake and how its changing structure resumes the strength required to generate more earthquakes. Even though aftershocks are one of the most predictable consequences of an earthquake, we do not understand the process that causes them. We also struggle to identify what aspects of fault geology separate seismogenic faults from faults that release stress by simply creeping. Immediately after a large earthquake, there is an opportunity to gain crucial information to fill these gaps in knowledge by observing the transient changes related to the rupture. For about two years, the fault is observably changed and a deep borehole can capture measurable signals germane to making progress in earthquake physics. Specifically, a deep borehole can address the following questions:

What Makes an Earthquake Big?

Earthquakes occur when the increasing local stress exceeds the frictional strength of a fault. A big earthquake occurs when the strength is pervasively low or the stress accumulation is high over a large region of the fault. Strain accumulation is relatively well understood and monitored through geodesy; however, the strength of faults and their time- and slip-dependence are much more poorly measured. At the large displacements and high velocities that characterize big earthquakes, even less is known about frictional strength, which may weaken dramatically during an earthquake. Understanding these currently unconstrained variations in frictional strength and their controlling mechanisms during large earthquakes is key to understanding the faulting process.

How Do Faults Heal for the Next Earthquake?

In the intervals between earthquakes, fault strength recovers slowly. Much of this process may not be measurable; however, recent work has shown that there is a window of opportunity after an earthquake when observable proxies for fault strength, such as seismic velocity, ground deformation, permeability, and aftershock rate, all change quickly (Brenguier et al., 2008; Kitagawa et al., 2002). Similarly, laboratory measurements document fault healing due to rate-and-state frictional and physicochemical processes, both of which evolve rapidly following slip (Hickman et al., 1995; Yasuhara et al., 2005). Scaling of laboratory observations to real faults is problematic but would be greatly aided by direct monitoring of a natural system.

How Do Earthquakes Interact?

Current data suggest that up to 80% of earthquakes are generally triggered by another earthquake (Marsan and Lengline, 2008). Some of these aftershocks themselves are large and damaging. However, while aftershocks are the most predictable earthquakes, we do not know if it is the dynamic stress of the seismic waves or the static stress change that activates the new earthquakes. How does stress change result in earthquakes that are delayed by weeks or months after the main shock? Determining the mechanism of triggered events is an important step on the road to earthquake prediction.

What are Important Material and Chemical Properties of the Seismogenic Zone?

Recent drilling projects on active faults show that fluids, mineralogy, and deformational structures all play important roles in fault behavior (Hickman et al., 1995; Ohtani et al., 2000; Ma et al., 2006). Specific combinations of these material properties may lead faults to episodically slip rapidly as opposed to gradually creep (Dixon and Moore, 2007). By sampling and instrumenting faults, we may be able to determine which combinations of properties lead to large ruptures. Furthermore, during healing following an earthquake, new surface area generated by fracturing is attacked by rapid chemical reactions that can destroy the original record of slip, so the sampling must be done before this process has advanced significantly.

How to Solve the Problems: What to Measure and Why

In order to learn about friction, healing, interaction, and nucleation, we need measurements of heat, stress, geological structures, frictional properties, permeability, seismic velocity, fluid chemistry, and local ground motion.

Temperature Measurements

Temperature profiles across the fault are the most direct way to quantify coseismic friction (Lachenbruch and Sass, 1980). Because most of the frictional resistance is dissipated as heat, any temperature increase on the fault at the time of the earthquake is potentially interpretable as a cumulative measure of frictional heat generation during slip. To measure the friction of an earthquake, the temperature must be measured in the vicinity of the seismogenic fault as quickly as possible and at depths where shear stress (a function of the effective normal stress and the effective coefficient of friction) is sufficiently large to generate an observable temperature anomaly (Fig. 1). The possibility of advection of frictionally generated heat by fluid flow following an earthquake must also be considered when interpreting downhole temperature data (Kano et al., 2006). Fulton et al. (2008) used a series of two-dimensional numerical models to simulate transient coupled fluid flow and heat transport to illustrate that temperatures across a fault zone should not be markedly affected by fluid flow driven away from the fault by locally elevated pore pressure (e.g., due to thermal pressurization or shear compaction). The results of these simulations illustrate that although the shape of a frictional heating anomaly will be modified if advective effects are substantial (e.g., in high-permeability cases), it would still be clearly resolvable after two years. Moreover, the use of repeated logging to monitor the evolution of thermal anomalies with time will allow separation of hydrologic (advective) and frictional heating signals.

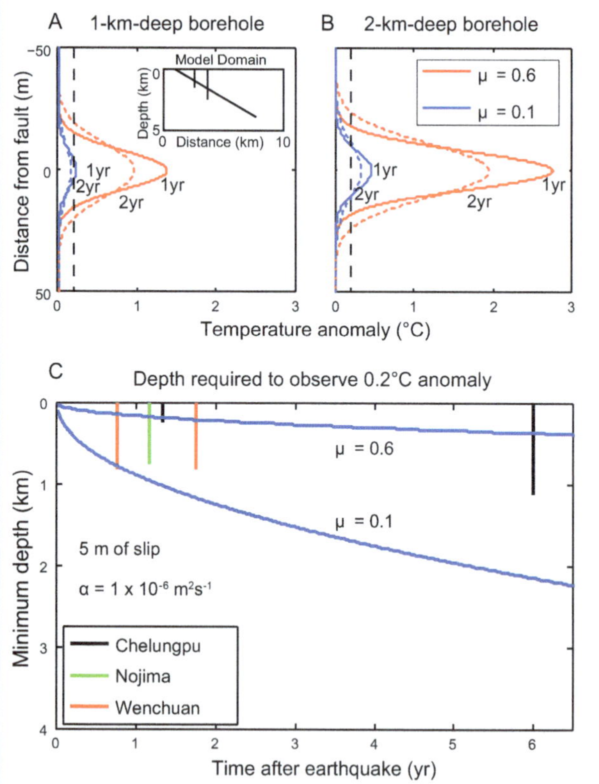

Figure 1. Frictional temperature anomalies resulting from a thrust earthquake with 5 m of slip and assuming a thermal diffusivity α of 10-6 m2s-1. (from Fulton et al., 2008) [A] Temperature anomaly for a borehole intersecting the fault at 1-km depth. Red and blue lines show frictional heating for effective friction coefficients of 0.6 and 0.1, representing strong and weak faults, respectively. Solid and dashed lines show the frictional heating anomaly one and two years after the earthquake, respectively. Dashed vertical black line shows the assumed detection threshold of 0.2°C. [B] Temperature anomaly for a borehole intersecting the fault at 2-km depth. [C] Blue lines show the minimum depth of borehole intersection with a fault to observe a temperature anomaly of 0.2°C as a function of time. The depth and timing of borehole completion of fault-zone drilling experiments are shown as vertical lines at the top of the panel [Figures from Fulton et al., 2008].

Direct Stress Measurements

An alternative approach to quantifying fault strength is to measure absolute stress directly. Determining the stress profile during rapid response fault-zone drilling, including the orientations and magnitudes of three-dimensional stresses, may tell us the stress change induced by fault rupture. Although it is difficult to estimate the magnitudes and orientations of three-dimensional stresses, the following quantities provide valuable information: vertical stress by downhole density logging, minimum horizontal stress by extended leak-off test/minifrac/hydraulic fracturing, and maximum horizontal stress by breakout width analyses (Zoback et al., 2003). A good complement to borehole data is core-based anelastic strain recovery (ASR) to produce complementary stress data (Lin et al., 2006). The longer the drilling is delayed, the more the stress will have decayed.

Geology of Cores

Some of the clearest evidence for the faulting process and frictional dissipation is preserved in fine-scale structures of cores. For instance, the presence of any melt rock (pseudotachylyte) is an immediate indicator of high friction early in the slip process, as heat is required to melt the rocks. Depending on the melt composition and permeability of the host, low friction during slip may be inferred later in the earthquake once melt has lubricated the surface. Geometric structures can also be used to assess the dynamic fluidization of the gouge and any likely rheology (Brodsky et al., 2009). If pseudotachylyte or fluidization structures are found, follow-up laboratory experiments are critical to determining the rheology of the melt or granular flow. Grain-size distribution also contains key information about dissipation. The energy that is absorbed in fracturing and surface creation is energy that is not dissipated by any other means, like friction. Similarly, measurements of fracture density in the borehole—using borehole imaging tools such as the Formation Micro Imager (FMI) and Resistivity-at-Bit (RAB)—and in the core itself constrain the dissipated energy, including that expended off the primary slip surface. Thus, observing the structure in a core provides some insight into frictional behavior. Fault-zone mineralogy changes rapidly during the healing process emphasizing the need for rapid sampling (Fig.2).

Laboratory Measurements

Most work on fault friction has been based on laboratory experiments on rocks sliding with limited velocities. However, unusual mineralogy, grain distribution, hydrological conditions, and structure in natural fault zones may significantly complicate the local frictional properties. Laboratory experiments can be manipulated to ascertain the exact effective stress and slip conditions to generate the observed stresses from the available materials. Furthermore, direct measurements of friction in the lab help to assess the applicability of the great body of rock mechanics to actual earthquake fault conditions. To diagnose the relative importance of various physical mechanisms during the earthquake, we must perform laboratory measurements of the frictional properties of gouge from the center of the fault zone (its core), and the results must be compared to the stresses inferred from the temperature and core-based stress measurements. Recent large-velocity experiments on natural gouge removed from fault cores show that significant weakening occurs as slip progresses (Tanikawa and Shimamoto, 2009).

Seismic Velocity

Seismic velocity has been shown to increase on a fault immediately after an earthquake (Li et al., 1998). A recent noise cross-correlation study showed velocity changes following the 2004 Parkfield earthquake that correlate with the slowly decaying strain transient (Brenguier et al., 2008). However, it is unclear if the seismic velocity change is an entirely shallow phenomenon or if it records a deeper process that is fundamental to the rheology of the fault; thus, measurements in a deep borehole are essential. Because changes often decay as 1/ time, measurements soon after an earthquake are important. Repeated vertical seismic profiles (VSPs) to measure velocity at depth in the nearfield of the fault could capture the progression of fault healing and changes in elastic moduli in the critical time period of the strong aftershock sequence. Borehole measurements should be supplemented with short-term surface deployments of short-period seismometers soon after the earthquake, with repeat deployments.

Hydrogeology

Dynamic fault-weakening mechanisms rely critically on the transient development of elevated (over ambient) pore fluid pressure via thermal or poro-elastic processes.

Figure 2. Slip lines observed on the slip surface of the Corinth core.

Permeability measurements in a fault zone are important in order to assess the sealing potential of the fault. The extent to which fluids may be trapped in a fault zone determines the shear stress during an earthquake and the locking strength before the next event. In addition, permeability can be treated as a proxy for fracturing and damage.

Permeability is known to evolve quickly after an earthquake rupture (Kitagawa et al., 2002). Pore pressure measurements are also important for evaluating static friction, since elevated pore pressure reduces the effective normal stress on faults and their overall shear strength (Rice, 1992). Preliminary interpretations based on drilling observations of the San Andreas Fault Observatory at Depth (SAFOD) (Zoback and Hickman, 2005) and the Chelungpu Fault in Taiwan (Doan et al., 2006) suggest that the fault zone is not overpressured at the depth where it is intersected by the borehole. These observations suggest that long-term weakening from high pore pressure may not be an adequate explanation for the apparent weakness of these fault zones.

Fluid Chemistry

Chemical precipitates close fractures and generate adhesion; therefore, the fluid chemistry must be sampled to develop a physical-chemical model of the healing process, and should include time series data from the fault core region (highly brecciated zone within the fault) as well as the damage zone. Real-time gas and hydrological monitoring will be critical for ascertaining the *in situ* chemical conditions.

What has been Learned from Prior Rapid Fault Zone Drilling Projects

Rapid drilling into a fault zone after an earthquake has already been done in the Nojima Fault, Japan, and the Chelungpu Fault, Taiwan. Future projects will build on this knowledge base. The full report also summarizes results from other fault zone drilling projects (SAFOD, Normal Aigion Fault in the Corinth Rift, Natural Earthquake Laboratory in South African Mines) that are not associated with recent large earthquakes.

Nojima Fault Drilling

Following the 1995 Kobe, Japan earthquake (Mw 6.9), the first rapid fault-drilling project was carried out on the Nojima Fault, which had surface rupture of about 1–2 m. Seven boreholes to depths of 500–1800 m were completed in fourteen months following the earthquake. Extensive geophysical measurements were made using standard logging techniques, along with borehole televiewer (BHTV), fullbore formation microimager (FMI), and dipole shear sonic imager (DSI) recordings. Resistivity, seismic velocity, and the various imaging techniques were useful for identifying fault zones and other physical properties throughout the drilled section. The borehole logging results were also correlated with the geological features of the core, such as fault gouge and cataclasites (Ikeda, 2001). Temperature measurements with relatively low resolution were made with the standard logging tools, but no signal associated with frictional heating was seen in the vicinity of the fault.

The boreholes provided nearly continuous core samples through the fault zone within granitic rocks and showed its complex geology. Distinct periods of seismic activity were recognized that were accompanied by intense hydrothermal alteration (Ohtani et al., 2000; Boullier et al., 2004; Tanaka et al., 2007). Ultracataclasites and pseudotachylytes that formed at 10-km depth or more were related to previous M6 to M7 earthquakes (Boullier et al., 2001; Murakami and Tagami, 2004). Estimates of the orientation of the local stress field were done with hydrofracturing and stress measurements on cores (Ikeda et al., 2001, Tsukahara et al., 2001; Yamamoto and Yabe, 2001). These stress direction studies showed that the maximum compressive stress was oriented nearly perpendicular to the fault, indicating low values for the static coefficient of friction.

Repeated water-injection experiments are being done in the 1800-m borehole to look for changes in fault permeability. Starting one year after the earthquake, and at subsequent intervals of a few years, water is pumped into the fault zone to induce small earthquakes. The timing and location of induced earthquakes, along with monitoring of resistivity, indicated permeabilities of 10^{-16} m^2 to 10^{-14} m^2 two years after the earthquake (Tadokoro, 2001). Subsequent injections suggest at least a 50% decrease in permeability over the subsequent three years (Kitagawa et al., 2002).

Taiwan Chelungpu Fault Drilling Project

An outstanding feature of the 1999 Chi-Chi, Taiwan earthquake (Mw 7.6) was the extensive surface rupture of the Chelungpu fault. The fault break was clearly visible for about 100 km, with surface displacements of 1–12 m. The regions of shallow slip provided the opportunity to access the seismogenic fault with drilling. The Taiwan Chelungpu Fault Drilling Project (TCDP) drilled two boreholes 40 m apart (completed 65 and 73 months after the earthquake) that penetrated the dipping thrust fault at depths of about one kilometer. The nearby surface fault had a displacement of about 12 m. There were also two shallow boreholes to depths of 330 m and 180 m on the northern and southern portions of the fault, respectively, which were completed 18 months after the earthquake (Tanaka et al., 2006).

The TCDP site near the town of Dakeng—where the earthquake had large displacement and slip velocity—was chosen in order to investigate the energy budget and the slip-weakening mechanisms of the earthquake (Ma et al., 2003). Identification of the fault zone in the core samples resulted from combining multiple techniques; they indicated a slip zone at 1111 m in Hole A and 1136 m in Hole B (Kuo et al., 2005; Ma et al., 2006; Hung et al., 2007; Sone et al., 2007; Yeh et al., 2007; Hirono et al., 2008). Evidence that the black gouge in these regions is in fact the slip surface comes from clay mineralogy, isotropic texture, absence of later reworking microstructures, and a very fine grain size indicative of high fracture energy in the Chi-Chi earthquake's energy budget (Ma et al., 2006).

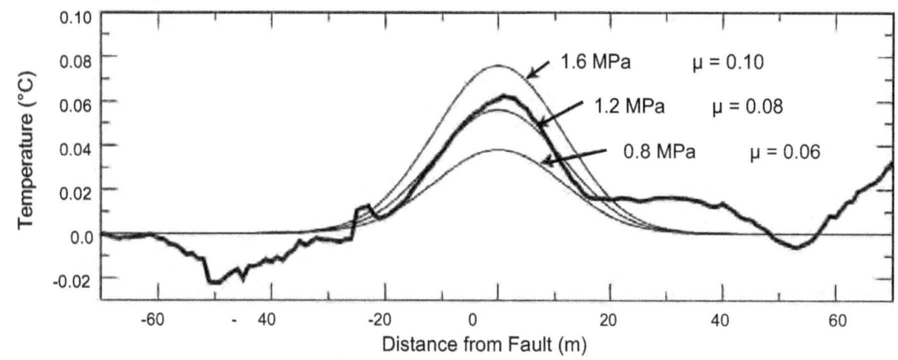

Figure 3. Temperature anomaly recorded in the Taiwan borehole and interpretation in terms of maximum friction coefficient (Kano et al., 2006).

Temperature measurements were made in the shallow and deep boreholes located in the northern region of large slip (Kano et al., 2006; Tanaka et al., 2006). Both studies found a consistent low level of residual heat close to the fault. In terms of frictional levels, these measurements imply a very low apparent coefficient of friction of around 0.1 to 0.2 (Fig. 3). Issues of water flow and determination of thermal diffusivity complicate the interpretation; however, these results show the potential importance of the temperature measurements at depth following a large earthquake.

Physical properties were measured continuously on cores using nondestructive methods (Hirono et al., 2007a). Clayclast aggregates and inverse grain size segregation were found in the Hole A principal slip zone and were interpreted as evidence for gouge fluidization as a result of frictional heating and thermal pressurization (Boutareaud et al., 2008; Boullier et al., 2009). Several other pieces of evidence—such as a magnetic susceptibility anomaly (Hirono et al., 2006; Mishima et al., 2006), clay mineralogy (Kuo et al., 2005; Hashimoto et al., 2008), whole-rock chemistry (Hirono et al., 2007b; Ishikawa et al., 2008), and gouge injections (Otsuki et al., 2005; Boullier et al., 2009)—show that the principal slip zone related to the Chi-Chi earthquake, as well as principal slip zones related to similar past earthquakes, have been heated to as much as 900°C and, consequently, are thermally pressurized and fluidized (Hirono et al., 2007b). Fluid circulation seems to be limited in the Chelungpu Fault, but healing processes and aseismic deformation by dissolution-diffusion-deposition mechanisms are important and efficient processes in the fault and damage zones (Boullier et al., 2009).

A cross-hole pumping test indicated a permeability of 10-16 m^2 to 10-18 m^2 along the fault, and simultaneous monitoring data suggested that the fault zone is moderately overpressured (Doan et al., 2006). Laboratory results on the shallow core samples show that the permeability of the northern region is an order of magnitude less than for the southern region (Tanikawa and Shimamoto, 2009).

TCDP samples have induced high international interest in friction experiments using soft rocks as initial material, and for slip-weakening processes in general. These studies have shown the lateral variability of the principal slip zone structure and thickness, along with the importance of clays and mineralogy of the gouge for slip weakening during coseismic slip. The analyses from FMI and DSI logging also indicate variations of the stress orientation near the principal slip zone, suggesting a complete stress drop during faulting with the exchange of the orientation of the SHmax to SHmin (Wu et al., 2007).

Wenchuan Fault Scientific Drilling Program

During the workshop, the Wenchuan Fault Scientific Drilling Program began drilling, with the first borehole begun only 178 days after the devastating Mw 7.9 earthquake in Sichuan, China on 12 May 2008. The project plans four deep holes through the faults in the Wenchuan region and represents one the most rapid drilling responses after an earthquake.

Proposed Earthquake Drilling Plan

Rapid drilling entails rapid decision making. Some of the discussion at the workshop centered on developing a drilling plan that can serve as a blueprint for future rapid response operations. All of the participants were cognizant of the variety of situations likely to be encountered in the real world and understood that no pre-earthquake drilling plan would be appropriate in detail for any particular situation. However, having a scenario plan that specifically addresses the key scientific questions helps provide a useful starting point for the work in the critical time following an earthquake.

Rapid response drilling should be considered whenever there is a large, on-land earthquake with sufficient surface slip (>1 m) to generate large geophysical and geological anomalies at depths accessible to a drilling program. The actual target earthquakes will probably be determined by the local scientific and technical infrastructure. Accessible onshore M≥7 earthquakes with at least 1 m of surface slip occur about every two to three years. The 1992 Landers, 1999 Izmit, 1999 Hector Mine, and 2002 Denali earthquakes would all have been appropriate targets, in addition to the 1995 Kobe, 1999 Chi-Chi, and 2008 Wenchuan earthquakes. Workshop participants reported on likely scenarios in China, Iran, Japan, New Zealand, Turkey, and the United States. Several of these countries already have fault zone drilling programs underway that could be accelerated and adapted when a large earthquake occurs.

For the plan, several basic selection requirements for drilling sites are aimed at addressing fundamental questions about earthquake energy budgets and fault zone processes. First, it is desirable to drill where the fault and regional geologic structure are well-known and relatively simple. A single major fault strand rather than multiple active traces will provide data that are more easily interpretable in the context of seismological information. Second, drilling should target a region of high co-seismic slip on the fault, as defined by seismological inversions and surface deformation data. Slip should be at least 1 m in order to ensure a measurable frictional signal. Ideally, the drilling would reach nucleation depths, although this is unlikely to be feasible in many cases, and thus most drilling projects will concentrate on the problems of friction, healing, and earthquake interaction. Third, crystalline rock is preferred over sedimentary rock because of its low permeability and its well-defined frictional properties. Unconsolidated sediments are particularly problematic for studying the seismogenic zone. Fourth, a dipping fault is ideal (though not required) because it would allow crossing the fault at a high angle with a vertical borehole. This scenario drilling plan assumes a dipping fault. Finally, it would be desirable to drill in a location where there are numerous pre-existing seismological and GPS stations, to provide additional needed information about the earthquake.

A significant challenge in designing and implementing a drilling plan is coordination of multiple downhole measurements and instrument installations, due to competition for limited space in the hole and potential interference. Multiple objectives must be spatially and temporally balanced, requiring care in the following: (a) design and completion strategy, including planning of measurement sequences to minimize interference; and (b) prioritization and compromise of some measurements (e.g., running only low volume perturbation tests for stress and hydrologic measurement objectives, or conducting only limited coring). Key observations of temperature logging, seismometers and strainmeters, hydrologic testing, and fluid sampling, all have specific hole configuration and/or timing requirements that are often conflicting.

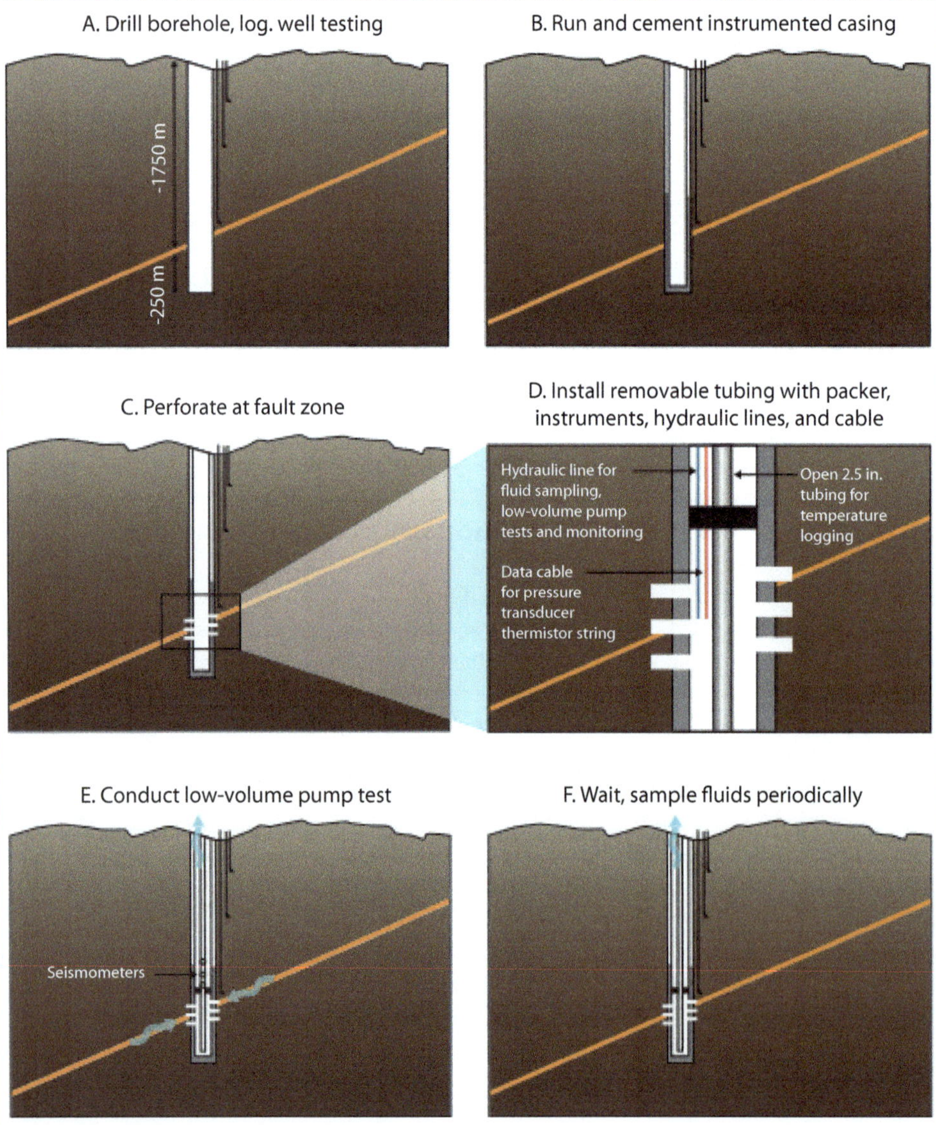

Figure 4. Schematic of a rapid drilling plan.

We provide a sample drilling plan (Fig. 4) to illustrate how to balance competing measurements in a realistic scenario. One key consideration is the minimum target depth required to obtain meaningful information about frictional heating, given instrument resolution and the magnitude of the expected thermal anomaly (based on average shear stress and the total amount of slip). There is a tradeoff between these considerations and the amount of time needed for drilling. For a suite of conservative assumptions about heat generation, a 2-km-deep hole drilled within six to twelve months following an earthquake is sufficient; deeper drilling is better, but the time and cost increase exponentially with increased depth. This depth range is also adequate for the desired geophysical measurements. This scenario drilling plan shows a 2-km-deep hole, penetrating a fault at 1750-m vertical depth. This plan and general completion strategy could be adjusted for any target fault depth, although the details of casing set points and sizes would need to be modified.

Recommendations

The key recommendation of the workshop is that rapid drilling should be a priority of the international drilling community following on-land earthquakes that occur with more than 1 m of slip in regions with sufficient infrastructure to support the technical operations. The optimal drill site is on crystalline rock with a dipping fault and well-characterized geology. The hole should be initiated within six months after the earthquake and should be drilled to a depth of at least 2 km.

To achieve this goal, the workshop participants recommend the following courses of action for ICDP and the scientific community (listed separately).

Recommendations for ICDP

- ICDP should form a standing committee on fault drilling.
- A mechanism should be created to facilitate rapid evaluation and funding of scientific drilling proposals immediately following an earthquake.
- Fault zone drilling projects should include a contingency plan for accelerating or reconfiguring the project in case of a major earthquake.
- An inventory of baseline data on and near major faults should be constructed under the supervision of the standing committee as soon as possible for reference information before an earthquake occurs. The inventory should include existing, accessible boreholes, current seismological and geodetic instrumentation, and paleoseismic constraints. As part of this process, geophysical logging should be conducted on extant boreholes in strategic locations near major faults.
- An emergency response kit should be built and maintained by ICDP, including material stockpiles and borehole instrumentation and including special downhole tools.
- Investments should be made into developing and improving sensors for deep, hot adverse fault zone environments.

Recommendations for the Scientific Community

- National and regional groups should form and work on scenario plans for their localities. Clear and accurate geological and geophysical information prior to a major earthquake are critical for optimally selecting a study site after the earthquake.
- Complementary seismological and geological studies at potential sites of large earthquakes should be undertaken. Existing boreholes should be logged. These baseline data are critical to post-earthquake interpretations.

Acknowledgements

The Rapid Response Fault Drilling Workshop and accompanying report were supported by the International Continental Scientific Drilling Program, Southern California Earthquake Center, University of California - Santa Cruz, and Kyoto University.

References

Boullier, A.M., Fujimoto, K., Ito, H., Ohtani, T., Keulen, N., Fabbri, O., Amitrano, D., Dubois, M., and Pezard, P., 2004. Structural evolution of the Nojima fault (Awaji Island, Japan) revisited from the GSJ drill hole at Hirabayashi. *Earth, Planets Space*, 56(12):1233–1240.

Boullier, A.M., Ohtani, T., Fujimoto, K., Ito, H., and Dubois, M., 2001. Fluid inclusions in pseudotachylytes from the Nojima fault, Japan. *J. Geophys. Res.* Solid Earth, 106(B10):21965–21977, doi:10.1029/2000JB000043.

Boullier, A.M., Yeh, E.-C., Boutareaud, S., Song, S.-R., and Tsai, C.-H., 2009. Microscale anatomy of the 1999 Chi-Chi earthquake fault zone. *Geochem. Geophys. Geosyst.*, 10:Q03016, doi:10.1029/2008GC002252.

Boutareaud, S., Boullier, A.M., Beck, P., Calugaru, D.-G., Han, R., and Tsutsumi, A., 2008. Clast-clay aggregates as new indicator of shallow crustal seismic slips. *Geophys. Res. Abstr.*, 10:1607-7962/gra/EGU2008-A-04694.

Brenguier, F., Campillo, M., Hadziioannou, C., Shapiro, N.M., Nadeau, R.M., and Larose, E., 2008. Postseismic relaxation along the San Andreas Fault at Parkfield from continuous seismological observations. *Science*, 321:1478–1481, doi:10.1126/science.1160943.

Brodsky, E.E., Rowe, C.D., Manighenni, F., and Moore, J.C., 2009. A geological fingerprint of extremely low viscosity fault fluids during an earthquake. *J. Geophys. Res.*, 114:B01303, doi:10.1029/2008JB005633.

Dixon, T.H., and Moore, J.C., 2007. *The Seismogenic Zone of Subduction Thrust Faults.* Irvington, N.Y. (Columbia University Press), 680 pp.

Doan, M.-L., Brodsky, E.E., Kano, Y., and Ma, K.-F., 2006. In situ measurement of the hydraulic diffusivity of the active Chelungpu Fault, Taiwan. *Geophys. Res. Lett.*, 33:L16317, doi:10.1029/2006GL026889.

Fulton, P.M., Saffer, D.M., and Brodsky, E.E., 2008. The role of advection on fault zone temperature after an earthquake: implications for rapid response drilling. *EOS Trans. AGU, Fall Meeting Suppl.*, 89(52), Abstract T53F-05.

Hashimoto, Y., Tadai, O., Tanimizu, M., Tanikawa, W., Hirono, T., Lin, W., Mishima, T., Sakaguchi, M., Soh, W., Song, S.-R., Aoike, K., Ishikawa, T., Murayama, M., Fujimoto, K., Fukuchi, T., Ikehara, M., Ito, H., Kikuta, H., Kinoshita, M., Masuda, K., Matsubara, T., Matsubayashi, O., Mizoguchi, M., Nakamura, N., Otsuki, K., Shimamoto, T., Sone, H., and Takahashi, M., 2008. Characteristics of chlorites in seismogenic fault zones: The Taiwan Chelungpu Fault Drilling Project (TCDP) core sample. *e-Earth*, 3:1–6, http://www.electronic-earth.net/3/issue1.html.

Hickman, S., Sibson, R., and Bruhn, R., 1995. Introduction to special section: Mechanical involvement of fluids in faulting. *J. Geophys. Res.*, 100:12831–12840, doi:10.1029/95JB01121.

Hirono, T., Lin, W., Yeh, E.-C., Soh, W., Hashimoto, Y., Sone, H., Matsubayashi, O., Aoike, K., Ito, H., Kinoshita, M., Murayama, M., Song, S.-R., Ma, K.-F., Hung, J.H., Wang, C.-Y., and Tsai, Y.-B., 2006. High magnetic susceptibility of fault gouge within Taiwan Chelungpu fault: nondestructive continuous measurements of physical and chemical properties in fault rocks recovered from Hole B, TCDP. *Geophys. Res. Lett.*, 33:L15303, doi:10.1029/2006GL026133.

Hirono, T., Sakaguchi, M., Otsuki, K., Sone, H., Fujimoto, K., Mishima, T., Lin, W., Tanikawa, W., Tanimizu, M., Soh, W., Yeh, E.-C., and Song, S.-R., 2008. Characterization of slip zone associated with the 1999 Taiwan Chi-Chi earthquake: X-ray CT image analyses and microstructural observations of the Taiwan Chelungpu fault. *Tectonophysics*, 449:63–84, doi: 10.1016/j.tecto.2007.12.002.

Hirono, T., Yeh, E.-C., Lin, W., Sone, H., Mishima, T., Soh, W., Hashimoto, Y., Matsubayashi, O., Aoike, K., Ito, H., Kinoshita, M., Murayama, M., Song, S.-R., Ma, K.-F., Hung, J.-H., Wang, C.-Y., Tsai, Y.-B., Kondo, T., Nishimura, M., Moriya, S., Tanaka, T., Fujiki, T., Maeda, L., Muraki, H., Kuramoto, T., Sugiyama, K., and Sugawara, T., 2007a. Nondestructive continuous physical property measurements of core samples recovered from hole B, Taiwan Chelungpu-Fault Drilling Project. *J. Geophys. Res.*, 112:B07404, doi:10.1029/2006JB004738.

Hirono, T., Yokoyama, T., Hamada, Y., Tanikawa, W., Mishima, T., Ikehara, M., Famin, V., Tanimizu, M., Lin, M., Soh, W., and Song, S.-R., 2007b. A chemical kinetic approach to estimate dynamic shear stress during the 1999 Taiwan Chi-Chi earthquake. *Geophys. Res. Lett.*, 34:L19308, doi:10.1029/2007GL030743.

Hung, J.-H., Wu, Y.-H., Yeh, E.-C., Wu, J.-C., and TCDP Scientific Party, 2007. Subsurface structure, physical properties, and fault zone characteristics in the scientific drill holes of Taiwan Chelungpu-fault Drilling Project. *Terr. Atmos. Ocean. Sci.*, 18(2):271–293, doi:10.3319/TAO.2007.18.2.271(TCDP).

Ikeda, R., 2001. Outline of the fault zone drilling project by NIED in the vicinity of the 1995 Hyogo-ken Nanbu earthquake, Japan. *Island Arc*, 10:199–205, doi:10.1046/j.1440-1738.2001.00318.x.

Ikeda, R., Iio, Y., and Omura, K., 2001. *In situ* stress measurements in NIED boreholes in and around the fault zone near the 1995 Hyogo-ken Nanbu earthquake, Japan. *Island Arc*, 10:252–260, doi:10.1046/j.1440-1738.2001.00323.x.

Ishikawa, T., Tanimizu, M., Nagaishi, K., Matsuoka, J., Tadai, O., Sakaguchi, M., Hirono, T., Mishima, M., Tanikawa, W., Lin, W., Kikuta, H., Soh, W., and Song, S.-R., 2008. Coseismic fluid–rock interactions at high temperatures in the Chelungpu fault. *Nature Geosci.*, 1:679–683, doi:10.1038/ngeo308.

Kano, Y., Mori, J., Fujio, R., Ito, H., Yanagidani, T., Nakao, S., and Ma, K.-F., 2006. Heat signature on the Chelungpu fault associated with the 1999 Chi-Chi, Taiwan earthquake. *Geophys. Res. Lett.*, 33:L14306, doi:10.1029/2006GL026733.

Kitagawa, Y., Fujimori, K., and Koizumi, N., 2002. Temporal change in permeability of the rock estimated from repeated water injection experiments near the Nojima fault in Awaji Island, Japan. *Geophys. Res. Lett.*, 29:1483, doi:10.1029/2001GL014030.

Kuo, L.-W., Song, S.-R., and Chen, H.-Y., 2005. Characteristics of clay mineralogy in the fault zone of the TCDP and its implication. *EOS Trans. AGU, Fall Meet. Suppl.*, 86(52), Abstract T43D-05.

Lachenbruch, A.H., and Sass, J.H., 1980. Heat flow and energetics of the San Andreas fault zone. *J. Geophys. Res.*, 85:6185–6222.

Li, Y.-G., Vidale, J.E., Aki, K., Xu, F., and Burdette, T., 1998. Evidence of shallow fault zone strengthening after the 1992 M7.5 Landers, California, earthquake. *Science*, 279(5348):217–219, doi:10.1126/science.279.5348.217.

Lin, W., Kwasniewski, M., Imamura, T., and Matsuki, K., 2006. Determination of three-dimensional in situ stresses from anelastic strain recovery measurement of cores at great depth. *Tectonophysics*, 426:221–238, doi: 10.1016/j.tecto.2006.02.019.

Ma, K.-F., Brodsky, E.E., Mori, J., Ji, C., Song, T.-R.A., and Kanamori, H., 2003. Evidence for fault lubrication during the 1999 Chi-Chi, Taiwan, earthquake (Mw7.6). *Geophys. Res. Lett.*, 30(5):1244, doi:10.1029/2002GL015380.

Ma, K.-F., Tanaka, H., Song, S.-R., Wang, C.-Y., Hung, J.-H., Tsai, Y.-B., Mori, J., Song, Y.-F., Yeh, E.-C., Soh, W., Sone, H., Kuo, L.-W., and Wu, H.-Y., 2006. Slip zone and energetics of a large earthquake from the Taiwan Chelungpu-fault Drilling Project. *Nature*, 444:473–476, doi:10.1038/nature05253.

Marsan, D., and Lengline, I., 2008. Extending earthquakes' reach through cascading. *Science*, 319:1076–1079, doi:10.1126/science.1148783.

Mishima, T., Hirono, T., Soh, W., Ikehara, M., Lin, W., Tanikawa, W., Yeh, E.-C., Song, S.-R., and Wang, C., 2006. Thermal history estimation of the Taiwan Chelungpu Fault using rock-magnetic methods. *Geophys. Res. Lett.*, 33:L23311, doi:10.1029/2006GL028088.

Murakami, M., and Tagami, T., 2004. Dating pseudotachylyte of the Nojima fault using the zircon fission-track method. *Geophys. Res. Lett.*, 31:L12604, doi:10.1029/2004GL020211.

Ohtani, T., Fujimoto, K., Ito, H., Tanaka, H., Tomida, N., and Higuchi, T., 2000. Fault rocks and past to recent fluid characteristics from the borehole survey of the Nojima fault ruptured in the 1995 Kobe earthquake, southwest Japan. *J. Geophys. Res.*, 105(B7):16161–16171, doi:10.1029/2000JB900086.

Otsuki, K., Uduki, T., Monzawa, N., and Tanaka, H., 2005. Clayey injection veins and pseudotachylyte from two boreholes penetrating the Chelungpu Fault, Taiwan: their implications for the contrastive seismic slip behaviors during the 1999 Chi-Chi earthquake. *Island Arc*, 14:22–36.

Rice, J.R., 1992. Fault stress states, pore pressure distributions, and the weakness of the San Andreas Fault. In: Evans, B., and Wong, T.-F., (Eds.), *Fault Mechanics and Transport Properties of Rocks*. San Diego, Calif., (Academic Press), 475–503.

Sone, H., Yeh, E.-C., Nakaya, T., Hung, J.-H., Ma, K.-F., Wang, C.-Y., Song, S.-R., and Shimamoto, T., 2007. Mesoscopic structural observations of cores from the Chelungpu fault system, Taiwan Chelungpu-Fault Drilling Project hole-A, Taiwan. *Terr. Atmos. Ocean. Sci.*, 18(2):359–377, doi:10.3319/TAO.2007.18.2.359(TCDP).

Tadokoro, K., 2001. Repeated water injection experiments at the Nojima fault zone, Japan: induced earthquakes and temporal change in permeability. Poster presentation. ICDP Workshop on Drilling the Chelungpu Fault. Investigating the Physics of Faulting for a Recent Large Earthquake, 27 –30 September 2001, Taiwan, Taipei.

Tanaka, H., Chen, W.-M., Wang, C.-Y., Ma, K.-F., Urata, N., Mori, J., and Anod, M., 2006. Frictional heat from faulting of the 1999 Chi-Chi, Taiwan, earthquake. *Geophys. Res. Lett.*, 33:L16316, doi:10.1029/2006GL026673.

Tanaka, H., Omura, K., Matsuda, T., Ikeda, R., Kobayashi, K., Murakami, M., and Shimada, K., 2007. Architectural evolution of the Nojima fault and identification of the activated slip layer by Kobe earthquake. *J. Geophys. Res.*, 112:B07304, doi:10.1029/2005JB003977.

Tanikawa, W., and Shimamoto, T., 2009. Frictional and transport properties of the Chelungpu fault from shallow borehole data and their correlation with seismic behavior during the 1999 Chi-Chi earthquake. *J. Geophys. Res.*, 114:B01402, doi:10.1029/2008JB005750.

Tsukahara, H., Ikeda, R., and Yamamoto, K., 2001. In situ stress measurements in a borehole close to the Nojima Fault. *Island Arc*, 10:261–265, doi:10.1046/j.1440-1738.2001.00324.x.

Wu, H.-Y., Ma, K.-F., Zoback, M., Boness, N., Ito, H., Hung, J.-H., and Hickman, S., 2007. Stress orientations of Taiwan Chelungpu-Fault Drilling Project (TCDP) hole-A as observed from geophysical logs. *Geophys. Res. Lett.*, 34:L01303, doi:10.1029/2006GL028050.

Yamamoto, K. and Yabe, Y., 2001. Stresses at sites close to the Nojima Fault measured from core samples. *Island Arc*, 10:266–281, doi:10.1046/j.1440-1738.2001.00325.x.

Yasuhara, H., Marone, C., and Elsworth, D., 2005. Fault zone restrengthening and frictional healing: the role of pressure solution. *J. Geophys. Res.*, 110:B06310, doi:10.1029/2004JB0033272005.

Yeh, E.-C., Sone, H., Nakaya, T., Ian, K.-H., Song, S.-R., Hung, J.-H., Lin, W., Hirono, T., Wang, C.-Y., Ma, K.-F., Soh, W., and Kinoshita, M., 2007. Core description and characteristics of fault zones from Hole-A of the Taiwan Chelungpu-Fault Drilling Project. *Terr. Atmos. Ocean. Sci.*, 18(2):327–357, doi:10.3319/TAO.2007.18.2.327(TCDP).

Zoback, M.D. and Hickman, S.H., 2005. Preliminary observations of stress and fluid pressure in and near the San Andreas Fault at depth in the SAFOD boreholes. *EOS Trans. AGU, Fall Meet. Suppl.*, 86(52), Abstract T21A-0438.

Zoback, M.D., Barton, C.A., Brudy, M., Castillo, D.A., Finkbeiner, T., Grollimund, B.R., Moos, D.B., Peska, P., Ward, C.D., and Wiprut, D.J., 2003. Determination of stress orientation and magnitude in deep wells. *Int. J. Rock Mech. Min. Sci.*, 40:1049–1076, doi:10.1016/j.ijrmms.2003.07.001.

Authors

Emily E. Brodsky, Deptartment of Earth and Planetary Sciences, UC Santa Cruz, 1156 High Street, Santa Cruz, Calif., 95060, U.S.A., e-mail: brodsky@pmc.ucsc.edu.
Kuo-Fong Ma, Department of Earth Sciences, Institute of Geophysics National Central University, No.300, Jhongda Road, Jhongli City, Taoyuan County 32001, Taiwan (R.O.C.).
Jim Mori, Disaster Prevention Research Institute, Kyoto University, Gokasho, Uji, Kyoto, 611-0011, Japan.
Demian M. Saffer, Department of Geosciences, The Pennsylvania State University, 310 Deike Building, University Park, Pa., 16802, U.S.A.
and participants of the ICDP/SCEC International Workshop on Rapid Response Fault Drilling

25

Deep Fault Drilling Project—Alpine Fault, New Zealand

by John Townend, Rupert Sutherland, and Virginia Toy

The Alpine Fault, South Island, New Zealand, constitutes a globally significant natural laboratory for research into how active plate-bounding continental faults work and, in particular, how rocks exposed at the surface today relate to deep-seated processes of tectonic deformation, seismogenesis, and mineralization. The along-strike homogeneity of the hanging wall, rapid rate of dextral-reverse slip on an inclined fault plane, and relatively shallow depths to mechanical and chemical transitions make the Alpine Fault and the broader South Island plate boundary an important international site for multi-disciplinary research and a realistic target for an ambitious long-term program of scientific drilling investigations.

Introduction

The mid-crust is the locus of several fundamental geological and geophysical phenomena. These include the transitions from brittle to ductile behavior and from unstable to stable frictional sliding; earthquake nucleation and predominant moment release; the peak in the crustal stress envelope; the transition from predominantly cataclastic to mylonitic fault rocks; and mineralization associated with fracture permeability. Current understanding of faulting, seismogenesis, and mineralization in this tectonically important zone is largely based on remote geophysical observations of active faults and direct geological observations of fossil faults.

The Alpine Fault, New Zealand, is a major dextral-reverse fault that is thought to fail in large earthquakes (M_W ~7.9) every 200–400 years and to have last ruptured in the year 1717. Ongoing uplift has rapidly exhumed a crustal section from depths of as much as 30 km, yielding a young (<~1Ma), well-preserved sample of mid-crustal structures comparable to those currently active at depth. Surface studies and geophysical experiments have provided valuable images of structures and associated deformation mechanisms, and contributed to archetypal models of continental faulting and orogenesis.

An International Continental Scientific Drilling Program-funded workshop attended by sixty-one researchers from seven countries was held on 22–28 March 2009 at Franz Josef, adjacent to the central portion of the Alpine Fault. The workshop addressed the state of knowledge of the Alpine Fault; the significance and feasibility of a multi-national program of drilling and allied science; and the preliminary steps required for site characterization, preparatory drilling, and longer-term science planning. The program involved seventeen technical presentations on the Alpine Fault, earthquake physics and faulting, and scientific drilling, as well as fourteen talks addressing related subjects and intended to stimulate discussion. The program also contained field trips to sites along the Alpine Fault to view fault rocks and geomorphic features as well as potential drill sites.

Three principal scientific themes emerged from the workshop: (1) evolution of a transpressive orogenic system; (2) ductile and brittle deformation mechanisms, and their interaction; and (3) seismogenesis and the habitat of earthquakes. The remarkable along-strike homogeneity of the Alpine Fault's hanging wall, the rapid rate of slip, and the dextral-reverse kinematics enable the progressive evolution of fault zone materials to be examined by linking rocks exposed at the surface to their *in situ* protoliths at depth along common exhumation trajectories. In the central portion of the fault, where exhumation rates and thermal gradients are highest,

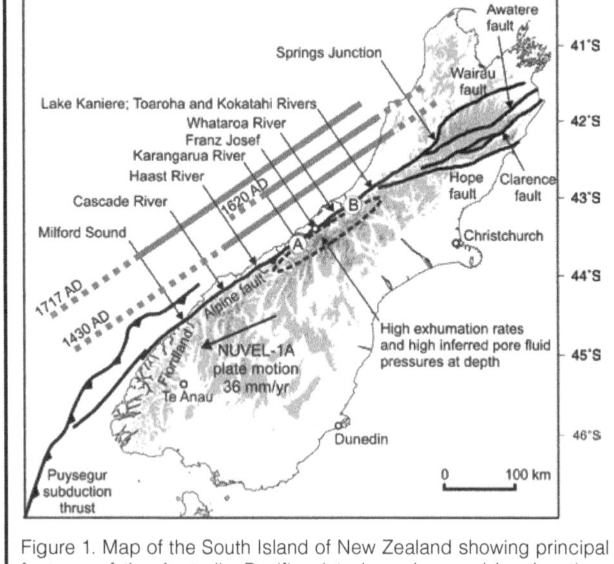

Figure 1. Map of the South Island of New Zealand showing principal features of the Australia–Pacific plate boundary and key locations referred to in the text. "A" and "B" mark the ends of the along-fault illustration in Fig. 4. The thick gray lines indicate the inferred extent of past Alpine Fault ruptures during historical times. Light gray shading demarcates topography higher than 800 m. Reproduced by permission of the American Geophysical Union after Sutherland et al., 2007. Copyright 2007 American Geophysical Union.

several sites can be identified at which rocks encountered at depth in boreholes would correspond to well-studied outcrops, enabling progressive geological deformation and petrological changes to be studied as functions of space and time along the transport path.

The Alpine Fault affords a rare opportunity to study, via scientific drilling and allied research, the physical character of tectonic deformation at depth within a major active continental fault that is late in its seismic cycle and which can be geophysically monitored in the coming decades.

Tectonic Setting

The Alpine Fault is the primary structure accommodating Australia–Pacific plate motion in New Zealand's South Island (Norris et al., 1990; Fig. 1). It is a mature dextral-reverse fault that offsets basement rocks laterally by ~470 km, with estimated late Quaternary horizontal displacement rates of 21–27 mm yr $^{-1}$ (Wellman, 1953; Norris and Cooper, 2001; Sutherland et al., 2006). Paleoseismic evidence near the Alpine Fault identifies earthquakes in the years 1717, 1620, and 1430, with estimated moment magnitudes of 7.9±0.3, 7.6±0.3, and 7.9±0.4, respectively (Sutherland et al., 2007).

In the central South Island, the Alpine Fault is dextral-reverse and bounds the western edge of the Southern Alps (Norris et al., 1990). Rapid exhumation (~6–9 mm yr $^{-1}$; Little et al., 2005) has also produced rapid cooling, indicated by decreasing metamorphic grade and increasing thermochronological age with distance from the fault (Koons, 1987; Batt et al., 2004; Little et al., 2005). It has produced a spectacular exposure of the crustal section from depths of as much as 20–30 km (Little et al., 2002; Norris and Cooper, 2003). The exposed section may provide a young proxy for material currently deforming at mid-crustal depths, allowing unprecedented ground-truthing of geophysical data and the interpretation of borehole observations in terms of recent fault products—and *vice versa*.

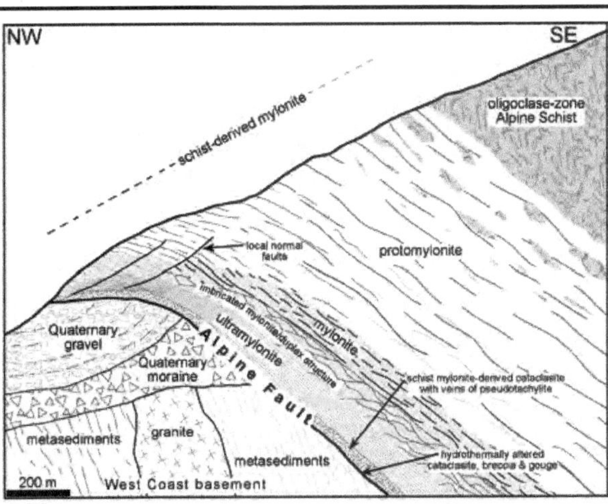

Figure 2. Schematic cross-section of an oblique thrust segment within the central section of the Alpine Fault. Reproduced by permission of the American Geophysical Union after Norris and Cooper, 2007. Copyright 2007 American Geophysical Union.

The exhumed 1–2-km-wide Plio-Pleistocene mylonite zone (illustrated schematically in Fig. 2) demonstrates that localized deformation occurs beneath the brittle part of the Alpine Fault (Sibson et al., 1979; Norris and Cooper, 2001, 2003; Toy et al., 2008 and references therein). Along its central segment, the fault juxtaposes a hanging wall of lithologically rather uniform high-grade (up to garnet-oligoclase facies) foliated Alpine schists against a footwall comprising a basement of Paleozoic greywackes (Greenland Group) intruded by Devonian–Cretaceous granites and overlain by a Cretaceous–Cenozoic cover sequence (see Cox and Barrel, 2007; Cox and Sutherland, 2007). Although portions of the fault zone are exposed in numerous river sections, no exposed complete fault-rock sequence from hard-rock hanging wall to hard-rock footwall has yet been located.

About 7 km structurally above the Alpine Fault lies a brittle-ductile shear array inferred to have formed at >20-km depths, with evidence for complex, transient semi-brittle behavior at high strain-rates and fluid pressure cycling on relatively short, possibly seismic, timescales (Wightman and Little, 2007). The base of the hanging wall seismogenic zone inferred from contemporary seismicity is relatively shallow (~8–12 km; Leitner et al., 2001). Geodetic studies are consistent with a shallow (5–10 km) depth for full fault locking (Beavan et al., 1999), though some degree of interseismic coupling may persist to as deep as ~18 km (Wallace et al., 2007). In the mid-crust, the fault zone exhibits low seismic wave speeds and high attenuation (Stern et al., 2001; Eberhart-Phillips and Bannister, 2002; Stern et al., 2007 and references therein) and high electrical conductivity (Wannamaker et al., 2002), suggesting interconnected saline fluids at high pressures within the ductile regime.

Why the Alpine Fault?

The Alpine Fault is a well-studied active continental fault that, unlike many other similar faults elsewhere, has not produced large earthquakes or measureable creep in historic times; however, paleoseismic data suggest that it has produced large earthquakes in the Holocene and that it is late in the earthquake cycle. The Alpine Fault's dextral-reverse kinematics, non-vertical dip, and rapid slip rates have exhumed to the ground surface a fresh sample of fault rocks inferred to have formed at depths of as much as 30 km within the last few million years. Due to its relatively steep dip, fluid-saturated state, and high near-surface geothermal gradient, the Alpine Fault serves further as an analogue for environments in which economically significant mesothermal mineralization has occurred (Koons and Craw, 1991; Weinberg et al., 2005; Sibson, 2007).

The following factors and related logistical considerations make the Alpine Fault a globally significant target of fundamental research into tectonic deformation, seismic hazard, and mineral resource formation:

- The >300-km along-strike exposure of a relatively uniform-lithology hanging wall and derived fault rocks that formed during the current tectonic regime, providing an exceptional reference for interpreting observations of contemporary processes and structure and older, exhumed structure
- Rapid uplift resulting in advection of crustal isotherms so that brittle-ductile transition processes can be studied at shallower depths than possible in other active transpressional settings
- Sequences of fault rocks developed during unidirectional exhumation on well-determined trajectories over relatively short time periods
- The opportunity to study on an ongoing basis the state of a locked fault thought to be late in the earthquake cycle
- Well-determined, rapid Quaternary slip rates
- An extensive body of geological and geophysical knowledge, and a modern, nationwide geophysical monitoring network (GeoNet; Fig. 3)
- A non-vertical fault orientation enabling fault penetration with either subvertical or incline holes
- A relatively benign political and physical environment in which to operate with existing petroleum industry onshore drilling activity and supporting infrastructure.

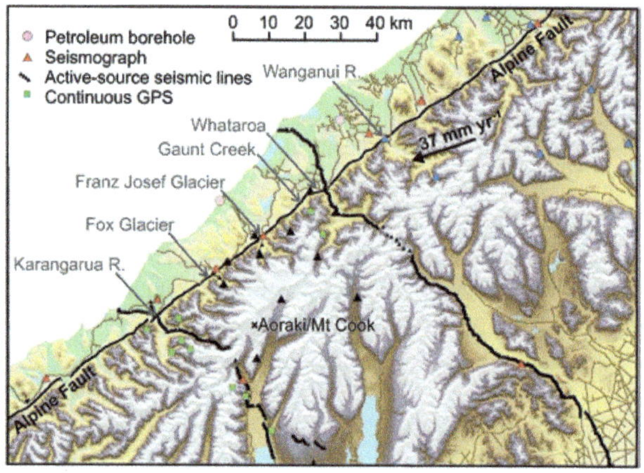

Figure 3. Map of the central section of the Alpine Fault showing existing infrastructure, the onshore portions of the South Island Geophysical Transect (SIGHT) active-source seismic lines, background seismicity, and the locations of possible drill sites.

Why Drill?

Drawing inferences about conditions and processes prevailing at seismogenic or greater depths based on outcrop observations is complicated by the fact that rocks exposed at the surface may have undergone modifications—structural, mineralogical, and geochemical—during their exhumation and at shallow depths.

One way of examining and accounting for the character and extent of upper crustal modifications is to examine a rock mass at depth whose future exhumation trajectory intersects a present-day surface outcrop (i.e., to treat the rock mass at depth as the protolith of the modified rocks now observed at the surface). The Alpine Fault kinematics are such that fault rocks evolve progressively on a path towards the surface, where they exit the system. This allows examination of progressive fault rock development using paired borehole and surface observations that is not possible on purely strike-slip faults where fault rocks may be continuously reworked at the same depth throughout the fault's history.

The second principal reason for drilling into the central Alpine Fault is to address the physics of faulting and seismogenesis by gaining access to the fault zone at depth and determining the temperature, fluid pressure and chemistry, bulk rock properties, and stress conditions prevailing at a late stage in the earthquake cycle, and establishing a long-term monitoring capability. Drilling enables continuous observations to be made of the fault zone in the immediate vicinity of the borehole itself and further afield using fault-zone guided waves, for instance (Li and Malin, 2008). It also provides a mechanism of calibrating and interpreting remote observations, such as those provided by the South Island Geophysical Transect project (SIGHT; Okaya et al., 2002; Stern et al., 2007), and linking these to surface data and rupture models. Laboratory measurements made on samples retrieved from depth, as well as measurements of the conditions under which those samples were collected, are required to more accurately describe the physical characteristics of fault rocks during coseismic rupture, and to account for strong ground motions (Beeler, 2006; Rice and Cocco, 2007).

Figure 4 illustrates the idealized geometry of the Alpine Fault in the central Southern Alps (inset stereogram) and a schematic view of the fault plane (as viewed normal to the plane), with the locations of key surface outcrops and fault-crossing roads marked. The regional fault plane strikes northeast and dips at approximately 50° to the southeast on

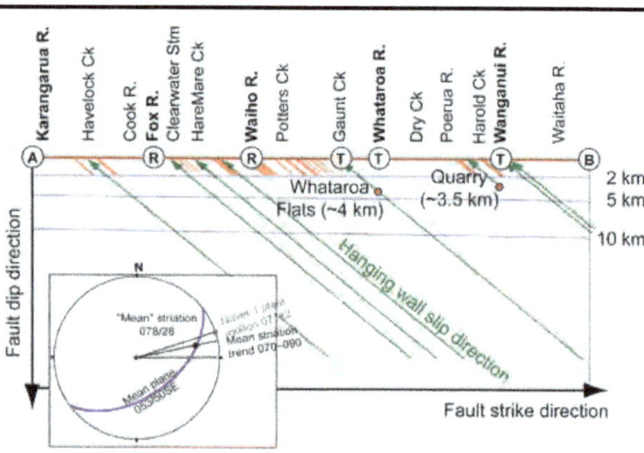

Figure 4. Schematic view normal to the mean fault plane in the central Alpine Fault region showing rock uplift trajectories and key outcrop locations. Place names shown in bold correspond to prospective drill sites. "A" (southwest) and "B" (northeast) are marked on Figure 1 for reference. "T" and "R" denote tracks and roads, respectively, crossing the fault trace. The short red lines mark strike-slip sections of the fault trace. The inset summarizes the mean fault geometry in stereographic projection.

average, and northeast-plunging striations define a mean rake of ~38° NE (Norris and Cooper, 1995; 1997; Little et al., 2002). Figure 4 demonstrates that there are several points along the fault at which several kilometers' access to the hanging wall is possible northeast of known surface outcrops. This means, for example, that a vertical borehole drilled approximately 3.5 km southeast of where the Alpine Fault crosses the Whataroa River would intersect the Alpine Fault at a depth of ~4 km, sampling material expected to breach the surface in ~0.4 Myr. Most importantly, however, the point at which the borehole intersected the fault plane would lie on the exhumation trajectory of the rocks now exposed in the Gaunt Creek outcrop.

Key Discussion Points

During the course of the workshop, three principal scientific themes and associated research goals emerged:

1. *Evolution of an orogenic system*—to determine how an active transpressional plate boundary system interacts with climate, landscape, and hydrological and thermal regimes;
2. *Ductile and brittle deformation mechanisms*—to determine via integrated surface and borehole observations what deformation mechanisms, mineralogical processes and conditions characterize the ductile and brittle regimes, and their interaction;
3. *Seismogenesis and the habitat of earthquakes*—to examine a major, locked, late-stage, continental fault at depth, determine the conditions under which earthquakes occur, and characterize the materials within which ruptures propagate.

The accompanying group discussions identified a number of common scientific issues related to the Alpine Fault and major continental faults in general, which are summarized below.

Ambient Conditions

A key theme to emerge from the group discussions was the vital importance of understanding the thermal and fluid flow regimes surrounding the Alpine Fault. Current thermal and hydrological models of the shallow to mid-crust in the vicinity of the Alpine Fault, particularly on the hanging wall, are limited by sparse data and consequent uncertainties in the maximum depth and pattern of topographically-induced fluid flow, the permeability structure, and shear heating effects.

The present-day shallow thermal regime west of the Alpine Fault can be inferred from petroleum exploration wells (Townend, 1999), but data are scarce in the immediate vicinity of the Alpine Fault's surface trace and further east. Models of the thermal structure of the Southern Alps orogen (Koons, 1987; Allis and Shi, 1995; Upton et al., 1995; Batt and Braun, 1999; Gerbault et al., 2003) differ quite markedly, and further work is required to reconcile these models with geodetic and seismological estimates of interseismic locking depths and the seismogenic thickness.

Temperature, fluid pressure and chemistry, and stress are all likely to be strongly perturbed at shallow depths by the pronounced topographic relief (as is the Alpine Fault's shallow structure itself; see below) and to differ more markedly from conditions prevailing at depth than has been the case in active fault drilling experiments elsewhere (Zoback et al., 2007). Detailed modeling of all three fields to determine how deep these effects persist is a high priority as plans for future drilling evolve.

Fluid and Rock Geochemistry

The full armory of elemental and isotopic techniques has yet to be brought to bear on fluids sampled in hot springs emanating from around the fault and trapped in exhumed veins within the fault zone and hanging wall. It is likely that more complete suites of geochemical data will aid the identification of fluid sources and flow-paths (Upton et al., 1995; Koons et al., 1998), and in particular enable more detailed analysis of progressive fluid-rock interaction. Among the outstanding questions related to fluid discharge are those of what factors control the number of hot springs along the Alpine Fault, their temperatures (<56°C), and the apparent gap in springs between the Cascade and Karangarua Rivers.

Alpine Fault Geometry

Key to the viability of any future drilling operations is improved knowledge of the Alpine Fault's shallow geometry. Seismic data indicate that the fault has a listric structure in the mid- and lower crust, with its southeastward dips decreasing from ~60° at 15 km depth to ~40° at 25 km depth and to almost zero at 30 km (Stern et al., 2007, and references therein). However, the fault's orientation at depths of hundreds of meters to several kilometers is still poorly constrained. Moreover, the central Alpine Fault (between Whataroa and Haast) exhibits "serial partitioning" into oblique thrust and strike-slip segments whose along-strike dimensions (typically <3 km) appear to be controlled by the spacing of major rivers (Norris and Cooper, 1995, 1997). Erosional control on the segmentation suggests that the segments likely merge into a single oblique structure at depths comparable to the topographic relief (~1000–1500 m), but detailed geophysical work is required to image these near-surface structures and determine their geometries at the depths that drilling might target.

Fault Zone Structure

Prevailing models of Alpine Fault zone structure at various depths have been compiled from partial data sets collected at different locations along the surface trace and in the hanging wall (Reid, 1964; Sibson et al., 1979; Sibson et al., 1981; Norris and Cooper, 1995, 1997, 2003, and 2007; Warr et al., 2007; Toy et al., 2008). Deconvolving near-surface or retrogressive effects that develop in response to depressurization, cooling, and interaction with progressively cooler fluids, from those resulting from cataclastic processes and frictional melting during deeper deformation (Warr and Cox, 2001) will be aided by analysis of suites of petrological and structural observations made at different depths along a single exhumation trajectory. Furthermore, borehole measurements are expected to provide information about changes in fault zone geometry and hence strain distribution, at different depths within the brittle and ductile regimes.

Seismogenesis and the Habitat of Continental Earthquakes

The Alpine Fault provides an excellent opportunity for elucidating the rheological properties and constitutive responses of real fault materials and host rocks. The overall mechanical behavior of a fault zone is likely to be controlled coseismically by thermal and hydraulic factors as well as the fault zone's intrinsic structure. Understanding the short- and long-term evolution of fault microstructures and fault zone fluids, their interaction and their temporal evolution should provide new insight into what mechanisms govern rupture propagation and termination (Sibson, 1985; Wesnousky, 2006; Rice and Cocco, 2007). Comparisons of fault rocks now exposed at the surface with samples retrieved from progressively greater depths and coincident measurements of *in situ* stress, fluid pressure, permeability, and elastic parameters would inform laboratory experiments and theoretical modeling. These issues provide a vital link between detailed structural and microstructural studies of fault rocks and rupture physics-based models of seismic radiation, strong ground motion, and transient interseismic processes. Further, they bear directly on questions of what governs the transition between brittle and ductile deformation processes, the generation and significance of pseudotachylytes, and the mechanical characteristics of a locked fault posing a major seismic hazard.

Interaction between Brittle and Ductile Deformation Processes

The transition in both space and time between continuous, temperature-controlled creep deformation at depth and episodic, pressure-sensitive frictional sliding in the upper-crust (commonly although controversially referred to as the "brittle–ductile transition") is fundamental to understanding the mechanics of crustal scale faults (Rutter, 1986). Characterizing this transition relies on being able to distinguish different grain-scale deformation mechanisms in samples of fault rocks, and determining the effects on those mechanisms of strain rate, temperature, pressure, and fluids (chemically and physically; e.g., Craw and Campbell, 2004).

Interaction of Climate, Landscape, and Tectonics

A great deal of scientific attention has focused on the role of tectonic processes in altering the Earth's topography and, more recently, the effects of surface processes on fault behavior (Koons and Kirby, 2007). The Alpine Fault is unusual in that it currently accommodates fault-parallel and fault-normal displacements on a single dipping structure (Norris et al., 1990; Stern et al., 2000). This has been suggested to reflect thermal weakening resulting from the highly asymmetric precipitation and erosion patterns of the Southern Alps (Koons et al., 2003) and serves as one example of the influence that orographic effects may exert on deep-seated faulting behavior (Upton et al., 2003, 2009). Better understanding of the rates and spatial variations in exhumation are required to more fully investigate climatic and physiographic effects on faulting.

Site Selection and Survey

Scientific Criteria

The key scientific factors that render the Alpine Fault a globally important focus for drilling—geometry, kinematics, rates, exposure—mean that the area of most relevance to drilling is the central section of the fault, between the Karangarua and Wanganui Rivers (Fig. 4), where dip-slip rates on the Alpine Fault and hanging wall exhumation rates are greatest. The central section is approximately bounded by the two longest deep-crustal seismic transects studied during the SIGHT project, providing the opportunity to tie various borehole observations to deeper remote measurements.

The key scientific criteria to be considered, therefore, are as follows: scientific relevance, representativeness, and significance; location in the central segment of the Alpine Fault; scope for multiple boreholes at different distances from the fault; and opportunities for aligned research.

Logistical Criteria

Much of the Alpine Fault lies within a region of very high conservation value; indeed, the central segment of the fault is also the area of highest conservation significance. Hence, one key logistical constraint on drilling operations and related surface studies is the need to minimize environmental impact within a high-profile conservation estate.

Steep topography, high rainfall, and active erosion make access to the hanging wall difficult (in fact, only three graded roads extend more than 1 km east of the fault trace in the central region of most rapid uplift), and field conditions are often challenging. In evaluating different potential drill sites, consideration of the following logistical factors will be particularly important: physical access to the hanging wall; permitting requirements; site conditions (e.g., vulnerability to flooding, ground stability); and visibility/impact.

Sites under Consideration

Figure 4 illustrates the five locations in the central segment of the Alpine fault where roads or farm tracks extend more than 1 km into the hanging wall. Also marked are five locations provisionally identified as possible drill sites requiring detailed site characterization studies—the Wanganui, Whataroa, Waiho (Franz Josef), Fox, and Karangarua River valleys. Of these sites, only the Wanganui, Whataroa, Waiho, and Karangarua River locations are closely linked to known surface outcrops along exhumation trajectories. The Waiho, Fox, and Karangarua Rivers are located in high-profile conservation areas, while the Wanganui River is somewhat removed from the northern SIGHT transect and the zone of most rapid uplift.

The experiment outlined here is well-suited to a multi-phase drilling program conducted over several years and involving a number of boreholes accompanied by detailed local and regional surveys and commencing with shallow (50–150-m-deep) core characterization and instrumentation holes and 1–2-km-deep exploratory holes, and going progressively deeper. This would enable the final and technologically most challenging phases of drilling and monitoring to be directly informed by sub-surface measurements in a similar manner to that used in projects elsewhere.

Surveys Required to Characterize and Define Drilling Targets

In order to adequately evaluate the suitability of any particular site for a long-term drilling operation, it will be necessary to conduct a coordinated multi-disciplinary site characterization program at each site. This program will focus on better defining the geometry of the fault and near-surface units, including alluvium; engineering site design, taking into account flood and landslide hazards; borehole design and stability analysis; and borehole instrumentation planning.

To take maximum advantage of downhole measurements and samples, a large program of aligned research that puts drilling results in context will be conducted. Some of this aligned research is being undertaken already (e.g., detailed geological and geomorphological mapping; paleoseismology; fault rock petrography and geochemistry; local seismicity analysis and tomography; active source seismic studies; numerical modeling). Other techniques (e.g., fault-zone guided wave studies; multi-system thermochronology; laboratory rock mechanics; permeability measurements; analysis of core retrieved from shallow boreholes) have yet to be fully applied to the Alpine Fault and will provide essential data for planning future deep boreholes.

Immediate Actions

Three working groups were convened at the March 2009 workshop to undertake preliminary studies in site selection and characterization, shallow drilling for core characterization and technical feasibility tests, and hydrogeological and thermal modeling and sampling. A fourth group will be convened in late 2009 to review and develop sampling protocols and coordinate permitting and funding applications.

The anisotropic, steeply-dipping fabrics of the hanging-wall Alpine Schist and the Alpine Fault mylonites may pose technical obstacles to drilling. Planning has commenced for a pilot drilling program, likely to take place in the 2010/2011 austral summer, during which one or more shallow (~50–150-m-deep) boreholes will be drilled using a mining rig to evaluate technical configurations, retrieve unweathered core from the fault zone, and conduct downhole logging in conjunction with surface measurements.

Ongoing programs of hot spring monitoring will be extended in 2010 with measurements of shallow subsurface flow and permeability in tunnels transecting the Alpine Fault hanging wall, and hot spring gas chemistry experiments. Temporary seismograph arrays currently deployed north and south of the Whataroa River will be augmented in 2010 to ensure that along-strike variations in seismicity near the Wanganui and Whataroa River prospective drill sites are recorded with a relatively homogeneous network with average instrument spacings of ~5 km. Planning will also begin for a higher-density (~1-km-spacing) transportable array with which to resolve fine-scale structures at each potential drill site in conjunction with active-source studies.

To facilitate the scientific and logistical activities, all activities will be coordinated via the project website (http://drill.gns.cri.nz/nzcdp/dfdp/), where up-to-date information regarding the working groups and the site characterization program may be found.

Acknowledgements

We gratefully acknowledge comments received on an earlier draft from Tim Davies, Tim Little, Peter Malin, Richard Norris, Dave Prior, Mark Quigley, Martha Savage, Damon Teagle, and Phaedra Upton. We thank all the workshop participants for their contributions to the discussions.

References

Allis, R.G., and Shi, Y., 1995. New insights to temperature and pressure beneath the central Southern Alps, New Zealand. *N. Z. J. Geol. Geophys.*, 38(4):585–592.

Batt, G.E., and Braun, J., 1999. The tectonic evolution of the Southern Alps, New Zealand: insights from fully thermally coupled dynamical modeling. *Geophys. J. Int.*, 136:403–420, doi:10.1046/j.1365-246X.1999.00730.x.

Batt, G.E., Baldwin, S.L., Cottam, M.A., Fitzgerald, P.G., and Brandon, M.T., 2004. Cenozoic plate boundary evolution in the South Island of New Zealand: new thermochronological constraints. *Tectonics*, 23(4):17, doi:10.1029/2003TC001527.

Beavan, J., Moore, M., Pearson, C., Henderson, M., Parsons, B., Blick, G., Bourne, S., England, P., Walcott, R.I., Darby, D., and Hodgkinson, K., 1999. Crustal deformation during 1994-98 due to oblique continental collision in the central Southern Alps, New Zealand, and implications for seismic potential of the Alpine fault. *J. Geophys. Res.*, 104(B11):25233–25255, doi:10.1029/1999JB900198.

Beeler, N.M., 2006. Inferring earthquake source parameters from laboratory observations and the scope of lab contributions to source physics. In: Abercrombie, R., McGarr, A., Kanamori, H., and Di Toro, G. (Eds.), *Earthquakes: Radiated Energy and the Physics of Earthquake Faulting. Geophysical Monograph Series.* Washington, DC (American Geophysical Union), 99–119.

Cox, S. and Barrel, D.J.A., 2007. Geology of the Aoraki area, Institute of Geological and Nuclear Sciences 1:250000 Geological Map. Lower Hutt, New Zealand (GNS Science), 71.

Cox, S. and Sutherland, R., 2007. Regional geological framework of South Island, New Zealand, and its significance for understanding the active plate boundary. In: D. Okaya, T.A. Stern and F. Davey (Eds.), *A Continental Plate Boundary: Tectonics at South Island, New Zealand. Geophysical Monograph Series.* Washington, DC (American Geoscience Union), 19–46.

Craw, D. and Campbell, J.R., 2004. Tectonic and structural setting for active mesothermal gold vein systems, Southern Alps, New Zealand. *J. Struct. Geol.*, 26(6-7):995–100, doi:10.1016/j.jsg.2003.11.012.

Eberhart-Phillips, D. and Bannister, S., 2002. Three-dimensional crustal structure in the Southern Alps region of New Zealand from inversion of local earthquake and active source data. *J. Geophys. Res.*, 107(10):20, doi:10.1029/2001JB000567.

Gerbault, M., Henrys, S., and Davey, F., 2003. Numerical models of lithospheric deformation forming the Southern Alps of New Zealand. *J. Geophys. Res.*, 108(7):18, doi:10.1029 2001JB001716.

Koons, P.O., 1987. Some thermal and mechanical consequences of rapid uplift; an example from the Southern Alps, New Zealand. *Earth Planet. Sci. Lett.*, 86(2):307–319, doi:10.1016/0012-821X(87)90228-7.

Koons, P.O. and Craw, D., 1991. Gold mineralization as a consequence of continental collision: an example from the Southern Alps, New Zealand. *Earth Planet. Sci. Lett.*, 103(1-4):1–9, doi:10.1016/0012-821X(91)90145-8.

Koons, P.O. and Kirby, E., 2007. Topography, denudation, and deformation: the role of surface processes in fault evolution. In: Handy, M. R., Hirth, G., and Hovius, N. (Eds.), *Tectonic Faults: Agents of Change on a Dynamic Earth. Dahlem Workshop Reports.* Cambridge, Mass. (The MIT Press), 205–230.

Koons, P.O., Craw, D., Cox, S.C., Upton, P., Templeton, A.S., and Chamberlain, C.P., 1998. Fluid flow during active oblique convergence; a Southern Alps model from mechanical and geochemical observations. *Geology*, 26(2):159–162.

Koons, P.O., Norris, R.J., Craw, D., and Cooper, A.F., 2003. Influence of exhumation on the structural evolution of transpressional plate boundaries: an example from the Southern Alps, New Zealand. *Geology*, 31(1):3–6.

Leitner, B., Eberhart-Phillips, D., Anderson, H., and Nabelek, J.L., 2001. A focused look at the Alpine fault, New Zealand: seismicity, focal mechanisms, and stress observations. *J. Geophys. Res.*, 106(B2):2193–2220, doi:10.1029/2000JB900303.

Li, Y.G. and Malin, P.E., 2008. San Andreas Fault damage at SAFOD viewed with fault-guided waves. *Geophys. Res. Lett.*, 35:L08304, doi:10.1029/2007GL032924.

Little, T.A., Cox, S., Vry, J.K., and Batt, G., 2005. Variations in exhumation level and uplift rate along the oblique-slip Alpine Fault, central Southern Alps, New Zealand. *Geol. Soc. Am. Bull.*, 117(5-6):707–723, doi:10.1130/B25500.1.

Little, T.A., Savage, M.K., and Tikoff, B., 2002. Relationship between crustal finite strain and seismic anisotropy in the mantle, Pacific-Australia plate boundary zone, South Island, New Zealand. *Geophys. J. Int.*, 151(1):106–116, doi:10.1046/j.1365-246X.2002.01730.x.

Norris, R.J. and Cooper, A.F., 1995. Origin of small-scale segmentation and transpressional thrusting along the Alpine Fault, New Zealand. *Geol. Soc. Am. Bull.*, 107(2): 231–240.

Norris, R.J. and Cooper, A.F., 1997. Erosional control on the structural evolution of a transpressional thrust complex on the Alpine Fault, New Zealand. *J. Struct. Geol.*, 19(10):1323–1324, doi:10.1016/S0191-8141(97)00036-9.

Norris, R.J. and Cooper, A.F., 2001. Late Quaternary slip rates and slip partitioning on the Alpine Fault, New Zealand. *J. Struct. Geol.*, 23(2-3):507–520, doi:10.1016/S0191-8141(00)00122-X.

Norris, R.J. and Cooper, A.F., 2003. Very high strains recorded in mylonites along the Alpine Fault, New Zealand: implications for the deep structure of plate boundary faults. *J. Struct. Geol.*, 25(12):2141–2157, doi:10.1016/S0191-8141(03)00045-2.

Norris, R.J. and Cooper, A.F., 2007. The Alpine Fault, New Zealand: surface geology and field relationships. In: Okaya, D., Stern, T.A., and Davey, F. (Eds.), *A Continental Plate Boundary: Tectonics at South Island, New Zealand. Geophysical Monograph Series.* Washington, DC (American Geophysical Union),157–175.

Norris, R.J., Koons, P.O., and Cooper, A.F., 1990. The obliquely-convergent plate boundary in the South Island of New Zealand: implications for ancient collision zones. *J. Struct. Geol.*, 12(5-6):715–725, doi:10.1016/0191-8141(90)90084-C.

Okaya, D., Henrys, S., and Stern, T., 2002. Double-sided onshore-offshore seismic imaging of a plate boundary: "super-gathers" across South Island, New Zealand. *Tectonophysics*, 355(1–4):247–263, doi:10.1016/S0040-1951(02)00145-2.

Reid, J.J., 1964. Mylonites, cataclasites, and associated rocks along the Alpine Fault, South Island, New Zealand. *N. Z. J. Geol. Geophys.*, 7:645–684.

Rice, J.R. and Cocco, M., 2007. Seismic fault rheology and earthquake dynamics. In: Handy, M. R., Hirth, G., and Hovius, N. (Eds.), *Tectonic Faults: agents of change on a dynamic Earth. Dahlem Workshop Reports.* Cambridge, Mass. (The MIT Press), 99–137.

Rutter, E.H., 1986. On the nomenclature of mode of failure transitions in rocks. *Tectonophysics*, 122(3-4):381–387, doi:10.1016/0040-1951(86)90153-8.

Sibson, R.H., 1985. Stopping of earthquake ruptures at dilational fault jogs. *Nature*, 316(6025):248–251, doi:10.1038/316248a0.

Sibson, R.H., 2007. Au-quartz mineralization near the base of the continental seismogenic zone. *Geol. Soc. London, Special Publ.*, 272(1):519–532.

Sibson, R.H., White, S.H., and Atkinson, B.K., 1979. Fault rock distribution and structure within the Alpine Fault Zone: a preliminary account. In: Walcott, R.I. and Cresswell, M.M. (Eds.), *Roy. Soc. N. Z.*, 18:55–65.

Sibson, R.H., White, S.H., and Atkinson, B.K., 1981. Structure and distribution of fault rocks in the Alpine fault zone, New Zealand. In: McClay, K.R. and Price, N.J. (Eds.), *Thrust and Nappe Tectonics*. London (Geological Society of London), 197–210.

Stern, T., Kleffman, S., Okaya, D., Scherwath, M., and Bannister, S., 2001. Low seismic-wave speeds and enhanced fluid pressure beneath the Southern Alps of New Zealand. *Geology*, 29:679–682.

Stern, T., Molnar, P., Okaya, D., and Eberhart-Phillips, D., 2000. Teleseismic P wave delays and modes of shortening the mantle lithosphere beneath South Island, New Zealand. *J. Geophys. Res.* 105(B9):21615–21631, doi:10.1029/2000JB 900166.

Stern, T.A., Okaya, D.A., Kleffmann, S., Scherwath, M., Henrys, S.A., and Davery, F.J., 2007. Geophysical exploration and dynamics of the Alpine fault zone. In: Okaya, D., Stern, T., and Davey, F. (Eds.), *A Continental Plate Boundary: Tectonics at South Island, New Zealand. Geophysical Monograph Series*, Washington, DC (American Geophysical Union), 207–233.

Sutherland, R., Berryman, K., and Norris, R., 2006. Quaternary slip rate and geomorphology of the Alpine fault: implications for kinematics and seismic hazard in southwest New Zealand. *Geol. Soc. Am. Bull.*, 118(3-4):464–474, doi:10.1130/ B25627.1.

Sutherland, R., Eberhart-Philllips, D., Harris, R.A., Stern, T.A., Beavan, R.J., Ellis, S.M., Henrys, S.A., Cox, S.C., Norris, R.J., Berryman, K.R., Townend, J., Bannister, S.C., Pettinga, J., Leitner, B., Wallace, L.M., Little, T.A., Cooper, A.F., Yetton, M., and Stirling, M.W., 2007. Do great earthquakes occur on the Alpine fault in central South Island, New Zealand? In: Okaya, D., Stern, T.A., and Davey, F. (Eds.), *A Continental Plate Boundary: Tectonics at South Island, New Zealand. Geophysical Monograph Series*, Washington, DC (American Geophysical Union), 235–251.

Townend, J., 1999. Heat flow through the West Coast, South Island, New Zealand. *N. Z. J. Geol. Geophys.*, 42:21–31.

Toy, V.G., Prior, D.J., and Norris, R.J., 2008. Quartz fabrics in the Alpine Fault mylonites: influence of pre-existing preferred orientations on fabric development during progressive uplift. *J. Struct. Geol.*, 30(5):602-621, doi:10.1016/j.jsg.2008.01.001.

Upton, P., Koons, P.O., and Chamberlain, C.P., 1995. Penetration of deformation-driven meteoric water into ductile rocks: isotopic and model observations from the Southern Alps, New Zealand. *N. Z. J. Geol. Geophys.*, 38(4):535–543.

Upton, P., Koons, P.O., and Eberhart-Phillips, D., 2003. Extension and partitioning in an oblique subduction zone, New Zealand: constraints from three-dimensional numerical modeling. *Tectonics*, 22(6):7–1, doi:10.1029/2002TC001431.

Upton, P., Koons, P.O., Craw, D., Henderson, C.M. and Enlow, R., 2009. Along-strike differences in the Southern Alps of New Zealand: consequences of inherited variation in rheology. *Tectonics*, 28:TC2007, doi:10.1029/2008TC002353.

Wallace, L.M., Beavan, J., McCaffrey, R., Berryman, K., and Denys, P., 2007. Balancing the plate motion budget in the South Island, New Zealand using GPS, geological and seismological data. *Geophys. J. Int.*, 168(1):332–352, doi:10.1111/j.1365-246X.2006.03183.x.

Wannamaker, P.E., Jiracek, G.R., Stodt, J.A., Caldwell, T.G., Gonzalez, V.M., McKnight, J.D., and Porter, A.D., 2002. Fluid generation and pathways beneath an active compressional orogen, the New Zealand Southern Alps, inferred from magnetotelluric data. *J. Geophys. Res.*, 107(B6):2117, doi:10.1029/2001JB000186.

Warr, L.N. and Cox, S., 2001. Clay mineral transformations and weakening mechanisms along the Alpine Fault, New Zealand, *Geol. Soc. London, Special Publ.*, 186:85–101.

Warr, L.N., van der Pluijm, B.A., and Tourscher, S., 2007. The age and depth of exhumed friction melts along the Alpine fault, New Zealand. *Geology*, 35(7):603–606, doi:10.1130/G23541A.1.

Weinberg, R.F., Van der Borgh, P., Bateman, R.J., and Groves, D.I., 2005. Kinematic history of the Boulder-Lefroy shear zone system and controls on associated gold mineralization, Yilgarn Craton, Western Australia. *Econ. Geology*, 100(7):1407–1426, doi:10.2113/100.7.1407.

Wellman, H.W., 1953. Data for the study of Recent and Late Pleistocene faulting in the South Island of New Zealand. *N. Z. J. Sci. Tech.*, 34(4):270–288.

Wesnousky, S.G., 2006. Predicting the endpoints of earthquake ruptures. *Nature*, 444(7117):358–360, doi:10.1038/nature 05275.

Wightman, R.H. and Little, T.A., 2007. Deformation of the Pacific Plate above the Alpine Fault ramp and its relationship to expulsion of metamorphic fluids: an array of backshears. In: Okaya, D., Stern, T.A., and Davey, F. (Eds.), *A Continental Plate Boundary: Tectonics at South Island, New Zealand. Geophysical Monograph Series*, Washington, DC, (American Geophysical Union), 177–205.

Zoback, M.D., Hickman, S., and Ellsworth, W., 2007. The role of fault zone drilling. In: Schubert, G. (Ed.), *Treatise on Geophysics*. Amsterdam (Elsevier), 649–674, doi:10.1016/B978-044452748-6.00084-5.

Authors

John Townend, School of Geography, Environment, and Earth Sciences, Victoria University of Wellington, P.O. Box 600, Wellington, New Zealand, e-mail:john.townend@vuw.ac.nz.

Rupert Sutherland, GNS Science, P.O. Box 30368, Lower Hutt, New Zealand, e-mail: r.sutherland@gns.cri.nz.

Virginia Toy, Department of Geology, University of Otago, P.O. Box 56, Dunedin, New Zealand, e-mail: virginia.toy@otago.ac.nz.

Related Web Links

http://drill.gns.cri.nz/nzcdp/dfdp/index.html
http://www.icdp-online.org

The ESF Magellan Workshop

ECORD
European Consortium for Ocean Research Drilling

"Beyond 2013—The Future of European Scientific Drilling Research", convened by G. Camoin, R. Stein, and M. Wagreich, was held at the Geocenter, University of Vienna, Austria on 24–25 April. Its main objectives were to define and document the European interests in an international scientific drilling program beyond 2013, and to prepare the INVEST Conference (Bremen, Germany, 23–25 September 2009). Eighty scientists from sixteen countries and from various expertise attended the workshop, which included general sessions and breakout group discussions. The topics that were discussed during the workshop included (1) the future of ECORD and IODP; (2) the new research initiatives and emerging fields in scientific drilling; (3) the relationships between IODP and other programs, and the collaboration between academia and industry; and (4) new technologies and the Mission Specific Platform approach. The major outcomes of the workshop and outlooks will be summarized in a white paper on the Future of European Ocean Drilling Research that will be edited prior to the INVEST Conference.

"Smart-Plug" Packer Systems Used for NanTroSEIZE

IODP
INTEGRATED OCEAN DRILLING PROGRAM

The multi-expedition IODP Nankai Trough Seismogenic Zone Experiment (NanTroSEIZE) will complete a transect of holes drilled through the Nankai accretionary prism. Coring, logging, and long-term monitoring experiments conducted in these drill holes will reach faults and wall rock at various depths, including the seismogenic part of the subduction thrust fault. During Stage 2 operations (*DV Chikyu* Expeditions 319 and 322), two holes will be drilled and cased along the transect in July–August 2009 to prepare for later installation of CORK observatories. To make use of these holes during the time between drilling and final CORK installation, a group of Canadian, U.S., German, and Japanese scientists have designed and built temporary "Smart-plug" packer systems to record formation pressures and temperatures. The data will document ambient conditions and provide proxies for strain and fluid flow related to tectonic and seismic activity. During NanTroSEIZE Stage 3, the "Smart-plugs" will be recovered and replaced by the permanent CORK observatories, and will ultimately be connected to the seafloor fiber-optic cable network DONET. These observatories will include seismometers, strain meters, thermistors, and pressure sensors.

Assembly of the interior of the "Smart-plug" (left); Smart-plug housing (center, with L-R: Martin Heesemann (University of Bremen), Bob Meldrum (Pacific Geoscience Center), Demian Saffer (Penn State University), Earl Davis and Bob Macdonald (both PGC).

Message from New President of IODP-MI, Dr. Kiyoshi Suyehiro

IODP
INTEGRATED OCEAN DRILLING PROGRAM

In my new role as IODP-MI President, I first would like to congratulate all the dedicated people who have contributed in many different ways bringing IODP to its current level of operations, achievement, and international regard. We have embarked on a truly unique international scientific undertaking and must continue to deliver exciting science, and thus attract young scientists who will lead future scientific ocean drilling investigations.

Because Earth's ocean is vast, a huge magnitude of unexplored space awaits your challenge. This 'inner space' invites you to understand Earth in four dimensions, in finer resolution, and to a wider extent. A dynamic system and supporting biosphere, Earth should genuinely arouse your curiosity in many ways. Confronting climate change and geohazards when the world population is rapidly growing, for example, immediately demands your greater understanding of the complex Earth system.

To meet such grand challenges, IODP will need to increase its interface with other scientific programs, each of which should consider IODP to be a vital partner. ICDP already demonstrates such a strategic partnership. And there are others who construct Earth in the cyber world or in labs, and who are developing observational networks, who are also natural partners.

As the scientific ocean drilling community gathers in Bremen to constructively look ahead constructively and build its future, I look forward to a successful INVEST meeting as a watershed event. Another step preparing for the future is the reorganization of IODP-MI into a single international office in Tokyo, starting in 2010.

Finally, I would like to emphasize that IODP-MI is about international access and mutual connection. We will strive further to gain your support for providing unbiased oversight and integration.

Continental Drilling Training Held in Sweden

A specific ICDP Training Course on conti-nental scientific drilling was organized in cooperation with the Swedish Deep Drilling Program. From 18 to 22 May 2009, members of this working group, other interested scientists, and lecturers invited by the Operational Support Group met at Skokloster, Sweden to discuss key aspects of scientific drilling projects. In addition to basic modules of the ICDP training from planning to drilling, other topics that included data handling, long-term experiments, and specific projects currently under devel-opment in Sweden were addressed (http://www.sddp.se/projects). Furthermore, the capabilities and limits of truck-mounted drill rigs were presented and discussed, as the Swedish group has decided to develop and deploy such comparatively low-priced drilling tools with slimhole wireline coring capabilities to 2.5 km depth.

Integration of Deep Biosphere Research into Continental Drilling Campaigns

An ICDP workshop on the integration of deep biosphere research into continental drilling will be held at the German Research Centre for Geosciences–GFZ in Potsdam on 27–29 September 2009. This workshop will address scientific, technical, administrative, and logistical prerequi-sites for the integration of deep biosphere research into land-based scientific drilling. The development of adequate technical protocols to recover appropriate contamination-controlled sample material and the integration of a newly developed mobile microbial laboratory (GFZ BUGLab) will be issues of paramount importance. Furthermore, identification of upcoming ICDP drilling projects for deep biosphere research and new targets for dedicated conti-nental deep biosphere drilling projects in the future will be in the focus of the meeting. Researchers are invited from disciplines such as micro-biology, bio-geochemistry, geology, and geochemistry, along with engineers with experience and interest in deep biosphere research and drilling. Interested scientists and engineers are requested to contact Kai Mangelsdorf: k.mangelsdorf@gfz-potsdam.de or Jens Kallmeyer: jens.kallmeyer@geo.uni-potsdam.de.

Scientific Drilling of the Heidelberg Basin (Germany)

The Heidelberg Basin, part of the northern Upper Rhine Graben, hosts one of the thickest successions of Pliocene and Quaternary sediments in Central Europe. Since 2006 the basin has been explored by new core drill-ings of up to 500 m depth. The drilling activities are embedded in an interdisciplinary project that aims to better understand the geological evolution of the basin, the control by climate change and tectonics, and the correlation of the Alpine and North European glacial evolution.

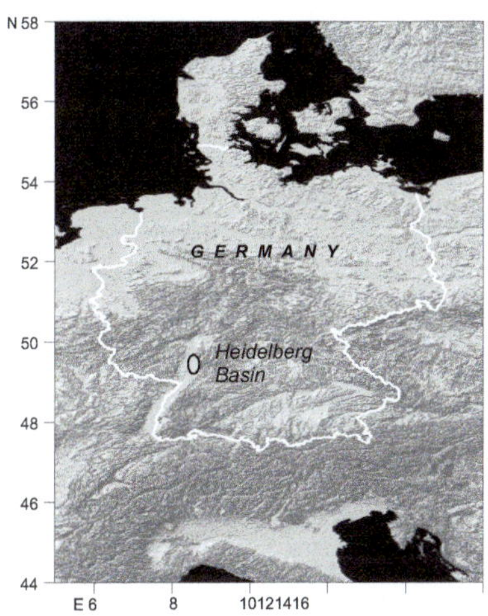

First results are published by Gabriel et al. (*Quaternary Science Journal*, 57:3–4, 2008). Beyond a summary of the project, eight papers deal with aspects such as the geophysical pre-site surveys, concise descriptions of the core material, first results from geophysical downhole logging, and palynological analysis. The volume also includes a discussion of Pleistocene mollusks and exemplary studies on luminescence dating of fluvial sediments. The data indicate that Pleistocene deposits are more than 500 m thick and well-resolved in terms of the early Pleistocene pattern of cold and warm periods.

Lake El'gygytgyn Scientific Drilling Successfully Completed

Between Octover 2008 and May 2009, an international team of scientists from Germany, the U. S., Russia, and Austria staged scientific drilling opera-tions at Lake El'gygytgyn, located in a 3.6-million-year-old impact crater in NE Russia. The project completed three holes in the center of the lake, reaching a depth of 517 m below lake floor, and one hole down to 141 m into permafrost deposits in the western lake catchment. The retrieved cores will provide new insights into the millennial-scale climate evolution of the Arctic, the formation of the meteorite crater, and the history of the permafrost. The successful completion of the project marks the end of years of challenging logistical planning (see previous issue of *Scientific Drilling*) and the develop-ment of international partnerships to share in the scientific goals.

Magma Drilled on Iceland!

An exciting event occurred on 24 June in the Iceland Deep Drilling Project (IDDP) at 2104 m depth. Suddenly the weight-on-bit declined while the rate of penetration and the torque shot up. After pulling up the drillstring a few meters, glassy rhyolitic cuttings followed by dark obsidian shards were circulated out, and it became clear that magma had been drilled.

The magma intrusion is presumably related to the nearby youngest eruptions in Northern Iceland (Krafla, 1975–1984). It may be a sill or dyke of unknown extension. However, due to the uncertainties it is now not feasible to continue drilling for supercritical fluids, the original goal of the IDDP. Nevertheless, the current surprise offers unique opportunities to study magma-fluid interaction at depth. Therefore, the well has been cased and cemented, and a slotted liner is set in the lowermost section. After surface valves are installed, tracer injection and flow tests will be performed to test connectivity to nearby wells and to produce superheated steam for research and eventually energy production from magma.

Depending on the results of these tests, future possibilities might include creating the world's highest-temperature Engineered Geothermal System (EGS). Details: http://iceland.icdp-online.org

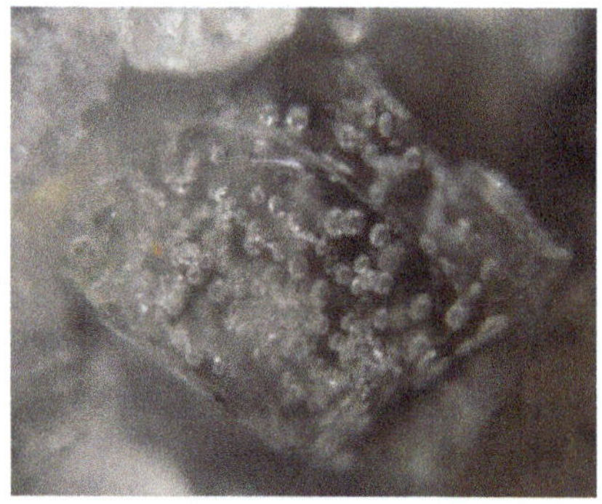

Colorless rhyolitic glass shard (~1 mm across) with spherical vesicles (photo from ISOR daily report).

Book Review: *Drill Me a Painting* by Christine Laverne

Reviewed by Catherine Mevel

Atlantica, 2008.
ISBN: 978-2-7588-0169-6

Understanding how the ocean crust is formed and evolves through time has been a major goal of the successive ocean drilling programs DSDP, ODP, and now IODP. A number of cruises over more than thirty years have been dedicated to drill as deeply as possible into the ocean crust, with the ultimate aim of penetrating the Mohorovicic Discontinuity (MOHO). However, drilling through the basaltic flows of the upper crust and the sheeted dike complex into the gabbros has proved as invaluable as it has proven difficult. Only by 2005 was the lower gabbroic crust reached below *in situ* upper crustal lavas and dikes (Wilson et al., *Science*, 312(5776):1016–1020, doi:10.1126/science.1126090, 2006). As an expert in the alteration of ocean crust, Christine Laverne participated in seven of the cruises on the *Glomar Challenger* and the *JOIDES Resolution* that contributed getting this far. She shares her fourteen months of experience in a book *Drill Me a Painting*, beautifully illustrated with her own water colors.

The book is written in a lively, personal diary style telling about life on board. This narrative is interwoven with sections explaining the scientific goals placed in the general context of Earth dynamics, and providing information on the drill ships, the drilling techniques, and the measurements made on board. Christine elucidates how a drilling cruise is a team effort, with all the participants focusing on a single aim—to drill deep and explore the unknown. She conveys how discouragement and excitement can alternate in concert with the progress of the drilling. Readers can also sense how she herself evolves from an inexperienced young one to a recognized scientist in her field—a well of knowledge for the younger generations. She also clearly explains the scientific goals of drilling and how the results contribute to our current understanding of the ocean crust that cover two-thirds of our planet. The combination of watercolors and scientifically very sound explanations is particularly attractive.

Participation in ocean drilling is not only a scientific challenge. It is also a human adventure. This is the message that Christine Laverne conveys to the reader with her enthusiasm and painting skills. If you want to convince a relative, a friend, or a neighbor that ocean drilling is an exciting endeavor, offer him or her this book. For young people, especially, it could even attract them to a career in science.

Welcome on board!

Order to: Atlantica éditions - 18, rue Marie-Politzer - BP 90041 - 64201 Biarritz cedex, France,
or online at www.atlantica.fr.

(French Version exists)

Joint IODP-ICDP Workshop Examines Challenges of Fault Zone Drilling

by Harold Tobin, Hisao Ito, Jan Behrmann, Steve Hickman, and Gaku Kimura

Introduction

This decade will likely be regarded in the geosciences as a time of an unprecedented explosion of fault zone drilling projects. The past few years have seen the initiation of projects including the Nojima Fault Drilling, San Andreas Fault Observatory at Depth (SAFOD), the Taiwan Chelungpu Fault Drilling Project (TCDP), the Gulf of Corinth Rift Laboratory (CRL) on land, and the Nankai Trough Seismogenic Zone Experiment (NanTroSEIZE) and Costa Rica Seismogenesis Project (CRISP) at sea (see Table 1 for references).

Each of these projects has unique objectives and scientific drivers, yet they share the fundamental goal of seeking direct access to zones of active faulting, particularly ones that generate earthquakes or have hosted seismic slip. This burgeoning interest in active fault drilling is taking place in the context of very rapidly growing research efforts on the mechanics and dynamics of faulting processes, integrating rock mechanics, seismology, geodesy, frictional physics, and fluid-fault interactions. There is a widespread recognition that our science lacks an integrated theory and understanding of the fundamental processes governing earthquakes and fault behavior, and the factors that control their natural variability.

On 23–26 May 2006, eighty scientists and engineers from around the world gathered for the *IODP-ICDP Workshop on Fault Zone Drilling: Developing a Global Perspective* in Miyazaki City on the island of Kyushu, Japan to explore the common ground—as well as the unique aspects—of all of these projects and to develop a consensus on the state of understanding of faults in the shallow crust and challenges to target in future work. The group represented a very broad spectrum in the Earth sciences, including seismologists, rock mechanicists, structural geologists, instrumentation specialists, geochemists, and many other specialties.

We have had over a century of detailed surface seismological observations, laboratory friction studies, geological observations, and case histories of earthquakes, yet processes acting at the fault interface remain elusive—because they occur deep in the subsurface, because faults are inherently heterogeneous materials, and because the initiation and propagation of fault rupture take place through complex and commonly non-linear physical processes. The physics of

Table 1. Major Currently-Active Fault Zone Drilling Projects.

Project	Location/Earthquake	Tectonic Setting	Principal Funding Agency	Overview Reference
SAFOD: San Andreas Fault Observatory at Depth	San Andreas Fault California, U.S.A.	Continental transform	NSF-EarthScope and ICDP	Zoback et al., 2006
TCDP: Taiwan Chelungpu Fault Drilling Project	Chelungpu Fault Taiwan ROC 1999 Chi-Chi M 7.6	Arc-continent collisional thrust	Taiwan, Japan-MEXT, and ICDP	Ma et al., 2006
CRL: Gulf of Corinth Rift Laboratory	Helike Fault Zone Corinth, Greece	Continental back arc rift	EESD-European Union and ICDP	Cornet et al., 2004
NanTroSEIZE: Nankai Trough Seismogenic Zone Experiment	Nankai Trough, Southwestern Japan	Subduction zone megathrust	IODP	Tobin and Kinoshita, 2006
Nojima Fault Drilling	1995 Kobe Earthquake, Kobe, Japan	Continental transform	NIED, GSJ, Universites	Ito et al., 1999, Oshiman et al., 2001
NELSAM and SeeSA	Mining-induced seismicity in South African Mines	Intraplate, reactivated faults	ICDP, NSF, NRF South Africa, MEXT	Reches et al., 2006, Ogasawara et al., this issue
CRISP: Costa Rica Seismogenesis Project	Costa Rica	Erosional subduction zone	IODP	Ranero et al., this issue
GONAF: Geophysical Observatory on the North Anatolian Fault	North Anatolian Fault, Marmara Sea, Turkey	Continental transform	ICDP (workshop)	Dresen et al., this issue

faulting in the crust, including the entire spectrum from aseismic creep through transient slow slip to rapid seismic slip, is a fundamental Earth process but one for which no integrated theory exists. Seismology, laboratory experiments in mechanics and friction, theoretical and modeling efforts, and fault zone geology and hydrology have all led to an understanding of the complexity of faulting and to hypotheses for many of the important processes. Testing of many of these ideas requires direct access to the actual active faults in the subsurface to collect new types of data that cannot be obtained by other means (e.g., Mori, this issue). This gap is the source of the need for the present rapid expansion of fault zone drilling.

Fault zone drilling projects are quite different in many respects from most other scientific drilling efforts. While a climate history project, for example, focuses mainly on acquiring a complete stratigraphy in cores and logs, fault zone projects target very specific zones for both coring and downhole measurements. In general, fault zone targets include *discontinuities* and small anomalous intervals in large rock volumes, rather than the stratigraphic column or broader rock volume of many other projects. The targets of interest are often the exact locations of poorest drilling conditions, such as a highly-fractured damage and gouge zone with potential for excess pore pressure above hydrostatic condition, anomalous stress state, weak rock, and other drilling hazards (Higuchi and Kobayashi; and Prevedel; this issue). The scientific approach commonly focuses on obtaining very extensive downhole measurements, logs, cores, and seismic structure in and around these fault zone environments. Measurements at the time of drilling as well as long-term observations are emphasized. All of these activities are technically very demanding, and in some cases they diverge from "standard" industrial borehole practice, so that creative new uses of existing technology or development of new technology is required.

Science Drivers for Fault Zone Drilling

Understanding the complex physics of tectonic faulting, earthquakes, and the generation of tsunami in the Earth's crust is among the grand remaining challenges of the geosciences. Because earthquakes take place deep below the Earth's surface, direct observation of processes acting in the faults themselves has generally not been possible; what we know about earthquake processes is derived from seismologic observations, geodetic deformation, and studies of exhumed rocks containing products of slip at depth long ago. Researchers have learned and inferred a great deal from these studies, but they are inherently limited. Much of fault zone drilling is aimed at understanding earthquake slip, but in fact fault physics encompasses a spectrum of slip and slip rates ranging from purely aseismic behavior, such as steady creep, through events such as slow earthquakes and creep, to true seismicity (Fig. 1).

Workshop Organization

The Miyazaki workshop provided an overview of past, active, and planned drilling projects in diverse tectonic settings. The range of plate boundary fault systems was very well represented, including convergent (both subduction and collision) thrust-fault systems, divergent (rift) normal-fault systems, and transform strike-slip fault systems. Intraplate, relatively small-scale but active (and in some cases very destructive), faults are also under intensive study in projects reported on and discussed at the workshop. The workshop comprised three meeting days, all in plenary session, and a full-day field trip between the second and third meeting day to visit spectacular seaside exposures of the Nobeoka thrust fault near the town of Nobeoka. This thrust fault is interpreted as an exhumed formerly active plate boundary thrust fault at seismogenic depths in a subduction-related accretionary complex (Kondo et al., 2005). The field trip vividly illustrated the challenges of fault zone drilling, as scientists and engineers examined a zone of chaotically folded, fractured, and damaged rock tens of meters thick surrounding a fault core only tens of centimeters thick with even thinner slip surfaces emebedded within it (see cover illustration; Tsuji et al.; and Ujiie; this issue). These narrow zones and wider unstable rock areas would present a challenge as targets of drilling, core recovery, and downhole measurement.

Participants (see back cover) contributed to an agenda organized around a series of themes. The first day of the workshop was devoted to several presentations discussing the motivations for fault zone drilling, including how to get seismologists, fault geologists, and those of other interested disciplines working together using common terminology and ideas. It also featured overviews of the main recent and active drilling projects, summarized below, and of key topics such as fault zone structure, seismology, hydrology, and modeling of fault physics. The second day began with a session devoted to seismology, strain measurements of several forms, and remote geophysical detection of physical properties in and around faults. This was followed by a session focused on the structure and composition of fault zones from outcrops and borehole samples and logging. Each session was accompanied by plenary discussion, which ultimately led to the consensus recommendations described below. The third day was devoted to the field trip to the Nobeoka thrust, where fault structures were observed in outcrop. Finally, on the fourth day of the meeting, several presentations described newly proposed or emerging future projects, and then the rest of the day was devoted to technological challenges for sampling, sample handling, drilling and completion, and downhole measurements and monitoring. The workshop was closed out with an extended period of plenary discussion, highlighting many of the areas of future opportunity as well as challenges. Posters on many specific projects and topics related to all of these sessions were displayed for the duration of the meeting, and the contents of many are reflected in the extended abstracts of this volume.

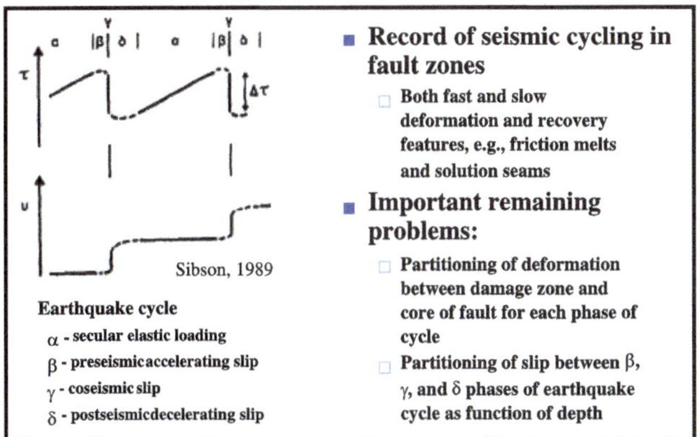

Figure 1. The earthquake cycle as a structural process (Chester, unpublished).

Significant recent progress in understanding the mechanics of earthquakes and fault slip has been achieved through new seismological and geodetic observations, fault geology, laboratory experiments, and theory (Fig. 2). However, the factors controlling stable vs. unstable slip on faults remain poorly understood, impairing our ability to evaluate associated hazards. At present, there is no unified theory of fault slip to account for earthquake nucleation and propagation, nor to explain the mechanisms of strain across the spectrum of different deformation rates now being observed with durations ranging from seconds to years. Our understanding of the mechanics and dynamics of plate boundary faulting is severely limited by a lack of information on ambient conditions and mechanical properties of active faults at depth. A major present goal in earthquake mechanics research is thus direct *in situ* sampling and instrumentation by drilling into the seismogenic zone of active faults, including both inter- and intra-plate fault systems in many tectonic settings.

Here, we briefly summarize some of the key scientific motivations for engaging in fault zone drilling that were especially highlighted at the workshop. Two recently published reviews of fault zone drilling activity by Reches and Ito (2007) and Zoback et al. (2007) both discuss the principal specific scientific goals of recent, active, and planned projects in detail, and it would be redundant to duplicate them here.

Zoback et al. (2007) summarized the fundamental motivation well, writing: "It is currently impossible to differentiate between—or even adequately constrain—the numerous conceptual models of active faults proposed over the years (Fig. 3). For this reason, the Earth science community is left in the untenable position of having no generally accepted paradigm for the mechanical behavior of faults at depth." This fundamental ignorance drives the research objectives of many of the projects touched on at this workshop. Some of the key questions include the following. What are the deformation mechanisms of fault slip in both aseismic and seismic conditions? Why is slip sometimes seismic, sometimes aseismic, and sometimes in an intermediate state, and what governs the transitions among these processes? How do earthquakes nucleate, and how does rupture grow and stop? Are there precursory phenomena in the fault or near field that indicate a preparatory phase to earthquake or slip nucleation? If so, can a prediction or early warning strategy based on borehole observation be developed? What are the processes governing tsunamigenic slip? What is the role of fluid content and fluid pressure in modulating faulting processes, especially during rapid seismic slip? How does permeability of the fault zone evolve interseismically and coseismically? What is the stress tensor in and around a fault zone throughout the earthquake cycle or other slip? What controls localization of slip vs. distributed strain in faulting? Fault zone drilling is justified by the need to shed light on these and related questions if we are to develop that paradigm for faults at depth.

Most, but not all, of the fault zone drilling efforts highlighted during the workshop focus on elucidating earthquake processes, though some have focused on aseismic faulting, particularly in accretionary prisms (Moore, this issue). The objectives and activities of the various fault zone drilling projects can be summarized as falling into three basic categories:

1. Direct sampling of fault zone and surrounding rocks and fluids: Acquiring samples (core, cuttings, and fluids) from the subsurface in a relatively pristine state for further study is a universal high priority (see Part 3 section this issue on Fault Zone Structure, Composition, and Physical Properties). The mineralogy, microstructure and fabric, mechanical properties, and fluid content and composition are all relevant to discovery of the physical processes involved in aseismic

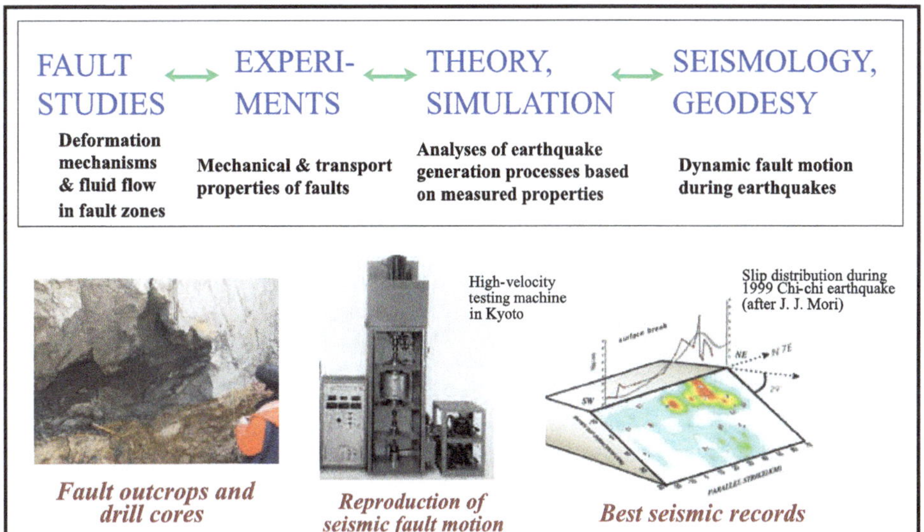

Figure 2. A concept of integrated fault studies in field,laboratory experiment, theory and simulation, and seismic/geodetic observations on faulting and earthquakes (T. Shimamoto, unpublished).

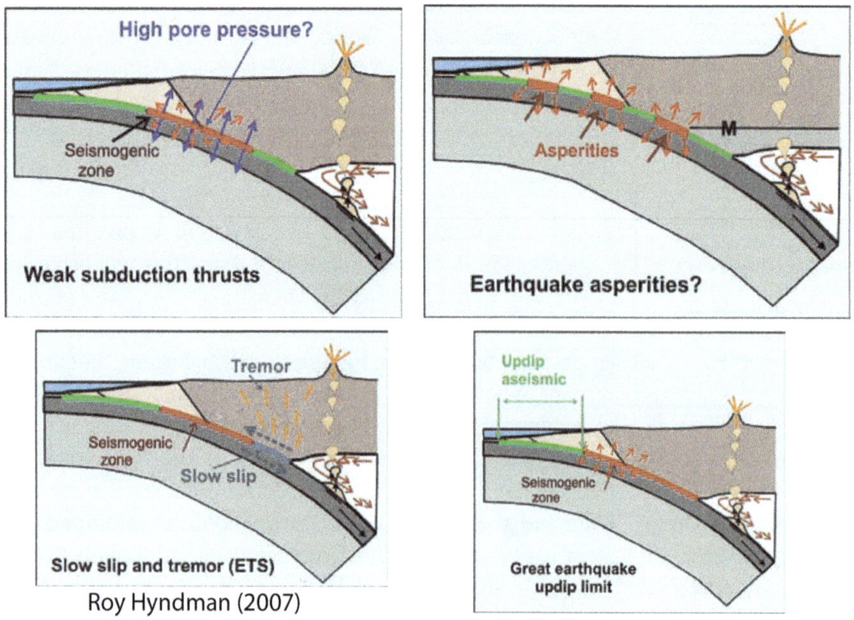

Figure 3 . Similar questions to fault zone drilling (Modified from Hyndman, 2007)

and seismic faulting. Only samples obtained from *in situ* positions can unequivocally be associated with fault processes uncontaminated by later alteration. Samples provide material for geological, geochemical, and petrophysical analyses to deduce the record of fault processes in the rocks (including ephemeral properties derived from very recent slip in some cases, such as TCDP or Nojima drilling). They also provide ground-truth to calibrate remote-sensed geophysical measurements based on seismic velocity and reflectivity, resistivity, gravity, and other observations over much larger volumes of the crust surrounding the borehole.

2. Conducting active tests or observations in situ in faults through borehole access: Besides the samples obtained through drilling, the borehole itself is a window into the fault zone and the faulting process. Borehole logging for detection of physical properties and imaging of stratigraphy, fractures, breakouts, and other features; downhole experimentation such as hydrologic testing through the use of borehole seals (packers); and collection of fluid samples are all emphasized to varying degrees in these projects (e.g., Hickman et al.; and Ito, this issue).

3. Monitoring the fault zone over time in borehole observatories: Long-term access to and use of the boreholes are key elements of most of these projects, though they are limited in many cases by technical challenges or budget. Monitoring of seismic activity, temperature, the strain field, pore fluid pressure, and seismic velocity structure of the fault zone and surroundings are consensus objectives across most of these programs. Borehole-based observations can be complemented by surface instruments such as land-surface or ocean-bottom broadband seismometers. A key question in fault zone studies is whether there is any evidence for a preseismic preparation phase—before nucleation of a major event—that might be detectable through near-field observations in the subsurface. Transient slip events, fault creep, and bursts of microseismicity may take place during the interseismic period on large faults and provide important information on mechanical process and constitutive properties. In other cases, such as deep mine fault drilling and seismicity studies, the monitoring is tied to active manipulation of the pore pressure and stress field, for a more explicit form of *in situ* experimentation (Ogasawara et al., this issue).

Current and Completed Projects

The workshop included presentations and exploration of virtually all of the major fault drilling projects worldwide in the past two decades. Most of these projects are represented by one or more papers in the accompanying extended abstract section. One interesting theme that emerges from this review is the following specific strengths associated with land-based and ocean-based drilling. ODP fault zone drilling at sea has been tightly integrated with high quality geophysical imaging (usually 3-D seismic reflection datasets that allow precise targeting), has made very productive use of logging-while-drilling (LWD) technology and cores by drilling multiple holes at the same location, and has deployed multiple sealed boreholes for hydrologic and other observations (Moore; Ranero; and Tobin and Kinoshita; this issue). The results include a great deal of quantitative information about the physical and hydrological properties of faults in relatively shallow (and aseismic) positions beneath the ocean floor in subduction zones. On land, by contrast, a major emphasis has been on sophisticated downhole experimentation and imaging, including seismology, hydraulic fracturing tests, breakout analysis, and more (Cornet; and Ellsworth et al., this issue). Land based boreholes have also obtained very high quality core in many cases, permitting detailed structural and petrological analyses of fault zone properties (Hung et al.; Lin et al.; Nielson; and Omura; this issue). It is therefore clear from the brief review below that oceanic and continental drilling studies of faults have been complementary operations to date, each advancing our understanding on different topics.

The section below is meant merely as a brief introduction to each area; an exhaustive review of the literature for each is beyond the scope of this paper. In roughly chronological order, the major projects discussed at the workshop were as follows.

KTB Project, Germany: The Kontinentales Tiefbohrprogramm der Bundesrepublik Deutschland (KTB), or German Continental Deep Drilling Program, was drilled from 1990 to 1994 in southeastern Germany. It is one of the deepest drilled holes on Earth, having reached 9101 m below the surface (Emmermann and Lauterjung, 1997). While not primarily intended as a fault zone project, KTB drilling did cross a tectonically inactive thrust fault system in metamorphic crystalline rock. Drilling activity, including fluid circulation and hydraulic fracturing tests and other perturbations of the borehole environment, caused microseismic activity in the near-field to the borehole, permitting analysis of the *in situ* state of stress (Zoback et al., 1993), presence of fluids and hydrologic properties, and thermal properties of the continental crust in this region. Many techniques used in continental drilling were pioneered or refined during KTB drilling.

Barbados Accretionary Wedge, Atlantic Ocean (ODP 110, 156, 171A): The Barbados wedge was the focus of substantial effort during DSDP and ODP, aimed at understanding the faulting conditions, ambient pore fluid pressure (Fig. 4), and mechanics in a very low angle (~2 degree dip) décollement thrust fault at the base of an accretionary wedge (see Moore, this issue). A high quality 3-D seismic survey (Shipley et al., 1994) imaged the décollement plane reflector in unprecedented detail, and served as the basis for targeting drilling into the ~30–40-m thick décollement zone in areas of contrasting reflection amplitude and character, all in the relatively shallow and presumed aseismic portion of the plate interface at the base of the accretionary wedge. Drilling and logging measurements made while drilling (LWD) demonstrated that the fault zone was a mechanically very weak interval with high excess pore fluid pressure and patchily distributed areas of anomalously high porosity (Moore et al., 1998). Installation of a borehole seal (CORK), downhole testing, and other studies at Sites 948 and 949 have shown that the fault zone has effective permeability that is strongly pressure dependent and varies over ~4 orders of magnitude with fluid pressure (Fisher, 2005; see also Saffer, this issue).

Nankai Trough Accretionary Wedge, Muroto Region, Western Pacific Ocean (ODP 131, 190, 196): Drilling by ODP of the central Nankai Trough region off Shikoku island also focused on the décollement zone at the outer toe of the accretionary deformation, and provided a detailed understanding of the mechanics of the fault zone (Moore et al., 2001). The 20–30-m-thick fault zone encountered at Sites 808 and 1174 is a zone of strong brittle fracture deformation that separates hydrologically distinct regimes above and below the fault; the footwall is at least partially sealed and pressurized with superhydrostatic pore fluid, limiting the effective normal stress that can act at the fault plane (Screaton et al., 2002). Microstructural and fabric analysis established that the décollement zone represents a virtually total strain decoupling of the footwall beneath the detachment, with basinal stress and strain state, and the accretionary wedge in the hanging wall, with heterogeneous and tectonically-enhanced strain (Byrne et al., 1993). These observations imply the décollement acts as a plane of very low shear strength during deformation. LWD measurements, especially resistivity and density data, suggest that the décollement fault has a core of compacted but fractured rock that is dilated in the high pore fluid pressure environment (Bourlange et al., 2003).

Nojima Fault Projects, Japan: Soon after the Kobe earthquake in 1995 (M6.9), several research teams drilled multiple boreholes into and across the shallow portion of the Nojima fault (the locus of that event) from ~750 m to ~1800 m depth (Ito et al., 1999; Omura, this issue,). Detailed compositional and structural analyses identified a narrow fault core of gouge and surrounding damage zone, with asymmetric structure in the footwall and hanging wall (Boullier et al., 2001; Fujimoto et al., 2001; Ito et al., 1996; Ohtani et al., 2001; Tanaka et al., 2001). Shear wave splitting studies of aftershocks showed a rapid evolution of fast direction after the earthquake, interpreted as evidence of fracture healing during a 12 month post-seismic period (Tadokoro and Ando, 2002). The preliminary results were collected in the special 2001 volume of *The Island Arc.* From repeated injection tests, permeability change with time was observed (Tadokoro et al., 2000). The structural properties (fault-zone width, shear wave velocity, and Qs) of the Nojima Fault were studied using the Love-wave-type fault-zone trapped waves (LTWs) recorded at two borehole seismic stations, which were con-

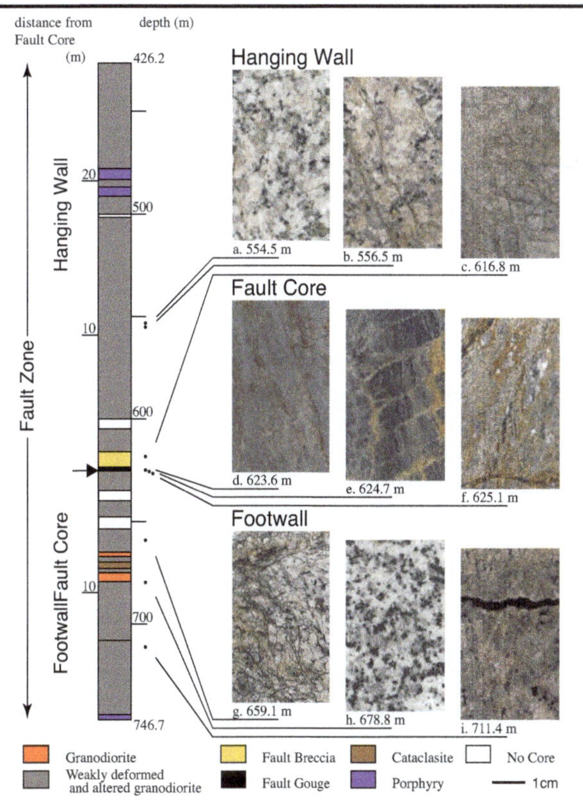

Figure 4. Fault zone structure of the Nojima fault. *Hanging wall:* Intensity of minor shear zones and the content of brown feldspar increase, and the mafic mineral content decreases toward the fault core. *Fault core:* Note three types of fault gouge. *Footwall:* These rocks are much less intensely deformed and altered than in hanging wall (Modified from Ohtani et al., 2000).

structed after the 1995 Kobe earthquake (Mizuno et al., 2007).

Costa Rica Convergent Margin, Nicoya Region, Eastern Pacific Ocean (ODP 170, 205): The Costa Rica margin off the Nicoya Peninsula in northwestern Costa Rica has been the site of two ODP expeditions that penetrated the décollement and other faults in two principal locations at depths of a few hundred meters below sea floor. Drilling demonstrated that the décollement fault overrides the subducting plate and all of the sedimentary section lying on top of it, such that the fault develops in a layer of extreme mechanical weakness made up of unconsolidated sediments (Silver et al., 2000). Structural studies and geochemical tracers of hydrological flow paths suggest that two strongly decoupled systems exist above and below the ~20-m-thick main fault zone, which acts simultaneously as a low-permeability barrier to cross fault flow and a high-permeability channel for along-fault flow (Tobin et al., 2001). Installation of a sealed borehole CORK during Leg 205 documented an apparently weakly overpressured décollement zone that exhibits temporal pressure excursions tied to apparent aseismic transients recorded geodetically on land (Davis and Villinger, 2006).

Corinth Rift Laboratory, Greece: The Gulf of Corinth is a location of back-arc extension within the Hellenic Arc system and is among the most seismically active areas of Europe (Cornet, this issue). The Corinth Rift Laboratory (CRL) Project drilled into the Aigion fault at 760 m depth and continued to 1000 m, encountering cataclastic fault rocks in a karstic carbonate host-rock environment. The fault zone forms a strong hydrologic barrier with an associated pressure differential across it. Transients in pore pressure have been detected associated with remote earthquakes (Cornet et al., 2004). Plans for upcoming activities at CRL include near-term deployment of a string of high-frequency three-component geophones, packers, pressure transducers, tiltmeters, and thermistance meters across the fault penetration, and also a proposed 4.5-km-deep borehole to monitor pore pressure and other features of the seismogenic zone (Cornet, this issue).

Taiwan Chelungpu Drilling Project, Taiwan: The devastating 1999 M7.6 Chi-Chi earthquake in Taiwan was a thrust event in a collisional tectonic setting that exhibited very large slip (>10 m), including large surface displacements, on the Chelungpu fault. The principal objective of drilling was to sample the main slip surface or zone, if one could be identified, in the relatively short post-seismic period. The fault zone was drilled in 2004 and 2005 in an area of high-velocity large slip during the main shock, and crossed the fault zone and several hundred meters of the footwall in two adjacent holes (Ma et al., 2006). The main slip zone and several subsidiary (or older) fault cores were identified in the borehole at ~1100 m depth (Hung et al., this issue). Studies of the fault gouge particle size, thickness, and composition have yielded quantitative estimates for the work done in forming gouge and damage (fracture energy), contributing to the understanding of energy budgets during earthquakes (Ma et al., 2006; Ma and Tanaka, this issue). Borehole hydrologic tests, temperature logging, and sample-based frictional experiments have all contributed to a tight integration of borehole derived in formation with seismological observations.

San Andreas Fault Observatory at Depth, California, U.S.A.: SAFOD targeted the San Andreas fault in a location at the transition from the locked zone that has ruptured repeatedly in M~6 earthquakes and the creeping segment to the north, where no large earthquakes have been observed. The location was chosen in part for the wealth of geophysical and other data generated during the decades-long Parkfield Experiment, where microseismic activity is very well located. The primary objectives included drilling into and across the San Andreas in a highly deviated (~55°) hole to sample and log the fault zone environment and materials, and to place instruments into the nucleation zone of one or more repeating earthquakes of magnitude 1–2. Fault rock properties, stress, pore pressure, and temperature *in situ*, and near-field observations of the earthquake process are key objectives of the research program (Ellsworth et al., this issue). A pilot hole was drilled in 2002 (Hickman et al., 2004), then the main hole crossed and logged the SAF in 2005. A zone ~230 m wide of low seismic velocity and density was detected at ~3100 m below the surface, in host rock of siltstones, shales, and sandstones, interpreted as a broad damage zone (Schleicher et al.; Solum et al.; and Wiersberg and Erzinger; this issue). Within this zone, several narrower bands of apparently more concentrated deformation, and velocity reduction were detected, including locations where the casing installed in the borehole has deformed since installation, interpreted as due to active fault creep. Temperature measurements to date and deduced stress orientation (Boness and Zoback, 2006) are consistent with the paradigm of a weak fault in a strong crust. Phase 3 drilling in 2007 targets continuous coring of three holes across the active fault traces as sidetracks from the main hole, and will finish with installation of a monitoring system to be placed within the rupture patch of M2 earthquakes (Ellsworth et al.; Hickman et al.; this issue).

Fault Drilling in South African Mines (NELSAM and SeeSA): Deep gold mines in South Africa, among the deepest in the world, contain Precambrian age faults that have been reactivated and accessed by recent mining operations. These faults have been extensively used as natural laboratories for research programs by several groups from Japan, the U.S.A., Europe, and South Africa. Ogasawara et al. (2002; this issue) describe installation of strainmeters and arrays of accelerometers in faults in several different mines as part of the project Semi-controlled Earthquake-generation Experiments in South African deep gold mines (SeeSA). Large strains were recorded co-seismically during events of magnitude ranging from -1 to 3.0. Even the smallest events showed source processes as complex as those of larger earthquakes, and hypothesized accelerating strain phenomena preceding seismic events were generally not observed. The Natural Earthquake Laboratory in South African Mines project (NELSAM) has taken the strategy of densely instrumenting a network of faults in the TauTona mine (Reches, this issue) with accelerometers, seismometers, strainmeters, temperature sensors, creep-meters, an electromagnetic radiation system, and acoustic emissions detection, in anticipation of magnitude ~2 events expected to be induced by mining operations.

Recent Progress in Technology

In response to scientific objectives, recent progresses in drilling (Higuchi and Kobayashi, this issue), logging, coring and observatory (Ito, this issue) were reported at the workshop. It was emphasized that operational issues (such as penetration rate, hole stability, direction control, over/under pressure) and quality control on core and borehole conditions should be carefully considered according to the scientific objectives.

For the logging, the following new technologies are discussed:

- LWD is crucial to make measurements in unstable formations, and provides extremely fast a full set of data allowing full characterization before coring.
- Wireline logging has higher resolution and, in particular sonic tools, offers measurements not available in LWD (anisotropy, full waveforms, permeability).
- Care must be taken when using LWD, wireline logging, and coring in different holes.
- Future IODP vessels will allow measurements that were not available in ODP.

Designing the wells for life is a paradigm change in scientific drilling and involves permanent monitoring and well completion technologies. One of the advantages of borehole observatory is noise reduction (Fig. 5; Ito, this issue), and new technologies on deep borehole observatory have been developed.

Planned Future Projects

In this section, we summarize a few additional projects that have been funded or for which planning and/or pilot drilling activity is underway and advanced. Additional projects that are currently at the proposal stage are highlighted in the accompanying abstracts (Kopf and Bohnoff, this issue).

Nankai Trough Seismogenic Zone Project (NanTroSEIZE): As the December 2004 Sumatra earthquake and Indian Ocean tsunami so tragically demonstrated, large subduction earthquakes represent one of the greatest natural hazards on the planet. Accordingly, drilling into and instrumenting an active interplate seismogenic zone are very high priorities in the IODP Initial Science Plan (2001). Through a decade-long series of national and international workshops, a consensus emerged that the Nankai Trough is an ideal place to attempt drilling and monitoring of the seismogenic plate interface. Factors in this selection include the relatively shallow extent of large slip and moment release in the 1944 M8.1 Tonankai earthquake, the good quality imaging of targets for drilling through multichannel seismic reflection surveys (including 3-D), and the vast amount of geophysical study that has gone into characterizing this subduction zone (Tobin and Kinoshita, 2006). NanTroSEIZE targets a series of thrust fault crossings at depths below the sea floor from <1000 m to >5000 m, to span the hypothesized up-dip transition from seismic to aseismic faulting processes. Coring, logging, downhole measurements, and the establishment of multiple instrumented boreholes spanning this zone are key objectives to address the questions of what controls the onset of seismogenic behavior with depth and how the faults behave spatially and temporally during interseismic and coseismic periods. The first phase of NanTroSEIZE drilling operations has now been scheduled for the fall of 2007 (Tobin and Kinoshita, this issue).

Costa Rica Seismogenesis Project (CRISP): CRISP is focused on access to the plate interface of a subduction zone where there is apparent "tectonic erosion" (the physical transfer of mass from the overriding plate to the downgoing plate at a subduction zone through thrust faulting). Erosional margins likely feature fault systems formed of protolith and fluid/heat conditions very different from accretionary margins such as Nankai (von Huene et al., 2004), because they may juxtapose crystalline crust of the upper plate with the downgoing basaltic slab and/or quantities of subducted sediment in the seismogenic zone. The CRISP-proposed project to IODP aims to drill first into the shallow portion of the plate

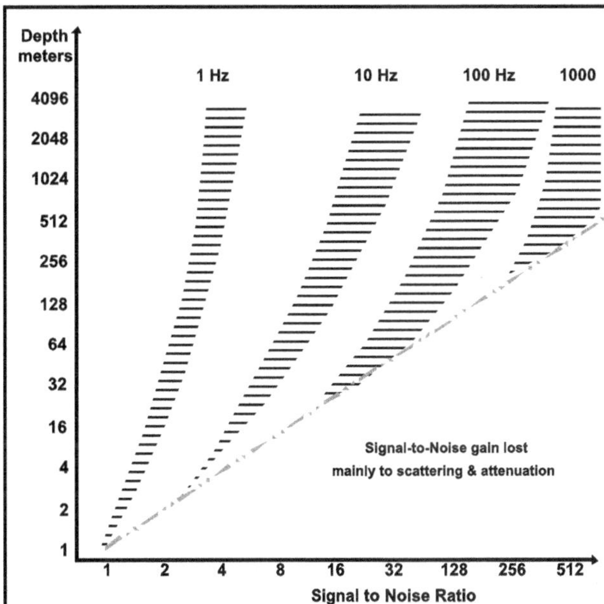

Figure 5 . Signal to noise improvement with depth (from Malin, unpublished).

interface at 1–3 km below the sea floor, to sample the rock and fluid inputs and outputs to and from the seismogenic plate boundary, then in a later phase to drill to depths of ~7 km below the sea floor, into the slip region of a 2002 magnitude 6.4 earthquake (Ranero et al., this issue).

Proposed North Anatolian Fault (NAF) Drilling: The North Anatolian fault is a strike-slip plate bounding continental fault that represents the greatest known seismic hazard to life and property in Europe (Dresen et al., this issue). This fault separates the Anatolian block from Eurasia, and has produced a devastating series of M~7 earthquakes in the past century, most recently the Izmit and Düzce quakes of 1999. The fault zone extends through the bed of the Marmara Sea near Istanbul, a location of identified seismic gap on the NAF. Coulomb stress analysis suggests that the 1999 earthquakes transferred positive stress change to this offshore segment, and an estimated probability of a major M~7 earthquake is 35%–70% in the next 30 years (Parsons, 2004). Dresen et al (this issue) propose an ICDP project involving a 2000-m-deep borehole to be drilled from an island in the near vicinity of the NAF to monitor seismicity, stress and stress changes, heat, and fluid flow. The goal is a monitoring system to better understand the temporal evolution and interseismic behavior of the NAF, with an eye also toward elucidating seismic hazards to the millions of inhabitants of the city of Istanbul and surrounding regions.

Outcomes of the Workshop—Recommendations for Future Efforts

Exploration at this workshop identified a host of topics and issues related to the science and technology of fault zone drilling. A common theme of these projects is that they seek to acquire three basic types of data: (1) studies of samples, including cores, cuttings, and fluids; (2) "spot" downhole measurements, including wireline and LWD/MWD logging, packer systems and other downhole testers, and borehole/casing surveys; and (3) long-term monitoring data, including seismicity, strain and tilt, temperature, and continuous fluid sampling.

All of these areas were the subject of extensive discussion during the workshop. The most essential conclusion was that the Fault Zone Drilling Workshop was in a sense an introductory meeting—individuals working on the many IODP, ICDP, national, and other projects came together to identify common themes and challenges. The dialogue has started and has already led to enhanced interaction among the various scientific constituencies in the time since the workshop was held. The most useful way forward would be for IODP and ICDP to take the sponsorship role to make sure that dialogues continue on a number of key, specific areas in the future. Before and after the 2006 workshop, the economic environment for mounting ambitious drilling efforts has steadily worsened, from the standpoint of business and funding of high-cost, high-risk science in several countries. Nevertheless, we suggest that IODP and ICDP can realize greater efficiency and success in future operations by leading efforts to combine force and expertise in new approaches to fault zone drilling.

Specific Recommendations: A number of recommendations for future efforts emerged during the numerous plenary discussions held throughout the workshop, as well as through the more informal interactions during poster sessions, evenings, and the field trip. These recommendations in some cases were for new driving questions and new strategies for subsurface access to faults; others were of a more technical or organizational nature.

The following overarching recommendations emerged:

- IODP and ICDP are urged to engage in joint efforts to develop optimal drilling techniques and drilling/casing/completion technologies specific to the fault zone environment.
- Through focused workshops, IODP and ICDP and the community should develop or document "best practices" statements for coring systems, core handling, and core orientation and for cuttings and fluids.
- Information exchange between IODP/ICDP/other drilling communities should be strengthened.
- The community should prioritize the development of plan(s) for a rapid drilling response to a suitable earthquake, on national and/or international levels.
- Sensor and deployment systems for monitoring should be identified/developed that are robust at temperatures of 120°C–160°C or greater (and have low power consumption for oceanic settings).
- New fault zone drilling proposals should be fostered through workshops networking with existing teams.

Grouping together several of these recommendations into themes, the motivation behind each and the specific recommendations as workshop outcomes are as follows.

1. Follow up with smaller working-group workshops on specific technical topics: The workshop identified many specific areas where new approaches and new technologies are either needed or have been developed for some projects but not yet widely adopted. We recommend that future joint efforts of IODP and ICDP include focused, small-team meetings to develop detailed "best-practices" for each area. Example topics for such working groups might be as broad as "Site Characterization" and as specific as "Core Orientation Methods". Other suggested areas for focus include: (1) improvements to core and fluid handling, such as developing more sophisticated approaches to capturing ephemeral properties in core, dealing with broken or fragile fault core, performing more scientific measurements using cuttings and drill mud, developing best practices for drilling fluid composition for scientific drilling, as deep land drilling and riser drilling at sea become more prevalent; (2) developing improved site characterization methods including optimal arrays and methods for earthquake location; (3) careful attention to downhole measurements such as hydrologic and stress testing, weighing value of data vs. risks to the borehole; (4) developing best practices for a host of drilling techniques/technologies for the fault zone environment, including casing strategies, mud weight balancing to optimize borehole stability, and directional control (vertical and deviated); and (5) development of completion and long-term monitoring best practices (see below in this section).

One challenge to developing these workshops is that many of the principal scientists and drilling/measurement experts are involved in one or more of the major current projects and are simply spread too thin to lead such efforts. This leads to an overwhelming recommendation—expand the scientific and technical interest group for fault zone drilling and foster new teams and new team members. As activity has increased, such a growth in the pool of scientists has been occurring in Japan, the U.S.A., Europe, and Taiwan. This must continue to be encouraged through high-profile meetings, publications, and other venues.

2. Enhance cross-communication among active and emerging projects: Because fault drilling projects are necessarily regional and often nationally supported, they require an explicit effort to keep lines of communication open. In particular, there are two principal organizations that support large-scale and international efforts at drilling into faults: IODP and ICDP. These two organizations are separated by sea level, an essentially meaningless boundary in terms of the scientific motivation for drilling faults, yet drilling on land vs. at sea requires different platforms and drilling technologies. Workshop participants perceived a continued division at times between the activities of these two organizations, in part because of different organizational structures and systems. Clearly, however, overcoming the considerable technical challenges of carrying out ambitious drilling and casing, logging, or especially long-term monitoring (for examples, see Prevedel, this issue) would greatly benefit from a much tighter cooperation between IODP and ICDP, as well as national organizations (such as DOSECC). One widely endorsed recommendation was to develop industry/academia fault zone drilling advisory panels for technical issues such as drilling and completion strategies, and development of long-term monitoring instrumentation. Further recommendations include standardization of the engineering process for development of new coring methods, logging instruments, and monitoring systems. For example, the benefits of standard borehole seismometer strings were highlighted by two presentations (e.g., Stephen, this issue), so that each team spends less time “reinventing the wheel.”

We endorse the recommendation for IODP and ICDP to intensify the exchange of information on technology, and to establish a common advisory or support panel for fault zone drilling.

3. Develop long-term monitoring systems and strategies capable of performing in the fault zone subsurface environment: The discussion of strategies for long-term monitoring system deployment was among the liveliest of the workshop. Participants generally agreed that monitoring over time is an absolutely essential aspect of studies in fault physics, since so many key processes are transitory or time-dependent in some way. Key parameters to monitor include seismicity, seismic velocity structure, strain and/or tilt, pore fluid pressure, temperature, and pore fluid chemistry. The fault zone environment at depth—where both temperature and pressure are greatly elevated, fluids may be corrosive, and return access is likely to be difficult or impossible—necessitates careful attention to monitoring schemes. Several key general practices emerged from this discussion. The principal one is to design for a long-term robust lifespan of the borehole as an observatory. Deep fault zone boreholes are large investments of capital and labor and must produce data for many years to realize their full potential. Thus, simple sensors are best; passive, non-powered instruments (e.g., geophones) should be preferred when possible. Some advocated maximizing use of passive instruments on the outside of installed casings, based on the logic that some form of measurement is better than none at all, and even if they fail, outside-casing instruments do not obstruct access to the interior of the borehole for other measurements. Redundancy was highlighted as an important objective, since individual sensors are prone to failure. The question of cementing in certain instruments, such as broadband seismometers and strainmeters, brought out a strong debate. On one hand, these instruments require very stable installations to work for years, favoring cementing in open holes. On the other hand, cementing them in blocks access to the hole for other instruments and may make it difficult or impossible to replace instruments that have failed. The choice depends on the balance of objectives unique to each project.

Workshop participants identified an urgent need for instruments and sensors robust for long periods of time at elevated temperatures and pressures. Lifespan is traded against stability at high temperatures in particular, such that a particular instrument might have a five-year life at 80°C, but only a six-month life at 120°C, for example. Broadband seismometers for installations above ~100°C do not exist, and accelerometers, fiber optic sensors, and other alternatives do not have acceptable resolution or absolute precision. These roadblocks are common to all projects, land or sea, so one recommendation from this workshop is that IODP and ICDP develop joint efforts in long-term monitoring system development. A joint working group or panel would be ideal.

4. Explore the potential for a rapid-response drilling plan: This is the most specific recommendation, but it is necessary because such a plan would require a new approach to planning for drilling. A number of workshop participants emphasized the scientific merit of drilling into a fault soon after slip during a large earthquake in order to measure the transient signal of frictional heating and other properties before they decay (Kano et al.; Mori; this issue). A goal of drilling in weeks to months (rather than years) post-event was recommended. Participants agreed that a candidate earthquake would have magnitude of roughly 6 to 7 or greater, would be well recorded locally, would have significant fault slip in the upper few thousand meters subsurface, and would be located in a region easily accessible for drilling. The most significant challenge is that a rapid-response drilling program would require (a) a well developed plan that is focused and has clear objectives, yet is generic enough to apply to any earthquake and (b) funding in place in advance, with an agreed-upon set of criteria for initiating the plan when a suitable earthquake occurs. In addition, there has to be a carefully thought out technical plan to mobilize a rig on relatively short notice. These requirements contrast sharply with the typical process of developing a specific drilling program over many years. One additional challenge to overcome is that such a plan would work best if candidate events in many, or at least more than one, country could be considered, so as to shorten the likely wait for a suitable event, but funding agencies' national interest and issues with permits might make it difficult to make this a strongly international effort. Despite these potentially formidable obstacles, the sense of the workshop was that this initiative should be explored. Since the Miyazaki workshop, an ICDP workshop proposal has been submitted to further explore the rapid response concept.

Extended Abstracts in This Volume

The collection of volunteered extended abstracts that follows this section represents a snapshot of many of the themes highlighted in both presentations and posters during the workshop. It is not intended to comprehensively capture the range of subjects discussed during the workshop, nor the full extent of current endeavor in this field. Nevertheless, the thirty-six abstracts here represent examples of current efforts in many areas. We have chosen to group them roughly into sections with the themes of *(1) Scientific Motivations for Fault Zone Drilling*, containing abstracts dealing with the broader issues that can be best or only addressed through drilling, and including *(2) overviews of many of the major projects*; *(3) Fault Zone Structure, Composition, and Physical Properties*, with papers principally on lithology, mineralogy, structural fabrics, and seismic and hydrologic properties; *(4) The Physics of Earthquake Rupture*, concerned with near-field seismology and fault physics as addressed through modeling and laboratory experimentation; and *(5) Technological Challenges of Drilling, Testing, Sampling, and Monitoring in Fault Zones*, which deals with new techniques for core handling and sampling, borehole instrumentation, downhole active testing of formation properties, logging, and drilling and completion operations in the fault zone environment. These reflect the key themes and challenges discussed in the section above. It is our intention that publication of this volume will stimulate new collaborations and links among developing research efforts on land and at sea, and also will encourage ambitious new programs to tackle these challenges. The improved understanding of fault processes will depend on a multi-pronged, multi-disciplinary approach using new tools and collecting novel types of data.

Concluding Remarks

A principal objective of the Miyazaki joint workshop was to bring together the land-based and ocean-based fault zone communities for an exchange of information and ideas, and an exploration of the commonalities and differences in working in these very different environments. On this aspect the workshop can only be judged a resounding success. New collaborations began during the four days of the meeting, and tighter integration of projects has begun. For example, an ICDP workshop has now been scheduled for summer 2007 to explore an on-land extension of the NanTroSEIZE offshore transect (see http://www.icdp-online.org/contenido/icdp/front_content.php?client=29&idcat=309&lang=28&idart=1931). The Mediterranean Ridge proposal (Kopf and Bohnoff, this issue) is another example of integrated land-sea drilling plans. All the projects, past or future, reviewed at this workshop have unique motivations, strategies for success, advantages and challenges, and it is likely that only by advancing on many of these simultaneously will the fault zone research community make substantial progress in coming years.

This workshop established a single global community of researchers on drilling-based fault studies. With two well established, highly visible and successful organizations in place (ICDP and IODP) to support scientists' efforts, it is our hope and expectation that the legacy of the workshop will be a dynamic, multi-disciplinary, and integrated research program in coming years.

Acknowledgements

We are very grateful for the support of both the Integrated Ocean Drilling Program (IODP) and the International Continental Drilling Project (ICDP) for their generous co-sponsorship of the costs of holding this workshop. In particular, we thank Dr. Manik Talwani of IODP for sparking the idea to hold a joint workshop, and Dr. Ulrich Harms of ICDP for unstinting support. The staff at the Phoenix SunGaia Hotel in Miyazaki provided excellent support in all respects. The most essential person for the success of the Fault Zone Drilling workshop was certainly Dr. Kelly Kryc of IODP-MI, who carried out the workshop organization with great professionalism, patience, and good humor. Production of this workshop report white paper and special volume of *Scientific Drilling* would not have taken place without the hard work and supreme patience of Emanuel Soeding, Mika Saido, and Hans-Christian Larsen of the journal, and we are indebted to them for their efforts.

References

Boness, N. and Zoback, M.D., 2006. A multi-scale study of the mechanisms controlling shear velocity anisotropy in the San Andreas Fault Observatory at Depth. *Geophysics*, 7(5):F131–F146, doi:10.1190/1.2231107.

Boullier, A-M., Ohtani, T., Fujimoto, K., Ito, H., and Dubois, M., 2001. Fluid inclusions in pseudotachylytes from the Nojima fault, Japan. *J. Geophys. Res.*, 106:21965–21977.

Bourlange, S., Henry, P., Moore, J.C., Mikada, H., and Klaus, A., 2003. Fracture porosity in the décollement zone of Nankai accretionary wedge using Logging While Drilling resistivity data. *Earth Planet. Sci. Lett.*, 209:103–112, doi:10.1016/S0012-821X(03)00082-7.

Byrne, T., Maltman, A., Stephenson, E., Soh, W., and Knipe, R., 1993. Deformation structures and fluid flow in the toe region of the Nankai accretionary prism. *In* Hill, I.A., Taira, A., Firth, J.V., et al., *Proc. ODP, Sci. Results, 131*: College Station, Texas (Ocean Drilling Program), 83–101.

Cornet, F.H., Bernard, P., and Moretti, I., 2004. The Corinth Rift Laboratory. *Comp. Rend. Geosci.*, 336:235–241, doi:10.1016/j. crte.2004.02.001.

Davis, E.E. and Villinger, H.W., 2006. Transient formation fluid pressures and temperatures in the Costa Rica forearc prism and subducting oceanic basement: CORK monitoring at ODP Sites 1253 and 1255. *Earth Planet. Sci. Lett.*, 245(1-2):232–244. doi:10.1016/j.epsl.2006.02.042.

Earth Oceans and Life: Integrated Ocean Drilling Program Initial Science Plan, 2003-2013, 2001. International Working Group Support Office, Washington, D.C., 110 p., Available from: http://www.iodp.org/isp

Emmermann, R. and Lauterjung, J., 1997. The German Continental Deep Drilling Program KTB: Overview and major results. *J. Geophys. Res.*, 102:18179–18201, doi:10.1029/96JB03945.

Fisher, A., 2005. Marine hydrogeology: recent accomplishments and future opportunities. *Hydrogeol. J.*, 13:69–97, doi:10.1007/s10040-004-0400-y.

Fujimoto, K., Tanaka, H., Higuchi, T., Tomida, N., Ohtani, T., and Ito, H., 2001. Alteration and mass transfer inferred from the Hirabayashi GSJ drill penetrating the Nojima fault, Japan. *The Island Arc*, 10:401–410.

Hickman, S., Zoback, M.D., and Ellsworth, W.E., 2004. Introduction to special section: Preparing for the San Andreas Fault Observatory at Depth. *Geophys. Res. Lett.*,. 31:L12S01, doi:10.1029/20040GL20688.

Hyndman, R.D., 2007. The seismogenic zone of subduction thrust faults: what we know and don't know. In *The Seismogenic Zone of Subduction Thrust Faults*, Eds.: Dixon, T.H. and Moore, C., Columbia University Press, 15–40.

Ito,H., Kuwahara, Y., Miyazaki, T., Nishizawa, O., Kiguchi, T., Fujimoto, K., Ohtani, T., Tanaka, H., Higuchi, T., Agar, S., Brie, A., and Yamamoto, H., 1996. Structure and physical properties of the Nojima fault by the active fault drilling, *Butsuri-Tansa*, 49, 522-535.

Ito, H., Fujimoto, K., Tanaka, H., and Lockner, D., 1999. Proceedings of the international workshop on the Nojima fault core and borehole data analysis. *Geological Survey of Japan Interim Report No. EQ/00/1*, USGS Open-File 895 Report 000-129.

Kondo, H., Kimura, G., Masago, H., Ohmori-Ikehara, K., Kitamura, Y., Ikesawa, E., Sakaguchi, A., Yamaguchi, A., and Okamoto, S., 2005. Deformation and fluid flow of a major out-of-sequence thrust located at seismogenic depth in an accretionary complex: Nobeoka Thrust in the Shimanto Belt, Kyushu, Japan. *Tectonics*, 24:TC6008, doi:10.1029/2004TC001655.

Ma, K.-F., Tanaka, H., Song, S.-R., Wang, C.-Y., Hung, J.-H., Tsai, Y.-B., Mori, J., Song, Y.-F., Yeh, E.-C., Soh, W., Sone, H., Kuo, L.-W., and Wu, H.-Y., 2006. Slip zone and energetics of a large earthquake from the Taiwan Chelungpu-fault Drilling Project. *Nature*, 444:473–476, doi:10.1038/nature05253.

Mizuno, T., Kuwahara, Y., Ito, H., and Nishigami, K., 2007. Spatial variations in fault-zone structure along the Nojima fault, central Japan, as inferred from borehole observations of fault-zone trapped waves. *Bull. Seism. Soc. Amer.*, in press.

Moore, G.F., Taira, A., Klaus, A., Becker, L., Boeckel, B., Cragg, A., Dean, A., Fergusson, C.L., Henry, P., Hirano, S., Hisamitsu, T., Hunze, S., Kastner, M., Maltman, A.J., Morgan, J.K., Murakami, Y., Saffer, D.M., Sánchez-Gómez, M., Screaton, E.J., Smith, D.C., Spivack, A.J., Steurer, J., Tobin, H.J., Ujiie, K., Underwood, M.B., and Wilson, M., 2001. New insights into deformation and fluid flow processes in the Nankai Trough Accretionary Prism: results of Ocean Drilling Program Leg 190. *Geochem. Geophys. Geosyst.*, 2, doi:10.129/2001GC000166.

Moore, J.C., Klaus, A., Bangs, N.L., Bekins, B., Buecker, C.J., Brueckmann, W., Erickson, S.N., Hansen, O., Horton, T., Ireland, P., Major, C.O., Moore, G.F., Peacock, S., Saito, S., Screaton, E.J., Shimeld, J.W., Stauffer, P.H., Taymaz, T., Teas, P.A., and Tokunaga, T., 1998. Consolidation patterns during initiation and evolution of a plate-boundary décollement zone: Northern Barbados accretionary prism. *Geology*, 26:811–814, doi:10.1130/0091-7613(1998)026<0811:CPDIAE>2.3.CO;2.

Ogasawara, H., Yanagidani, T., and Ando, M., 2002. Seismogenic process monitoring. *Proceedings of a Joint Japan-Poland Symposium on Mining and Experimental Seismology*, Kyoto, Japan (A.A. Balkema), 414 pp.

Ohtani, T., Fujimoto, K., Ito, H., Tanaka, H., Higuchi, T., Tomida, N., 2000. Fault rocks and past to recent fluid characteristics from the borehole survey of the Nojima fault ruptured in the 1995 Kobe earthquake, southwest Japan. *J. Geophys. Res.*, 105, 16161–16171.

Ohtani, T., Tanaka, H., Fujimoto, K., Higuchi, T., Tomida, N., and Ito, H., 2001. Internal structure of the Nojima fault zone from the Hirabayashi GSJ drill core. *The Island Arc*, 10:392–400.

Oshiman, N., Shimamoto, T., Takemura, K., and Wibberley, C.A.J., 2001. Nojima Fault Zone Probe. *The Island Arc*, 10:195–505.

Parsons, T., 2004. Recalculated probability of $M \geq 7$ earthquakes beneath the Sea of Marmara, Turkey. *J. Geophys. Res.*, 109: B05304, doi:10.1029/2003JB002667.

Reches, Z. and Ito, H., 2007. Scientific drilling of active faults: Past and future. *In* Harms, U., Koeberl, C., and Zoback, M.D. (Eds.), *Continental Scientific Drilling: A Decade of Progress and Challenges for the Future*, Heidelberg, Germany

(Springer), doi:10.1007/978-3-540-68778-8_6.

Screaton, E., Saffer, D., Henry, P., Hunze, S., and the Leg 190 Scientific Party, 2002. Porosity loss within the underthrust sediments of the Nankai accretionary complex: Implications for overpressures. *Geology*, 30(1):19–22, doi:10.1130/0091-7613(2002)030<0019:PLWTUS>2.0.CO;2.

Shipley, T.H., Moore, G.F., Bangs, N.L., Moore, J.C., and Stoffa, P.L., 1994. Seismically inferred dilatancy distribution, northern Barbados Ridge décollement: Implications for fluid migration and fault strength. *Geology*, 22:411–414, doi:10.1130/0091-7613(1994)022<0411:SIDDNB>2.3.CO;2.

Silver, E.A., Kastner, M., Fisher, A.T., Morris, J., McIntosh, K.D., and Saffer, D.M., 2000. Fluid flow paths in the Middle America Trench and Costa Rica margin. *Geology*, 28:679–682, doi:10.1130/0091-7613(2000)28<679:FFPITM>2.0.CO;2.

Tadokoro, K. and Ando, M., 2002. Evidence for rapid fault healing derived from temporal changes in S wave splitting, *Geophys. Res. Lett.*, 29(4), 1047, doi:10.1029/2001GL013644.

Tadokoro, K., Ando, M., and Nishigami, K., 2000. Induced earthquakes accompanying the water injection experiment at the Nojima fault zone, Japan: Seismicity and its migration. *J. Geophys. Res.*, 105:6089–6104, doi:10.1029/1999JB900416.

Tanaka, H., Fujimoto, K., Ohtani, T., and Ito, H., 2001. Structural and chemical characterization of shear zones in the freshly activated Nojima fault, Awaji Island, southwest Japan. *J. Geophys. Res.*, 106:8789–8810.

Tobin, H. and Kinoshita, M., 2006. NanTroSEIZE: the IODP Nankai Trough seismogenic zone experiment. *Sci. Drill.*, 2:23–27, doi:10.2204/iodp.sd.2.06.2006.

Tobin, H., Vannuchi, P., and Meschede, M., 2001. Structure, inferred mechanical properties, and implications for fluid transport in the décollement zone, Costa Rica convergent margin. *Geology*, 29:907–910, doi:10.1130/0091-7613(2001)029<0907:SIMPAI>2.0.CO;2.

von Huene, R., Ranero, C.R., and Vannucchi, P., 2004. A model for subduction erosion. *Geology*, 32:913–916, doi:10.1130/G20563.1.

Zoback, M., Hickman, H., and Ellsworth, W., 2006. Structure and properties of the San Andreas fault in central California: Preliminary results from the SAFOD experiment. *Geophys. Res. Abstr.*, 8:02474.

Zoback, M.D., Apel, R., Baumgartner, J., Brudy, M., Emmermann, R., Engeser, B., Kessels, W., Fuchs, K., Rischmuller, H., Rummel, F., and Vernik, L., 1993. Upper-crustal strength inferred from stress measurements to 6 km depth in the KTB borehole. *Nature*, 365:633–634, doi:10.1038/365633a0.

Zoback, M.D., Hickman, S., and Ellsworth, W., 2007. The role of fault zone drilling, 2007, *In* Kanamori, H., and Schubert, G. (Eds.), *Earthquake Seismology - Treatise on Geophysics.* Amsterdam, The Netherlands (Elsevier), in press.

Authors

Harold Tobin, Associate Professor, Department of Geology and Geophysics, University of Wisconsin-Madison, 1215 West Dayton Street, Madison, Wis. 53706, U.S.A., e-mail: htobin@wisc.edu.

Hisao Ito, Center for Deep Earth Exploration (CDEX) – Japan Agency for Marine-Earth Science and Technology (JAMSTEC), 3173-25 Showa-machi, Kanazawa-ku, Yokohama, Kanagawa, 236-0001, Japan.

Jan Behrmann, Marine Geodynamics, IFM-GEOMAR, Wischhofstr. 1-3, 24148, Kiel, Germany.

Steve Hickman, U.S. Geological Survey, MS977, 345 Middlefield Road, Menlo Park, Calif., 94025, U.S.A.

Gaku Kimura, Department of Earth and Planetary Sciences, Graduate School of Science, The University of Tokyo, 7-3-1 Hongo, Bunkyo-ku, Tokyo, 113-0033, Japan.

27

Fault Zone Drilling in Subduction Zones: The Accomplishments of Ocean Drilling and Associated Studies

by J. Casey Moore

Subduction zones produce the world's largest earthquakes and cumulatively account for 90% of the global seismic moment release. The production of such large earthquakes is probably due to the shallow inclination (typically <20°) of the major plate-boundary subduction thrust. Thus, in comparison to vertically-oriented transform faults, subduction thrusts have a large area in the seismogenic zone, or the depth interval mechanically capable of producing earthquakes.

Beginning with the later stages of the Deep Sea Drilling Program (DSDP) and continuing through the Ocean Drilling Program (ODP), Earth scientists have concentrated on drilling through the subduction thrust. Restrictions of the riserless drilling process required that the drilling focus on relatively shallow portions of the subduction thrust with burial depths limited to about 1 km below the seafloor.

Subduction thrusts convey the oceanic crust and overlying sedimentary deposits to the seismogenic zone and beyond. Hence, we can know the material that is entering the subduction thrust and trace its evolution with depth, both by direct sampling and experimental studies of its changing mechanical properties and mineral phases. Where drilled, subduction thrusts consist of zones of scaly mudstone 20–40 m thick. They are characterized by consolidative deformation where intergranular porosity is declining; these fault rocks include a fracture porosity and permeability that may be intermittently dilated by high fluid pressure. Hydrologic tests show that permeability varies with fluid pressure. Geochemical evidence suggests that separate fluid systems may exist around the subduction thrust, in the underthrust sediments, and in the subjacent oceanic crust. Geochemical evidence further suggests fluids associated with the subduction thrusts are typically far-traveled along this shallowly inclined fault.

Logging while drilling (LWD) has been an effective tool for determining *in situ* physical properties, establishing borehole structure, and inferring stress orientations in ODP boreholes in subduction zones. Physical property results indicate impedance values (velocity × density) that are consistent with the seismic reflection images. Structural features from analysis of LWD resistivity images are similar in orientation and type to those measured from reoriented cores. Borehole breakouts indicate stress orientations mirroring those determined from structural features of the borehole and also consistent with plate convergence vectors.

Because subduction thrusts dip shallowly, they may be well imaged with seismic reflection techniques. Seismic reflection images, coupled with drilling results, indicate that the fluids being expelled from the consolidating, dehydrating subduction zone deposits may be channelized during transport along the subduction thrust. Seismic data also suggest progressive underplating of the underthrust sediments with depth; the quantity of water in the underthrust sediments also diminishes with depth. Models, also well linked to drilling results, suggest the fluid pressure initially rises tens of kilometers landward of the deformation front and then diminishes towards the upper aseismic-seismic transition along the subduction thrust.

A combination of seismological information and thermal models suggests that the upper aseismic to seismic transition begins at temperatures of about 125°C. Here, the combined effects of consolidation and a series of diagenetic-metamorphic reactions apparently foster the onset of stick-slip behavior. At some localities incoming sediment may have been underplated by the depth of the seismogenic zone, leaving the basaltic oceanic crust as the lower plate of the subduction thrust. The fluid system of the oceanic crust, documented by ocean drilling, may be important here in controlling the mechanics of faulting.

Subduction thrusts probably transition from thick consolidating zones at shallow depths to more localized fault surfaces in the seismogenic zone. Candidates for such seismogenic faults in exhumed subduction complexes include layers of frictional melts (pseudotachylytes), zones of fluidized but unmelted fault rock, and intervals of conspicuous veining. A major challenge in drilling into the seismogenic zone will be recognition of the active fault surface, especially when it is included in a large volume of previously deformed rocks. Careful inspection of borehole images and cores, as well as monitoring of the borehole, will be essential in unequivocal specification of the active subduction thrust at depth.

Author

J. Casey Moore, Earth and Planetary Sciences, University of California Santa Cruz, Santa Cruz, Calif., 95064, U.S.A., e-mail: cmoore@es.ucsc.edu.

Fault Characteristics, Energy Estimates, and Earthquake Recurrence: What *One* Seismologist Wants from Fault Drilling

by Jim Mori

There are some significant differences between the way seismologists and geologists typically look at faulting from large earthquakes. Seismologists tend to analyze waveform data which can usually resolve structures on the order of tens of kilometers; possibly down to one kilometers there is very good near-field data. Also, the derived physically properties are averaged over this scale length. Although it is difficult to resolve small scale features, seismologic studies have the advantage of seeing all depths of the fault. In contrast fault zone geologists can make direct observations of the fault zone structures and properties, which are quite different from the remote sensing techniques of seismologists. Geologists tend to look at structures on the scale of microns to meters when examining the cores obtained from drilling into faults. These observations are spot measurements and it is often difficult to assess how representative they are of the entire fault. Also, fault zone sample studies are usually limited to a few kilometers depth, which may barely be in the range of the seismogenic zone that produces the seismic waves. These differences mean that seismologists and geologists are often looking at quite different aspects of the earthquake process. The approaches are different, but can also be complementary.

Drilling Depth

The depth of drilling is an important point for fault zone investigations. To compare physical properties obtained in cores with seismological results obtained from waveform analyses, it is necessary to reach depths of the seismogenic zone, that is the regions of fault slip that have significant stress change to produce seismic waves. The shallow regions of faults, where materials have low rigidity, likely slip sympathetically with the seismogenic zone but do not produce seismic waves. The depth of the beginning seismogenic zone depends on local geology and probably is a transition zone. Waveform inversion studies suggest that the seismogenic zone may be as shallow as 1 km for the Chi-Chi, Taiwan earthquakes (Ji et al., 2001). It would be interesting if geologic studies could distinguish between seismic and non-seismic slip of the fault.

Temperature Measurements

One of the fundamental issues in understanding the physical mechanisms of earthquakes is clarifying the level of friction on the fault. One way to estimate the frictional levels during the faulting is to measure the heat produced. Measurements of the heat flow associated with the San Andreas fault have long been discussed over the past decades (e.g., summarized in Scholz, 2002, section 3.4.4). A more direct estimate would be to measure the fault zone temperatures immediately after a large earthquake. There were informal discussions to measure fault-zone temperatures following the large 1992 Landers and 2001 Denali earthquakes, although measurements were not done. Currently, the only available data of fault-zone temperatures following an earthquake are for the 1999 Chi-Chi, Taiwan earthquake (Kano et al., 2006). This result infers a very low level of dynamic friction during the earthquake; however, the measurement was made 5 years after the earthquake, when the temperature signal was quite small, and it is difficult to make clear interpretations. To obtain better estimates of the fault-zone heat, measurements more quickly (a few months) after a large earthquake are necessary (Fig.1).

Asperities

Fault slip during large earthquakes is usually heterogeneous with regions of large and small slip. The areas of large slip are often termed 'asperities'. These areas of large slip dominate the energy radiation and may control the rupture process. There are also suggestions that the stress accumulation during the interseismic period is different on the asperities, compared to the rest of the fault. The actual physical process that controls the stress accumulation and large slip of the asperities is currently an active topic of discussion in seismology. Yamanaka and Kikuchi (2004)

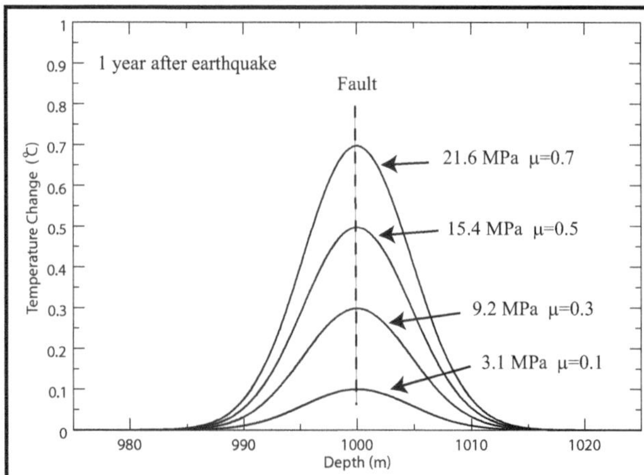

Figure 1. Expected temperature changes at 1000 m depth one year following a fault slip of one meter. The various curves show the changes for different levels of friction on the fault.

Table 1. Complementary differences between typical seismology studies and fault zone investigations.

Seismological Observations—Macroscopic	Fault Zone Investigations—Microscopic
Scales of km to tens of km, average properties of the fault zone	Scales of nanometers to meters for spot measurements
Remote sensing of entire depth range of fault zone	Direct samples from shallow depths of the fault
Estimates of total seismic radiated energy	Estimates of heat generated by temp. measurements after earthquakes
Estimates of non-radiated (fracture?) energy	Determination grain size, crack distributions for energy estimates
Seismic determination of slip weakening distance	Estimates of fault frictional properties from core samples

suggest that asperities are persistent features that are characteristic of a fault zone. However, results from the two recent Parkfield earthquakes and also large subduction earthquakes along the New Britain trench (Park and Mori, 2007) indicate that asperity distributions can be different for repeated ruptures of the same zone. Some related seismological issues that might be answered by detailed analyses of fault zone structures from drilling include the following:

- Are repeating earthquakes on a fault occurring on the same exact fault plane?
- Are the locations of asperities (areas of large slip) the same in repeating earthquakes?
- What is the physical cause of asperities?
- How does each earthquake contribute to making the fault zone?
- What is the fracture/healing process on the fault?

Macro- and Microscopic Observations

Kanamori and Heaton (2002) emphasize the importance for understanding and integrating small and large scale results for fault zone studies. The physical processes that occur on microscopic scales of the fault zone are reflected in the macroscopic observations made by seismologists.

Table 1 lists some of the different and complementary aspects of seismological and borehole studies of earthquakes and fault zones. For example, detailed analyses of near-field seismograms for large earthquakes can give estimates of both the radiated and non-radiated energy. The non-radiated energy is thought to be dissipated by processes such as fracture formation and heat production. These dissipated processes can be studies with direct measurements of temperature, crack distributions, and grain sizes obtained from borehole observations.

Onshore and Offshore Drilling

The logistics of drilling crustal faults on land and drilling faults offshore, such as large megathrust subduction zones, are quite different and are addressed by the differences in the programs of the International Continental Scientific Drilling Program (ICDP) and the Integrated Ocean Drilling Program (IODP). Crustal onshore faults tend to be shallower and more accessible to drilling investigations. For large offshore subduction zones, identifying and reaching the fault surfaces with drilling is a difficult endeavor. Drilling both faulting environments is important for understanding the physical structures of the fault zones. There are also opportunities for cooperative onshore/offshore drilling projects that can combine observations of the same fault zone.

Wish List

Finally, to address the issues mentioned above, I would recommend that a fault zone drilling program would include the following (roughly in order of difficulty).

- Design projects to combine microscopic and macroscopic observations
- Drill to sufficient depth to reach the seismogenic zone (region that produces seismic waves)
- Drill soon after a large earthquake to measure temperature changes caused by faulting
- Use deep boreholes for measurements of earthquakes and strain (borehole observatories)
- Sample a fault in the same place before and after an earthquake

References

Ji, C., Helmberger, D.V., Song, T.R., Ma, K.-F., and Wald, D.J, 2001. Slip distribution and tectonic implication of the 1999 Chi-Chi, Taiwan, earthquake. *Geophys. Res. Lett.* 28:4379–4382, doi:10.1029/2001GL013225.

Kanamori, H. and Heaton, T.H., 2002. Microscopic and macroscopic physics of earthquakes. *In* Rundle, J., Turcotte, D.L., and Klein, W. (Eds.), *Geocomplexity and the Physics of Earthquakes, Geophysical Monograph no. 120*, Washington, DC (American Gephysical Union), 147p.

Kano, Y., Mori, J. Fujio, R., Ito, H., Yanagidani, T., Nakao, S., and Ma, K.F., 2006. Heat signature on the Chelungpu fault associated with the 1999 Chi-Chi, Taiwan earthquake. *Geophys. Res. Lett.*, doi:10.1029/2006GL026733.

Park, S. and Mori, J., 2007. Are asperity patterns persistent? Implication from large earthquakes in Papua, New Guinea, *J. Geophys. Res.*, doi:10.1029/2006JB004481.

Scholz, C.H., 2002. *The Mechanics of Earthquakes and Faulting.* Cambridge (Cambridge University Press), 471 p.

Yamanaka, Y. and Kikuchi, M., 2004. Asperity map along the subduction zone in northeastern Japan inferred from regional seismic data. *J. Geophys. Res.*, 109, B07307, doi:10.1029/2003JB002683.

Author

Jim Mori, Disaster Prevention Research Institute, Kyoto University, Gokasho, Uji, Kyoto, 611-0011, Japan, e-mail: mori@eqh.dpri.kyoto-u.ac.jp.

29

Pore Pressure at Plate Boundaries: Insights from Modeling

by Demian Saffer

Introduction

Fluids play a key role in a wide range of mechanical and transport processes in the Earth's crust. Fluid flow can efficiently redistribute heat and solutes; as a result, fluid assisted transport and fluid rock interaction are important controls over the rate and extent of metamorphism, precipitation, and sealing of fractures. Fluid-rock reaction and precipitation impact permeability and fault strength (Chester et al., 1993). In addition to its chemical effects, pore fluid mediates the strength of brittle rocks because pore pressure controls effective stress (Hubbert and Rubey, 1959). In particular, elevated pore fluid pressure has been invoked as a cause of mechanical weakness along plate boundary faults and within the surrounding crust, at subduction zones (Davis et al., 1983) and along continental transforms (Irwin and Barnes, 1975). Because fluid pressure is central to a host of hypotheses explaining apparent fault weakness, direct measurements and indirect constraints on fluid pressure and flow are high priorities for fault zone drilling.

Numerical models provide key tools to link observations at shallow depths or the surface with borehole observations, and to test hypotheses generated on the basis of such data. These efforts can complement deep drilling by helping to direct measurement strategies, and by estimating pressures and flow patterns at depths and over a larger spatial extent than reached by drilling. In the context of understanding fluid pressure at plate boundaries, modeling efforts have provided insights into many fundamental questions, including: What are the magnitudes and spatial patterns of suprahydrostatic pore pressure? What controls pore pressure, and how should it vary—systematically between margins or with depth? Are hypothesized fluid sources sufficient to generate fault-weakening overpressures? What role do permeable zones (faults, sands) play in drainage, and in localizing pressure or flow?

Here I summarize three studies in which hydrologic models are combined with available data to better understand pore pressure and its mechanical effects at plate boundaries (Table 1). In the first example, field and laboratory data constrain a simple analytic model, which elucidates fluid flow patterns and controls on pore pressure buildup during initial loading at subduction zones. In the second example, regional scale numerical models are used to quantify the influence of fundamental geologic factors on pore pressure, and consequently on rock strength, in subduction zones. In the third example, models are used to evaluate hypotheses for pore pressure generation and compartmentalization invoked to explain mechanical weakness along the San Andreas fault system in central California.

Example 1: Simple Model for Dewatering at the Toe of Subduction Margins

At subduction zones, as incoming sediments are offscraped or underthrust at the trench, elevated pore pressures result from rapid loading of low-permeability sediment (Table 1). Pore pressure within the sediment underthrust at the trench is especially important for the mechanical strength of the plate boundary fault system, because the main décollement localizes immediately above this sediment, and at many margins downsteps into the section within a few to tens of kilometers from the trench. Along transects where Ocean Drilling Program (ODP) drilling has penetrated the underthrust section at the Nankai, northern Barbados, and Costa Rican subduction zones, it maintains anomalously high porosity for its burial depth, suggesting under-consolidation and associated high fluid pressure. Quantitative estimates of *in situ* pore pressure have been obtained from laboratory consolidation tests on core samples, and from observed compaction trends both in boreholes and inferred from seismic data (Screaton et al., 2002; Tobin and Saffer, 2003). At all three margins, the different methods yield consistent results and indicate development of overpressures that increase systematically with distance from the trench. Estimated pore pressures suggest nearly undrained conditions at the base of the section (excess pressures equal to the load emplaced by subduction burial) and partially drained conditions at the top.

To first order, the data can be explained by a simple model describing a layer subjected to a continuously increasing load, in which fluid escapes upward to a free surface (Wissa et al., 1971; Fig. 1). In the model, pore pressure varies systematically with distance from the trench and depends upon the thickness of the consolidating sediment layer (H), the permeability and compressibility of the sediment layer (described by a single variable, C_v), and the plate convergence rate (v_p). For each margin, laboratory measurements constrain sediment permeability and compressibility. Notably, v_p, H, and C_v are different for each of the three locations, yet all of the data are consistent with a generic model in which degree

Table 1. Parameters of 2-D numerical models of subduction zones.

Example	Model Source Terms	Scale	Permeability
Subduction zones (toe)	Loading by underthrusting (virtual)	100's of m	Sediment permeability measured in laboratory, function of porosity
Subduction zones (regional)	Prescribed porosity loss based on drilling and seismic data	10's of km	Sediment permeability measured in laboratory, function of porosity; range of fault permeabilities explored
San Andreas Fault System	Fluid release by prograde metamorphism	~10km	Function of depth; range of fault permeabilities explored

of under-consolidation depends on the relative magnitudes of loading rate, drainage path length, and sediment hydraulic properties (Fig. 1). The agreement between observation and model implies that upward drainage to a highly permeable décollement dominates dewatering to distances of at least ~5–10 km landward of the trench. The model results also suggest that severe under-consolidation should persist for tens of kilometers from the trench (Saffer, 2007).

Example 2: Regional Scale Hydrologic Models of Subduction Complexes

In several studies, 2-D numerical models of subduction zones have been used to estimate pore pressures and test hypotheses about the nature of fluid flow at the regional scale (tens of kilometers; Bekins et al., 1995; Screaton et al., 1990). In these models, pore pressure and flow are driven by sediment compaction (Table 1). Fluid sources from compactive dewatering are incorporated at each node within the model domain, and are constrained by porosity data from boreholes and inferred from seismic velocity data (these models are described in detail by Bekins and Dreiss, 1992; see also Screaton et al., 1990). Sediment bulk permeability in the models is constrained by laboratory data, and a range of fault zone permeability has been investigated. Fault permeability is typically assumed to be higher than that of the sediment matrix, based on observations of (1) fluid expulsion at the seafloor and (2) chemical and thermal anomalies centered on faults. For several well studied margins, modeled pore pressure is in excellent agreement with that measured in borehole observatories (Barbados; Bekins and Screaton, 2007) and estimated by methods described in the previous example (Nankai and Costa Rica; Spinelli et al., 2006; Tobin and Saffer, 2003).

Regional scale models have also been used to evaluate the relative importance of factors that control pore pressure and ultimately the taper angle of subduction accretion complexes. The factors include sediment and fault permeability, incoming sediment thickness, convergence rate, and partitioning of sediment between accretion and underthrusting. By isolating and quantifying the influence of each of these factors on pore pressure, a comprehensive sensitivity analysis provides a basic framework for considering observations from active subduction complexes. As in the case of burial at the toe of subduction systems described above, the results show that pore pressure in accretionary wedges can be viewed as a balance between factors which drive pore pressure (source terms) and those that limit flow (permeability and drainage path length). Sediment permeability and incoming sediment thickness are most important, whereas fault permeability and the partitioning of sediment have small effects (Fig. 2).

By combining the hydrologic model with critical taper theory, systematic relationships emerge between the fundamental factors that govern fluid pressure and stable taper angle (Saffer and Bekins, 2006). For example, as sediment permeability is increased, modeled pore pressure decreases and allows taper angles to steepen. With increased sediment thickness, increased pore pressure drives a decrease in taper angle (Fig. 2). In general, low permeability and thick incoming sediment sustain high pore pressures consistent with shallowly tapered geometry, whereas high permeability and thin incoming sediment should result in steep geometry. The model results compare favorably with available data from active accretionary complexes—active margins characterized by a significant proportion of fine-grained sediment within the incoming section (such as the northern Barbados Trench and the eastern Nankai Trough) exhibit thin taper

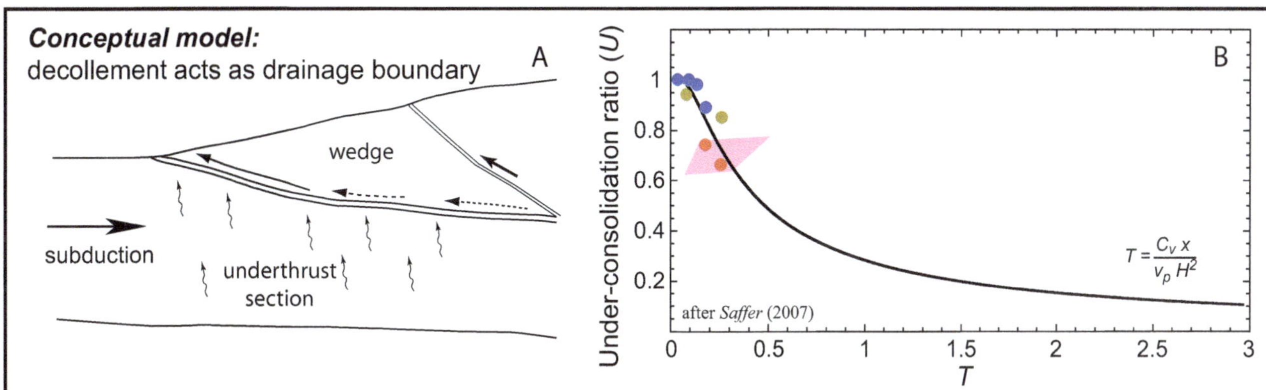

Figure 1. [A] Schematic showing drainage pattern for 1-D consolidation model. [B] Modeled under-consolidation ratio, U, vs. dimensionless time (solid black curve); mean values of U from ODP sites at Barbados (blue), Nankai (red) and Costa Rica (tan), and inferred from seismic velocity data at Nankai (pink shaded region).

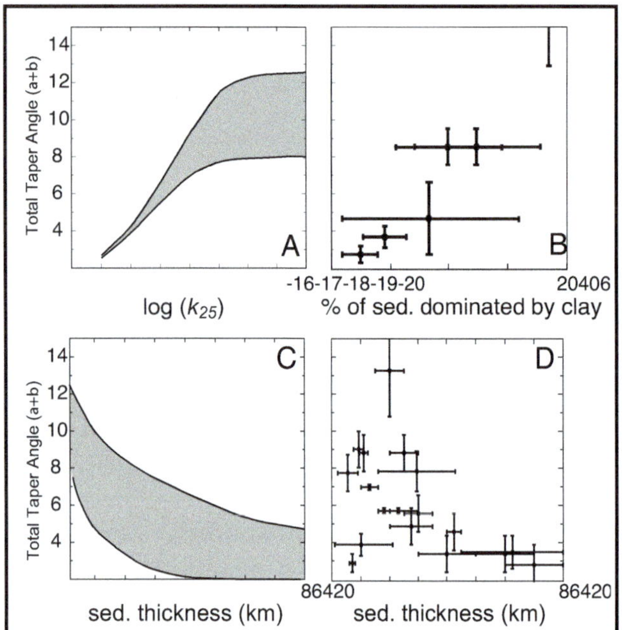

Figure 2. [A] Model results showing influence of permeability (plotted as log permeability at 25% porosity) on taper angle of accretionary wedges. [B] Data from active accretionary complexes where drilling has documented lithologies of incoming sediment, showing relationship between proportion of clay in sediment and taper angle. [C] Model results. [D] Data from active accretionary complexes, showing influence of sediment thickness on taper angle. After Saffer and Bekins (2006).

angles, whereas those characterized by a higher proportion of sandy turbidites (such as Cascadia, Chile, and Mexico) have steep taper angles. Observations from active margins also indicate a strong trend of decreasing taper angle with increased sediment thickness. One key implication is that hydrologic properties should strongly influence the strength of the crust in a wide range of geologic settings.

Example 3: A Test of Hypotheses Along the San Andreas Fault System

Along the San Andreas fault in California, heat flow observations and inferred stress orientations have been interpreted to indicate that the fault moves under considerably smaller shear stresses than predicted by laboratory-derived values for rock friction (Zoback et al., 1987). Elevated fluid pressures are one likely candidate to explain the apparent fault weakness—such overpressures may be either regionally extensive resulting in a weak fault embedded within a weak crust (Hardebeck and Hauksson, 2000), or localized along the fault resulting in a weak fault embedded within a strong crust (Zoback et al., 1987). Several hypotheses have been proposed to explain the generation of fluid pressure in the vicinity of the fault. In the study summarized here, Fulton et al. (2005) used numerical models to investigate (1) whether fluid produced by prograde metamorphism of the Franciscan assemblage is a likely candidate for generating significant and lasting overpressures along the fault system, and (2) if so, whether overpressures can be localized along the fault.

The apparent correlation between CO_2 discharge, ^{18}O enrichment, geologic structure, and observed seismicity patterns originally led Irwin and Barnes (1975) to suggest that fluids of metamorphic origin sourced from dehydration of Franciscan rocks are trapped and forced into the fault zone, causing fluid overpressures in areas where overlying low permeability Great Valley sediments and serpentinite act as a barrier to upward fluid expulsion. Fulton et al. (2005) quantitatively tested this hypothesis using a two-step approach. First, the spatial and temporal distribution of fluid sources from prograde dehydration was estimated from theoretical values of whole-rock H_2O content and P-T histories for the Franciscan crust in the wake of northward migration of the Mendocino Triple Junction (MTJ) (Table 1). Second, the fluid sources were incorporated into transient 2-D models of fluid flow within cross sections perpendicular to the fault. In the models, permeability of the Salinian and Franciscan crust was constrained by the work of Ingebritsen and Manning (1999) and assumed to decrease logarithmically with depth. In addition to a 500-m-thick low-permeability serpentinite body ($k=10^{-20}$ m^{-2}) extending across the eastern half the model domain at a depth of 2 km, model simulations also explore a range of fault permeability architectures including a fault barrier, a fault conduit, and a combined conduit-barrier.

For models that incorporate both a serpentinite cap and fault barrier, overpressures as large as 62% of lithostatic pressure develop within 4 My of MTJ passage (a region extending ~200 km south of the present day location of the MTJ), where sources are abundant. However, simulated pressures dissipate within several thousand years because fluid sources die out due to stabilization in temperature and the low remaining H_2O content in lower crustal rocks. Overall, the presence of a low permeability cap can enhance regional overpressure development; however, none of the structures investigated allowed overpressures to localize along the fault (Fig. 3). The results suggest that metamorphic dehydration within the Franciscan crust has the potential to cause regional overpressure and weakening of the crust for

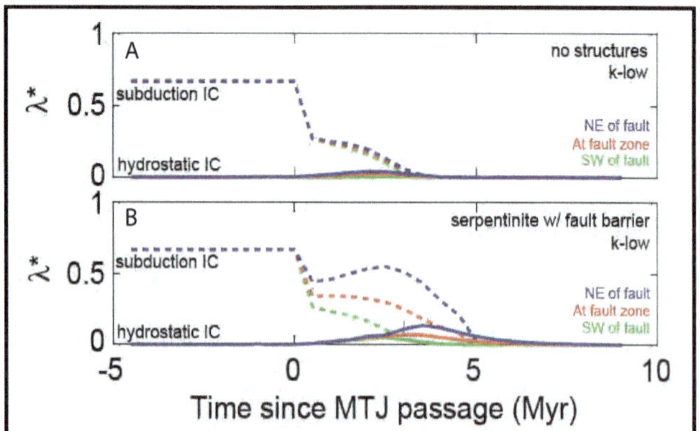

Figure 3. Model results showing the development of overpressures in the crust to the NE (blue), to the SW (green), and within (red) the San Andreas fault, for [A] a simple model with no geologic structures and [B] models incorporating a serpentinite "cap" and a fault barrier. Overpressures are plotted using the normalized pore pressure ratio $\lambda^*=(P_{fluid} - P_{hydrostat})/(P_{lithostat} - P_{hydrostat})$. In both panels, solid curves show hydrostatic initial condition, and dashed curves show an elevated pore pressure initial condition associated with subduction north of the MTJ.

~4 Ma after MTJ passage, currently encompassing a region ~200 km south of the MTJ, but is limited as a sole mechanism for long-term weakening of the San Andreas fault along much of its length because overpressures cannot be sustained or localized along the fault.

Summary

In the examples summarized above, the integration of hydrologic modeling with datasets enabled by drilling—including laboratory studies on core samples and down-hole measurements—has provided insight into first-order questions about the magnitude, distribution, and mechanisms of elevated pore pressure in tectonically active environments. However, the utility of such models is closely tied to the availability of key datasets. Although data from the land surface (or seafloor) are important, drilling provides the most powerful and direct constraints on pore pressure and fluid flow in the subsurface. In particular, the hydraulic properties of faults and wall rock at relevant scales are key unknowns in most models; down-hole and long-term perturbation tests are one effective way to constrain these quantities. Likewise, rock sampling and geochemical data are critical toward better defining the fluid source terms that drive flow in these dynamic environments. Down-hole pressure and temperature measurements provide key data to test flow models. In this context, fault zone drilling in conjunction with surface measurements will provide vital data to clarify the role of fluids in faulting and to more definitively test hypotheses for fluid pressure development and its mechanical and chemical effects.

References

Bekins, B. and Screaton, E., 2007. Pore pressure and fluid flow in the northern Barbados accretionary complex. *In* Dixon, T., Moore, C. J., Eds. *The Seismogenic Zone of Subduction Thrusts*, New York (Columbia University Press), 148–170.

Bekins, B.A. and Dreiss, S.J., 1992. A simplified analysis of parameters controlling dewatering in accretionary prisms. *Earth Planet. Sci. Lett.*, 109:275–287, doi:10.1016/0012-821X(92)90092-A.

Bekins, B.A., McCaffrey, A., and Dreiss, S.J., 1995. Episodic and constant flow models for the origin of low chloride waters in a modern accretionary complex. *Water Resour. Res.*, 31:3205–3215, dodi:10.1029/95WR02569.

Chester, F.M., Evans, J.P., and Biegel, R.L., 1993. Internal structure and weakening mechanisms of the San Andreas fault. *J. Geophys. Res.*, 98:771–786.

Davis, D., Suppe, J., and Dahlen, F.A., 1983. Mechanics of fold-and-thrust belts and accretionary wedges. *J. Geophys. Res.*, 88:1153–1172.

Fulton, P.M., Saffer, D.M., and Bekins, B.A., 2005. Crustal dehydration and overpressure development on the San Andreas fault. *Eos Trans. AGU*, 86(52), Fall Meet. Suppl., Abstract T43C-02.

Hardebeck, J.L. and Hauksson, E., 2000. The San Andreas fault in southern California: A weak fault in a weak crust. *In* Kovach, R. (Ed.), *Proceedings of the 3rd Conference on Tectonic Problems of the San Andreas Fault System*, September 6–8, 2000, Stanford University.

Hubbert, M.K. and Rubey, W.W., 1959. Role of fluid pressure in mechanics of overthrust faulting. *Geol. Soc. Am. Bull.*, 70:115–166, doi:10.1130/0016-7606(1959)70[115:ROFPIM]2.0.CO;2.

Ingebritsen, S.E. and Manning, C.E., 1999. Geological implications of a permeability-depth curve for the continental crust. *Geology*, 27:1107–1110, doi:10.1130/0091-7613(1999)027<1107:GIOAPD>2.3.CO;2.

Irwin, W.P. and Barnes, I., 1975. Effect of geologic structure and metamorphic fluids on seismic behavior of the San Andreas Fault system in central and northern California. *Geology*, 3:713–716, doi:10.1130/0091-7613(1975)3<73:EOGSAM>2.0.CO;2.

Saffer, D.M., 2007. Pore pressure within under thrust sediments in subduction zones. *In* Dixon, T., Moore, C. J., Eds., *The Seismogenic Zone of Subduction Thrusts*, New York (Columbia University Press), 692 p.

Saffer, D.M. and Bekins, B.A., 2006. An evaluation of factors influencing pore pressure in accretionary complexes: Implications for taper angle and wedge mechanics. *J. Geophys. Res.*, 111:B04101 , doi:10.1029/2005JB003990.

Screaton, E.J., Wuthritch, D.R., and Dreiss, S.J., 1990. Permeabilities, fluid pressures, and flow rates in the Barbados ridge complex. *J. Geophys. Res.*, 95 :8997–9007.

Screaton, E.J., Saffer, D.M., Henry, P., Hunze, S., and the Leg 190 Shipboard Scientific Party, 2002. Porosity loss within underthrust sediment of the Nankai accretionary complex: Implications for overpressures. *Geology*, 30:19–22, doi:10.1130/0091-7613(2002)030<0019:PLWTUS>2.0.CO;2.

Spinelli, G., Saffer, D.M., and Underwood, M.B., 2006. Hydrogeologic responses to three-dimensional temperature variability, Costa Rica subduction margin. *J. Geophys. Res.*, 111:B04403, doi:10.1029/2004JB003436.

Tobin, H.J. and Saffer, D.M., 2003. Fluid pressure in the shallow plate interface at the Nankai Trough subduction zone. *Eos Trans. AGU*, 84(46), Fall Meet. Suppl., Abstract T41A-04.

Wissa, A.E.Z., Christian, J.T., Davis, E.H., and Heiberg, S., 1971. Consolidation at constant rate of strain, *J. Soil Mech. Found. Div., ASCE*, 97(SM10):1393–1413.

Zoback, M.D., Zoback,M.L., Mount, V.S., Suppe , J., Eaton, J.P., Healy, J.H., Oppenheimer, D., Reasenberg, P., Jones, L., Raleigh, C.B., Wong, I.G., Scotti, O., and Wentworth, C., 1987. New evidence on the state of stress of the San Andreas Fault, *Science*, 238:1105–1111, doi:10.1126/science.238.4830.1105.

Author

Demian Saffer, Department of Geosciences, The Pennsylvania State University, 310 Deike Building, University Park, Pa. 16802, U.S.A., e-mail: dsaffer@geosc.psu.edu.

30

Dynamic Rupture Propagation Modeling

by Satoshi Ide

Resolving earthquake dynamic rupture is essential to understand the physics behind earthquake rupture. This paper first reviews the current status of the slip inversion, and then discusses the scaling of earthquake dynamic rupture.

Slip inversion methods have been developed since the 1980s to study the heterogeneity of spatio-temporal slip distribution of earthquakes. Several methods have been applied to many earthquakes using various data, such as near-field strong-motion records, far-field broadband records, and geodetic observations including GPS and InSAR. Some attempts have been made to compile a catalog of slip models to study their general features. Increasing the amount of data and communication speed allows even quasi-real time analysis of slip inversion, which gives an impression that these analyses are easy and straightforward. However, each slip model is based on some assumptions that are not always described well in research papers and can be a reason for apparent large discrepancies between models. For example, the models presented for the 1999 Izmit, Turkey earthquake (Mw 7.6) were quite different (Fig. 1) as shown in Ide et al. (2005). Basic characteristics, such as the locations of large slip and even the starting point of the rupture, were quite different. The data seem not to be insufficient; some strong motion stations are located close to the surface trace of the source fault, and some geodetic measurements are available. However, mainly due to uneven distribution of stations and inaccurate timings of seismometers, the data cannot resolve the growth process of earthquake rupture and the assumptions made by each research group to determine the characteristics of their slip model.

We have to use many model-dependent assumptions to obtain slip models. These include data selection and preprocessing, model settings such as fault plane geometry and parameterization, and inversion method including the assumption of probability function of error. Among them, the most uncertain factor is how to calculate synthetic data that we compare with observed data. Usually we assume layered structures to calculate seismic waveforms. It has been demonstrated that a reliable 3-D structure can increase the resolution of slip inversion, but such structure information is practically unavailable in many cases. The method of Empirical Green's functions, which uses the records of small earthquakes as theoretical waves, provides a good approximation of waves in complex structures, but to find a set of good small events is not easy. Computation of seismic waveform is the bottleneck of current slip inversion.

Even now, the number of well resolved earthquakes is limited to about ten. Nevertheless, we find some general features of slip models. Earthquakes generally consist of successive ruptures of subevents, which are sometimes referred to as asperities with vague definition. The pulse-like rupture propagation is now widely accepted. As reported for some earthquakes, such as the 1906 San Francisco and the 2002 Denali, rupture pulses can propagate faster than the S-wave speed, but the average propagation velocity does not exceed it. Well resolved slip models can be used to discuss dynamic properties of earthquake rupture. Fault constitutive relationships and their characteristic parameters, slip weakening distance D_c and fracture energy G_c, have been estimated for some well resolved earthquakes, such as the 1992 Landers, the 1995 Kobe, and the 2000 Tottori. While D_c tends to be overestimated, G_c is a relatively stable parameter controlled mainly by rupture propagation velocity.

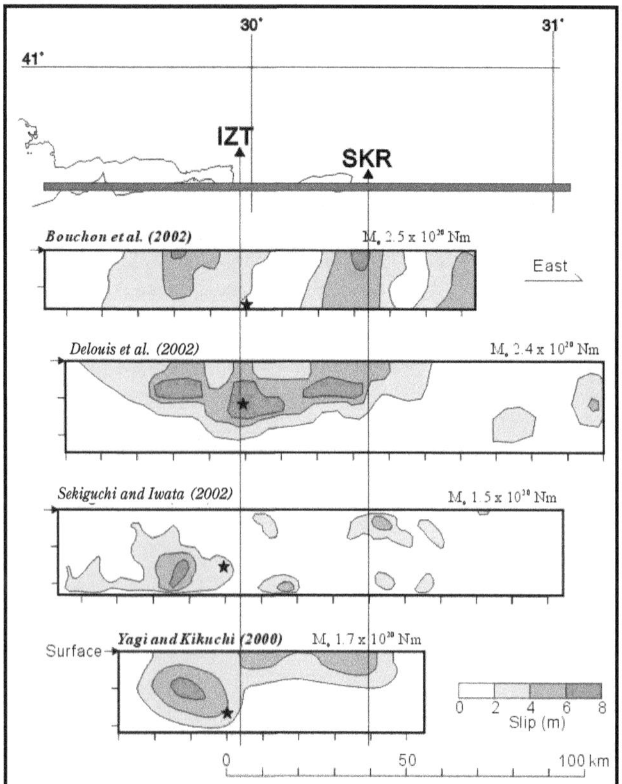

Figure 1. Slip models for the 1999 Izmit earthquake (Ide et al., 2005).

Earthquake faults and slip generally satisfy geometrical similarity; in other words, earthquakes have similar stress drop. If earthquakes are also temporally similar having similar rupture velocities, we expect to observe an almost constant ratio between seismic energy and seismic moment. Although there is still room for discussion whether the ratio is strictly constant, its change is small compared to the wide difference in seismic moment (about 15-fold).

While the scaling for these macroscopic parameters is well documented, the study for the scaling of earthquake complexity has started only recently. For example, Somerville et al. (1999) pointed out that the size of asperity (large slipped areas) scales as the square of the cubic root of seismic moment, suggesting geometrical similarity. Using a large catalog of slip models, Mai and Beroza (2000) also found similar scaling for seismic moment and fault length and width. However, they also showed that the scaling is different for the width of strike slip events, which is an evidence of a break of self-similarity. Sometimes, small earthquakes are assumed to be simple compared to large ones; however, this concept is not so obvious. Moreover, recently estimated slip models for microearthquakes (~ Mw 1) in a South Africa gold mine (Yamada et al., 2005) do have a subevent sequence, and characteristic time, length, and slip amount are all smaller by a factor of 1000 compared to those of the Kobe earthquakes. The information about scaling laws and their limitation are quite useful to construct hypothetical earthquake models for strong motion prediction.

Slip models can be used to scale dynamic parameters, too. Fracture energy G_C is a relatively reliable parameter calculated for many slip models. The average fracture energy estimated for earthquakes of M_w 5 to 7 is almost proportional to seismic moment, although the examples are limited.

All the previously mentioned studies for earthquake scaling compare small and large earthquakes using different measures that are comparable to the final size of events, but there are other ways to compare. One is the scaling of complexity during rupture growth of one earthquake—namely, comparison between the initial and final stages of one event. Once we recognize this scaling law, we can recognize whether one small rupture is the initial part of a large event or just a small event. Unfortunately, traditional slip inversion cannot solve this problem. Recently, we developed a new method suitable to this problem, which we refer to as the multiscale slip inversion method (Uchide and Ide, 2006). The initial stage of an earthquake can be investigated using high-frequency seismic waves near the onset of signal, while traditional inversion details overall rupture propagation. We can prepare a set of observation equations on these two scales, or with some additional intermediate scales. These observation equations are unified into a multiscale observation equation using a renormalized multiscale slip model. By solving this equation in the same way as traditional slip inversion problem, we obtain a multiscale slip model with changing resolution as the rupture grows.

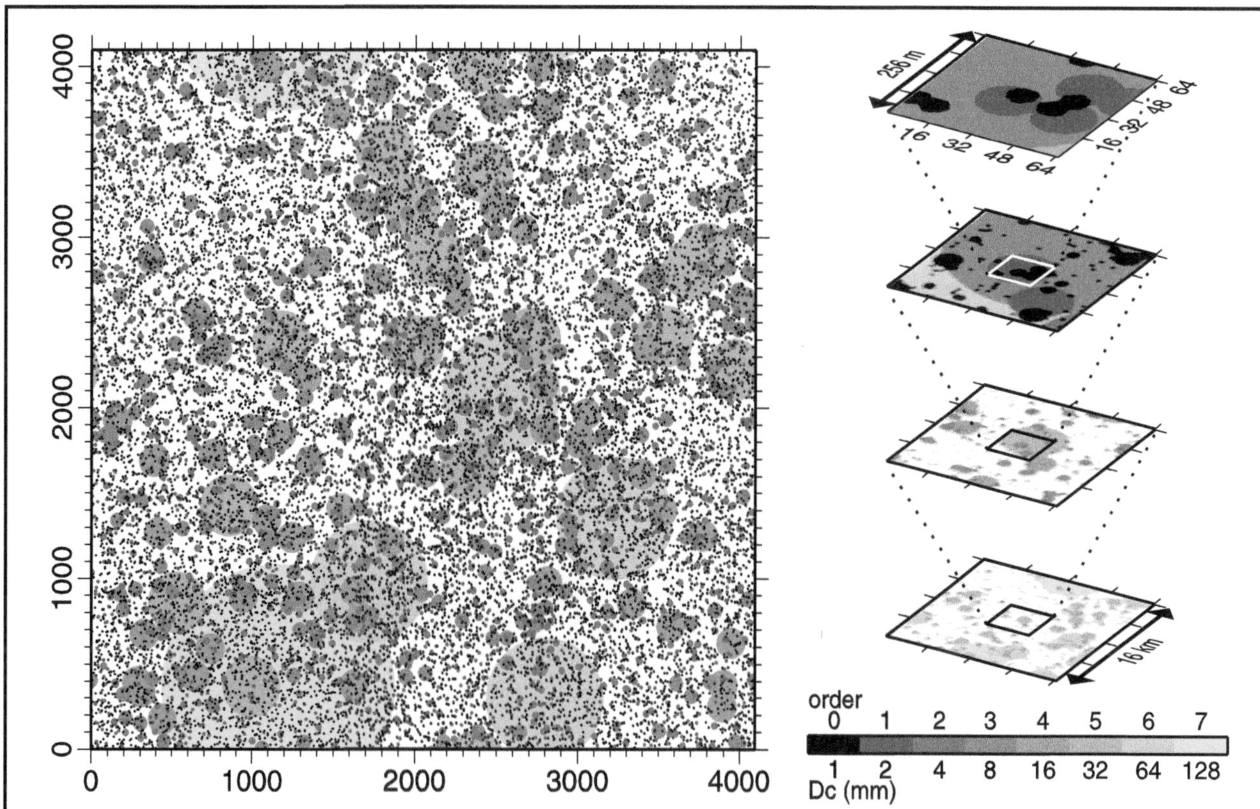

Figure 2. Distribution of circular patches (from Ide and Aochi, 2005).

We first applied this method to the mid-Niigata prefecture earthquake in 2004 in Japan. The spatio-temporal slip distribution is obtained as a nested spatio-temporal slip distribution on an assumed fault plane. Unlike ordinary slip models, this model contains a detailed rupture process in the first 1 s. The resolution in the next 1 s is worse, but the model area increases. Finally, we obtain the full scale rupture process with low resolution, which is similar to the result of traditional slip inversion. We observe several slip pulses in different directions even in the first 1 s. In every scale, slip rate is about 1 m s^{-1}, suggesting similarity.

When the earthquake is a self-similar process, how does a rupture propagate? Ide and Aochi (2005) have modeled self-similar dynamic ruptures assuming a heterogeneous distribution of circular patches on a planar crack (Fig. 2). Two points of the model are that 1) slip weakening distance D_c scales with the size of patch, and 2) the size-number distribution of patches satisfies a power law. Assuming that a rupture starts from one of the smallest patches, we solve the elastodynamic equation using the boundary integral equation method with the renormalization technique. Some ruptures grow into very large events, while most ruptures remain small. The moment rate function of a large event begins with a so-called initial phase which may also contain an initial phase in a smaller scale. Therefore, the rupture process is statistically self-similar. The rupture speed is subshear in average, but it can locally exceed S-wave speed, which is consistent with recent observations and slip models.

Whether a self-similar dynamic rupture growth is a good approximation of an earthquake remains to be proven. The answer to this question is important, not only because it is fundamental for the physics behind earthquakes, but also because it can be the basis of various practical usages of small earthquakes, such as assessment of earthquake predictability and predication of strong motion. Slip inversion is a useful tool to assist in resolving this question, and we have to overcome many problems to obtain sufficient numbers of reliable slip models, which are required for a statistical discussion.

References

Bouchon, M., Toksöz, N.M., Karabulut, H., Bouin, M.-P., Dietrich, M., Aktar, M., and Edie, M., 2002. Space and time evolution of rupture and faulting during the 1999 Izmit (Turkey) earthquake. *Bull. Seismol. Soc. Am.*, 92:256–266, doi:10.1785/0120000845.

Delouis, B., Giardini, D., Lundgren, P., and Salichon, J., 2002. Joint inverstion of InSAR, GPS, teleseismic, and strong motion data for the spatial and temporal distribution of earthquake slip: Application to the 1999 Izmit mainshock. *Bull. Seismol. Soc. Am.*, 92:278–299, doi:10.1785/0120000806.

Ide, S. and Aochi, H., 2005. Earthquakes as multiscale dynamic ruptures with heterogeneous fracture surface energy. *J. Geophys. Res.*, 110:B11303, doi:10.1029/2004JB003591.

Ide, S., Beroza, G.C., and McGuire, J.J., 2005. Imaging earthquake source complexity. *Seismic Earth: Array Analysis of Broadband Seismograms. Geophys. Monogr. Ser.* 157:117–135.

Mai, P.M. and Beroza, G.C., 2000. Source scaling properties from finite-fault-rupture models. *Bull. Seismol. Soc. Am.*, 90:604–614, doi:10.1785/0119990126.

Sekiguchi, H. and Iwata, T., 2002. Rupture process of the 1999 Kocaeli, Turkey, earthquake estimated form strong-motion waveforms. *Bull. Seismol. Soc. Am.*, 92:300–311, doi:10.1785/0120000811.

Somerville, P., Irikura, K., Graves, R., Sawada, S., Wald, D., Abrahamson, N., Iwasaki, Y., Kagawa, T., Smith, N., and Kowada, A., 1999. Characterizing crustal earthquake slip models for the prediction of strong ground motion. *Seismol. Res. Lett.*, 70:59–80.

Uchide, T. and Ide, S., 2006. Development of multiscale slip inversion method and its application to the 2004 mid-Niigata prefecture earthquake. *Eos Trans. AGU*, 87(52), Fall Meet. Suppl., Abstract S42C-01.

Yagi, Y. and Kikuchi, M., 2000. Source rupture process of the Kocaeli, Turkey, earthquake of August 17, 1000, obtained by joint inversion of near-field data and teleseismic data. *Geophys. Res. Lett.*, 27:13–1969.

Yamada, T., Mori, J.J., Ide, S., Kawakata, H., Iio, Y., and Ogasawara, H., 2005. Radiation efficiency and apparent stress of small earthquakes in a South African gold mine. *J. Geophys. Res.*, 110:B01305, doi:10.1029/2004JB003221.

Author

Satoshi Ide, Department of Earth and Planetary Science, University of Tokyo, 7-3-1, Hongo, Bunkyo, Tokyo, 113-0033, Japan, e-mail: ide@eps.s.u-tokyo.ac.jp

31

Fault Zone Complexity and Earthquake Ruptures

by Ryosuke Ando and Teruo Yamashita

Earthquake ruptures are propagated inside fault zones that consist of structures spanning a wide spectrum of lengths. A natural fault is known to consist of hierarchy structure ranging from the scale of fine grains forming a gauge or ultracataclasite zone along the principal slip plane (Fig. 1, top) up to that of fault system consisting of discrete fault segments (Fig. 1, bottom) and/or branched fault segments (Fig. 1, middle and bottom). Although the physics of earthquakes has been well studied on microscopic and macroscopic scales (Fig. 1, top and bottom, respectively) based on laboratory experiments and seismic inversion analysis, respectively, the effects on a mesoscopic scale (Fig. 1, middle) are poorly understood. Since these structures interact with earthquake ruptures, one of the seismologically fundamental problems is to understand how they are formed and evolved associated with the earthquakes that are the most intensive activities on fault zones.

In this study we focus on the investigation of the formation of branches in mesoscopic and macroscopic scales, and their effects on dynamic ruptures. In our model, branch fault segments are assumed to be nucleated at some discrete points on the main fault, which implies that the main fault is weak locally at these points. Each branch is assumed to be nucleated at each prescribed point on the main fault once the main fault tip passes, and the following fracture criterion is satisfied there. We also assume branches are extending in the direction of the maximum shear traction at each branch. Our studies rely on a simple slip-weakening law as the fracture criterion and the friction law:

$$\tau(S) = \frac{\tau_p - \tau_r}{D_c}(D_c - S)H(D_c - S) + \tau_r \quad (1)$$

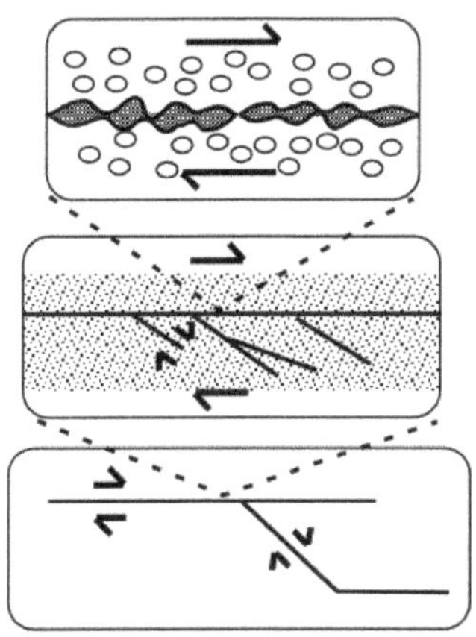

Figure 1. Schematic illustration of hierarchical structure of earthquake fault. Bottom (macroscopic scale): Scale of entire length of a fault system composed of planar segments; Middle (mesoscopic scale): length scale of fault zone involving damage zone (shaded) and branches; Top (microscopic scale): scale of frictional surface involving gouge and microcracks (after Ando and Yamashita,, 2007).

where τ_p, τ_r, S, D_c, and H (s) denote the peak shear strength, residual shear stress, slip, critical slip displacement, and a unit step function, respectively. The boundary integral equation method (BIEM) is employed to numerically solve the elastodynamic response of the medium. This method has advantages (in the treatment of non-planar fault geometry) and problems (where rupture paths are not specified but dynamically determined as the ruptures are extended).

Figure 2 shows snapshots of a typical example of fault growth. First, the main fault starts its extension and begins to nucleate branches. However, the nucleated branches do not begin their growths when the extension of the main fault is not long enough because of insufficient enhancement of stress (Fig. 2A). As the size of the main fault increases, the shear stress is magnified near its extending tip. If such stress enhancement becomes sufficiently large, branches can begin their growths while their extensions are arrested after attaining certain lengths (Fig. 2B and C). The spatial distribution of these arrested branches forms a roughly triangular zone on one side of the main fault. The lengths of these branches increase with main fault size. As the main fault extends further and its size exceeds a certain critical length L_m, a limited number of branches continue to grow together

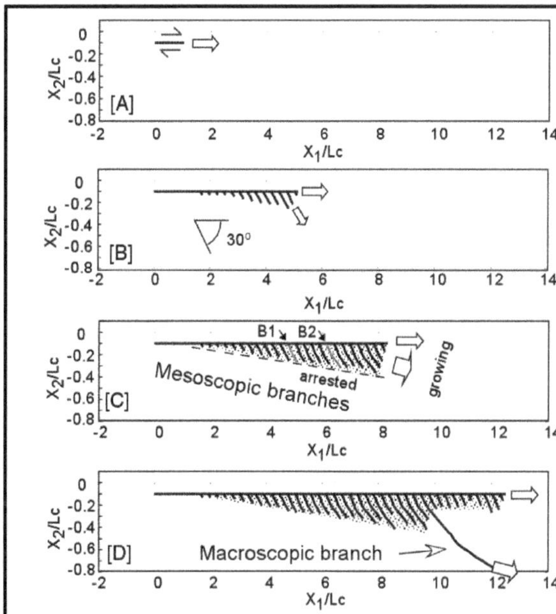

Figure 2. Snapshots of dynamic growth of fault with multiple branches; the value of $\Delta B/Lc$ = 0.30 is assumed. Note that the scale is magnified in the x2 direction. (after Ando and Yamashita, 2007).

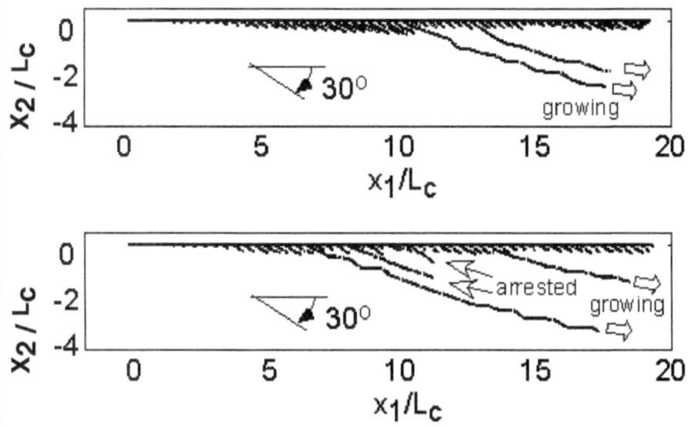

Figure 3. Examples of longer computation. We assume $\Delta B/Lc$ = 0.31 and 0.38 in the upper and lower graphs, respectively (after Ando and Yamashita, 2007).

with the main fault without arresting. This suggests that complicated fault geometry can be generated dynamically. It is also shown in the figure that the emergence of such a large-scale branch tends to suppress the growth of the next nucleated branches in its neighborhood, which is due to the stress shadow effect as will be discussed below. Such phenomena are also observed in Figure 3, which shows an example of longer simulation time.

Figure 4 shows the extending velocity of the tip of main fault soon attains a constant value with the extension. Comparing the velocity at each position between this multiple branch model and the case of a planar fault, we can clearly see that the dynamic branching reduces the rupture velocity of the main fault; the assumed physical parameters are identical between them (except the geometry). This data will contribute to the formation of approximately self-similar shapes for the spatial distribution of meso-branches.

We further investigate the effect of mesoscopic branches as the stress-slip relationship on a macroscopic scale, neglecting the thickness of the fault zone. Figure 5 shows this relationship obtained at each point along the main fault by calculating shear stress and displacement immediately outside the fault zone. The critical slip weakening distance D_c and the fracture energy G_c become proportional to the

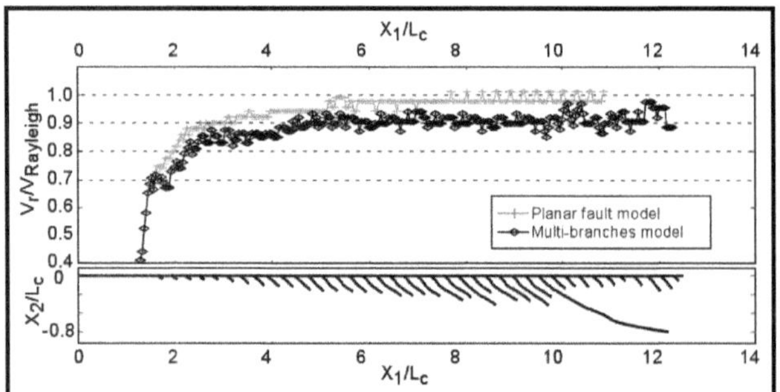

Figure 4. Reduction of extending velocity of the tip of main fault to approximately a constant sub-Rayleigh value due to multiple branching; the model shown in Figure 4 is assumed. (Upper) Black curve denotes the fault tip velocity in the multi-branch model. Gray curve shows the case of planar fault model with no branches; the other assumptions are the same as in the multi-branch model. (Lower) Thick curves denote the corresponding fault geometry of the multi-branch model (after Ando and Yamashita, 2007).

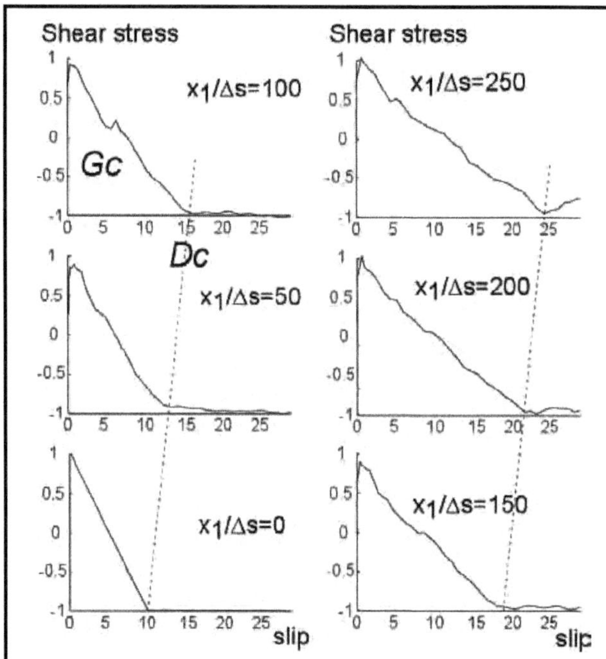

Figure 5. Slip-shear stress relationship in macroscopic scale. Each panel shows the shear stress as the function of slip obtained for equally Ispaces sampling points along main fault. The dimension of the triangle corresponds the fracture energy Gc (after Ando, 2005).

distance of rupture propagation. This explains why the observed constant rupture velocity (Fig.4) decreased under a sub-Rayleigh speed. Note that this simple scaling is satisfied as long as the rupture propagation distance is smaller than a critical length L_m.

References

Ando, R. and Yamashita, T., 2007. Effects of mesoscopic-scale fault structure on dynamic earthquake ruptures: Dynamic formation of geometrical complexity of earthquake faults, *J. Geophys. Res.*, 112, B09303, doi:10.1029/20062006JB004612.

Ando, R., 2005. Development of efficient spatio-temporal boundary integral equation method and theoretical study on dynamics of fault zone formation and earthquake ruptures (in Japanese), Univ. of Tokyo, Tokyo.

Authors

Ryosuke Ando, Natonal Research Institute for Earth Science and Disaster Prevention, 3-1 Tennodai, Tsukuba, Ibaraki, 305-0006, Japan, e-mail:ando@bosai.go.jp.

Teruo Yamashita, Earthquake Research Institute, The University of Tokyo, 1-1-1 Yayoi, Bunkyo-ku, Tokyo, Japan.

Structure and Properties of the San Andreas Fault in Central California: Recent Results from the SAFOD Experiment

by Stephen Hickman, Mark Zoback, William Ellsworth, Naomi Boness, Peter Malin, Steven Roecker, and Clifford Thurber

The San Andreas Fault Observatory at Depth (SAFOD) is a 3.2-km-deep borehole observatory drilled into the San Andreas fault zone at seismogenic depths to study directly the physical and chemical processes controlling faulting and earthquake generation. SAFOD forms one component of EarthScope, a major Earth Science facility program of the U.S. National Science Foundation; the other two funded components are U.S. Array, a continental-scale seismic network, and the Plate Boundary Observatory, an extensive network of GPS receivers and strainmeters deployed across the western U.S. SAFOD is being conducted in collaboration with the U.S. Geological Survey and the International Continental Scientific Drilling Program (ICDP). A detailed scientific rationale for drilling into active faults (including an overview of past and ongoing fault zone drilling projects) is presented in Zoback et al. (2007).

SAFOD is located in central California within the transition between the creeping and locked sections of the San Andreas Fault, 9 km NW of Parkfield, Calif. and just NW of the rupture zone of the 2004 M6.0 Parkfield earthquake. The fault displays a range of behaviors at this site. At the surface, it is creeping at a rate of 1.8 cm yr^{1}, with most of the fault displacement localized to a zone no more than 10 m wide (Burford and Harsh, 1980; Schulz, 1989). Numerous earthquakes occur directly on the San Andreas fault at this location, and they define the fault to be a narrow, near-vertical seismically active zone with a top at about 3-km depth (Fig. 1).

An important feature of microearthquakes along the San Andreas fault at SAFOD is that they occur in families of repeating events. The University of California - Berkeley High Resolution Seismic Network has observed individual earthquakes to recur every few years at precisely the same location and with the same magnitude (Nadeau and McEvilly, 1997; Nadeau et al., 2004). Repeating sources of up to M2 are located at depths as shallow as 3 km at the SAFOD site, and they play a key role in guiding SAFOD drilling.

In preparation for drilling of the main SAFOD borehole, a 2.2-km-deep vertical pilot hole was funded by the ICDP and rotary drilled in 2002 at the SAFOD site, located 1.8 km to the southwest of the surface trace of the San Andreas fault (Fig. 2). Results from the pilot hole were instrumental in defining the background geological and geophysical setting of the fault at this location, and in providing information on geologic structure and earthquake locations needed for planning of the main SAFOD experiment. Results from the SAFOD Pilot Hole and related site characterization studies were presented by Hickman et al. (2004) and in associated papers published in two special issues of Geophysical Research Letters.

During Phases 1 and 2 in the summers of 2004 and 2005, the SAFOD main hole was rotary drilled vertically and then deviated across the entire San Andreas fault zone through the zone of repeating microearthquakes, passing beneath the surface trace of the San Andreas fault at a depth of 3.1 km (Fig. 2C). From its inception, a major goal of SAFOD has been to drill into, sample, and monitor one or more of the repeating M2 earthquake sources (green, red, and blue earthquakes in Figs. 1 and 2) to follow the buildup and release of strain through multiple earthquake cycles.

Extensive drill cuttings were obtained during Phases 1 and 2, and a real-time analysis of gases in the drilling mud was carried out (Wiersberg and Erzinger, 2007 a, b). In addition, about sixty meters of spot core was acquired from the SAFOD main borehole at vertical depths of 1.5 km, 2.5 km, and 3.2 km, and sidewall cores were collected in the lower section of the hole. An extensive suite of open-hole

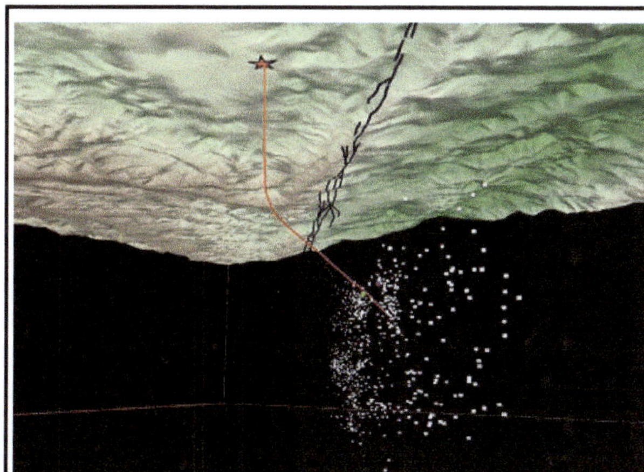

Figure 1. Seismicity of the San Andreas fault as seen from a vantage point in the earth looking to the northwest. Hypocenters were determined using tomoDD by Haijiang Zhang (University of Wisconsin - Madison) utilizing data from the Parkfield Area Seismic Observatory, supplemented by data from local USGS and University of California - Berkeley network stations. The San Andreas fault is made visible by more than 1000 microearthquakes. The SAFOD main hole is shown in red, extending downward from the surface facility (star). The surface trace of the fault is shown in black draped over the topography (3-D EarthVision plot by Luke Blair, USGS).

geophysical logs was obtained along almost the entire length of the well. For a complete description of drill cuttings, core samples and geophysical logs acquired during SAFOD Phases 1 and 2 see the ICDP Web site on SAFOD, link to Public Data (safod.icdp-online.org). Preliminary results from SAFOD Phases 1 and 2 were presented in two special sessions of the December 2005 American Geophysical Union meeting, and numerous publications are in print or under review detailing early results from SAFOD. Results from analyses of drill cuttings and cores from SAFOD are discussed in Solum et al. (2007), Schleicher et al. (2007), and references cited therein.

Four major geologic units were encountered along the trajectory of the SAFOD main hole. In the vertical section of the wellbore, the near-surface Quaternary and Tertiary sediments were found to be underlain by Salinian granite at a depth of ~700 m. After the borehole was deviated toward the fault, arkosic sediments (most likely locally-derived from Salinian granite) were encountered about 300 m NE of the drill site, perhaps after crossing the Buzzard Canyon fault, a NW-trending strike-slip fault exposed at the surface that trends sub-parallel to the San Andreas (M. Rymer, personal comm.). Approximately 1200 m NE of the drill site, a possibly ancestral trace of the San Andreas was crossed as the lithology changed abruptly to claystones and siltstones of the Great Valley Formation, found throughout central California on the east side of the San Andreas.

Geophysical logs and cuttings analysis indicate that the San Andreas fault is a zone of anomalously low P- and S-wave velocity and resistivity that defines a relatively broad (~200-m-wide) damage zone (Fig. 3). The location of a currently active trace of the San Andreas fault is revealed by casing deformation (indicated by the red line in Fig. 3) detected through repeated multifinger caliper logs. This active fault is associated with a narrower (~30-m-wide), more highly localized zone of low P- and S-wave velocity and resistivity embedded within the broader damage zone. Serpentinites and talc were also found in cuttings immediately adjacent to this deforming zone (Moore and Rymer, 2007; Solum et al., 2006, 2007), which might be responsible for the predominantly creeping behavior of the San Andreas fault at this location as well as its low long-term strength. Work is now underway to determine if this actively deforming fault trace is associated with repeating microearthquakes in the SAFOD target region.

Stress measurements in the SAFOD pilot hole and main hole provide strong evidence in support of the hypothesis that the San Andreas is a weak fault imbedded in an otherwise strong crust. Horizontal differential stress magnitudes in the pilot hole were found to be high adjacent to the fault, while the orientation of maximum horizontal stress at depth in the hole indicates low levels of shear stress resolved onto the fault itself (Hickman and Zoback, 2004). Shear-wave anisotropy measurements in the SAFOD main hole further indicate that the direction of maximum horizontal compression remains nearly perpendicular to the fault to within a few hundred meters of the active fault trace at depth (Boness and Zoback, 2006). In addition, heat flow measurements in the SAFOD pilot hole and main hole (to 3.1 km depth) are consistent with shallower data in the region which show no evidence of frictionally-generated heat (Williams et al., 2004; 2006).

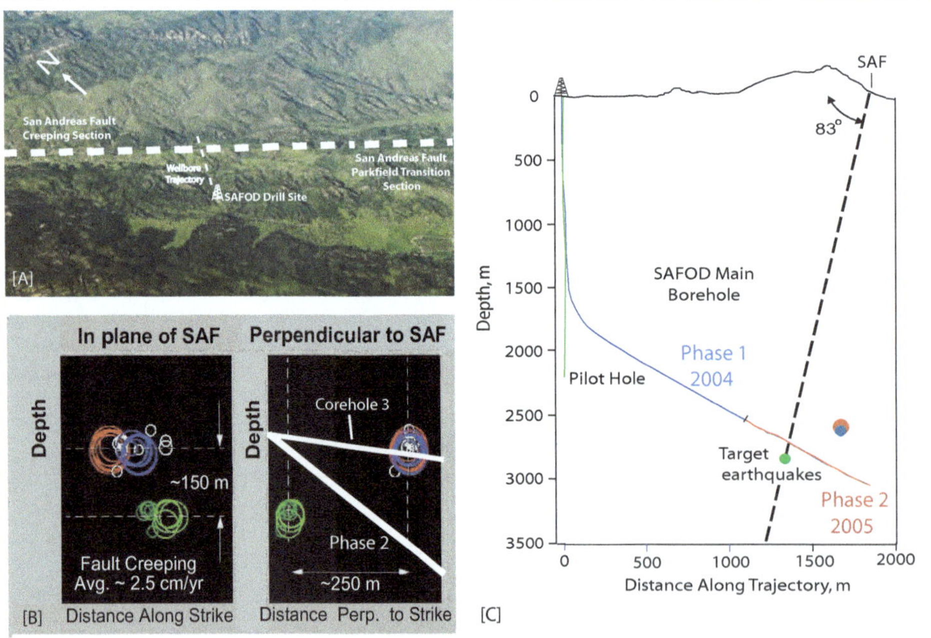

Figure 2. [A] The SAFOD drill site is located at the southeastern end of the creeping section of the fault, where the surface creep rate begins to gradually decrease in the Parkfield section. [B] There are three groups of repeating earthquakes in the target area. As shown in the cross-section in a plane parallel to the fault, the groups of repeaters shown in red, blue, and green correspond to M~2 earthquakes that recur every ~2.5 years. As shown in the cross-section perpendicular to the fault, the red and blue repeaters occur on a different strand of the fault than the green repeaters. Hence, there are multiple active fault traces at depth. The white lines schematically illustrate the location of the SAFOD main borehole—drilled in the summers of 2004 and 2005—and a core hole targeting the microearthquakes to be drilled in 2007. [C] Trajectories of the SAFOD pilot hole and main hole. The microearthquake locations are to the southwest of the surface trace of the fault.

Beginning in February 2006, we conducted a series of instrumental investigations in the SAFOD main hole in close proximity to where fault movement is actively deforming the borehole casing. A digitally telemetered pair of 3-component 15-Hz geophones was mechanically clamped inside the casing to record very local earthquakes at a sample rate of 4000 sps. The tool also contains a biaxial tiltmeter. We succeeded in

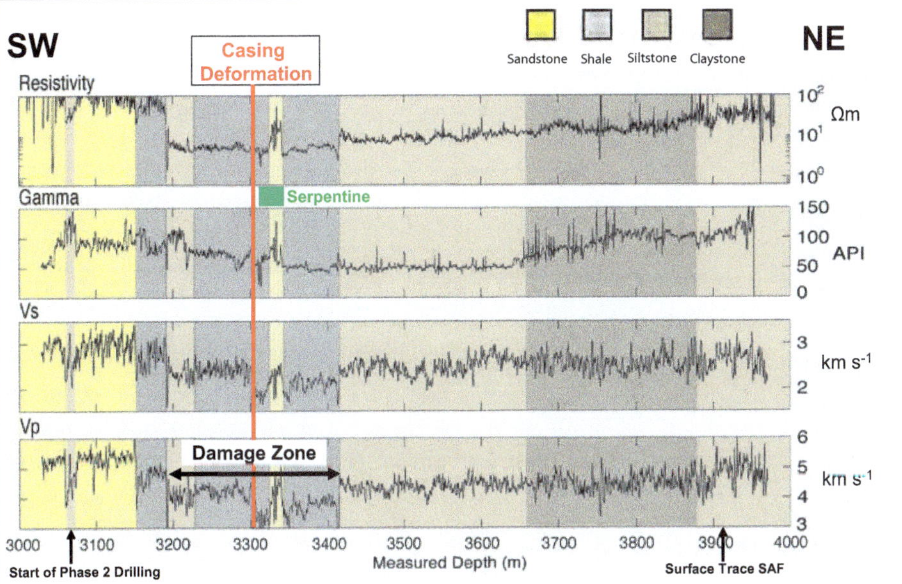

Figure 3. Geophysical well logs acquired across the San Andreas fault zone during SAFOD Phase 2, superimposed on major lithologies derived from cuttings analysis. Depths are as measured along the borehole, which is inclined at an angle of 54°–60° from the vertical (see Figure 2C for reference). Also shown are zones of casing deformation revealed by repeat multi-finger caliper logging (corresponding to an active trace of the fault at a measured depth of ~3300 m) and a ~200-m-wide damage zone defined by anomalously low P- and S-wave velocities and resistivity. Analyses of cuttings reveal the occurrence of serpentine (shown in green) and talc just to the northeast of the casing deformation zone.

recording a wide variety of seismic phenomena with this instrumentation, which are discussed in more detail by Ellsworth et al (2007). Notably, several of the M2 target earthquakes were recorded in the very near-field, at distances down to ~400 m. SAFOD downhole seismometers also recorded a new type of seismic wave that arrives between the P- and S-waves and appears to be a "leaky mode" PSV fault guided wave; the well known SH-type fault guided waves were also seen. If borne out by more complete modeling, these observations provide strong support for the hypothesis that continuous narrow, low-velocity fault zones extend deep into the seismogenic crust. We also successfully conducted several "virtual earthquake" experiments in which small explosive charges were detonated next to surface seismograph stations to measure the travel time from the station to the borehole seismometer. Using travel time reciprocity, these data are being used to determine with greater accuracy the absolute locations of the SAFOD target earthquakes in preparation for Phase 3.

During Phase 3 of SAFOD, planned for the summer of 2007, continuous cores will be acquired from three multilateral holes branching off the main SAFOD borehole to directly sample the damage zone (Fig. 3) and both the creeping and seismically active faults at depth (Fig. 2B). These core samples will be extensively studied in the laboratory to compare and contrast the composition, deformation mechanisms, physical properties, and rheological behavior of creeping and seismogenic fault rocks at realistic *in situ* conditions. At the conclusion of SAFOD Phase 3, an array of instruments will be deployed in the cased borehole in the immediate vicinity of the repeating M2 target earthquakes for long-term monitoring of near-field radiated seismic energy, deformation, fluid pressure, and seismicity through multiple earthquake cycles (Fig. 4). This will allow scientists to test—for the very first time—a variety of hypotheses related to earthquake rupture nucleation, propagation, and arrest, as well as the possible role of fluid pressure in controlling fault strength and earthquake periodicity.

References

Boness, N. and Zoback, M.D., 2006. A multi-scale study of the mechanisms controlling shear velocity anisotropy in the San Andreas Fault Observatory at Depth. *Geophysics*, 7(5):F131–F146, doi:10.1190/1.2231107.

Burford, R.O., and Harsh, P.W., 1980. Slip on the San Andreas fault in central California from alinement array surveys. *Bull. Seismol. Soc. Am.*, 70:1233–1261.

Ellsworth, W., Malin, P.E., Imanishi, K., Roecker, S.W., Nadeau, R., Oye, V., Thurber, C.H., Waldhauser, F., Boness, N.N., Hickman, S.H., and Zoback, M.D., 2007. Seismology inside the fault zone: applications to fault-zone properties and rupture dynamics. *Scientific Drilling Special Issue*, 1:85–88.

Hickman, S. and Zoback, M.D., 2004. Stress measurements in the SAFOD pilot hole: implications for the frictional strength of the San Andreas Fault. *Geophys. Res. Lett.*, 31:L15S12.

Hickman, S., Zoback, M.D., and Ellsworth, W.E., 2004. Introduction to special section: Preparing for the San Andreas Fault Observatory at Depth. *Geophys. Res. Letters,. 31,* L12S01, doi:10.1029/20040GL20688.

Moore, D.E. and Rymer, M.J. , 2007, Talc-bearing serpentinite and the creeping section of the San Andreas fault. Nature 448, 795–797, doi:10.1038/nature06064

Nadeau, R.M. and McEvilly, T.V., 1997. Seismological studies at Parkfield V: Characteristic microearthquake sequences as fault-zone drilling targets. *Bull. Seismol. Soc. Am.*, 87:1463–1472.

Nadeau, R.M., Michelini, A., Uhrhammer, R.A., Dolenc, D., and McEvilly, T.V., 2004. Detailed kinematics, structure and recurrence of micro-seismicity in the SAFOD target region. *Geophys. Res. Lett.*, 31:L12S08, doi:10.1029/2003GL019409.

Schleicher, A.M., van der Pluijm, B.A., Warr, L.N., and Solum, J.G., 2007. Electron Microscopy of Clay Minerals in Mudrocks from the San Andreas Fault Observatory at Depth (SAFOD). *Scientific Drilling Special Issue* 1:68–70.

Schulz, S.S., 1989. Catalog of creepmeter measurements in California from 1966 through 1988. *U.S. Geological Survey Open-File Report 89-650.*

Solum, J.G., Hickman, S.H., Lockner, D.A., Moore, D.E., van der

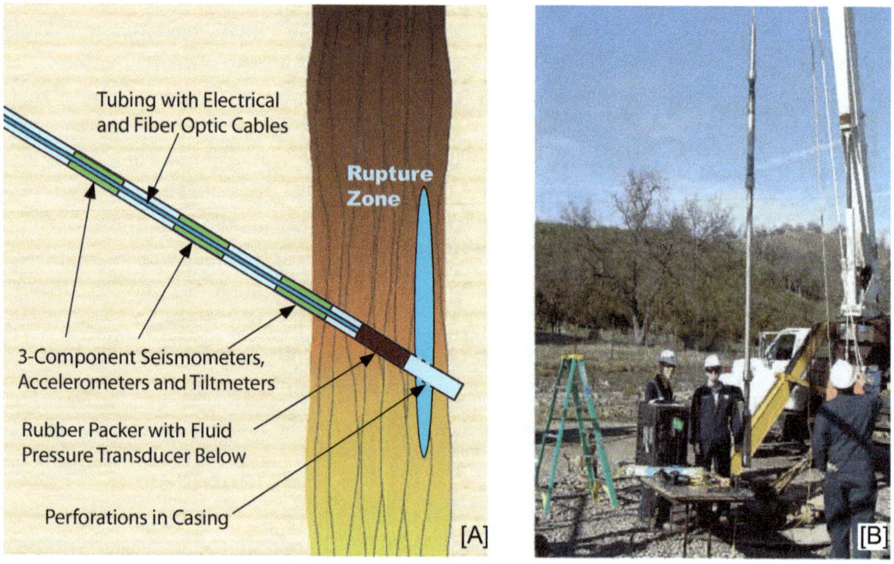

Figure 4. [A] Schematic of the long-term monitoring array for SAFOD. After casing is cemented in place from the bottom of corehole 3 to the surface (see Fig. 2B), the casing will be perforated in the fault zone, and the array will be emplaced on rigid tubing close to the target earthquake rupture zone. This array is designed so that it can be periodically removed for maintenance and repair. Unidirectional bow-springs (not shown) decentralize the seismic sensors to insure good coupling with the casing, and a rubber packer is inflated just above the perforations to monitor variations in fluid pressure during the earthquake cycle. [B] Engineers from Pinnacle Technologies testing a prototype for the seismic and tilt sensors in the SAFOD borehole following Phase 2 drilling. Photo taken by Peter Malin.

Pluijm, B.A., Schleicher, A.M., and Evans, J.P., 2006. Mineralogical characterization of protolith and fault rocks from the SAFOD main hole. *Geophys. Res. Lett.*, 33:L21314, doi: 10.1029/2006GL027285.

Solum, J.G., Hickman, S., Lockner, D.A., Tembe, S., Evans, J.P., Draper, S.D., Barton, D.C., Kirschner, D.L., Chester, J.S., Chester, F.M., van der Pluijm, B.A., Schleicher, A.M., Moore, D.E., Morrow, C., Bradbury, K., Calvin, W.M., and Wong, T.-F., 2007. San Andreas fault zone mineralogy, geochemistry and physical properties from SAFOD cuttings and core. *Scientific Drilling Special Issue* 1:64–67.

Wiersberg, T. and Erzinger, J., 2007a. A helium isotope cross-section study through the San Andreas Fault at seismogenic depths. *Geochem. Geophys. Geosyst.*, 8(1):Q01002, doi:10.1029/2006GC001388.

Wiersberg, T. and Erzinger, J., 2007b. Real-time mud gas monitoring: A technique to obtain information on the composition and distribution of gases at depth while drilling. *Scientific Drilling Special Issue* 1:71–72.

Williams, C.F., Grubb, F.V., and Galanis, S.P., 2004. Heat flow in the SAFOD pilot hole and implications for the strength of the San Andreas Fault. *Geophys. Res. Lett.*, 31:L15S14, doi:10.1029/2003GL019352

Williams, C.F., Grubb, F.V., and Galanis, Jr, S.P., 2006. Heat flow measurements across the San Andreas Fault near Parkfield, California - preliminary results from SAFOD. *Eos Trans. AGU*, 87(52), Fall Meet. Suppl., Abstract S33B-0241.

Zoback, M.D., Hickman, S., and Ellsworth, W., 2007. The role of fault zone drilling. *In* Kanamori, H., and Schubert, G. (Eds.), *Earthquake Seismology - Treatise on Geophysics*. Amsterdam (Elsevier), in press.

Related Web Link

safod.icdp-online.org

Authors

Stephen Hickman, U.S. Geological Survey, 345 Middlefield Road, Mail Stop 977, Menlo Park, Calif., 94025, U.S.A., e-mail:hickman@usgs.gov.

Mark Zoback, Stanford University, Mitchell Building, Room 359, Stanford, Calif., 94305-2215, U.S.A.

William Ellsworth, U.S. Geological Survey, 345 Middlefield Road, Mail Stop 977, Menlo Park, Calif., 94025, U.S.A.

Naomi Boness, Chevron Energy Technology Company, 6001 Bollinger Canyon Road, San Ramon, Calif., 94583, U.S.A.

Peter Malin, Duke University, 109A Old Chem, Box 90227, Durham, N.C., 27708, U.S.A.

Steven Roecker, Department of Earth and Environmental Sciences, Rensselaer Polytechnic Institute, Troy, N.Y., 12180, U.S.A.

Clifford Thurber, Department of Geology and Geophysics, University of Wisconsin, 1215 West Dayton Street, Madison, Wis., 53706, U.S.A.

33

Drilling of the Chelungpu Fault after the 1999 Chi-Chi, Taiwan Earthquake (Mw7.6): Understanding Physics of Faulting

by Kuo-Fong Ma and Hidemi Tanaka

The 21 September 1999 Chi-Chi, Taiwan earthquake (Mw 7.6) ruptured the N-S trending Chelungpu fault with a major 90-km structure that dips shallowly (30°) to the east, and principally slips within and parallel to bedding of the Pliocene Chinshui shale. A shallow borehole was drilled into the northern portion of the fault in the year 2000 (Fig. 1) in order to sample the slip zones that had large slip (>10 m) during the earthquake. The borehole was drilled at an angle of about 50° to the west, penetrating the fault at a drilling depth of around 300 m in January, 2001 (Tanaka et al., 2002). The success of the shallow hole drilling led to the deeper holes drilling named TCDP (Taiwan Chelungpu-fault Drilling Project; Ma et al., 2006), which drilled two vertical holes 40 m apart (Hole A to a depth of 2 km, and Hole B to a depth of 1.3 km) and a side-track from Hole B (Hole C, depth 0.95–1.2 km) about 2 km east of the surface rupture and 3 km to the south of the shallow hole, near the town of DaKeng (Fig. 1). TCDP carried out continuous coring for depths of 500 m to 2000 m, 950 m to 1300 m, and 950 m to 1200 m for Hole A, Hole B, and Hole C, respectively. Geophysical well logs were carried out in Hole A to collect seismic velocities, densities, and digital images.

For the shallow hole, the distributed fracture zone ranges from 285 m to 330 m depth, which contained nine slip zones. The neutron geophysical logs show that the porosity abruptly increases in the drilling interval between 285 m and 330 m, suggesting that this zone would be the newest, being fractured during co-seismic slip by the Chi-Chi earthquake. Slip zones are basically composed of fault gouge and fault breccia. The fault gouge is brighter gray than host rocks. The boundary between fault gouge and host rock is gradual and is associated with a thin transition zone (<2 mm) from host rock to fault gouge. The mineral grains are more disaggregated nearer to the slip zone. Grains in fault gouge are of extremely fine cryptocrystalline materials (<2 μm diameter), and the maximum size is about 25 μm in diameter. Microscopic textures were not indicative of frictional melting, such as

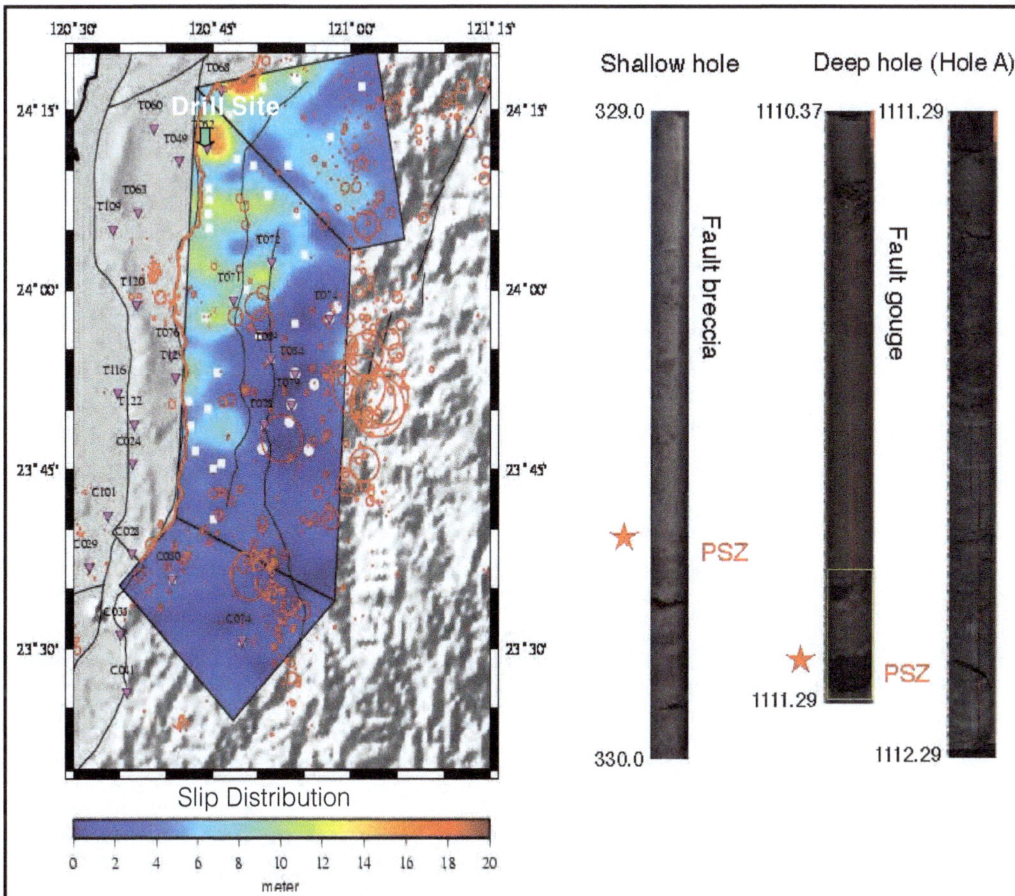

Figure 1. Location (green arrow, top) of the shallow and deeper drill holes of the Taiwan Chelungpu fault (Ji et al., 2003). The shallow hole is 4 km to the north of the deep holes (TCDP) at the town of DaKeng. The amount of slips near the drill site is about 10 m, as shown in the spatial slip distribution (shaded colors). The ruptured Chelungpu fault associated with the 1999 Chi-Chi earthquake is shown by red line. The triangles indicate the distribution of the strong-motion stations near the fault. The circles indicate the distribution of the aftershocks. Occurrence of slip zones from shallow and deep boreholes were presented from the retrieved cores. PSZ with red star is principal slip zone inferred from geological and geophysical logs of the cores. Both PSZ's are located close to the bottom of fracture zones.

spherulitic or dendritic crystal morphology, microlites or grassy matrix in the slip zones. In the deep holes, the Chelungpu fault zone from the Hole A core is observed within the Chinsui shale at depths from 1105 m to 1115 m, and consists of fractured rocks, fault breccia, and thin fault gouge layers. The degree of fracturing increases from the top to the bottom of the zone. Near the bottom of the broad zone of deformation, a 12-cm-thick primary slip zone (PSZ) can be identified based on the presence of ultra-fine grained fault gouge and increased fracture density at depths of 1111.23 m to 1111.35 m. A corresponding feature was also found in Hole B core at the depths of 1136.50 m to 1136.62 m, confirming the fault dip of 30°E. The geophysical logging measurements of low seismic velocities and low electrical resistivity around 1111 m depth also confirm that this is the main fault zone (Wu et al., 2007). The PSZ observed in the core from Hole C after splitting and polishing shows several layers of slip zones associated with several repeating earthquakes. The individual slip zone has a thickness of about 2–3 cm with a 5-mm ultra-fine grain zone at the bottom. Among the slip zones, the bottom zone in the PSZ, 2-cm in thickness, suffered the least disturbance from later events, suggesting that this narrow band might be the major slip zone (MSZ) that corresponds to the Chi-Chi earthquake. Other estimates of the thickness for the slip zone from nearby sites are 50–300 μm observed at the surface near the DaKeng drill site, and 1 cm from a fault core at a depth of 330 m in shallow drilling. These determinations of slip zone thicknesses are all from layers located near the bottom of the fracture zone. The variation of thickness of the slip zone at different depths might correspond to differences in normal stress as observed in Nojima fault core and other exhumed faults (Gratier et al., 2003; Tanaka et al., 2007).

The grain size distribution of the major slip zone was observed to estimate the surface fracture energy to compare with the seismic fracture energy to understand the energy partition of the earthquake. Direct measurements of the heat generated by this large earthquake were made for the shallow and deep holes. The local increases in the temperature profile across the fault observed both in shallow and deep holes were interpreted to be the residual heat generated during the earthquake. The frictional heat was calculated from temperature anomaly data observed at the depth associated with the identified slip zones of the Chi-Chi earthquake. Using the thermal parameters measured on-site with non-destructive methods as constraints, the analyses of the temperature anomalies lead to low estimates of the dynamic shear stress and to small values of the frictional coefficient. These results suggest the involvement of dynamic weakening mechanisms of faulting, such as thermal pressurization, during the large earthquake.

References

Gratier, J.-P., Favreau, P., and Renard, F., 2003. Modeling fluid transfer along California faults when integrating pressure solution crack sealing and compaction process. *J. Geophys. Res.*, 108(No. B2):2104, doi:10.1029/2001JB000380.

Ji, C., Helmberger, D.V., Wald, D.J., and Ma, K.-F., 2003. Slip history and dynamic implications of the 1999 Chi-Chi, Taiwan, earthquake, *J. Geophys. Res.*, 108(B9), 2412, doi:10.1029/2002JB001764.

Ma, K.-F., Tanaka, H., Song, S.-R., Wang, C.-Y., Hung, J.-H., Tsai, Y.-B., Mori, J., Song, Y.-F., Yeh, E.-C., Soh, W., Sone, H., Kuo, L.-W., and Wu, H.-Y., 2006. Slip zone and energetics of a large earthquake from the Taiwan Chelungpu-fault drilling project. *Nature*, 444:473–476, doi:10.1038/nature05253.

Tanaka, H., Wang, C.-Y., Chen, W.-M., Sakaguchi, A., Ujiie, K., Ito, H., and, Ando, M., 2002. Initial science report of shallow drilling penetrating into the Chelungpu fault zone, Taiwan. *Terr. Atm. Oceanic Sci.*, 13(3):227–251.

Tanaka, H., Omura, K., Matsuda, T., Ikeda, R., Kobayashi, K., Murakami, M., and Shimada, K., 2007. Architectural evolution of the Nojima fault and identification of the activated slip layer by Kobe earthquake. *J. Geophys. Res.*, 112 (B7), B07304, doi:10.1029/2005JB003977.

Wu, H.-Y., Ma, K.-F., Zoback, M., Boness, N., Ito, H., Hung, J.-H., and Hickman, S., 2007. Stress orientations of Taiwan Chelungpu-Fault Project (TCDP) hole-A as observed from geophysical logs. *Geophys. Res. Lett.*, 34:L01303, doi:10.1029/2006GL028050.

Authors

Kuo-Fong Ma, Department of Earth Science and Institute of Geophysics, National Central University, No.300, Johngda Rd., Johngli City, Taoyuan County 32001, Taiwan (R.O.C.), e-mail: fong@earth.ncu.edu.tw.

Hidemi Tanaka, Solid Earth Science Group, Department of Earth and Planetary Sciences, The University of Tokyo, Faculty of Science Building #1, Hongou 7-3-1, Bunkyo-ku, Tokyo, 113-0033, Japan.

34

The Corinth Rift Laboratory or an *in situ* Investigation on Interactions between Fluids and Active Faults

by François Henri Cornet

Objectives

Earthquakes result from the sudden acceleration of a preliminary long-term slow deformation process. The objective of the Corinth Rift Laboratory (CRL) is to investigate *in situ* this quasistatic deformation process and mechanisms leading to a sudden catastrophic acceleration. Of particular interest is the characterization of the structure of the deforming zones and of the progressive localization of deformation. Special attention is given to the role of fluids but also on the influence of faults on regional fluid flow.

Corinth Rift Laboratory

The Gulf of Corinth, in western Greece, is one of the most seismically active regions in Europe. It is possibly the fastest continental rift in the world and provides an ideal site for an *in situ* investigation of the physics of earthquake sources and for developing efficient seismic hazard reduction procedures. It was selected several years ago, after an Intenational Continental Scientific Drilling Program (ICDP) supported workshop (Cornet et al., 1997), as a site for developing an *in situ* laboratory, the CRL.

The Corinth Rift (Fig. 1), which separates the Peloponnesus from continental Greece, is a 110-km-long, N 110° E-oriented graben, bound by systems of very recent steeply dipping normal faults less than 2 Myr old (Armijo et al., 1996; Jolivet et al., 2004; Le Pichon et al., 1995). This structure is the site of continental break-up, with up to 1.5 cm yr^{-1} of north-south extension and a few millimeters per year uplift of the southern shore (Avallone at al., 2004; Briole et al., 2000; Pantosti et al., 2004). This rapid opening is associated with a shallowly north-dipping seismic zone located at depths ranging from 6 km to 12 km (Bernard et al., 2006; Gautier et al., 2006; Rigo et al., 1996). Five events with magnitude larger than 5.8 have been recorded in this region within the last forty years. The rifting is associated with the subduction to the southwest of Peloponnesus but also possibly to the propagation of the North Anatolian Fault.

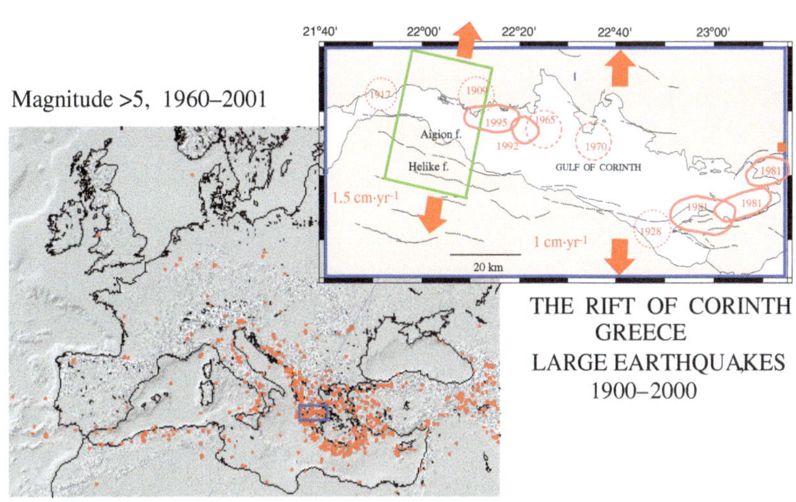

Figure 1. Geodynamic context of the Corinth Rift zone.

CRL (www.corinth-rift-lab.org; Comptes Rendus Geosciences, 2004) is located in the vicinity of the city of Aigion, some 40 km east of Patras. It covers an area about 30 km × 30 km, extending across the gulf between Aigion on the southern shore and Eratini on the northern shore. This area was chosen for several reasons:

- The local strain rate and the microseismicity are highest (1.5×10^{-6} per year);
- The faults in this area have not been the site of any earthquake of magnitude larger than 5.5 for more than a century (three centuries for the western part). Further, from GPS data analysis, these faults are considered to be in the final stage of their seismic cycle (a few decades) and in a state of accelerated strain (Bernard et al. 2006);
- These faults were recently loaded by the 1995, magnitude 6.2, so-called Aigion earthquake, at the north eastern edge of the area (Bernard et al., 1997);
- The gulf is less than 8 km wide there, allowing on-land access to instrumented sites.

The CRL involves the monitoring of seismic activity with a network of twelve three-component (2-Hz) seismic stations and three broadband stations; the monitoring of surface deformation by continuous GPS, high resolution surface tiltmeters, and downhole strainmeters; and the monitoring of water level or flow rate variations of natural springs and their changes in chemical composition, including rare gas content. In addition, it involves deep boreholes (presently two) for better characterization of the geological structure and for *in situ* monitoring of fluid-fault interactions.

One of the goals of the CRL is to understand the geometric connections between the outcropping steeply dipping fault planes and the deeper shallowly dipping seismogenic zone. Another goal is to explore the relationship between seismic and non-seismic deformation, with particular attention to the role of circulating fluids, for only about 40% of observed deformations (last five years) can be accounted for by the cumulated seismic activity.

More than five years of observation are presently available (Fig. 2). Of particular interest is the microseismic crisis (maximum magnitude 3.5) along the Aigion fault, that started about one month after the 1000-m-deep AIG10 borehole (Cornet et al., 2004) modified the hydraulic conditions in the upper part of the fault (October 2004). The destabilization of the fault in its upper portion is evidenced by minute (40–80 Pa) pressure drops (Fig. 3) dynamically triggered by large distant earthquakes (magnitude ≥8 with epicenters located more than 10,000 km away). The crisis, which lasted about five months, migrated westward and downward and generated more than 2000 events, many of which are multiplets. Preliminary results outline the

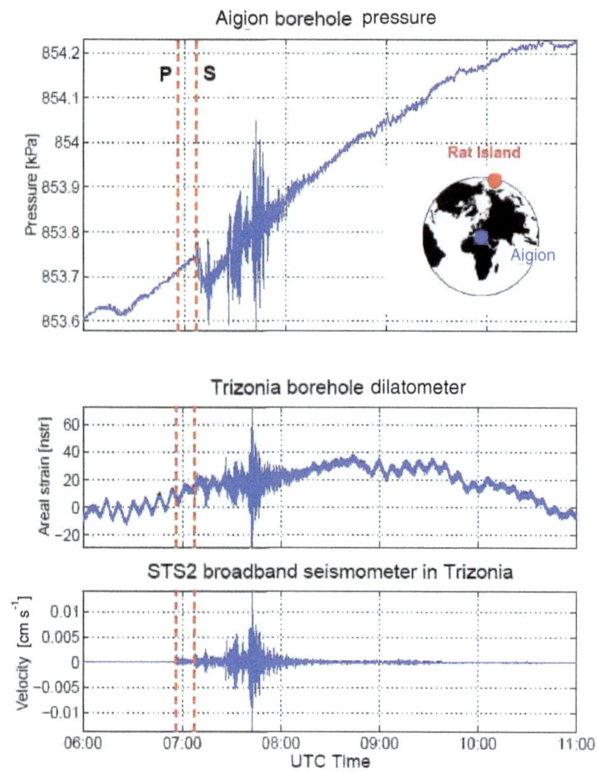

Figure 3. Pressure gauge readings in the 1000-m-deep AIG10 well that intersects the active Aigion fault at 760 m. Response to the S-waves generated by the Mw7.9 17 November 2003 Rat Island earthquake (Aleutian Isles). Seismic waves are recorded because of S-P conversion (Doan and Cornet, 2007). We also present data acquired 10 km north of the site. The borehole dilatometer and the broadband seismometer record the seismic waves, but do not display any anomaly, indicating that the pressure drop is induced in the vicinity of the AIG10 borehole.

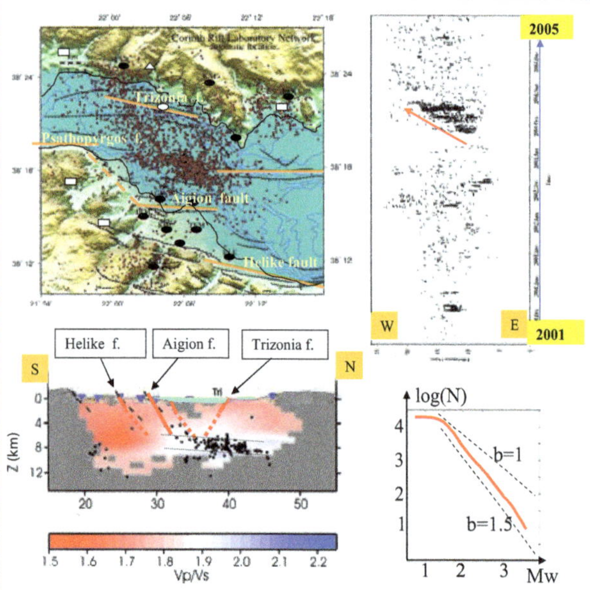

Figure 2. Seismicity at CRL. Top left: seismicity map 2000–2005, from automatic picking of CRL stations (black ovals). white oval: Trizonia STS2 station with real-time satellite telemetry. white triangle: broad band Czeck seismometer. white squares: broad band seismometers with real-time telemetry, planned summer 2006. orange lines: outcrop of main presently active normal faults. Top right: Time-space evolution of seismicity from 2000 to 2005 on an ESE-WNW axis. red arrow: migration of seismicity possibly linked with water diffusion (Cornet and Bourouis, and CRL Team, 2005). Bottom left: Vp/Vs ratio and 2000–2005 seismicity. red lines: major active faults (Gautier et al., 2006). Bottom right: red curve: Gutenberg-Richter law for the 2000–2005 period, with moment magnitude Mw.

existence of subvertical slip planes, some of which are strike slip (Bourouis et al., 2005; Cornet et al., 2005) as well as low dip fault planes. This is in contrast with the other most significant seismic crisis (2001) observed during the five-year life span of CRL existence that occurred in a zone that was thought to be inactive. The 2001 crisis is much more localized and is thought to have been associated with upward fluid motion (Lyon-Caen et al., 2004).

The 1000-m-deep AIG10 well intersects around 760 m the 10-km-long Aigion normal fault that was reactivated during the 1995 magnitude 6.2 Aigion earthquake. Below the Aigion fault, the AIG10 well entered a karstic limestone with a 0.9 MPa overpressure, while the aquifer just above the fault was found also to be artesian with a 0.45 MPa overpressure. The fault is about 10 m thick and involves an argileous 60-cm-thick core. Within this clay zone, observed slip direction is consistent with the maximum principal stress component being normal to the main fault plane (Sulem, 2007).

Also significant is the observation of a transient creeping process (Fig. 4), preceding by about half an hour a local magnitude 3.5 earthquake in December 2002, which involved the shallow part of the Psathopyrgos Fault, at about 2 km in

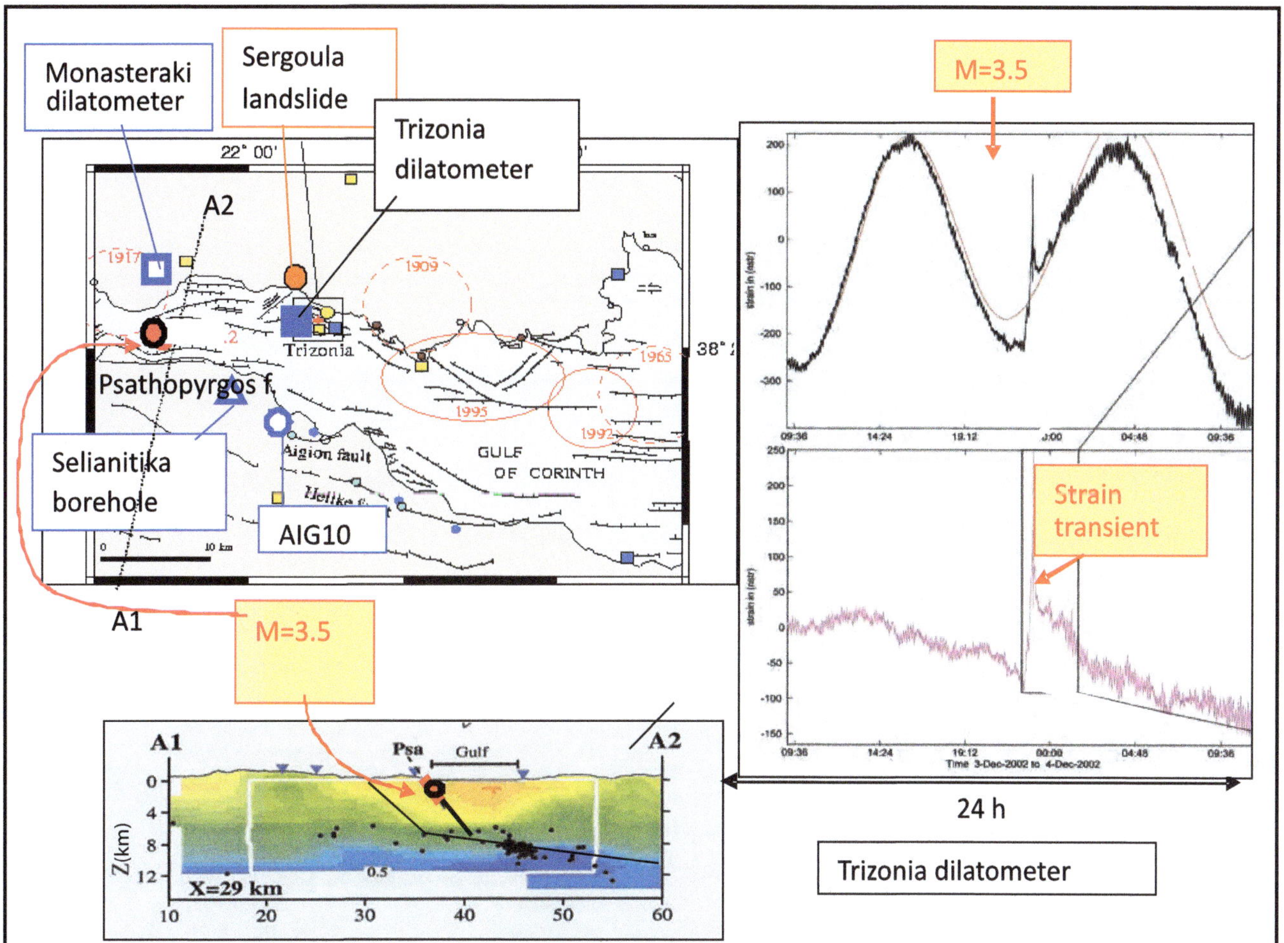

Figure 4 . Strain transients at CRL. Top left: Continuous strain-related observation array. Blue open circle: AIG10 borehole; Blue square: Trizonia dilatometer borehole; open blue square: Monasteraki borehole dilatometer, planned June 2006; Open blue triangle: Selianitika borehole; Yellow squares: continuous GPS; Blue dots: pore pressure in aquifers; Green square: geochemical station in aquifer; Green circle: radon probes on fault scarp. Black/Red circle: shallow 2002 earthquake coincident with the recorded strain transient; Top right: 24-hour record of the Trizonia dilatometer (3-12-2002). Red curve, top: synthetic of the tidal effect; bottom: data after tide removed. Note the one-hour-long transient dilatation, peaking with the M3.5 event (Bernard et al., 2006). Short period fluctuations are the pressure effect due to the free oscillations of the gulf. Bottom left: NS cross-section of the Psathopyrgos fault. Red and black circle: M3.5 event; Red segment: possibly creeping shallow Psathopyrgos fault, likely source of the strain transient.

depth (equivalent magnitude 5), i.e., well above the seismogenic zone of the area (Bernard et al., 2006). This earthquake was also the largest of a seismic swarm located about 15 km west of the CRL seismic network that lasted six weeks and involved deeper parts of the same fault.

The present plan is to deploy a complete suite of packers, high frequency (2.5 KHz sampling rate) 3C geophones, hydrophones (2.5 KHz sampling rate), high precision pressure transducers, thermistors, and tilt meters on both sides of the Aigion fault (February 2007). These data will be part of a feasibility study for drilling down to 5 km for sampling circulating fluids and monitoring high frequency signals within the seismogenic zone.

Acknowledgements

CRL was developed from 1999 to 2003 mostly through European funding and with the support of the International Continental Drilling Program for the drilling of AIG10. It has benefited greatly also from continuous support from the French National Centre for Scientific Research (CNRS) since 1999.

References

Armijo, R., Meyer, B., King, G.C.P., Rigo, A., and Papanastassiou, D., 1996. Quaternary evolution of the Corinth Rift and its implications for the late Cenozoic evolution of the Aegean. *Geophys. J. Int.*, 126:11–53.

Avallone, A., Briole, P., Agatza-Balodimou, A.M., Billiris, H., Charade, O., Mitsakaki, C., Nercessian, A., Papazissi, K., Paradissis, D., and Veis, G., 2004. Analysis of eleven years of deformation measured by GPS in the Corinth Rift Laboratory area. *C. R. Geoscience*, 336(4–5):301–312.

Bernard, P., Briole, P., Meyer, B., Lyon-Caen, H., Gomez, J.M., Tiberi, C., Berge, C., Cattin, R., Hatzfeld, D., et al., 1997. The Ms=6.2, June 15, 1995 Aigion earthquake (Greece): evidence for low angle normal faulting in the Corinth rift. *Journal of Seismology*, 1:131–150.

Bernard, P., Lyon-Caen, H., Briole, P., Deschamps, A., Boudin, F., Macropoulos, K., Papadimitriou, P., Lemeille, F., Patau, G.,

Billiris, H., Paradissis, D., Papazissi, K., Castarède, H., Charade, O., Nercessian, A., Avallone, A., Pacchian, F., Zaradnick, J., Sacks, S., and Linde, A., 2006. Seismicity, deformation and seismic hazard in the western rift of Corinth: new insights from the Corinth rift laboratory (CRL). *Tectonophysics*, 426:7–30.

Bourouis, S., Bouin, M.P., Patau, G., Bernard, P., Lyon-Caen, H., and Deschamps, A., 2005. The Corinth Rift laboratory - Multiplet analysis of the January 2004 microseismic crisis. European Geosciences Union Annual Meeting, Vienna.

Briole, P., Rigo, A., Lyon-Caen, H., Ruegg, J.C., Papazissi, K., Mitsakaki, C., Balodimou, A., Veis, G., Hatzfeldand, D., and Deschamps, A., 2000. Active deformation of the Corinth Rift, Greece : results from repeated Global Posiontioning Systems surveys between 1990 and 1995. *J. Geophys. Res.*, 105:25605–25625.

Comptes Rendus Geoscience, 2004. 336(4/5); (special issue on Corinth Rift, with 25 papers).

Cornet, F.H., Bourouis, S., and the CRL Team, 2005. Estimate of hydraulic diffusivity at 8 km depth in the Corinth Rift (Greece) from seismic activity migration. AGU Fall Meeting, San Francisco.

Cornet, F.H., Borm, G., McEvilly, T., and Vardoulakis, I., (Orgs.), 1997. Development of a multi-borehole observatory at the Gulf of Corinth, Athens. *In* Proceedings of the 1st workshop on the development of a multiborehole observatory at the Gulf of Corinth. 26–28 October 1997, Athens. 1997 ICDP 1996–1997 report available at http://www.corinth-rift-lab.org.

Cornet, F.H., Doan, M.L., Moretti, I., and Borm, G., 2004. Drilling through the active Aigion Fault: The AIG10 well observatory. *C. R. Geoscience*, 336:395–406.

Doan, M.L. and Cornet, F.H., 2007. Small pressure drop triggered near a fault by small teleseismic waves. *Earth Planet. Sci. Lett.*, in press.

Gautier, S., Latorre, D., Virieux, J., Deschamps, A., Skarpelos, S., Sotiriou, A., Serpetsidaki, A., and Tselentis, A., 2006. A new passive tomography of the Aigion Area (Gulf of Corinth, Greece) from the 2002 data set. *Pageoph.*, 163(2–3);431–453, doi:10.1007/s00024-005-0033-7.

Jolivet, L., Famin, V., Mehl, C., Parra, T., Avigad D., and Aubourg, C., 2004. Progressive strain localisation, crustal scale boudinage and extensional metamorphic domes in the Aegean sea. *American Geological Society, Special paper 380*, 185–210.

Le Pichon, X., Chamot-Rooke, N., Lallement, S., Noomen, R., and Veis, G., 1995. Geodetic determination of the kinematics of central Greece with respect to Europe: implications for Eastern Mediterranean tectonics. *J. Geophys. Res.*, 100(B7):12675–12690.

Lyon-Caen, H., Papadimitriou, P., Deschamps, A., Bernard, P., Makropoulos, K., Pacchiani, F., and Patau, G., 2004. First results of the CRL seismic network in the western Corinth Rift: evidence for old fault reactivation. *C. R. Geoscience*, 336:343–352.

Pantosti, D., De Martini, P.M., Koukouvelas, I., Stamatopoulos, L., Palyvos, N., Pucci, S., Lemeille, F., and Pavlides, S., 2004. Palaeoseismological investigations of the Aigion Fault (Gulf of Corinth, Greece). *C. R. Geoscience*, 336:335–342.

Rigo, A., Lyon-Caen, H., Armijo, R., Deschamps, A., Hatzfeld, D., Makropoulos, K., Papadimitriou, P., Kassaras, P., and Kassaras, L., 1996. A microseismic study of the western part of the Gulf of Corinth (Greece): implications for the large-scale normal faulting mechanisms. *Geophys. J. Int.*, 126:663–688.

Sulem, J., 2007. Stress orientation evaluated from strain localisation analysis in Aigion fault. *Tectonophysics*, 442(1):3–13.

Related Web Link

http://www.corinth-rift-lab.org

Author

François Henri Cornet, Institut de Physique du Globe de Strasbourg, 5 rue René Descartes, 67000 Strasbourg, France, e-mail: francois.cornet@eost.u-strasbg.fr.

35

Drilling the North Anatolian Fault–Part I

by Georg Dresen, Marco Bohnhoff, Mustafa Aktar, and Haluk Eyidogan

An international workshop entitled "GONAF: A deep Geophysical Observatory at the North Anatolian Fault", was held 23–27 April 2007 in Istanbul, Turkey. The aim of this workshop was to refine plans for a deep drilling project at the North Anatolian Fault Zone (NAFZ) in northwestern Turkey. The current drilling target is located in the Marmara Sea offshore the megacity of Istanbul in the direct vicinity of the main branch of the North Anatolian Fault on the Prince Islands (Figs. 1 and 2).

The NAFZ represents a 1600-km-long plate boundary that slips at an average rate of 20–30 mm·yr^{-1} (McClusky et al., 2000). It has developed in the framework of the northward moving Arabian plate and the Hellenic subduction zone where the African lithosphere is subducting below the Aegean. Comparison of long-term slip rates with Holocene and GPS-derived slip rates indicate an increasing westward movement of the Anatolian plate with respect to stable Eurasia. During the twentieth century, the NAFZ has ruptured over 900 km of its length. A series of large earthquakes starting in 1939 near Erzincan in Eastern Anatolia propagated westward towards the Istanbul-Marmara region in northwestern Turkey that today represents a seismic gap along a ≥100-km-long segment below the Sea of Marmara. This segment did not rupture since 1766 and, if locked, may have accumulated a slip deficit of 4–5 m. It is believed being capable of generating two M≥7.4 earthquakes within the next decades (Hubert-Ferrari et al., 2000); however, it could even rupture in a large single event (Le Pichon et al., 1999).

The most recent devastating earthquakes in the region occurred in 1999 near Izmit and Düzce with magnitudes >7.

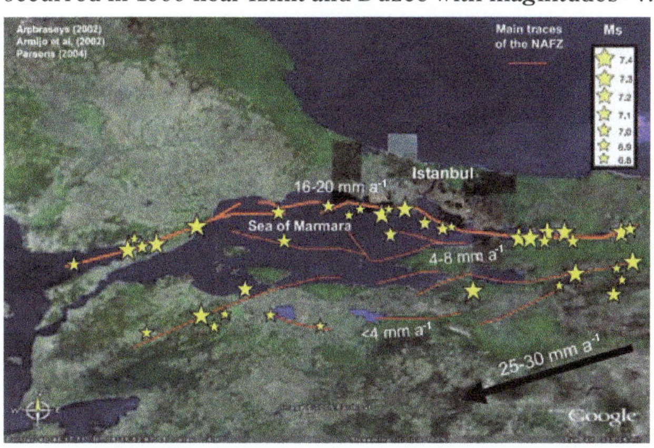

Figure 1. Google Earth view of the Marmara Sea / Istanbul region. Red lines indicate major segments of the North Anatolian Fault Zone (NAFZ). Stars indicate major events that occurred in the last 2000 years.

Their western termination of rupture is located offshore below the eastern Sea of Marmara possibly extending to just south of the Princes Islands (Özalaybey et al., 2002) within ~20 km of Istanbul.

Current seismic activity in the eastern Marmara Sea indicates a complex fault network at the transition between the western end of the Izmit earthquake rupture and the assumed seismic gap south of Istanbul. The majority of focal mechanism solutions indicate dominant strike-slip motion with minor normal faulting activity (Örgülü and Aktar, 2001; Karabulut et al., 2002). However, existing seismic observations lack the spatial and temporal resolution required to accurately distinguish between locked and creeping segments of the NAFZ. This is due to the threshold (magnitude >2) of the existing seismic networks. The knowledge of the stress state at the NAFZ is rudimentary at best. Stress orientation (World Stress Map) with respect to the fault zone is mainly based on a small number of focal mechanisms of larger seismic events (Heidbach et al., 2004) and aftershocks (Bohnhoff et al., 2006). Maximum compressive stress is generally oriented at 35°–45° with respect to the fault trend and in agreement with predictions from Coulomb friction theory. In contrast to other major plate bounding faults, the NAFZ does not appear to be a weak fault. However, no data exist on stress magnitudes and on heat flow close to the NAFZ in the Marmara Sea region.

The GONAF initiative focuses on the installation of a borehole observatory in a deep borehole at the NAFZ. This will conduct long-term monitoring of seismic activity, stress, heat and fluid flow. The target area is located offshore Istanbul in the Marmara Sea close to the main branch of the NAFZ on the outermost island of Sivriada (Fig. 2). The projected observatory is located at the transition between the western end of the 1999 Izmit rupture and the 150-km-long seismic gap along the western NAFZ that may have accumulated a 4–5 m slip deficit within the past 250 years.

Figure 2. Proposed drilling location on the island of Sivriada that is located in direct vicinity to the main branch of the NAFZ.

Presentations at the workshop included an overview of existing fault drilling projects, a session on seismotectonics, seismology, and the geological setting of western part of the NAFZ and the Marmara Sea region, and deep borehole monitoring results and technology. The potential drill site on Sivriada island was visited during a one-day field trip. A summary of suggested pre-site and drilling-phase studies concluded the workshop. Key scientific and technical aspects of a deep drillhole and long-term geophysical observatory were discussed in order to prepare for a full drilling proposal to be submitted to the International Scientific Continental Drilling Program (ICDP).

Reports on fault mechanics and earthquake processes as well as technical and logistical challenges of a drilling presented included projects in California (San Andreas Fault Observatory at Depth; Hickman, et al., 2007), Taiwan (Taiwan Chelungpu-Fault Drilling Project; Ma and Tanaka, 2007), Japan (Nojima Fault; Ito et al., 2003), and Greece (Corinth Rift Laboratory; Cornet, 2007). All fault drilling projects produced new and ground-breaking results with regards to physics of faulting, slip distribution at hypocentral depth, source mechanisms, and the earthquake energy budget. Other presentations detailed the structure, kinematics, and seismotectonics of the western NAFZ emphasizing recent and partly unpublished field data. Unresolved fault structure and hypocenter locations in the Marmara Sea underscored the need for high-resolution long-term seismological observations and detailed wide-angle seismic profiling close to the planned drill site.

A final series of presentations focussed on state-of-the-art monitoring strategies for a borehole observatory including instrumentation and other technical aspects. In particular, detailed introductions presented new scientific results and technical achievements covering high-resolution earthquake monitoring, strain monitoring, velocity measurements, stress measurements, heat and fluid flow measurements, and borehole logging techniques.

References

Bohnhoff, M., Grosser, H., and Dresen, G., 2006. Strain partitioning and stress rotation at the North Anatolian fault zone from aftershock focal mechanisms of the 1999 Izmit Mw=7.4 earthquake. *Geophys. J. Int.*, 160:373–385.

Cornet, F.H., 2007. The Corinth Rift Laboratory or an *in situ* investigation on interactions between fluids and active faults), *Sci. Drill. Special Issue*, 1:35–38.

Heidbach, O., Barth, A., Connolly, P., Fuchs, F., Müller, B., Reinecker, J., Sperner, B., Tingay, M., and Wenzel, F., 2004. Stress maps in a minute: The 2004 world stress map release. *Eos Trans.*, 85(49):521–529.

Hickman, S., Zoback, M., Ellsworth, W., Boness, N., Malin, P., Roecker, S., and Thurber, C., 2007. Structure and properties of the San Andreas Fault in central California: Recent results from the SAFOD experiment. *Sci. Drill. Special Issue*, 1:29–32.

Hubert-Ferrari, A., Barka, A., Jacques, E., Nalbant, S.S., Meyer, B., Armijo, R., Tapponnier, P., and King, G.C.P., 2000. Seismic hazard in the Marmara Sea region following the 17 August 1999 Izmit earthquake. *Nature*, 404:269–273, doi:10.1038/35005054.

Karabulut, H., Bouin, M.-P., Bouchon, M., Dietrich, M., Cornou, C., and Aktar, M., 2002. The seismicity in the Eastern Marmara Sea after the 17 August 1999 Izmit Earthquake. *Bull. Seismol. Soc. Am.*, 92(1):387–393, doi:10.1785/0120000820.

Le Pichon, X., Şengör, A.M.C., and Taymaz, T., 1999. The Marmara fault and the future Istanbul earthquake. In Karaca, M., and Ural, D.N. (Eds.), *ITU-IAHS International Conference on the Kocaeli earthquake 17 August 1999*, Istanbul (Istanbul Technical University), 41–54.

Ma, K.-F., and Tanaka, H.,2007. Drilling of the Chelungpu Fault after the 1999 Chi-Chi, Taiwan earthquake (Mw7.6): Understanding physics of faulting). *Sci. Drill. Special Issue*, 1:33–34.

McClusky, S., Balassanian, S., Barka, A., Demir, C., Ergintav, S., Georgiev, I., Gurkan, O., Hamburger, M., Hust, K., Kahle, H., Kastens, K., Kekelidze, G., King, R., Kotzev, V., Lenk, O., Mahmoud, S., Mishin, A., Nadariya, M., Ouzounis, A., Paradissis, D., Peter, Y., Prilepin, M., Reilinger, R., Sanli, I., Seeger, H., Tealeb, A., Toksöz, M.N., and Veis, G., 2000. Global positioning system constraints on plate kinematics and dynamics in the eastern Mediterranean and Caucasus. *J. Geophys. Res.*, 105:5695–5719, doi:10.1029/1999JB900351.

Ito, H., Roeloffs, E., Matsumoto, N., Kuwahara, Y., 2003. Has the Nojima Faulthhealed Rapidly? – Results from seismogenic Zone Drilling and Monitoring at the Nojima Fault, Japan. European Geophysical Society, *Geophys. Res. Abstr.*, Vol. 5, 06047.

Örgülü, G., and Aktar, M., 2001. Regional Moment Tensor Inversion for Strong Aftershocks of the August 17, 1999 Izmit Earthquake (M_w=7.4). *Geophys. Res. Lett.*, 28(2), 371–374, doi:10.1029/2000GL011991.

Özalaybey, S., Ergin, M., Aktar, M., Tapirdamaz, D., Biçmen, F., and Yörük, A., 2002. The 1999 Izmit earthquake sequene in Turkey: seismological and tectonic aspects, *Bull. Seismol. Soc. Am.* 92: 376–386, doi:10.1785/0120000838.

Authors

Georg Dresen, GeoForschungsZentrum Potsdam, Telegrafenberg, D-14473, Potsdam, Germany, e-mail: dre@gfz-potsdam.de.

Marco Bohnhoff, GeoForschungsZentrum Potsdam, Telegrafenberg, D-14473, Potsdam, Germany.

Mustafa Aktar, Bogazici University, Kandilli Observatory and Earthquake Research Institute (KOERI), Cangelkoy, Istanbul, 81220, Turkey.

Haluk Eyidogan, Istanbul Technical University, Mining Faculty, Department of Geophysics, 34469, Maslak, Istanbul, Turkey.

Related Web Link

http://www.gonaf.de/

Drilling the North Anatolian Fault–Part II

by Georg Dresen, Mustafa Aktar, Marco Bohnhoff, and Haluk Eyidogan

The North Anatolian Fault Zone (NAFZ) represents a ~1400-km-long plate boundary that slips at an average rate of 20–30 mm yr^{-1}. It separates the westward moving Anatolian block in the south from Eurasia in the north (McClusky et al., 2000). Comparison of long-term slip rates (6.5 mm yr^{-1} for the last 13 Ma, 17 mm yr^{-1} for the last 5 Ma,) with Holocene and GPS-derived slip rates (20–25 mm yr^{-1}) indicate an accelerated movement in the recent geological past. During the twentieth century, the NAFZ has ruptured over 900 km of its length. A series of large earthquakes starting in 1939 near Erzincan in eastern Anatolia propagated westward towards the Istanbul-Marmara region in northwestern Turkey. A large part of the Sea of Marmara today represents a seismic gap along a ≥100-km-long segment which did not rupture since 1766 and may have accumulated a slip deficit of 4–5 m. It is believed to be capable of generating two M≥7.1 earthquakes within the next decades or could even rupture in a large single event.

The most recent devastating earthquakes in the region occurred in 1999 near Izmit and Düzce with magnitudes greater than 7. Their western termination of rupture is located offshore below the eastern Sea of Marmara possibly extending to just south of the Princes Islands within ~20 km of Istanbul (Fig.1).

Below the Sea of Marmara, the NAFZ branches in two prominent offshore fault segments (Fig. 1) that bound a pull-apart depocenter (Cinarcik Basin) in the NE and SE containing 16–20 mm yr^{-1} GPS-derived surface displacement at present. The NAFZ is again exposed on the western Dardanelles Peninsula merging with the Ganos Fault. Here, the last M>7 earthquake occurred in 1912 rupturing approximately 60 km into the Sea of Marmara.

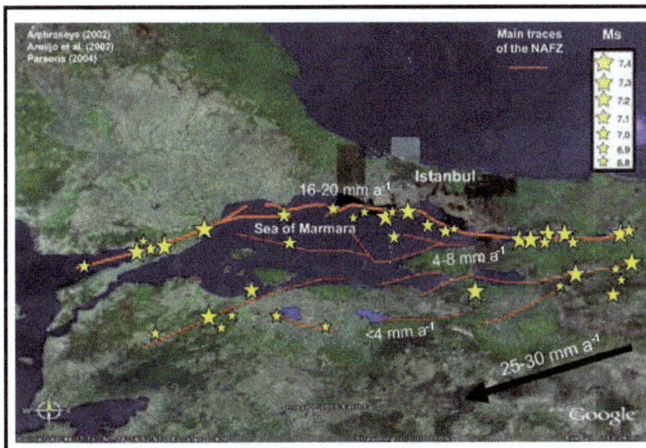

Figure 1. Google Earth® view of the Sea of Marmara, Istanbul region. Red lines indicate major segments of the North Anatolian Fault zone. Stars indicate main shocks M>6.5 that occurred in the last 2000 years.

In the eastern Sea of Marmara, five seismic events with M>6 have occurred since 1509. The events in 1509, 1754, 1766, and 1894 generated tsunamis with waves exceeding 6 m in height in some cases. Also, the 1999 Izmit earthquake, reflecting pure strike-slip faulting, generated a minor tsunami with local wave heights >2 m.

Current seismic activity in the eastern Marmara Sea indicates the seismic gap south of Istanbul and a complex fault network active at the transition between the western end of the Izmit earthquake rupture (Fig. 2; Aktar et al., 2004; Karabulut et al., 2002). The majority of focal mechanism solutions indicate dominant strike-slip motion with minor normal faulting activity; however, existing seismicity images of the region lack the spatial and temporal resolution required for an accurate definition of seismic versus aseismic zones. This is due to the magnitude threshold of the existing seismic networks that is currently limited to about magnitude 2.

Our knowledge of the stress state at the NAFZ is rudimentary (Heidbach et al., 2004). Stress orientation with respect to the fault zone as given in the World Stress Map is mainly based on a small number of focal mechanisms of larger seismic events. Maximum compressive stress is generally oriented at 35°–45° with respect to the fault trend, so the NAFZ does not appear to be a weak fault in contrast to other major plate bounding faults. No data exist on stress magnitudes and on heat flow close to the NAFZ for the entire Marmara Sea area.

Positive Coulomb stress changes in the Marmara region suggest that the Izmit earthquake significantly increased the seismic hazard for the Istanbul-Marmara region. The Izmit event has increased the 2004–2034 probability for an earthquake M≥7 by 2–6 times. Estimated thirty-year probability for an event M≥7 is 35%–70% (Parsons., 2004; Parsons et al., 2000; Wright et al., 2001).

We currently investigate aftershock focal mechanisms of the Mw=7.4 Izmit earthquake of 17 August 1999, on the western North Anatolian Fault Zone (Bohnhoff et al., 2006). Spatial clustering and orientation of 446 fault plane solutions were analyzed. Aftershock clusters define four individual

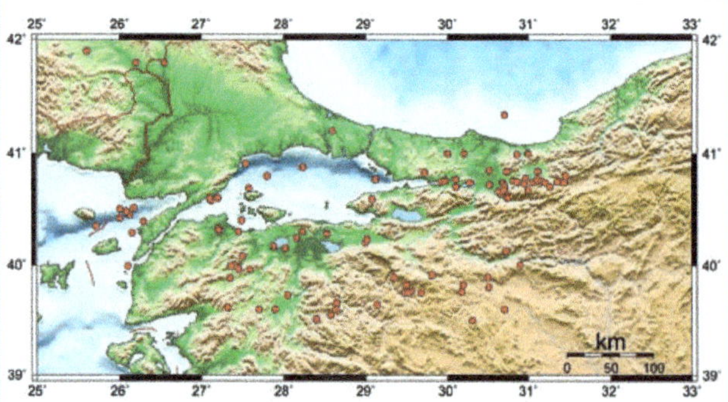

Figure 2. Seismicity of the Sea of Marmara region since 1900 for event magnitudes M>5 (Kandilli Observatory and Earthquake Research Institute, National Earthquake Monitoring Center (KOERI-NEMC), 2005).

fault segments. The distribution of aftershock focal mechanisms corresponds to fault segmentation of the NAFZ in the Izmit-Düzce region produced by coseismic slip. Areas with high coseismic slip show aftershocks that are dominantly strike-slip, but low-slip barriers show mostly normal faulting aftershocks.

Focal mechanisms surrounding epicenters of the Izmit and subsequent Düzce main shock (Mw=7.1, 12 November 1999) indicate dominantly strike-slip but also normal faulting. Below the easternmost portion of the Sea of Marmara, alignments of aftershocks suggest branching of the NAFZ into three or more active segments that differ significantly in orientation of focal mechanisms.

Stress tensor inversions of the aftershock focal mechanisms show systematic rotations of the local stresses following the Izmit main shock. We conclude that the Izmit earthquake caused significant stress partitioning along the rupture. The direction of stress rotation is related to the orientation of the individual fault segments along the NAFZ.

We propose to drill a deep borehole close to the main branch of the NAFZ on the outermost Princes Islands close to the Istanbul metropolitan region and to install a multiparameter deep borehole observatory (Fig. 3). Regarding recent seismic activity, the Princes Islands are located on a transition between the western end of the Izmit rupture and a fault segment showing very low activity, suggesting that this part of the fault is currently locked.

The deep-borehole observatory will allow an enhanced monitoring of microseismic activity, stress, heat and fluid flow near the NAFZ. The location of the observatory represents the only possible long-term monitoring site of seismic activity in the vicinity of Istanbul, and it is located at a NAFZ segment that is considered a seismic gap. Owing to post-seismic stress redistribution after the Izmit earthquake, the area is likely subjected to enhanced stresses.

The observatory will be installed at the bottom of a 2000-m-deep vertical borehole. The vertical seismometer array will be combined with an array of seismometers in shallow boreholes. A deep borehole observatory will permit observing earthquake activity close to the seismically active region of the fault zone at a depth between 4 km and 15 km. A chain of borehole seismometers will enhance the accuracy of hypocenter determinations and significantly decrease the magnitude-detection threshold. This would allow studying the rupture process of small-magnitude earthquakes close to their source region in unprecedented detail.

The orientation and magnitude of local stresses and strains at the fault and their change with depth can be determined. This will assist estimating the *in situ* strength of a major plate boundary fault. Fluid pressures in close proximity to active fault strands of the NAFZ will be monitored. The proposed borehole observatory will allow us to address the following fundamental problems, partly in combination of with a shallow borehole array and offshore campaign measurements:

- Is the North Anatolian Fault Zone in northwestern Turkey a weak or a strong plate boundary?

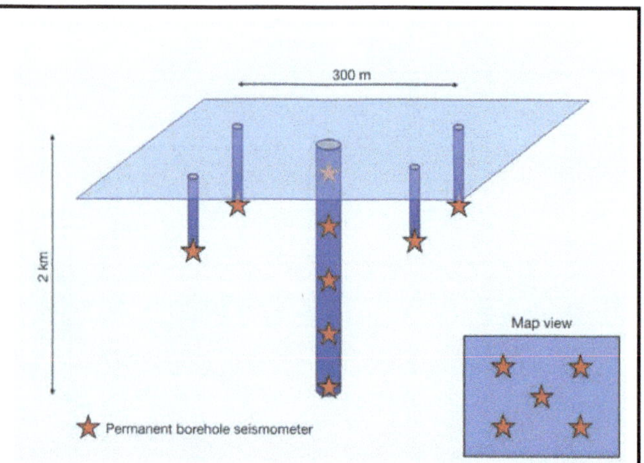

Figure 3. [Left] Proposed location for a deep drill hole on the Prince Islands near (~2–3 km from) the main fault of the NAFZ. View of Prince Islands (insert: Sivriada). [Right] Schematic view of the seismometer array proposed for a deep drill hole on the Prince Islands.

- What is the orientation of the principal stresses at the fault zone? Do stresses at the fault zone vary with depth and time?
- How does the structural heterogeneity of the fault affect the earthquake nucleation process?
- How is the dynamic propagation of a shear rupture affected by the fault structure?
- How is deformation partitioned between aseismic creep and seismic slip at the termination of a major earthquake rupture?
- What is the continuation of the faults at depth? Is there a deep-seated master fault in the Sea of Marmara south of Istanbul?

Currently, discussion of fault kinematics in the Marmara segment of the NAFZ has been controversial (Armijo et al., 2002, 2005; Le Pichon et al., 2001, 2003). Fault kinematics have important consequences for the seismic hazard assessment pertinent to the Istanbul area. For example, Armijo et al. (2005) suggested a maximum seismic risk for a well defined 70–100-km-long fault segment between the 1912 and 1999 ruptures where a slip deficit of 4–5 m might have accumulated (this is capable of generating a M~7 event). This is in agreement with recurrence intervals of 270 years for this region with regard to the 1766 event. Other studies suggest that the fault is less segmented and might rupture through its entire length of up to 150 km. The proposed deep-borehole geophysical observatory close to the NAFZ will permit us to examine some of the key questions pertinent to faulting activity and earthquake mechanics in general and seismic hazard assessment for the Istanbul area in particular.

References

Aktar, M., Özalaybey, S., Ergin, M., Karabulut, H., Bouin, M.P., Tapýrdamaz, C., Biçmen, F., Yörük, A., and Bouchon, M., 2004. Fault zone heterogeneity and variations of seismicity parameters across 1999 Ýzmit-Düzce earthquake sequence. *Tectonophysics*, 391:325–334, doi:10.1016/j.tecto.2004.07.020.

Ambraseys, N., 2002. The seismic activity of the marmara sea region over the last 2000 years. *Bull. Seismol. Soc. Am.*, 92:1:1–18, doi:10.1785/0120000843.

Armijo, R., Meyer, B., Navarro, S., King, G., and Barka, A., 2002. Asymmetric slip partitioning in the Sea of Marmara pull-apart: A clue to propagation processes of the North Anatolian fault? *Terra Nova*, 14(2):80–86, doi:10.1046/j.1365-3121.2002.00397.x.

Armijo, R., Pondard, N., Meyer, B., Uçarkus, G., Mercier de Lépinay, B., Malavieille, J., Dominguez, S., Gustcher, M.-A., Schmidt, S., Beck, C., Çagatay, N., Çakir, Z., Imren, C., Eris, K., Natalin, B., Özalaybey, S., Tolun, L., Lèfevre, I., Seeber, L., Gasperini, L., Rangin, C., Emre, O., and Sarikavak, K., 2005. Submarine fault scarps in the Sea of Marmara pull-apart (North Anatolian fault): Implications for seismic hazard in Istanbul. *Geochem. Geophys. Geosyst.*, 6(6):Q06009, doi:10.1029/2004GC000896.

Bohnhoff, M., Grosser, H., and Dresen, G., 2006. Strain partitioning and stress rotation at the North Anatolian fault zone from aftershock focal mechanisms of the 1999 Izmit Mw=7.4 earthquake. *Geophys. J. Int.*, in press.

Heidbach, O., Barth, A., Connolly, P., Fuchs, F., Müller, B., Reinecker, J., Sperner, B., Tingay, M., and Wenzel, F., 2004. Stress maps in a minute: The 2004 world stress map release. *Eos Trans.*, 85(49):521–529.

Karabulut, H., Bouin, M.-P., Bouchon, M., Dietrich, M., Cornou, C., and Aktar, M., 2002. The seismicity in the Eastern Marmara Sea after the 17 August 1999 Izmit Earthquake. *Bull. Seismol. Soc. Am.*, 92(1):387–393, doi:10.1785/0120000820.

LePichon, X., Þengör, A.M.C., Demirbað, E., Rangin, C., Ýmren, C., Armijo, R., Görür, N., Çaðatay, N., Mercier de Lepinay, B., Meyer, B., Saatçýlar, R., and Tok, B., 2001. The active Main Marmara Fault. Earth Planet. *Sci. Lett.*, 192:595–616, doi:10.1029/2002JB001862.

LePichon, X., Chamot-Rooke, N., Rangin, C. and Sengör, A.M.C., 2003. The North Anatolian fault in the Sea of Marmara. *J. Geophys. Res.*, 108(B4):2179, doi: 10.1029/2002JB001862.

McClusky, S., Balassanian, S., Barka, A., Demir, C., Ergintav, S., Georgiev, I., Gurkan, O., Hamburger, M., Hurst, K., Kahle, H., Kastens, K., Kekelidze, G., King, R., Kotzev, V., Lenk, O., Mahmoud, S., Mishin, A., Nadariya, M., Ouzounis, A., Paradissis, D., Peter, Y., Prilepin, M., Reilinger, R., Sanli, I., Seeger, H., Tealeb, A., Toksöz, M.N., and Veis, G., 2000. Global positioning system constraints on plate kinematics and dynamics in the eastern Mediterranean and Caucasus, *J. Geophys. Res.*, 105:5695–5719, doi:10.1029/1999JB900351.

Parsons, T., 2004. Recalculated probability of M≥7 earthquakes beneath the Sea of Marmara, Turkey. *J. Geophys. Res.*, 109: B05304, doi:10.1029/2003JB002667.

Parsons, T., Toda, S., Stein, R.S., Barka, A., and Dieterich, J.H., 2000. Heightened odds of large earthquakes near Istanbul: An interaction-based probability calculation. *Science*, 288:661–664, doi:10.1126/science.288.5466.661.

Wright, T., Fielding, E., and Parsons, B., 2001. Triggered slip: observations of the 17 August 1999 Izmit (Turkey) earthquake using radar interferometry. *Geophys. Res. Lett.*, 28(6):1079–1082, doi:10.1029/2000GL011776.

Authors

Georg Dresen, GeoForschungsZentrum Potsdam, Sektion 3.2 Deformaton und Rheologie, Telegrafenberg, D-14473, Potsdam, Germany, e-mail: dre@gfz-potsdam.de.

Mustafa Aktar, Bogazici University, Kandilli Observatory and Earthquake Research Institute (KOERI), Cangelkoy, Istanbul, 81220, Turkey.

Marco Bohnhoff, GeoForschungsZentrum Potsdam, Haus D, Projectbereich 3.2, Telegrafenberg, D-14473, Potsdam, Germany.

Haluk Eyidogan, Istanbul Technical University, Mining Faculty, Department of Geophysics, 34469, Maslak, Istanbul, Turkey.

37

Fault Zone Drilling in the Backstop to the Mediterranean Ridge Accretionary Complex off Crete, Greece

by Achim Kopf and Marco Bohnhoff

Regional Geological Setting

Scientific drilling and the installation of downhole seismic instrumentation to observe seismicity at a low-magnitude detection threshold are powerful tools to monitor densely populated areas in Southern Europe, such as the Athens metropolitan region, to minimize societal and environmental dangers, which millennia ago destroyed entire cultures, two of the seven wonders of the world (Colossus of Rhodes 224 BC, Lighthouse on Pharos 365 AD) and several historical sites in the circum-Mediterranean. The Eastern Mediterranean hosts one of the most prominent retreating convergent margins worldwide that was capable of generating M>8 earthquakes (365 AD, western Crete) and exhaustive volcanic eruptions (1650 BC, Santorini) in historic times.

The Hellenic Subduction Zone (HSZ) in the Eastern Mediterranean Sea represents a mature subduction zone where African crust is thrust (i.e., subducted) beneath Eurasia and where a wide backstop area of partly old accreted strata (marine), partly HP/LT rocks (the exhumed island of Crete as forearc high), an extensional submerged landward forearc (Cretan Sea), and a volcanic arc and backarc basin (Aegean Sea) evolved. The HSZ is an ideal natural laboratory to study collisional processes which are well recorded over the past ca. 35 million years, including an intermittent stage of micro-continent collisions between about 30 Ma and 20 Ma, followed by breakoff of the subducting slab, and incipient collision with the passive African margin today (Thomson et al., 1998). The island of Crete represents a horst structure developed within the last 5 million years in the central forearc, and it provides excellent onshore access to the internal structure of the forearc at various levels (for overall geodynamic situation in map view, see Fig. 1).

Scientific Drilling

An "amphibic" scientific drilling approach is planned for the HSZ, comprising a northern (International Continental Scientific Drilling Project (ICDP)-driven) and a southern (Integrated Ocean Drilling Program (IODP)-driven) domain. Having been kicked off at an ICDP workshop on Crete in 1998, proposals for backstop drilling in the Cretan Sea (ICDP), on Crete (ICDP), and in the Mediterranean Sea (IODP) were developed and later combined to an onshore-offshore deep drilling transect along the HSZ (Fig. 2). With moderately deep marine drill sites (ca. 1 km depth) in the backstop region south of Crete, one continental 3–4-km-deep drill hole onshore Crete, and further ICDP (Glad800) offshore holes within the Cretan Sea (i.e., the forearc-backarc transitional zone) and the volcanic arc (i.e., the submerged Kolumbo volcano as part of the Santorini volcanic complex), we anticipate a substantial contribution to the understanding of earthquake hazard and mitigation, tectonic evolution, and rheology of active collision zones. In this paper, we focus on the marine drilling project where three drill sites are anticipated to penetrate active fault zones.

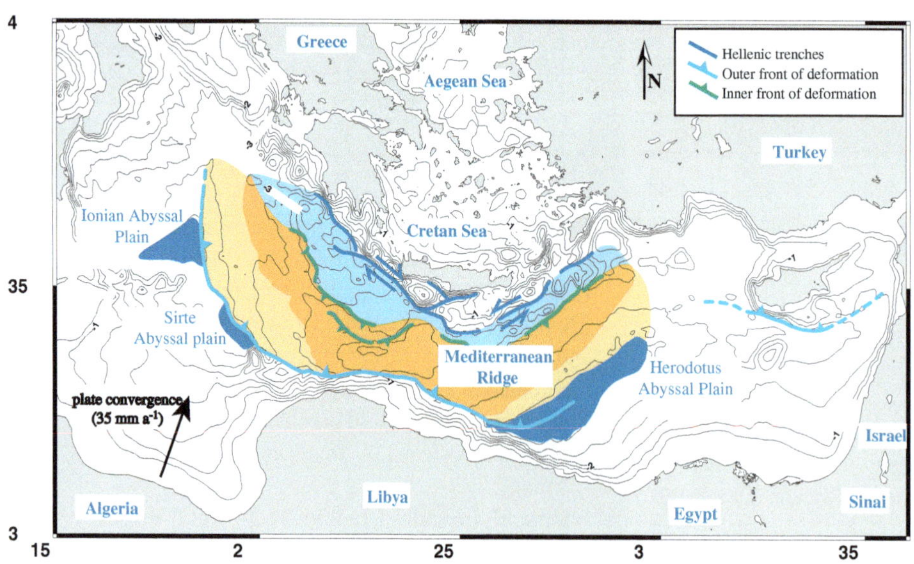

Figure 1. The Cretan Margin as part of the Hellenic subduction zone (HSZ); heavy colored lines with triangles = deformation front, accretionary complex in gray, inner ridge backstop domain in light blue.

Fault Zone Drilling in the Backstop

Within ODP (Ocean Drilling Program) and its

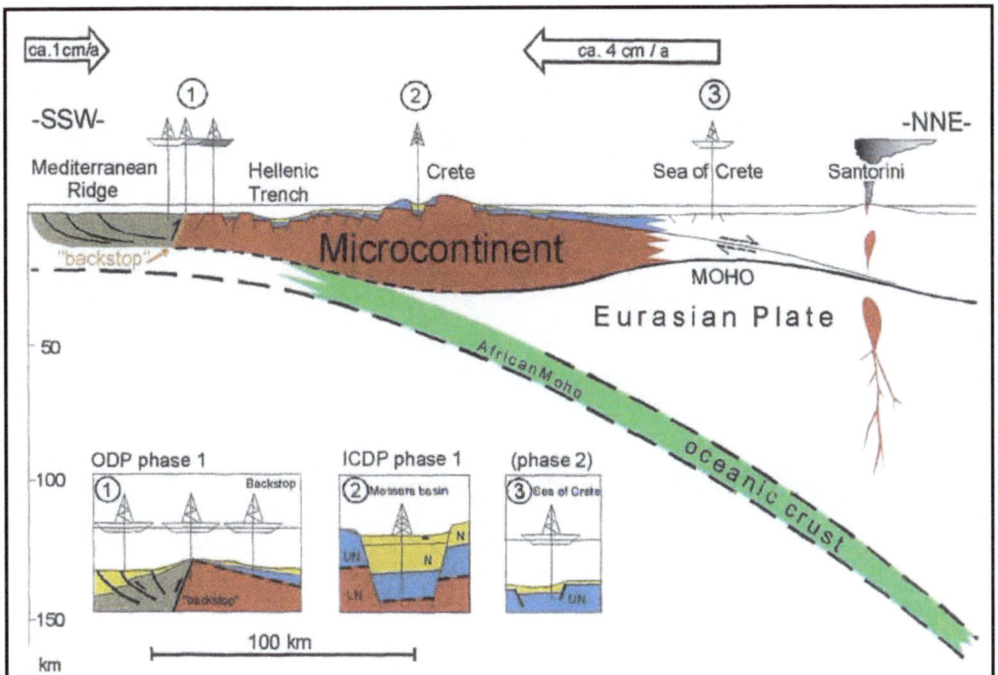

Figure 2. Transect of ICDP and marine targets for the investigation of the Hellenic Subduction Zone .

predecessor DSDP (Deep Sea Drilling Project), several convergent margins have been investigated; however, all these regions have in common that drilling focused entirely on the frontal portion of the forearc prism. Due to this shortcoming, Kopf et al. (1999) proposed to drill a transect of three sites south of the island of Crete (E Mediterranean) from the distal part of the Mediterranean Ridge accretionary prism (MedRidge) across its backstop. Following geophysical surveys and analog modeling, drilling a backstop (or buttress) setting is essential to elucidate fundamental deep fluid flow processes and deformation mechanisms within an accretionary prism and its buttress. Variability in fluid flux across the wedge to its buttress must influence fluid budgets of the accretionary complex, and earlier work attested that flux rates in the backstop domain exceed those near the toe of the accretionary complex (Kopf et al., 2001). The marine transect aims to illuminate (1) mass and fluid transfer at an accreting convergent margin, (2) the significance of spatial variability of fluids from mineral dehydration and diagenetic alteration at depth, and their interaction with the rock, and (3) their possible effect on seismicity. Each of the three 1-km-deep drill holes in the backstop domain will penetrate deep-seated thrusts and backthrust faults whose fluids act as windows down to several kilometers depth. Fluids will be indicative of enhanced dewatering reactions and possibly cause earthquake swarms in the area. Monitoring of fluid-flow and pressure variations would have profound effects on the understanding of fluid budgets and global mass balances, and seismogenesis.

The three fault zones to be penetrated by ocean drilling are:

(i) An active backthrust fault within the distal part of the accretionary complex, which is juxtaposed by a small (1 km across) mud volcano;

(ii) The backthrust contact between the Inner Ridge backstop and the accretionary complex; and

(iii) A reverse fault in the higher Cretan submarine forearc, which separates the exhumed HP/LT units from the overlying thrust nappes forming the bulk part of the island of Crete.

Seismic reflection data suggest fluid migration and activity of each of these fault zones, the latter being the most exciting one since the former detachment fault plane will provide us a window to the depth of slab break off.

References

Kopf, A., Klaeschen, D., and Mascle, J., 2001. Extreme efficiency of mud volcanism in dewatering accretionary prisms. *Earth Planet. Sci. Lett.*, 189(3-4):295–313, doi:10.1016/S0012/821X(01)00278-3.

Kopf, A., Robertson, A.H.F., Screaton, E.J., Mascle, J., Parkes, R.J., Foucher, J.P., DeLange, G.J., Stöckhert, B., and Sakellariou, D., 1999. Backstop hydrogeology of a wide accretionary complex south of Crete, Eastern Mediterranean Sea. *Full drilling proposal (#555) for the Ocean Drilling Program*, 25p.

Thomson, S.N., Stöckhert, B., and Brix, M.R., 1998. Thermochronology of the high-pressure metamorphic rocks of Crete, Greece: Implications for the speed of tectonic processes. *Geology*, 26:259–262, doi:10.1130/0091-7613(1998)026<0259;TOTHPM>2.3.CO;2.

Authors

Achim Kopf, Research Center Ocean Margins (RCOM), University of Bremen, Leobener Strasse, 28359 Bremen, Germany, e-mail: akopf@uni-bremen.de.

Marco Bohnhoff, GeoForschungsZentrum Potsdam, Telegrafenberg D424, 14473 Potsdam, Germany.

Generation Depth of the Pseudotachylyte from an Out-of-Sequence Thrust in Accretionary Prism—Geothermobarometric Evidence

by Shin'ya Okamoto, Gaku Kimura, Asuka Yamaguchi, Haruka Yamaguchi, and Yoko Kusuba

Introduction

Major out-of-sequence thrusts (OOSTs) in accretionary prisms are faults which generate great earthquakes in subduction zones, such as the 1944 Tonankai and 1946 Nankai earthquakes in the Nankai Trough (Cummins and Kaneda, 2000; Park et al., 2002), the 1964 Alaska earthquake in the Aleutian Trench (Plafker, 1972), and the 2005 Sumatra earthquake in the Sunda Trench (Ammon, 2006). The 1999 Chi-Chi Earthquake in Taiwan along the Chelungpu Fault (Chen et al., 2001) also occurred in the same setting as other OOSTs, even though the fault is exposed on land.

Because great earthquakes of Mw>8.0 dominantly take place in accretive margins (Ruff and Kanamori, 1983), understanding the faulting mechanisms of a major OOST is one of the keys in determining the earthquake generation process as such.

An on-land, fossilized, major OOST thrust exhumed from seismogenic depth can be observed in the early Tertiary Shimanto Belt, Kyushu, Japan (Kondo et al., 2005). The shear zone associated with the thrust is thicker than 100 m and is characterized by the development of crack-fillings and crack-seal veins, which suggest that the shear zone was filled by geothermal fluids at depth. Velocity analysis of the rocks in the shear zone shows high and low velocities in the hanging wall and footwall, respectively. This is consistent with the impedance contrast expected from negative polarities of the modern out-of-sequence thrust deeper than 10 km in the Nankai Trough (Tsuji et al., 2006).

A pseudotachylyte-bearing subsidiary fault was recently discovered from the shear zone in the hanging wall (Okamoto et al., 2006). The pseudotachylyte is characterized by ubiquitous association of fluid implosion breccias and mineral veins of quartz and carbonates. This occurrence is very different from that of continental pseudotachylytes developed under dry conditions and suggests abundant fluid association, thermal pressurization, and hydrofracturing preceding frictional melting (Okamoto et al., 2006).

It is important to constrain the exact pressure-temperature conditions of the setting for the pseudotachylyte, because its occurrence is a key to understanding the faulting mechanism if the slip took place in depth. From this point of view, we estimated the pressure-temperature conditions on the basis of fluid inclusion analysis from the mineral veins.

Geological Setting of the Nobeoka Thrust

The Nobeoka thrust is a major OOST bounding the northern and southern Shimanto belts of the Cretaceous-Tertiary accretionary complex in Kyushu, southwest Japan (Fig. 1A). The thrust is traceable for more than 300 km parallel to the modern Nankai Trough.

Hanging wall rocks of the thrust are composed of the Eocene Kitagawa group mainly consisting of plastically deformed phyllites and sandstones. Footwall strata of the Eocene Hyuga group are composed of a shale matrix mélange with sandstone and basaltic blocks deformed mainly by brittle mechanisms. Deformation fabrics of both of these rocks are consistent with the sense of shear along the Nobeoka thrust (Kondo et al., 2005). Geothermometry using vitrinite reflectance and CH_4-H_2O fluid inclusion data indicates that the Kitagawa and the Hyuga groups experi-

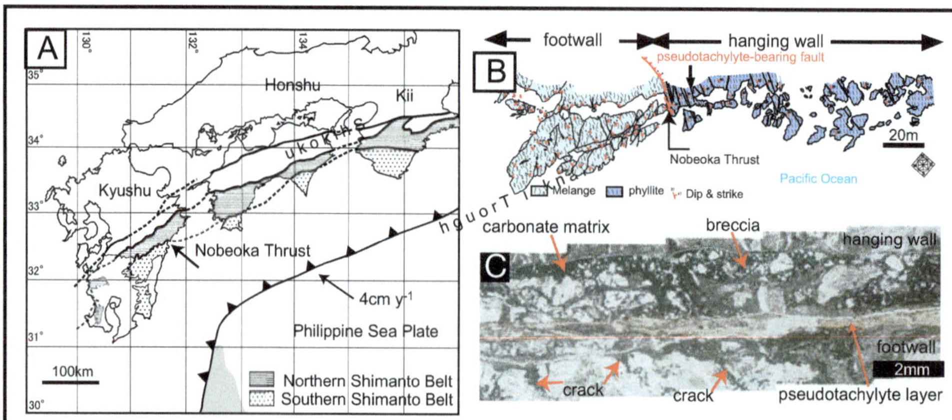

Figure 1. [A] Geological setting of the Nobeoka thrust and distribution of the Shimanto belt in southwest Japan. [B] Sketch map of horizontal exposure of the Nobeoka thrust and shear zone on the tidal flat (addition to Kondo et al., 2005). [C] Photomicrograph of the slipped part of the pseudotachylite-bearing subsidiary fault. Translucent fault core (pseudotachylite) is bounded by breccia in the dark carbonate matrix (modified from Okamoto et al., 2006).

enced heating up to maximum temperatures of ~320°C and ~250°C, respectively (Kondo et al, 2005).

The Nobeoka thrust is characterized by a cataclastic fault core about 20 cm thick and by a plastic and brittle damage zone whose thickness is several tens of meters in the hanging wall and about 100 meters in the footwall. Ubiquitous subsidiary shear fractures, with orientations parallel to that of the fault core of the Nobeoka thrust, are present in the damaged zones.

Kondo et al. (2005) pointed out from the viewpoint of thermal models (Hyndman and Wang, 1993; Hyndman et al., 1997) that the Nobeoka thrust was formed at seismogenic depths, in comparison with the OOST in the modern Nankai Trough (Park et al., 2002).

Experimental velocity measurements of fault rocks from the damaged zone of the Nobeoka thrust show 4–5 km s^{-1} for the hanging wall and 3–4 km s^{-1} for the footwall, presenting a clear impedance contrast, and thus a good analog of the deep seismogenic part of a negative polarity of the major OOST in the modern Nankai Trough (Tsuji et al., 2006).

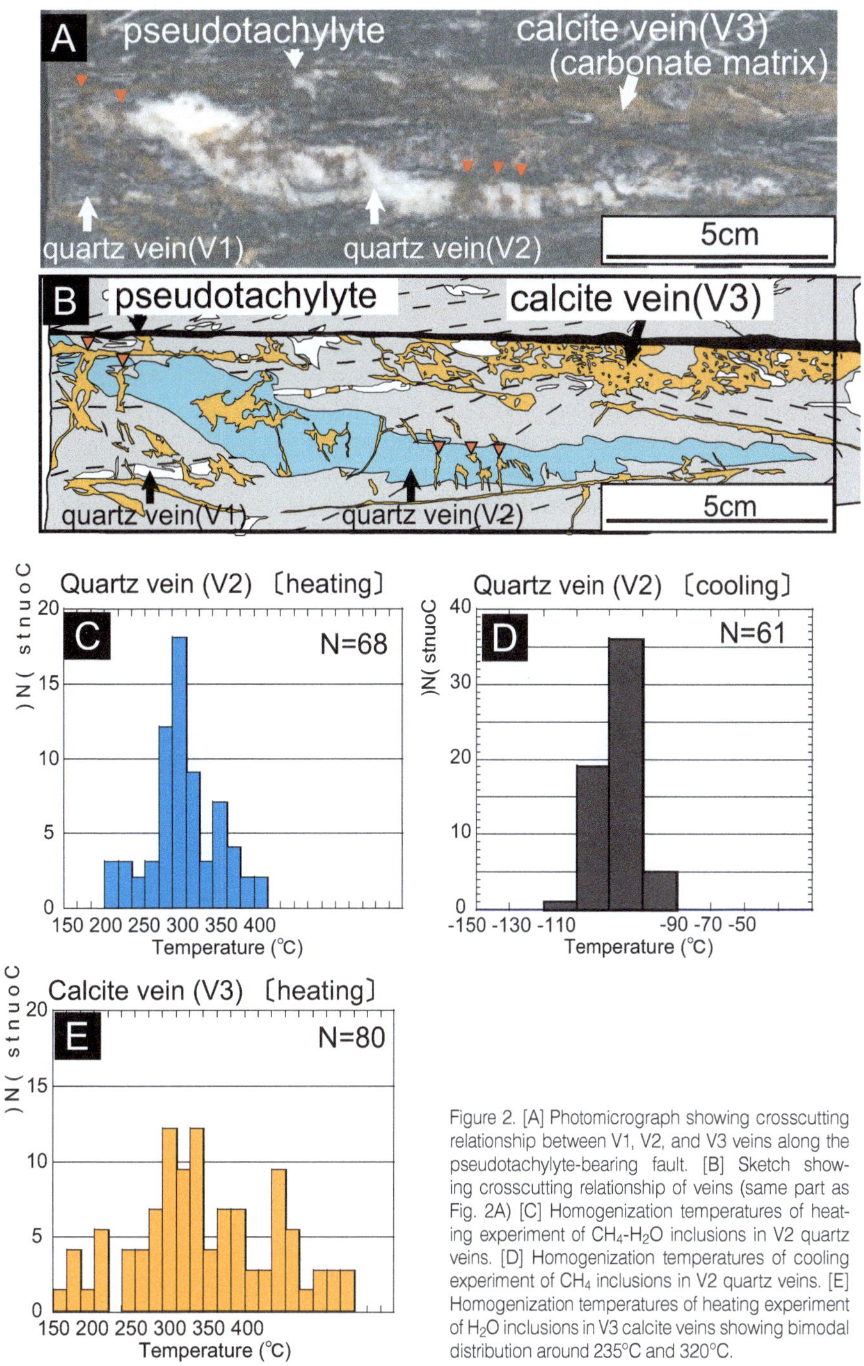

Figure 2. [A] Photomicrograph showing crosscutting relationship between V1, V2, and V3 veins along the pseudotachylyte-bearing fault. [B] Sketch showing crosscutting relationship of veins (same part as Fig. 2A) [C] Homogenization temperatures of heating experiment of CH_4-H_2O inclusions in V2 quartz veins. [D] Homogenization temperatures of cooling experiment of CH_4 inclusions in V2 quartz veins. [E] Homogenization temperatures of heating experiment of H_2O inclusions in V3 calcite veins showing bimodal distribution around 235°C and 320°C.

Occurrence of Pseudotachylyte and Veins

A pseudotachylyte-bearing fault was found in one of the subsidiary faults in the hanging wall, about 10 m above the fault core of the Nobeoka thrust (Fig. 1B; Okamoto et al., 2006). The fault is composed of two different parts along the fault surface: narrow, planar slip sections and wider zones along dilation jogs. The planar slip section shows two types of fault rock. One is a carbonate matrix-supported breccia, and the other is the pseudotachylyte (Fig. 1C). Lines of evidence for frictional melting are discussed by Okamoto et al. (2006). The dilation jogs are filled with carbonate matrix-supported breccia (Okamoto et al., 2006), and the breccias are composed of fragmented host rocks of phyllites several tens of microns to millimeters in size. The carbonate matrix constitutes micron-scale calcite and ankerite, filling the inter-calcite grain boundaries (Fig. 1C).

Crosscutting relationships of veins are as follows (Fig. 2A, B). The oldest veins (V1) occur in the phyllites. Most of them are composed of quartz and are involved in the plastic deformation accommodated by pressure solution (Kondo et al., 2005). Younger quartz veins (V2) are not deformed by such mechanisms, but keep precipitation-related idiomorphic crystal textures. Carbonate veins (V3) fill tension cracks cutting V1 and V2 quartz veins (Fig. 2A, B). The carbonates are connected to the matrix carbonates of the breccias

(Fig. 2A, B). All vein generations are cut by pseudotachylyte (Fig. 2A, B) in slipped parts of the fault. As described in Okamoto et al. (2006), the pseudotachylyte pinches out at dilation jogs filled by carbonate matrix-supported breccia. These occurrences suggest that the breccia formation and carbonate matrix precipitation were preceded by the frictional melting and maintained during the slip.

Occurrences and Compositions of Fluid Inclusions

V2 quartz and V3 calcite veins contain abundant fluid inclusions. V2 quartz veins include single-phase and two-phase inclusions. The single-phase inclusions are composed only of methane, which is shown by a remarkable peak of 2919.2 cm^{-1} using Raman micro-spectroscopy. Two-phase inclusions consist of vapor and liquid phases. A broad peak from 3100 cm^{-1} to 3500 cm^{-1} indicates water, and a peak of 2917.2 cm^{-1} coincides with the methane peak. The results suggest that water is saturated with methane, and water and methane co-exist as immiscible fluids (Vrolijk et al., 1988). No solid carbon dioxide was observed during the cooling experiments (see next section), suggesting that carbon dioxide content is less than 10% (Burruss, 1981).

V3 calcite veins contain only two-phase inclusions. They consist of vapor and liquid phases. Only the water peak from 3100 cm^{-1} to 3600 cm^{-1} is detected by Raman micro-spectroscopy. Therefore, the vapor and liquid phases are both composed of water.

Results of Fluid Inclusion Analysis

Sixty-eight and sixty-one inclusions, respectively, were analyzed by heating and cooling experiments for V2 quartz veins, while eighty inclusions were analyzed by heating experiments for the V3 calcite veins. Results are shown in Fig. 2C, D, E.

Homogenization temperatures of fluid in V2 quartz veins range from 180°C to 296°C (Fig. 2C) with a modal peak at 230°C–240°C. Cooling experiments represent a range from -130°C to -90°C (Fig. 2D). Pressure estimation on the basis of the methane saturation curve (Mullis, 1979) and the results of heating and cooling experiments shows a range of 102–272 MPa. The homogenization temperatures of fluids in the V3 calcite veins were 154°C–362°C and present bimodal peaks. The primary peak is at 230°C–260°C, and the secondary one is at 310°C–320°C (Fig. 2E). The primary peak coincides with that of the V2 quartz veins, and the secondary peak is higher.

Deformation History and its Thermal Conditions

The crosscutting relationships suggest that the deformation history of the hanging wall was as follows:

1. Ductile deformation related to phyllitic cleavage formation. V1 quartz veins are involved in this deformation. Vitrinite reflectance suggests that the temperature was about 320°C (Kondo et al., 2005)

2. Tension cracking and V2 quartz vein precipitation. No more phyllitic cleavages overprint this vein generation, and fluid inclusion analysis suggest that the hanging wall rock was uplifted at about 235°C.

3. Slip initiation, breccia formation, and carbonate precipitation. These events may have started at about 235°C, reaching about 320°C by frictional heating, and ending with finally cooling to about 235°C.

4. Continuing slip resulting in frictional melting. Temperatures along the fault may have been more than 1700°C locally, as quartz grains show signs of plastic deformation and melting (Okamoto et al., 2006).

Background Geothermal Gradient and Depth at the Initial Stage of Slip

The paleo-geothermal gradient in the northern Shimanto belt is estimated at 20–30°C km^{-1} based on metamorphic parageneses (Toriumi and Teruya, 1988), while the paleo-geothermal gradient in the footwall of the Nobeoka thrust is estimated at 28–47°C km^{-1} based on fluid inclusion analysis (Kondo et al., 2005). However, no studies of the thermal conditions of the hanging wall of the Nobeoka thrust have been reported so far. Here, the geothermal gradient is estimated as follows based on the fluid inclusion analyses described above.

It is assumed that the densest methane fluid inclusions were trapped under nearly lithostatic conditions, because the veins are interpreted to be deposited soon after fracturing under the near-lithostatic fluid pressures (Hashimoto et al., 2003; Matsumura et al., 2003; Vrolijk, 1987). Densities are assumed to be 2700 kg m^{-3} and 1000 kg m^{-3}, respectively for solid and fluid.

Thus, fluid pressure (P) is expressed as a function of depth (h) and porosity (θ) as

$$P = \int [2700(1-\theta) + 1000\theta] g \, dh$$

where g is the gravity acceleration, θ is a porosity at depth h expressed by the empirical equation $\theta = Ae^{-bh}$ of Bray and Karig (1985). The equation can be rewritten as

$$P = \int [2700(1 - Ae^{-bh}) + 1000Ae^{-bh}] g \, dh$$

where A and b are constants, 0.65 and 1.54, respectively. The maximum P value of V2 quartz veins is estimated to be about 240 MPa when a temperature of 235°C is assumed.

Then, solving the equation for h, the depth is calculated as 8.6 km. This yields 27°C km^{-1} of geothermal gradient, which is consistent with previous results from the northern Shimanto belt. Geologic occurrences of the Nobeoka thrust, therefore, present excellent information for faulting mechanisms of seismogenic OOST in depth.

Acknowledgements

This study was supported by the Research Program for Plate Dynamics of the Japan Agency for Marine-Earth Science & Technology, and the 21st Century Center of Excellence Program of the University of Tokyo. Our field work was supported by T. Kanda, University of Miyazaki. We are grateful to all the participants of the IODP-ICDP Workshop on Fault Zone Drilling held in Miyazaki for fruitful discussions and comments.

References

Ammon, C.J., 2006. Megathrust investigations. *Nature*, 440:31–32, doi:10.1038/440031a.

Bray, C.J. and Karig, D.E., 1985. Porosity of sediments in accretionary prisms and some implications for dewatering processes. *J. Geophys. Res.*, 90:768–778.

Burruss, R.C., 1981. Analysis of fluid inclusions - Phase-equilibria at constant volume. *Am. J. Sci.*, 281:1104–1126.

Chen, Y.-G., Chen, W.-S., Lee, J.-C., Lee, Y.-H., Lee, C.-T., Chang, H.-C., and Lo, C.-H., 2001. Surface rupture of 1999 Chi-Chi earthquake yields insights on active tectonics of central Taiwan. *Bull. Seismol. Soc. Am.*, 91:977–985.

Cummins, P.R. and Kaneda, Y., 2000. Possible splay fault slip during the 1946 Nankai earthquake. *Geophys. Res. Lett.*, 27:2725–2728.

Hashimoto, Y., Enjoji, M., Sakaguchi, A. and Kimura, G., 2003. *In situ* pressure-temperature conditions of a tectonic melange: Constraints from fluid inclusion analysis of syn-melange veins. *Island Arc*, 12:357–365, doi:10.1046/j.1440-1738.2003.00405.x.

Hyndman, R.D. and Wang, K., 1993. Thermal constraints on the zone of major thrust earthquake failure – the Cascadia Subduction Zone. *J. Geophys. Res.*, 98:2039–2060.

Hyndman, R.D., Yamano, M., and Oleskevich, D.A., 1997. The seismogenic zone of subduction thrust faults. *Island Arc*, 6:244–260, doi:10.1111/j.1440-1738.1997.tb00175.x.

Kondo, H., Kimura, G., Masago, H., Ohmori-Ikehara, K., Kitamura, Y., Ikesawa, E., Sakaguchi, A., Yamaguchi, A., and Okamoto, S., 2005. Deformation and fluid flow of a major out-of-sequence thrust located at seismogenic depth in an accretionary complex: Nobeoka Thrust in the Shimanto Belt, Kyushu, Japan. *Tectonics*, 24:TC6008, doi:10.1029/2004TC001655.

Matsumura, M., Hashimoto, Y., Kimura, G., Ohmori-Ikehara, K., Enjohji, M., and Ikesawa, E., 2003. Depth of oceanic-crust underplating in a subduction zone: Inferences from fluid-inclusion analyses of crack-seal veins. *Geology*, 31:1005–1008, doi:10.1130/G19885.1.

Mullis, J., 1979. The system methane-water as a geologic thermometer and barometer from the external part of the central Alps, *Bull. Mineral.*, 102:526–536.

Okamoto, S., Kimura, G., Takizawa, S., and Yamaguchi, H., 2006. Earthquake fault rock indicating a coupled lubrication mechanism. *e-Earth*, 1:23–28.

Park, J.O., Tsuru, T., Kodaira, S., Cummins, R., and Kaneda, Y., 2002. Splay fault branching along the Nankai subduction zone. *Science*, 297:1157–1160, doi10.1126/science.1074111.

Plafker, G., 1972. Alaskan earthquake of 1964 and Chilean earthquake of 1960 — implications for arc tectonics. *J. Geophys. Res.*, 77:901–925.

Ruff, L. and Kanamori, H., 1983. Seismic coupling and uncoupling at subduction zones. *Tectonophysics*, 99(2–4):99–117, doi:10.1016/0040-1951(83)90097-5.

Toriumi, M. and Teruya, J., 1988. Tectono-metamorphism of the Shimanto Belt. *Mod. Geol.*, 12:303–324.

Tsuji, T., Kimura, G., Okamoto, S., Kono, F., Mochinaga, H., Saeki, T., and Tokuyama, H., 2006. Modern and ancient seismogenic out-of-sequence thrusts in the Nankai accretionary prism: Comparison of laboratory-derived physical properties and seismic reflection data. *Geophys. Res. Lett.*, 33:L18309, doi:10.1029/2006GL027025.

Vrolijk, P., 1987. Tectonically driven fluid-flow in the Kodiak accretionary complex, Alaska. *Geology*, 15:466–469.

Vrolijk, P., Myers, G., and Moore, J.C., 1988. Warm fluid migration along tectonic mélanges in the Kodiak accretionary comprex, Alaska. *J. Geophys. Res.*, 93:10313–10324.

Authors

Shin'ya Okamoto, Department of Earth and Planetary Sciences, Graduate School of Science, The University of Tokyo, 7-3-1 Hongo, Bunkyo-ku, Tokyo, 113-0033, Japan.

Gaku Kimura, Department of Earth and Planetary Sciences, Graduate School of Science, The University of Tokyo, 7-3-1 Hongo, Bunkyo-ku, Tokyo, 113-0033, Japan; and Institute for Research on Earth Evolution (IFREE), Japan Agency for Marine-Earth Science and Technology, 2-15 Natsushima-cho, Yokosuka, Kanagawa, 237-0061, Japan, e-mail: gaku@eps.s.u-tokyo.ac.jp.

Asuka Yamaguchi, Department of Earth and Planetary Sciences, Graduate School of Science, The University of Tokyo, 7-3-1 Hongo, Bunkyo-ku, Tokyo, 113-0033, Japan.

Haruka Yamaguchi, Department of Earth and Planetary Sciences, Graduate School of Science, The University of Tokyo, 7-3-1 Hongo, Bunkyo-ku, Tokyo, 113-0033, Japan; and Institute for Research on Earth Evolution (IFREE), Japan Agency for Marine-Earth Science and Technology, 2-15 Natsushima-cho, Yokosuka, Kanagawa, 237-0061, Japan.

Yoko Kusuba, Department of Earth and Planetary Sciences, Graduate School of Science, The University of Tokyo, 7-3-1 Hongo, Bunkyo-ku, Tokyo, 113-0033, Japan.

Drilling the Seismogenic Zone of an Erosional Convergent Margin: IODP Costa Rica Seismogenesis Project CRISP

by César R. Ranero, Paola Vannucchi, Roland von Huene and the CRISP proponents

Introduction

Interplate subduction zone earthquakes occur in the seismogenic zone, downdip of the transition from stable slip to unstable slip along the plate boundary megathrust. Currently, mineral transformations and fluid migration are thought to govern the stable to unstable slip transition. In the seismogenic zone, subducting clastic material in the subduction channel changes structure and physical properties and releases fluid as it migrates toward the mantle. Where sediment accretion dominates a convergent margin, the subducted clastic input is trench sediment, and the subducted plate is likened to a conveyor belt traveling from the trench into conditions of elevated temperature and pressure. Trench sediment has been sampled, and the diagenetic and metamorphic changes can be inferred. Moore and Saffer (2001) describe the decrease in porosity and increase in temperature with depth, which results in a transition from opal to quartz, the transformation of clay minerals, cementation by carbonate, zeolites, and quartz, all accompanied by pressure solution. However, where subduction erosion dominates a convergent margin, material input to the seismogenic zone is unknown because it is derived from the base of the upper plate, beyond the sampling capabilities of past scientific ocean drilling (von Huene et al., 2004). Mineral transformations involved in seismogenesis at erosional margins may be very different from those at accretionary margins. Zones of upper plate subduction erosion have never been drilled, nor does geophysics resolve their structure, lithology, and physical properties (Fig. 1). The Japanese riser drill ship *Chikyu* in the Integrated Ocean Drilling Program (IODP) provides the opportunity to drill the seismogenic zone of an erosional margin and this is the objective of CRISP (Costa RIca Seismogenesis Project).

Subduction erosion along the plate boundary of the Middle America convergent margin: During CRISP the seismogenic zone of the Middle America erosional convergent margin will be sampled and instrumented. A key feature is the unconformable surface across the top of crystalline basement of the upper plate, a regional unconformity imaged in seismic records. Drilling on DSDP Legs 67 (1979) and 84 (1981) revealed that the inferred accretionary prism of Guatemala is instead an extension of the upper plate igneous basement covered unconformably by Eocene to Miocene shallow water sediment. Drill samples from the lower slope contain microfossils that are now more than 3 km deeper than the water depth in which they lived. Microfossil depth indicators in the

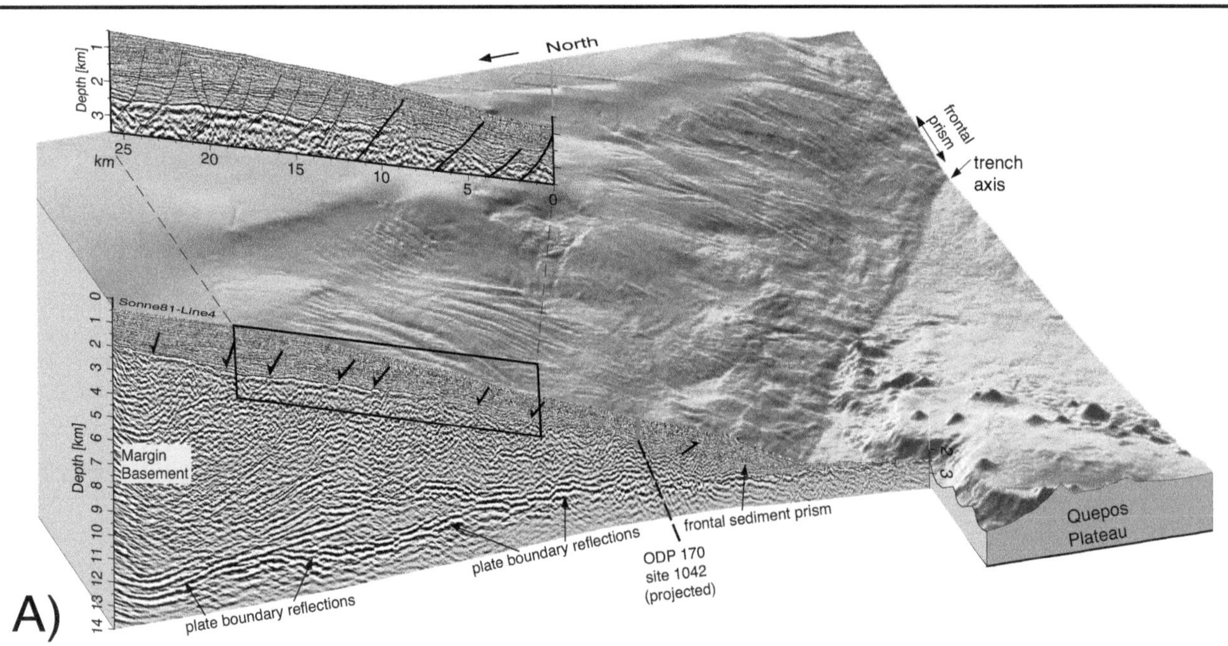

Figure 1. [A] Seismic depth image (pre-stack depth migration) of the Costa Rica margin near the Osa Peninsula (Ranero and von Huene, 2000). Margin basement is formed by igneous rock; major normal faults are emphasized with lines. Inset on the left shows detail of normal faults displacing the top of basement uncorformity and the slope sediment section. Seafloor topography was acquired with multibeam surveying.

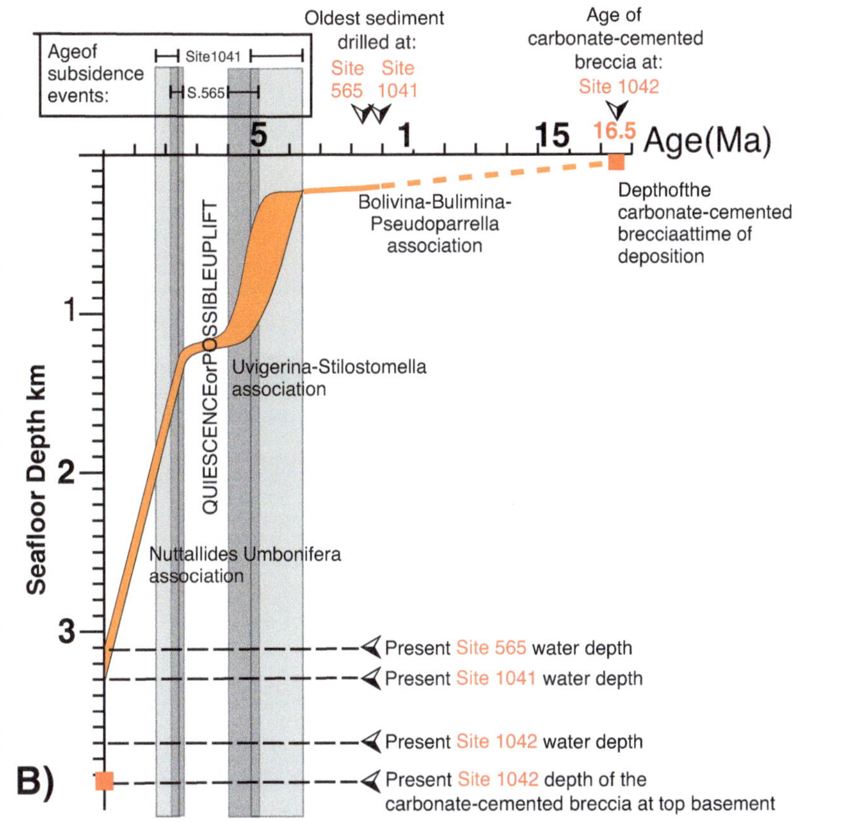

Figure 1. [B] Long-term subsidence history of the Costa Rica margin composed with data from several drill sites offshore the Nicoya Peninsula. Paleodepth information was obtained from studies of benthic foraminifera in the slope sediment, indicating a non-linear subsidence (two main pulses) of about 4 km for the middle continental slope (Vannucchi et al., 2003).

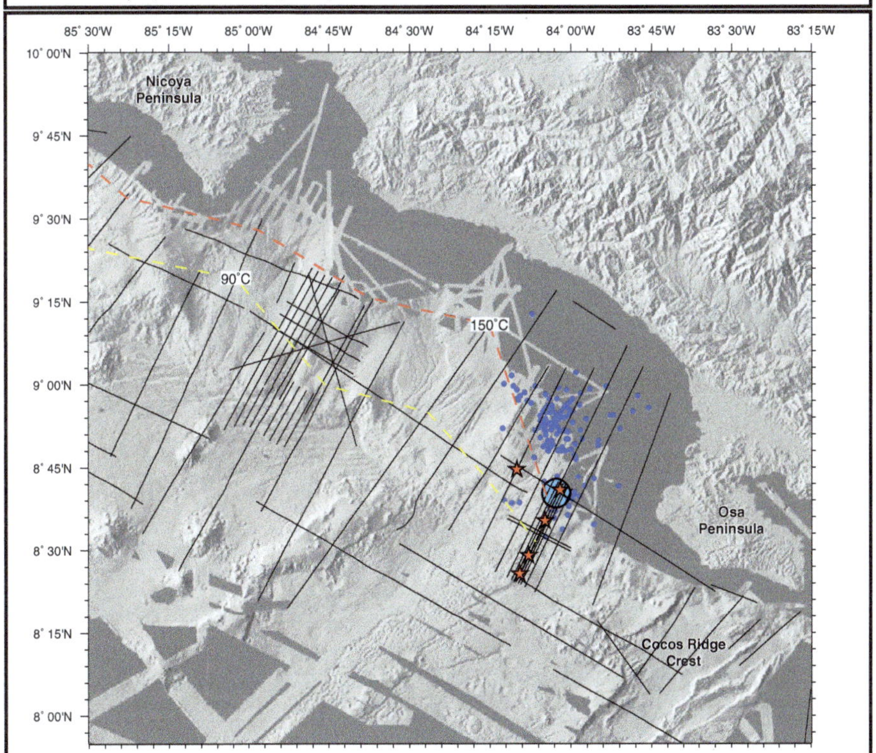

Figure 2. Multibeam bathymetry and topography of the Costa Rica margin between Nicoya and Osa Peninsula. Dashed lines delineate the 90°C and 150°C temperature isotherms along the plate boundary. Large blue circle marks nucleation area of the 2002 Mw6.4 earthquake, and small blue circles are aftershocks occurring during 6 days after the main shock (I. Arroyo, unpublished). Drill site in the upper slope is located on the nucleation area of the 2002 Mw6.4 earthquake (see Fig. 4). Black lines indicate tracks of multichannel seismic reflections profiles collected during SONNE 81 and BGR99 cruises.

overlying sediment record the progressive margin subsidence since Miocene times (Vannucchi et al., 2004). Subsidence from the surf zone to trench depths requires crustal thinning by erosion of material along the underside of the upper plate.

Drilling on ODP Leg 170 (1996) off Nicoya confirmed that, like Guatemala, the Costa Rican unconformity surface has subsided from shallow water in the past few Myr (Vannucchi et al., 2003) similarly to the neighbor segment off Nicaragua (Ranero et al., 2000). Additionally, the igneous basement subsequently dredged from the lower continental slope of Nicaragua indicated an extension of the continental framework to the trench there. In seismic records the regional unconformity can be followed continuously from the ODP Leg 170 drill transect off Nicoya to the proposed Osa Peninsula drill transect.

Ongoing erosion is indicated by morphology of the slope (Fig. 2). Osa bathymetric high forms the southern flank of a major embayment where erosion and slope retreat are accelerated by subduction of Cocos Ridge and its associated seafloor ridges and seamounts. The regional erosion surface was ground truth during Leg 170, and the current retreat of the continental slope provides evidence of vigorous tectonic erosion where CRISP drilling is proposed.

Resultsfrom recent work: Researchers in the German SFB 574 (Kiel University) have used high-resolution geophysical tools to map the mounds at a series of mid-slope vent sites. In all, about 125 vent sites were located along ~600 km of the margin off Costa Rica and Nicaragua, and they seem to be common from Panama to Mexico. Material and fluids were sampled from several mounds, and probe heat flux measurements were collected across a number of sites. Cameras just above the seafloor displayed abundant chemosynthetic carbonates and fauna at the vent sites.

The mounds are a few hundreds to one thousand meters across with a very hummocky topography. They occur in the area of normal faulting and locally appear at fault scarps. Heat flux over and around the mounds is high compared to the regional background values, indicating venting of deeply-sourced fluid (Grevemeyer et al., 2004). Pore fluid chemistry documents low chlorinity fluids, and chemical modeling indicates that they come from dehydration of clays (Hensen et al., 2004). First-order estimates of fluid flow budget across the forearc indicate that most fluid released during dehydration of sediment in the subduction channel traverses the upper plate and is released across the mid-slope vent sites (Ranero et al., 2007). Dehydration of the sediment subducting with the lower plate is proposed as one of the significant causes for the transition to stick-slip behavior and seismogenesis. The massive fluid release across the upper plate of erosional margins is one of the unique processes investigated during CRISP.

Also unique to CRISP is the possibility to drill the nucleation area of the 20 November 2002, Mw6.4 interplate thrust earthquake. The earthquake was recorded on an array of ocean bottom seismometers as well as on the Costa Rican National Seismological Network on land (Fig. 3). The full array provides a database for accurate relocation of the earthquake and its aftershocks (I. Arroyo, in preparation). The main shock is essentially located in the position of the proposed 5-km-deep riser drill hole to the seismogenic zone. Downhole instrumentation, including seismometers, will provide a 3-D seismic image around the drill hole with a radius of several kilometers including the area of the Mw6.4 nucleation. Attribute analysis of the seismic data will permit an estimation of the general physical properties of the upper plate and plate boundary.

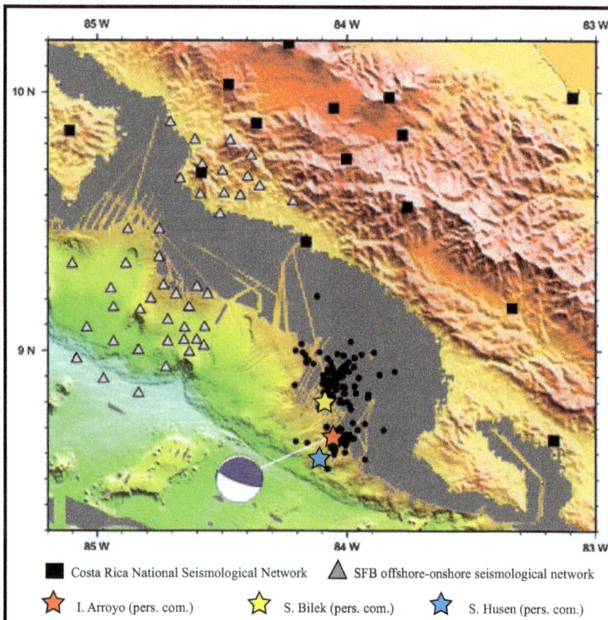

Figure 3. Location of the main shock and aftershocks of the 2002 Mw6.4 earthquake (red star). Triangles indicate the array of ocean bottom seismometers that recorded the quake; black squares are seismographs of the local network. Previous locations without the offshore seismograph array are shown in yellow and blue stars.

CRISP: The shallow dip of the subduction zone off southern Costa Rica and the relatively high temperature of the subducting ocean crust bring materials and processes involved in seismogenesis to depths reachable by drilling with the riser ship *Chikyu*. CRISP is structured in non-riser (Program A) and riser (Program B) stages that systematically lead from shallow non-riser to deep riser drilling (Fig. 4). Program A provides background information for Program B, which reaches the unstable slip of the seismogenic zone. Program A drilling allows the characterization of lower plate oceanic igneous rock and hydrology before subduction, as well as under the front of the margin. With standard ODP drilling technology it also provides the samples to characterize upper plate basement rock. Logging and instrumentation will provide accurate information on the upper plate thermal structure, interplate earthquake nucleation, and rupture propagation, which are key for a successful planning of deeper riser drilling. In Program B, the riser sites locations are constrained with the results from downhole instruments and conventional geophysical experiments, rock mechanics laboratory experiments, and forward modeling. Program B drilling reaches 3 km below seafloor to sample for the first

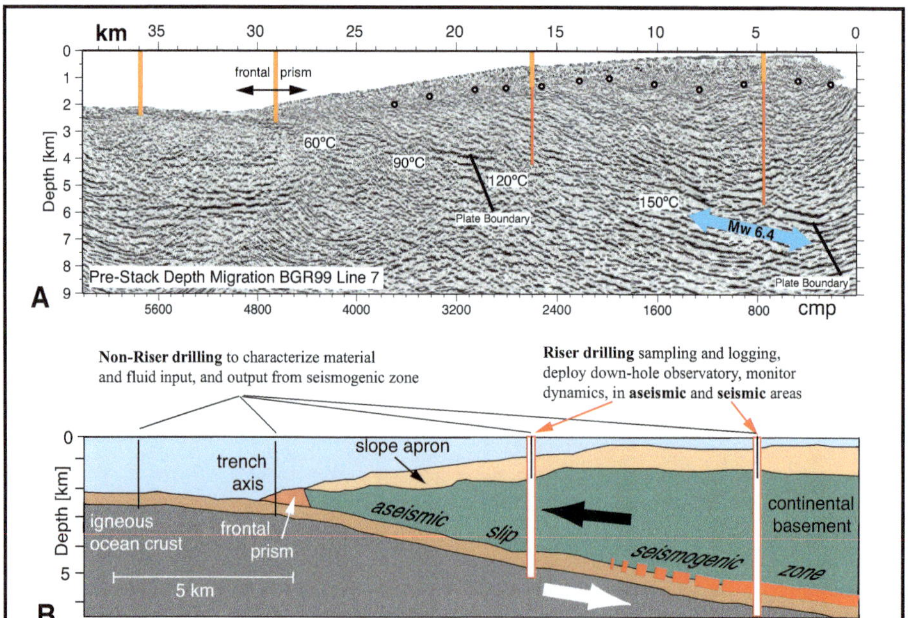

Figure 4. [A] Depth image across the location of proposed drill sites (see Fig. 2). Shallow non-riser sites are shown in orange, and deeper riser holes in red. Note that drilling of the deepest hole will penetrate the nucleation area of the 2002 Mw6.4 earthquake (approximate location shown by blue arrow). [B] Diagram of main geological units, based on seismic image of [a], showing the plan to drill on either side of the updip limit of the seismogenic zone.

Subsurface Structure, Fault Zone Characteristics, and Stress State in Scientific Drill Holes of Taiwan Chelungpu Fault Drilling Project

by Jih-Hao Hung, Kuo-Fong Ma, Chien-Yin Wang, Shen-Rong Song, Hisao Ito, Weiren Lin, and En-Chao Yeh

Introduction

The 1999 Chi-Chi earthquake (M_w7.6) produced a >90-km-long surface rupture zone along the north-south trending, west-vergent Chelungpu fault. The most striking feature of the coseismic displacement field is that areas of large surface displacement lie above the footwall ramp of the thrust and at the northern termination (up to 12 m). An important question that needs to be addressed is what physical properties or dynamic processes within the fault zone cause large coseismic displacements in the northern segment. Hypotheses that have been proposed include 1) change of the fault-plane geometry (Yue et al., 2005); 2) static (long-term) physical properties such as intrinsic low coefficient of friction, high pore-pressure, and solution-transport chemical processes; and 3) dynamic change of physical properties during slip. To address the above questions two holes (A and B) were drilled for the Taiwan Chelungpu Fault Drilling Project (TCDP) during 2004–2005 at Dakeng, west-central Taiwan, where large surface slip (~10 m) was observed. Continuous coring and geophysical downhole logging in two holes 40 meters apart were completed from a depth of 500–2003 m (Hole A) and 950–1350 m (Hole B). Data from the drilled holes provide a unique opportunity to understand deformation mechanisms and physical properties of the Chelungpu fault where large slip occurred in the Chi-Chi earthquake.

Subsurface Structure and Fault-Zone Characteristics

Subsurface structure, stratigraphy, and corresponding log depth encountered in Hole A are shown in Fig. 1. Regional bed attitude above FZA1712, identified from cores and FMI/FMS images in Hole A and from correlation of fault zones between Hole A and Hole B, is trending N15°-E21°, dipping 20°–40° (30° on average) toward SE. Nonetheless, intervals of increasing (from 30° to 75°) or decreasing (from 70° to 20°) dip, as well as changes of dip azimuth, appear across fault zones. A gradual increase of bedding dip with depth starts from FZA1712, and a drastic change of dip from 20°–40° to 60°–80° occurs across FZA1855 where steep to overturned beds extend to the bottom hole.

A total of twelve fault zones identified in Hole A are located in the Plio-Pleistocene Cholan Formation, Pliocene Chinshui Shale, and Miocene Kueichulin Formation. Common fault rocks in the cores include intensely deformed fault core (clayey gouge) and adjacent highly fractured damage zones (fault breccia). The fault gouge is composed of ultra-fine-grained clay minerals and massive to foliated fabrics; occasionally, thin layers of indurate black material appear within the gouge zone. A typical example is the Chelungpu fault zone, FZA1111 (Fig. 2). The fault is bedding-parallel consisting of fault breccia and fault gouge 1109 m to 1112 m.

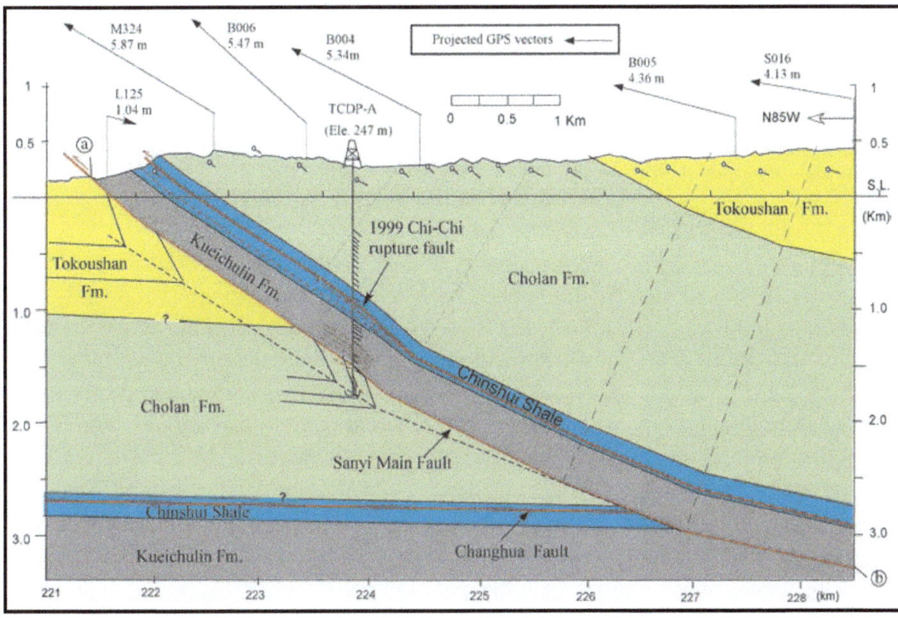

Figure 1. Interpreted structural profile across the Dakeng well (Hole A) based on the surface and subsurface drilled data. Measured depth (from rig floor, 3 m above the ground level) intervals and true thickness (in parenthesis) of formations in this borehole includes: Cholan, 0–1013 m (877 m); Chinshui, 1013–1300 m (249 m); and Kueichulin, 1300–1712 m (357 m). Underlying the Sanyi fault is the repeated section of Cholan formation from 1712–2003 m. Locations of Chelungpu (Chi-Chi rupture), Sanyi, and interpreted Changhua faults are shown by solid lines. Older faults formed prior to the Chi-Chi earthquake are shown by dashed lines. The displacement of the Sanyi fault is greater than 9 km (from eroded base of hanging-wall cutoff point "a" to footwall cutoff point "b"). The GPS vectors are projected components of the coseismic displacement in the plane of section (station number and displacement are given). Note that the displacement vectors, except in the footwall of the rupture fault, are approximately parallel to the fault at depth.

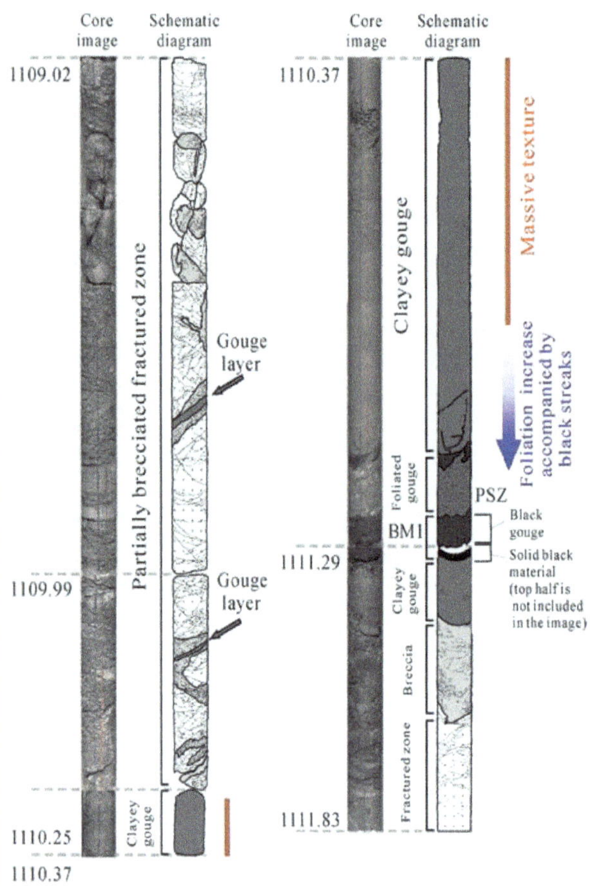

Figure 2. The core image and sketch diagram of fault zone FZA1111 from depth 1109.02 m to 1111.83 m of Hole A with descriptive comments. Note that the degree of deformation increases to the bottom of clayey gouge zone, and a 12-cm primary slip zone develops at depths of 1111.23–1111.35 m (from Ma et al., 2006).

The degree of fracturing increases from the top of the damage zone towards the gouge zone in which the fabrics changed from massive to foliate between 1110.25 m and 1111.35 m. The Chi-Chi major slip zone (MSZ, about 2 cm thick) is contained within the 12-cm-thick primary slip zone (PSZ), which is located near the bottom of this broad gouge zone (Ma et al., 2006).

In spite of large surface displacements, no temperature anomaly was observed near FZA1111 due to circulation of mud immediately after the drilling. Nevertheless, Kano et al. (2006) reported a heat anomaly of 0.06°C during repeated temperature measurements 6 months after the completion of drilling.

In situ Stress Measurements

Leak-off test: A standard commercial procedure of open-hole, extended leak-off tests was conducted in Hole B at depths between 940 m and 1350 m to determine *in situ* magnitudes of maximum (S_{Hmax}) and minimum (S_{Hmin}) horizontal stresses. Successful leak-off tests have been done at four locations in Hole B—1279.6, 1179.0, 1085.0, and

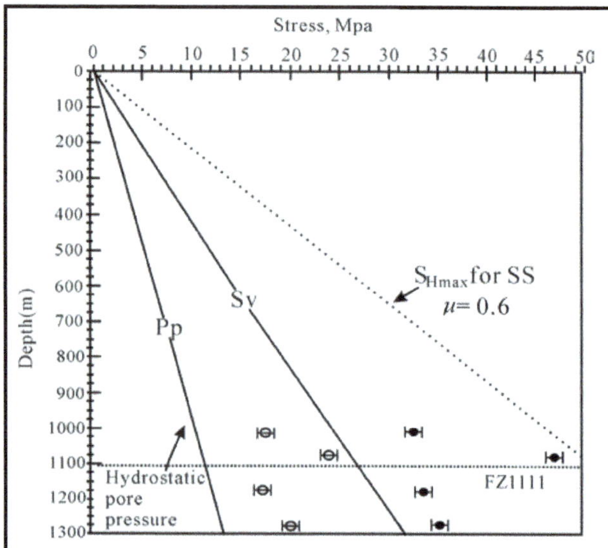

Figure 3. Plot of overburden stress (Sv) from integration of density logs, pore pressure, and S_{Hmax} (filled symbol) and S_{hmin} values (open symbol) determined from leak-off tests at different depths. The calculated upper bound S_{Hmax} (dashed line) for a strike-slip faulting stress regime assuming coefficient of friction μ=0.6 and location of FZA1111 are also shown.

1019.5 m—with two above and two below the FZB1137 (equivalent to FZA1111). Given the depth of overburden, we can calculate the vertical stress from integration of density logs and compile with pore pressure and hydrofracturing data from various depths (Fig. 3). The measurements clearly indicate a strike-slip fault regime after the Chi-Chi earthquake in this area.

Wellbore failure: *In situ* stresses S_{Hmax} determined from borehole breakouts and drilling-induced tensile fractures from Hole A and Hole B (Fig. 4) show that a significant change of S_{Hmax} azimuth occurs across the depth of 1300 m (also a stratigraphic boundary between the Chihsui shale and the Kueichulin Formation). The S_{Hmax} was oriented at 103°–138° with an average of 123° in the section of 700–1300 m, as opposed to 137°–164° (154° on average) from 1300 m to 1700 m. Borehole breakouts are relatively better developed in the Kueichulin Formation than in other places. This observation agrees with stronger anisotropy (stress magnitude) in the Kueichulin Formation, as shown by the shear wave anisotropy.

Shear Seismic Wave Anisotropy

Data from Dipole-Shear Sonic Imager (DSI™, Schlumberger) logs acquired over the interval of 508–1870 m in Hole A were used to assess shear wave velocity anisotropy. Analyses at these depths are shown by scatter plots and rose diagrams for nine discrete intervals of similar orientations of fast shear wave polarization (Fig. 5). A prominent NW-SE fast shear polarizing direction was generally observed except in a few depth zones, such as 738–770 m, 785–815 m, and 1517–1547 m. In particular, a very consistent mean direction with small dispersion of 115°±1°–2° (95% confidence interval) appears in the strongly anisotropic Kueichulin Formation at 1300–1650 m. Relatively consistent fast shear polarization directions appear across FZA1111 (average 165° between 1105 m and 1115 m) compared to the interval of 1078–1190 m with trending in a much broader range of 130°–170°. Thus, there is no observable systematic change of trend on fast shear polarization across the Chi-Chi slip zone. On the other hand, from the change of fast shear azimuth at depth 1000 m and contrasting dgree of anisotropy across the depth of 1300 m, the perturbation of regional stresses may have occurred within the upper and lower boundaries of the Chinshui Shale as suggested from detailed study of borehole breakouts (Wu et al., 2007).

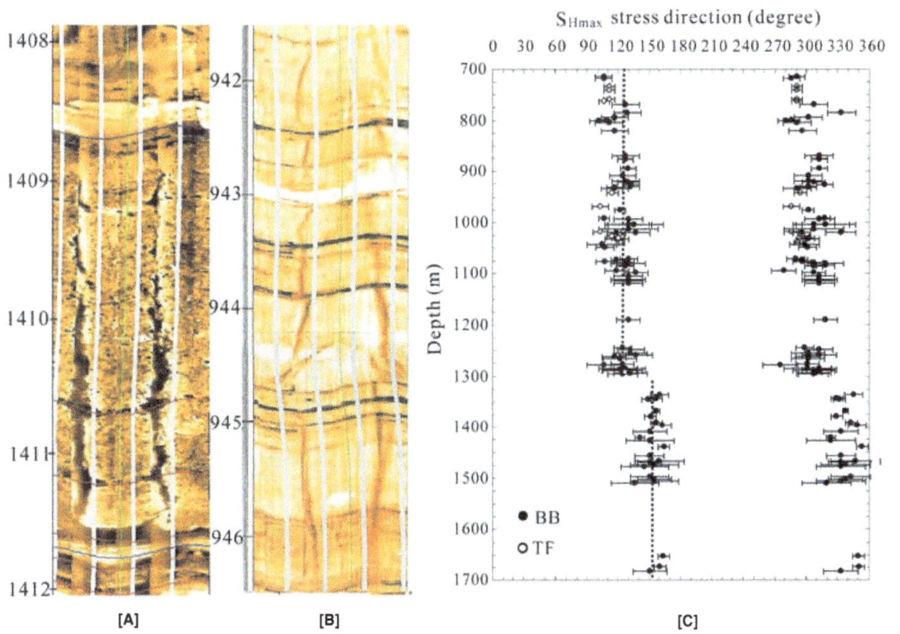

Figure 4. [A] Borehole breakouts as dark bands in opposite sides of the wellbore wall in FMI image logs of Hole A between 1409 m and 1411 m, and [B] drilling-induced tensile fractures in the Hole B at 941–947 m. [C] The plot of azimuth of S_{hmzx} determined from breakouts (BO) and tensile fractures (TF) in the TCDP wells. Width of bar shows an opening angle of BO and TF with dark and open circles, respectively, as mid-point. Dotted lines are average of mid-points between sections of 700–1300 m and 1300–1700 m.

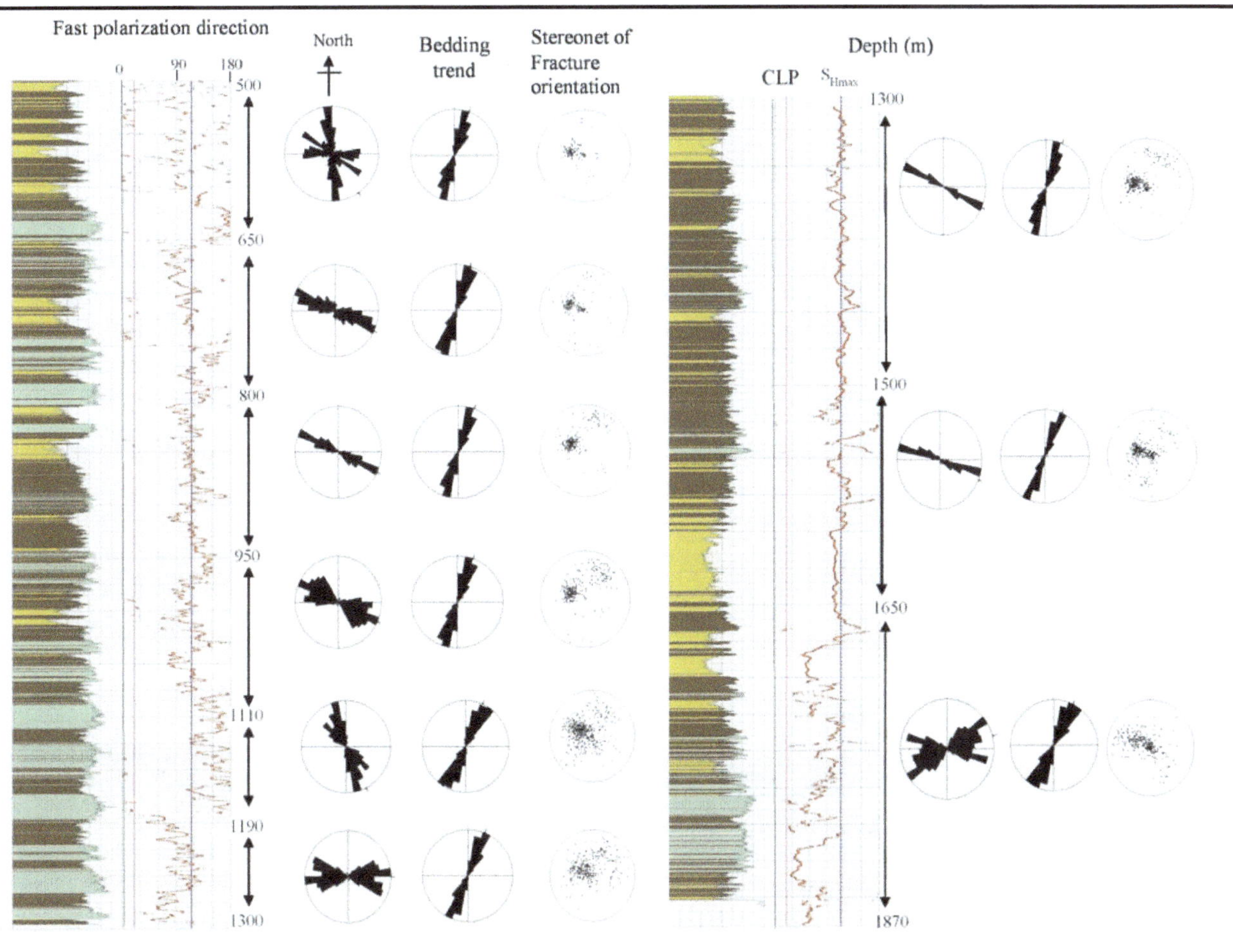

Figure 5. Comparison of the fast-shear polarization direction with bedding trend determined from borehole images and fracture orientations measured from core images. Azimuths of the Chelungpu fault (CLP, 020) and maximum horizontal stress (S_{hmax}, 115) in central Taiwan determined from earthquake focal mechanisms (Yeh et al., 1991) are shown for references.

41

Drilling Investigations on the Mechanics and Structure of Faults

by Kentaro Omura

Fault zone drilling provides valuable insights into understanding the dynamics, physical properties, and structure of an active fault. We drilled into the four major active faults in central and western Japan to directly access mechanics, physical properties, and fault rock distributions in and around the fault zones—the Nojima fault (which appeared by the 1995 Hyogo-ken Nanbu Earthquake, M7.2, also called the Kobe Earthquake); the Neodani fault, which appeared by the 1891 Nobi earthquake (M8.0); the Atera fault of which some parts seemed to have been dislocated by the 1586 Tensho earthquake (M7.9); and the Gofukuji Fault, which is said to have activated more than a thousand years ago. We measured *in situ* stress magnitude and direction in and around a fault zone by hydraulic fracturing and borehole breakout method. Additionally, integrating *in situ* measurements in a deep borehole and ground surface geophysical survey across a fault zone, we observed physical properties and structures of fault zones. The stress state and structure of different faults may reflect different stages in the earthquake pre-event process in the earthquake recurrence cycle.

By the *in situ* stress measurements, we found the differential stress magnitude decreases in the depth close to fracture zones in the Neodani fault (1300-m-deep well; Ikeda et al., 1996) and the Nojima fault (1800-m-deep well; Ikeda et al., 2001a). We also found the orientations of the maximum horizontal compressive stress were nearly perpendicular to the fault strikes of the Nojima Fault (Ikeda et al., 2001a). As for the Atera Fault, the orientation of the maximum horizontal compressive stress was the reverse of the fault moving direction (Ikeda et al., 2001b). These results support the idea that the differential stress is small at narrow zone adjoining fracture zones and the fault is quite "weak" after the earthquake. Additionally, outside the narrow zone the frictional strength of rock is still high. The level of shear stress adjacent to the fault seems to be controlled principally by the frictional strength of rock in and around the fault.

As for the Nojima fault, the pre-shock stress field adjacent to (about 300 m distance from) the fault was estimated by adding the stress change due to the co-seismic slip from the post-shock stress field measured by the hydraulic fracturing method (Yamashita et al., 2004). The reconstructed stress state just before the earthquake was different from that just after the earthquake and shows that the maximum horizontal principal stress direction is oblique to the fault, which is consistent with the background tectonic stress direction (East-West) before the earthquake. As for the Gofukuji fault, the orientation of the maximum horizontal principal stress adjacent to (about 300 m distance from) the fault was oblique to the fault trace (Omura et al., 2003a). The Headquarters for Earthquake Research Promotion, Japan determined that the Gofukuji fault has not activated for a time longer than the mean earthquake recurrence interval. The strength of the Gofukuji fault may be recovered enough to sustain shear stress on the fault, which is the same situation as estimated pre-shock stress state of the Nojima. The findings on the stress states of Nojima and Gofukuji fault suggest that the fault recovered as hard as the host rock surrounding the fault from the weak strength just after the earthquake, and that the orientation of the stress near the fault changes during the inter-seismic period of the earthquake recurrence cycle.

Downhole physical logging, and macro- and microscopic observation of recovered cores show complicated fault rock distribution in the fault fracture zones. In a schematic view, narrow strongly fractured sub-zones and broad weakly fractured sub-zones are distributed within a fracture zone. The strongly fractured sub-zones are characterized as having low electrical resistance, low density, low P-wave velocity, high porosity, and enlargement of borehole radius (Omura et al., 2003b). In particular, in the Nojima Fault cores, we observed three distinct fracture zones, each including strongly and weakly fractured sub-zones (Kobayashi et al., 2001). X-ray analyses of the Nojima fault cores show that the three fracture zones have different distribution pattern of chemical elements and alteration minerals which were produced by hydrothermal changes associated with fault activities (Matsuda et. al., 2004). The fault seems not to have slipped on a specific plane, but on more than one fault plane during earthquake cycles. The complicated fault zone structure should have formed progressively during repeating earthquake cycles.

The drilling method is of great advantage to study the mechanics and structure of fault zones because we can access the materials and internal structure of the fault zones directly. On the other hand, as the depth and the number of drillings are limited, it is important to include significant integrated studies such as geophysical prospect, geophysical observations, laboratory experiments, and numerical simulation.

References

Ikeda, R., Iio, Y., and Omura, K., 2001a. *In situ* stress measurements in NIED boreholes in and around the fault zone near the 1995 Hyogoken-Nanbu earthquake, Japan. *The Island Arc*, 10(3/4):261–265.

Ikeda, R., Omura, K., Iio, Y., and Tsukahara, H., 1996. Scientific drilling and *in situ* stresses in active fault zones at Neodani, central Japan. *Proc. 8th Int'l. Symp. On the Observation of the Continental Crust Through Drilling*, 30–35.

Ikeda, R., Omura, K., Matsuda, T., Mizuochi, Y., Uehara, D., Chiba, A., Kikuchi, A., and Yamamoto, T., 2001b. Central Japan's Atera active fault's wide-fractured zone: An examination of the structure and in-situ crustal stress. *EOS. Trans. Am. Geophys. Union*, 82(47):Abstract T51A-0848.

Kobayashi, K., Arai, T., Ikeda, R., Omura, K., Sano, H., Sawaguchi, T., Tanaka, H., Tomita, T., Tomida, N., Hirano, S., Matsuda, T., and Yamazaki, A., 2001. Distribution of fault rocks in the fracture zone of the Nojima Fault at a depth of 1140m: Observations from the Hirabayashi NIED core. *The Island Arc*, 10(3/4):411–421, doi:10.1016/j.1440-1738.2001.00339.x.

Matsuda, T., Omura, K., Ikeda, R., Arai, T., Kobayashi, K., Shimada, K., Tanaka, H., Tomita, T., and Hirano, S., 2004. Fracture-zone conditions on a recently active fault: Insights from mineralogical and geochemical analyses of the Hirabayashi NIED drill core on the Nojima fault, southwest Japan, which ruptured in the 1995 Kobe earthquake. *Tectonophysics*, 378:143–163, doi:10.1016/j.tecto.2003.09.005.

Omura, K., Ikeda, R., Iio, Y., Arai, T., Kobayashi, K., Shimada, K., Tanaka, T., Hirano, S., and Matsuda, T., 2003b. Logging in a fault fracture zone –A case study at Hirabayashi NIED Borehole drilled through Nojima fault. *BUTSURI-TANSA*, 56:401–414. (Japanese with English abstract)

Omura, K., Ikeda, R., Matsuda, T., and Mizoguchi, Y., 2003a. Active fault drilling research on stress, electrical and base rock structure – Fault zone structure of Gofukuji fault, central Japan. 2003 General Assembly of the International Union of Geodesy and Geophysics (IUGG), Poster B468.

Yamashita, F., Fukuyama, E., and Omura, K., 2004. Estimation of fault strength: Reconstruction of stress before the 1995 Kobe earthquake. *Science*, 306:261–263, doi:10.1126/science.1101771.

Author

Kentaro Omura, National Research Institute for Earth Science and Disaster Prevention, Tennoudai 3-1, Tsukuba, Ibaraki, 305-0006, e-mail: omura@bosai.go.jp.

42

Clay Mineral Composition and Diagenesis: Effects on the Location and Behavior of Faults in the Frontal Portions of Subduction Zones

by Michael B. Underwood

One of the more ambitious goals of the Seismogenic Zone Experiment (SEIZE) is to discriminate between the effects of changing intrinsic frictional strength (i.e., gouge versus wall-rock) and the effects of gradual or transient increases in pore pressure within fault zones. In subduction zones, both variables are likely to change with increasing distance down the plate boundary. In addition to sediment's cohesion and coefficient of internal friction, the shear strength of any given stratigraphic interval depends upon fluid pressure and total normal stress. The release of fluids from mudstones during diagenesis (including volatiles generated from organic matter) is modulated by a combination of mechanical compaction, tectonic consolidation, and mineral reactions. Thus, it is essential to determine how permeability, sediment-derived fluid sources, and fluid flow contribute to overpressure conditions within the subduction inputs, beginning seaward of the subduction front at depths of 2500 m or less.

Primary lithostratigraphic architecture exerts a first-order influence over the material properties and tectonic behavior of subduction zones, at least near the toe of the margin wedge. Generic sedimentary facies models for trenches (Piper et al., 1973) show upward thickening and coarsening trends (from basalt at the base through pelagic ooze, hemipelagic mud, silty turbidites, and sandy turbidites). As glaring exceptions to that norm, however, there are substantial thicknesses of highly permeable sand (abyssal-plain deposits) that continue to move down the subduction path (i.e., beneath the frontal décollement) in Nankai (Ashizuri transect), Barbados, and Cascadia (Underwood and Moore, 1995; Underwood, 2007). Until those sands are lithified by cementation, their high net permeability will control three-dimensional patterns of fluid circulation. The amount and types of clay-size particles within each stratigraphic interval are also important because they affect any sediment's coefficient of friction and permeability. Empirical and experimental studies demonstrate that clay-rich fault gouges are weaker than surrounding wall rocks, typically because of their constituent particle size and mineralogy. Expandable minerals of the smectite group (e.g., montmorillonite) are notorious in the eyes of geotechnical engineers for affecting strata's coefficient of internal friction, compressibility, and permeability (Lupini et al., 1981; Morrow et al., 1992; Saffer and Marone, 2003).

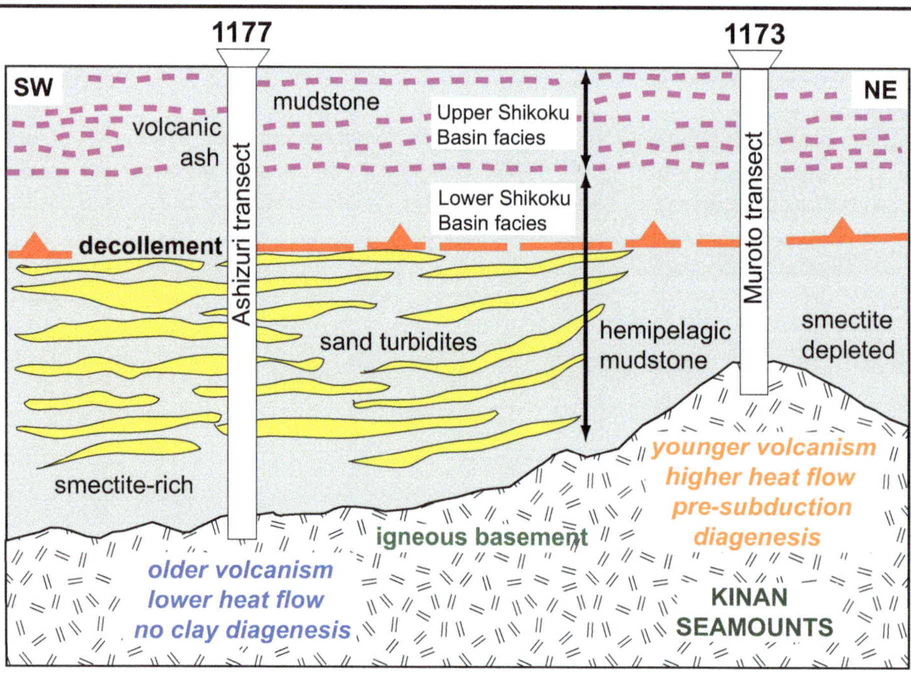

Figure 1. Schematic cross-section through ODP Sites 1177 and 1173 near the subduction front of the Nankai Trough. The plane of the section is oriented parallel to the strike of the margin and shows pinch-out of Miocene to Pliocene turbidites against the paleo-seafloor and basement relief of the Kinan seamounts.

Common hydrous minerals (e.g., smectite and opal) also increase fluid production during diagenesis. Fluids produced from dehydration reactions become increasingly important as the burial temperature increases and the pore volume is reduced by compaction. To complicate matters further, the thermal structure of subducting lithosphere varies with crustal age, basement relief, and patterns of hydrothermal circulation. The zones of maximum fluid production, therefore, should shift spatially in response to variations in two independent input parameters: heat flow and abundance of hydrous minerals. Moreover, pathways of focused fluid migration will track the architecture of high-permeability sand bodies inherited from the depositional

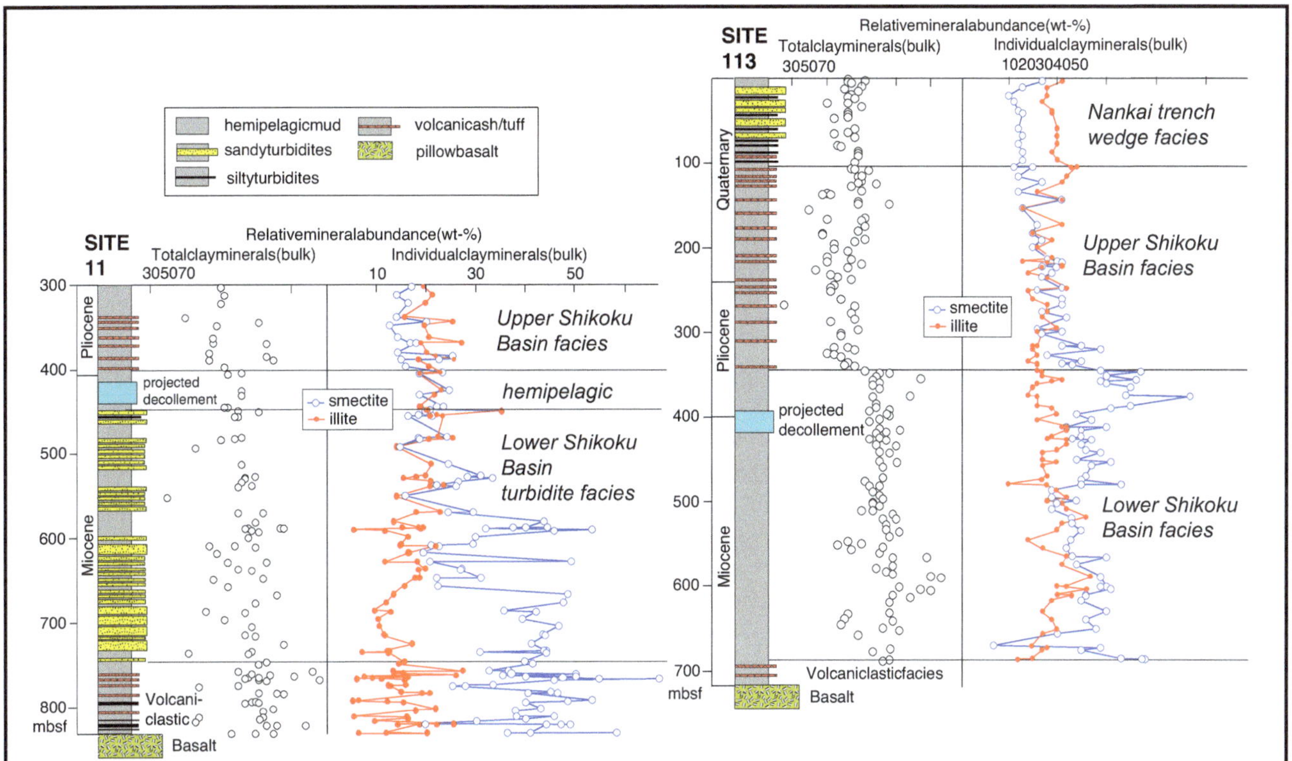

Figure 2. Stratigraphic columns for ODP Sites 1177 and 1173, Nankai Trough, with projected position of the frontal décollement in blue. Also shown are calculated values of total clay minerals (from bulk-powder x-ray diffraction) and estimated weight percentages of smectite and illite in bulk mudstone (from clay-size XRD). See Steurer and Underwood (2003) for XRD data.

environment (i.e., isolated shoe-string channel sands, stacked channel-levee complexes, unconfined sand sheets, depositional lobes). Compartments of excess pore-fluid pressure are, in many depositional systems, set up initially by those sand-body geometries (Bredehoeft et al., 1988; Flemings et al., 2002). Conversely, if subduction inputs consist of low-permeability mud throughout a system, then diffuse drainage will be retarded, and focused flow is more likely to follow structural conduits of fracture-enhanced permeability.

When investigating the change from stable sliding to stick-slip behavior, the "smectite-to-illite hypothesis", in its elementary form, provides scientists with a tractable idea to test (Hyndman et al., 1997; Vrolijk, 1990). But that influential concept is oversimplified because it does not take into account three-dimensional variations in lithostratigraphy, either within or among individual subduction margins. Similarly, the hypothesis fails to account adequately for local and global variability in total clay abundance, the absolute amount of smectite, grain fabric, pore pressure, heat flow, and reaction progress prior to subduction. A more holistic view, stressed by Moore and Saffer (2001), is that several interrelated factors and processes (e.g., precipitation of carbonate and silicate cements, progressive growth of phyllosilicates, clay-mineral dehydration, changes in frictional properties of clay) act collectively to modulate the strength and mechanical behavior of strata on opposite sides of the updip limit of the seismogenic zone. The recipe for blending each factor and process almost certainly changes from one subduction margin to another, within each individual margin, and through time. This natural variability needs to be taken into account when drilling projects target subduction thrusts.

Variability in subduction inputs can lead to substantial differences in how quickly diagenetic reactions progress, both outboard of a subduction front and downdip along the subduction path. This effect is particularly obvious in the Nankai Trough, where heat flow varies with distance from the fossil spreading ridge of the Shikoku Basin (Yamano et al., 2003). Inputs of sediment to the Nankai Trough change in three dimensions because irregular basement highs deflected turbidity currents during early stages of abyssal-floor sedimentation. Sandy turbidites are the norm, rather than the exception, beneath the frontal décollement where smooth basement enters the subduction zone (Fig. 1). The content of smectite also changes along strike in Nankai because of local perturbations of heat flow and pre-subduction clay diagenesis (Fig. 2). By the time strata pass beneath the toe of the warmer Muroto accretionary prism, the smectite-to-illite reaction advances nearly to completion at the sediment-basalt interface (Fig. 2). In contrast, coeval strata within the nearby (cooler) Ashizuri transect remain virtually unaltered. Accurate predictions of where and how much fluid volume might be released through deeper dehydration reactions demand accurate three-dimensional knowledge of the starting materials and thermal structure.

Each subduction margin around the world displays its own idiosyncrasies. The frontal décollement in the Barbados accretionary prism occupies a stratigraphic interval that is

San Andreas Fault Zone Mineralogy, Geochemistry, and Physical Properties from SAFOD Cuttings and Core

by John G. Solum, Stephen Hickman, David A. Lockner, Sheryl Tembe, Jim P. Evans, Sarah D. Draper, D.C. Barton, David L. Kirschner, Judith S. Chester, Frederick M. Chester, Ben A. van der Pluijm, Anja M. Schleicher, Diane E. Moore, Carolyn Morrow, Kelly Bradbury, Wendy M. Calvin, and Teng-fong Wong

Introduction

The San Andreas Fault Observatory at Depth (SAFOD), drilled near the town of Parkfield, California (Fig. 1) (Hickman et al., 2004) as part of the U.S. National Science Foundation's EarthScope Project (see http://www.earthscope.org), provides a continuous set of samples through the active San Andreas Fault (SAF) zone. These samples can help address decades-old questions such as apparent weakness of the SAF (Zoback, 2000) providing the unparalleled opportunity to constrain the parameters that control the behavior of plate bounding faults like the San Andreas.

Samples collected from SAFOD also complement studies of exhumed fault zones. While studies of exhumed fault zones have benefit from regionally extensive exposures that allow for detailed mapping and analyses (Chester and Chester, 1998; Evans and Chester, 1995), such studies also have inherent limitations including uncertainty about stress states and styles of deformation (i.e., seismogenic vs. creeping). Moreover, mineral assemblages in exhumed fault zones can be altered during exhumation, obscuring fault-related mineral assemblages and textures (Solum et al., 2003; Solum and van der Pluijm, 2004).

The SAFOD target zone contains two parallel strands that generate repeating M2 earthquakes in addition to having a component of creep (Nadeau et al., 2004). Therefore, SAFOD also offers the opportunity to obtain samples from seismogenic and aseismic faults and to constrain the rheology of the active SAF system.

Figure 1. Location map of SAFOD. SAF denotes active surface trace of the San Andreas Fault.

Samples Collected

Three types of samples were collected during drilling of the SAFOD main hole: rotary drill cuttings, spot cores, and percussive sidewall cores. A complete list of samples is available at: http://safod.icdp-online.org/.

The vast majority of samples collected to date are cuttings (>99% of the length of the hole). Two groups of rotary cuttings were collected at 3-m intervals over the entire length of the hole. One set was composed of unwashed cuttings (cuttings mixed with drilling mud collected as they come off the shale shaker), while the other was composed of washed cuttings, which are equivalent to unwashed cuttings that were gently rinsed on-site to remove drilling mud.

Three spot cores were collected following casing runs—one in a hornblende-biotite granodiorite at 1462–1470 m measured depth (MD), one in an arkose/siltstone at the bottom of the Phase I hole at 3056-3067 m MD, and one at the bottom of the Phase II hole at 3990–3998 m MD. Fifty-two percussive side wall cores were collected from 3081–3953 m MD preceding casing of the Phase 2 hole in August 2005. Most of these samples were in siltstone/shale; however, two were in arkose.

Each type of sample has potential problems that must be addressed individually. For cuttings, these include loss of formation clays during on-site washing; difficulty resolving thin features like thin beds or faults (since cuttings represent an average over a 3-m-long interval); mixing as cuttings travel from the drill bit to the surface; contamination of cuttings derived at the drill bit by borehole collapses and abrades further up the hole; and contamination from materials from the drill string (e.g., steel shavings, paint) or from drilling additives (e.g., nut plug). Problems associated with side wall cores include disruption of natural textures and inducement of fractures during the collection process; contamination of the core from drilling mud (as the tool shoots through the borehole wall into the formation); and the potential that the core will sample the cuttings bed and not the formation. In addition, side wall cores can only be collected from regions of the hole that are relatively in gauge, which can eliminate the sampling weak areas such as fault zones, that are prone to washouts. Side wall cores also provide only a 2.5-cm snapshot of lithology. Drilling mud

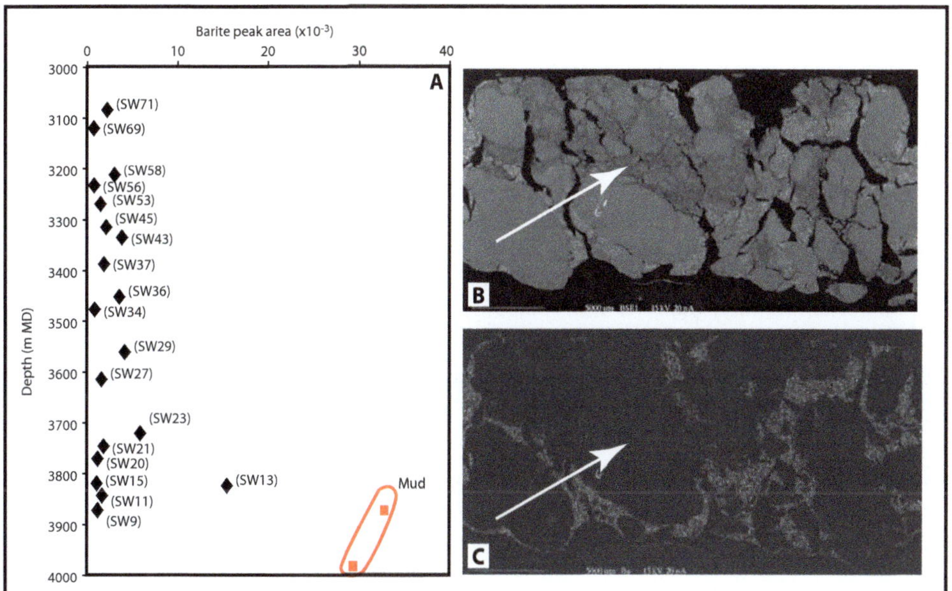

Figure 2. [A]. Relative barite concentrations in side wall cores from XRD patterns. Drilling mud samples are shown in red. [B] BSE image of SW61 (10461' MD). [C] Barium scans of SW61. The breccia with the bright (barite-rich) matrix is drilling-related, while the breccia indicated by the white arrow is formation-related.

contamination of side wall cores can be constrained using XRD and SEM (Fig. 2). SEM characterizations of the side wall cores allow textures that are drilling-induced to be distinguished from natural textures. Spot cores present fewer analytical difficulties than rotary cuttings or side wall cores; however, spot cores sampled less than 0.5% of the length of the SAFOD main hole. It is therefore vital to place spot and side wall cores into a broader geologic context, and this must be done using analyses of rotary cuttings (Fig. 3).

Sample Mineralogies

The mineral assemblages of samples were characterized using x-ray diffraction (XRD) (Solum et al., submitted), optical microscopy (Almeida et al., 2005; Barton et al., 2005; Draper et al., 2005; Evans et al., 2005), infrared reflectance (IR) (Calvin and Solum, 2005), scanning electron microscopy (SEM) (Hickman et al., 2005) and transmission electron microscopy (TEM) on select spot core samples (Schleicher et al., 2006). A summary of cuttings and core mineralogy is shown in Fig. 2.

Changes in mineralogy inferred from cuttings strongly correlate with geophysical logs (Solum et al., submitted), indicating that cuttings analyses reveal real lithologic boundaries and faults. These changes define mineralogical zones that are likely fault-bounded.

After penetrating the sedimentary cover, the hole entered Salinian granitic rocks. At 1926 m MD the hole crossed the Buzzard Canyon Fault (Rymer et al., 2003) and entered a series of arkosic sediments composed of an upper and a lower unit separated by a large fault at ~2560 m MD (Solum et al., submitted; Draper et al., in prep.), and containing a potentially active member of the SAF system at 3067 m MD. The hole entered a series of mudrocks (Great Valley sequence) at 3158 m MD, which are separated from a mineralogically distinct series of Great Valley mudrocks by an actively creeping member of the SAF system that is causing casing deformation (Zoback et al., 2005). This formation extends to the bottom of the hole at 3998 m MD.

At least five major faults were crossed during drilling (Figs. 2 and 4), including an actively creeping member of the SAF system that is causing casing deformation at a depth of 3295–3313 m MD (Zoback et al., 2005). The creep occurs within a low-velocity zone interpreted as the damage zone of the active SAF (Zoback et al., 2005). Between 3322 m and 3353 m MD, this zone contains up to ~2 wt% serpentine, preliminarily identified as lizardite with minor chrysotile, based on crystal morphologies, (Solum et al., submitted); however, given the limitations inherent in the use of cuttings, the relationship of this serpentine to faulting cannot be definitively established until Phase III drilling in the summer of 2007. Two major faults, one at 2545–2560 m MD and one at 3067 m MD, are clay-rich (>50 wt%) containing a neoformed mixed-layer clay (Solum et al., submitted; Schleicher et al., 2006). The Buzzard Canyon Fault at 1926–1235 m MD is rich in quartz and feldspars (>80 wt%) and contains relatively little clay (<3 wt%). These preliminary results highlight the

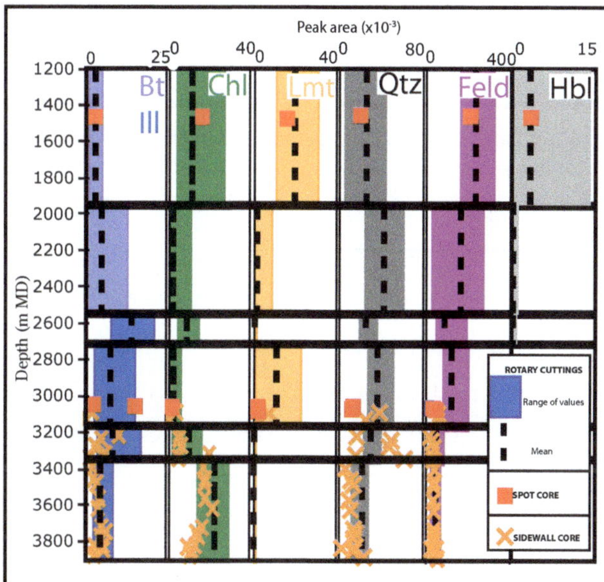

Figure 3. Major mineralogical zones of the SAFOD main hole as defined by XRD analyses of rotary cuttings. The mineralogies of spot and side wall cores are also shown.

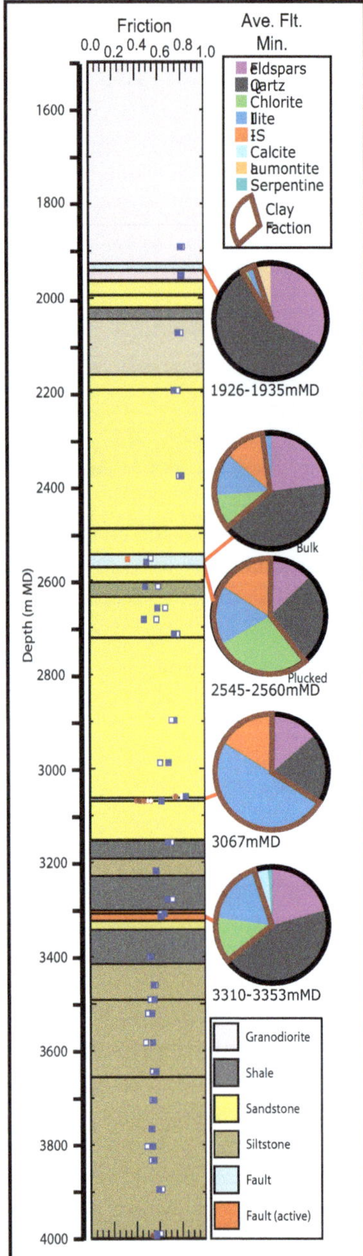

Figure 4. Coefficients of friction for samples from the SAFOD hole (Tembe et al., submitted). Average fault rock mineral assemblages are shown at right (Solum et al., in review). Lithologies are from Boness and Zoback (2006).

compositional and rheological heterogeneity of the faults of the SAF system at SAFOD.

Alteration phases such as clays, zeolites, and some feldspars may record alteration by fluids that were confined by sealing faults. In addition, Kirschner et al. (2005) found that the stable isotopic composition of carbonates plucked from cuttings at 3100–3950 m MD fell into two partially overlapping populations. One occurred more frequently in fault zones, while the other occurred more frequently in protolith. This suggests that in the region of SAFOD faults are tapping a fluid reservoir that is not seen in the protolith.

Constraining the frictional properties of the SAF is a key objective of SAFOD. Tembe et al. (2005) showed that cuttings preserve meaningful information about the frictional properties of the rocks penetrated during drilling, with faults generally weaker than adjacent protolith (μ=0.40–0.55 vs. μ=0.55–0.80). Plucked grains of clay-rich fault rocks from 2556 m MD have μ=0.33, while bulk samples from that depth have $\mu \approx 0.45$. This indicates that analyses of bulk samples provide an upper bound on the coefficient of friction, and that the properties of fault rocks can be better constrained by plucking grains of fault rocks from bulk cuttings. Analyses of plucked fault rocks from the section of the hole that contains the actively creeping fault are ongoing.

Conclusions

The ongoing studies briefly discussed here will help to increase the understanding of the behavior of the active SAF. Further, comprehensive studies of all of the faults penetrated during drilling of SAFOD provide the opportunity to constrain the rheological, structural, and geochemical evolution of this section of the SAF. Analyses of cuttings and Phase I and II cores define the broad geologic framework of the SAFOD area and will help guide the selection of targets for Phase III drilling in the summer of 2007.

References

Almeida, R., Chester, J.S., Chester, F.M., Kirschner, D.L., Waller, T.D., and Moore, D.E., 2005. Mesoscale structure and lithology of the SAFOD Phase 1 and II Core Samples. *Eos Trans. AGU*, 86(52), Fall Meet. Suppl., Abstract T21A-0454.

Barton, D., Bradbury, K.K., Solum, J.G., and Evans, J., 2005. Structural and lithologic characterization of the SAFOD pilot hole and phase one main hole. *Eos Trans. AGU*, 86(52), Fall Meet. Suppl., Abstract T21A-0451.

Boness, N.L. and Zoback, M.D., 2006. A multiscale study of the mechanisms controlling shear velocity anisotropy in the San Andreas Fault Observatory at Depth. *Geophysics*, 71:131–146, doi:10.1190/1.2231107.

Calvin, W.M. and Solum, J.G., 2005. Drill hole logging with infrared spectroscopy. *Geothermal Resources Council Trans.*, 29:565–568.

Chester, F.M. and Chester, J.S., 1998. Ultracataclasite structure and friction processes of the Punchbowl fault, San Andreas system, California. *Tectonophysics*, 295:199–221, doi:10.1016/S0040-1951(98)00121-8.

Draper, S.A., Boness, N.L., and Evans, J.P., 2005. Source and significance of the sedimentary rocks in the SAFOD borehole: Preliminary analysis, *Eos Trans. AGU*, 86(52), Fall Meet. Suppl., Abstract T24B-02.

Evans, J., Moore, D., Kirschner, D., and Solum, J.G., 2005. Lithologic characterization of the deep portion of the SAFOD drill hole. *Eos Trans. AGU*, 86(52), Fall Meet. Suppl., Abstract T21A-0450.

Evans, J.P. and Chester, F.M., 1995. Fluid-rock interaction in faults of the San Andreas system; inferences from San Gabriel Fault rock geochemistry and microstructures. *J. Geophys. Res.*, 100:13007–13020, doi:10.1029/94JB02625.

Hickman, S., Solum, J., Moore, D., Chester, J., Guillemette, R., Draper, S., Tembe, S., Boness, N., and Summers, R., 2005. Photographs and scans of SAFOD phase 2 sidewall sores, SAFOD phase 2 sample party, Menlo Park, Calif., U.S.A., http://geoweb.tamu.edu/Faculty/chesterj/safod/products/Phase_2_Sidewall_Cores.ppt

Hickman, S., Zoback, M., and Ellsworth, W., 2004. Introduction to special section: Preparing for the San Andreas Fault Observatory at Depth. *Geophys. Res. Lett.*, 31:L12S01, doi:10.1029/2004GL020688.

44

Electron Microscopy of Clay Minerals in Mudrocks from the San Andreas Fault Observatory at Depth (SAFOD)

by Anja M. Schleicher, Ben A. van der Pluijm, Laurence N. Warr, and John G. Solum

Introduction

The San Andreas Fault Observatory at Depth (SAFOD) drill site situated near Parkfield, California (Hickman et al., 2004) offers the opportunity to assess the interplay between clay formation, faulting, and fluid migration in an active fault zone through direct sampling from depth and comparison with exhumed fault strands (Chester et al., 2004; Evans and Chester, 1995; Solum and van der Pluijm, 2004). Recent electron microscopy and x-ray diffraction (XRD) studies of rock fragments and cuttings from this drill site show abundant clay minerals in the host rock, on fracture surfaces, and within mineralized veins (Schleicher et al., 2006; Solum et al., 2006). It is, therefore, of importance to determine the timing of clay formation in relation to pre-, syn-, or post-faulting activities, and to establish their influence on the seismic behavior of the San Andreas Fault (Wu, 1975). Within this context, the formation of clays at depth is a potential weakening mechanism of crustal-scale faults (Warr and Cox, 2001; Wintsch et al., 1995; Zoback, 2000).

In this study of mineral transformations, the scanning electron microscope (SEM) and transmission electron microscope (TEM) are used to provide new insights into the nano-scale characteristics of clay minerals, and to evaluate the mineralogical characteristics of very fine-grained particles in fault rocks from the SAFOD drill hole. Using high-resolution TEM, the following aspects are being investigated: i) the microstructural characteristics of the clay mineral phases down to the nanometer scale, ii) the clay polytype structures by selected-area diffraction analysis (SAED), iii) the microchemistry of clay particles using analytical electron microscopy (AEM) and elemental geochemistry (ICP), and iv) the hydration state of smectite using wet-cell XRD and TEM. Based on these investigations, different events of clay formation can be assessed in order to characterize diagenetic or hydrothermal fluid-driven growth processes.

Sampling and Methods

Fine-grained, clay-rich rock fragments up to 1 cm in average length were collected from a spot-core of a clay-rich shear zone at 3067 m measured depth (MD, upper star in Fig. 1). This part of the borehole represents a potentially active section of the San Andreas fault zone, although it is currently not considered to contain the main fault trace (Zoback et al., 2005). The samples show distinct polishing and striations on particle surfaces and on fracture surfaces. Some of these carefully washed rock chips were ultrasonically treated to remove surface particles and to study their nature.

Other rock fragments with similarly polished surfaces were found at 3436 m MD from inside a core catcher after a failed coring attempt (lower star in Fig. 1). In an attempt to preserve the mineralogical structure, natural hydration state, and textural properties, specimens were bottled and embedded in resin directly on site. The travel-time of rock-chips from the bottom of the hole to the surface is about 7.5 hours. The fractured core chips were cleaned of drilling mud using dry paper towels and placed into small sample bottles containing 50% methanol and 50% LR White resin. In the laboratory, the solvent was progressively removed by using increasing concentrations of LR White resin following the procedure of Kim et al. (1995). Polymerization was achieved by transferring to casting molds, and the bubbles were removed in a vacuum desiccator. Finally, the samples were hardened in a vacuum oven at 60°C for 24 h. Backup samples were stored in humidity-controlled vessels to avoid drying and any contact with liquid water.

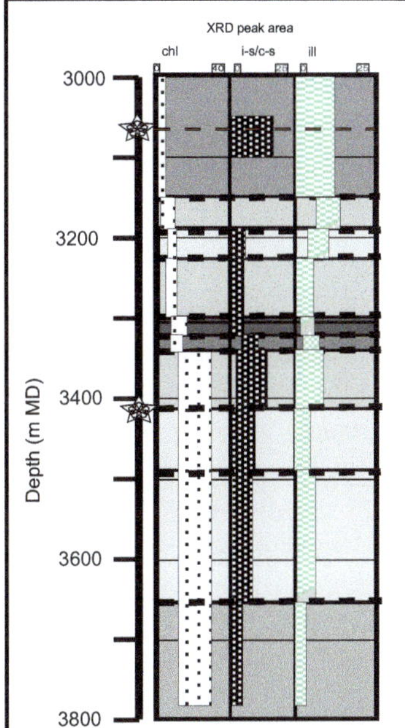

Figure 1. Mineralogic characteristics of the SAFOD main hole obtained from XRD powder patterns. Main minerals are chlorite (chl), illite (ill), mixed-layer illite-smectite and chlorite-smectite (i-s/c-s). Major boundaries are shown by dashed lines and bands correspond to lithologic units (Solum et al., 2006).

SEM of thin sections prepared from the impregnated rock fragments allowed pre-selection of locations of interest that were subsequently investigated by TEM. Small Cu-washers were glued on these areas and thinned using an ion mill apparatus. All

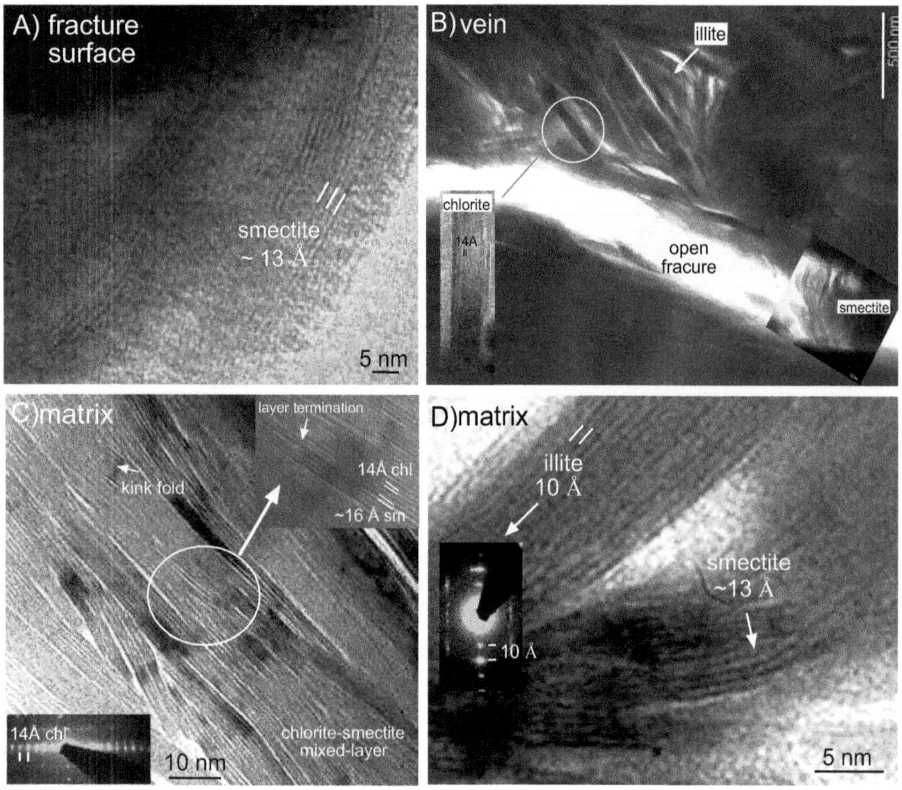

Figure 2. Low and high resolution TEM images of [A] smectite ultrasonically separated from fracture surfaces, [B] smectitic vein fillings that are surrounded by packets of chlorite and illite crystals, [C] rock-matrix samples with interstratifications of chlorite and smectite (note microstructures such as kink folds and layer terminations), and [D] packets of illite growth with illite-smectite mixed-layering and smectite.

electron microscopy work was conducted in the EMAL at the University of Michigan, using the methods described in Warr and Nieto (1998). Microscopy was undertaken using a Philips CM12 scanning transmission electron microscope at an operating voltage of 120 kV and a beam current of 20 μA. High-resolution imaging was conducted at magnifications between 50,000× and 350,000×. Chemical analysis was performed in scanning mode by energy-dispersive spectrometry analysis over areas of 100 nm^2 using a Kevex Quantum solid-state detector.

Results and Ongoing Work

At least three different structural types of smectite-occurence could be distinguished in the SAFOD mud-rock fragments: i) fracture surfaces with thin films and slicken-fibers (Fig. 2A, Schleicher et al., 2006), ii) vein mineralization in the rock matrix (Fig. 2B), and iii) the matrix containing chlorite, smectite, chlorite-smectite, and illite-smectite mixed-layerings (Figs. 2C and 2D). The thin films that cover the fracture surfaces at 3067 m and 3436 m MD reveal smectitic aluminosilicate minerals with a variable cation content of Na, K, Ca, Mg, and Fe. These compositions differ from the smectite minerals of the fresh drilling mud. A smectite phase with a composition similar to the fracture coatings was detected in the mineralized veins at 3436 m MD. Here, chlorite and illite crystals occur as mixed-layerings together with a discrete smectite phase covering the margins of the veins.

In contrast, the matrix shows small areas with discrete smectite phases; however, most smectites occur as mixed-layers of illite-smectite and chlorite-smectite. They usually occur together with large packets of illite and chlorite. SEM investigations of these minerals reveal large irregular detrital grains with curved and damaged particle shapes. They are surrounded by a compacted fabric (presumably bedding parallel), perpendicular to the polished fractures and containing notably higher concentrations of K, Mg, and Fe. Illite, chlorite, and interstratifications of smectite with chlorite and illite have also been recognized by Solum et al. (2006) in the whole-rock XRD patterns. Based on first analysis of illite diffraction patterns (SAED), ordered and disordered 1M polytypes can be distinguished, whereas a higher-temperature $2M_1$ illite polytype has not yet been recognized.

Preliminary observations of samples impregnated on site also indicate the coexistence of authigenic illite-smectite and discrete smectite particles that were not disturbed during embedding and polymerization. However, the characterization of microstructures and comparison with non-impregnated material are not yet fully determined and require more detailed study.

Discussion

Based on the ongoing electron microscopy study of SAFOD drill rocks, illite, chlorite, and smectite in combination with diverse mixed-layered minerals occur in the argillaceous lithologies. However, the TEM analysis in particular shows that the minerals were formed in texturally different microstructural sites, suggesting different formation processes.

Among the key observations is a natural smectite phase on the fracture surfaces that occurs as oriented platy and fibrous minerals, forming thin film coatings with slicken-fibers (Fig. 2A). This clay phase is interpreted as an authigenic phase formed by dissolution-precipitation reactions during the movement of aqueous fluids along permeable fractures and veins at temperatures possibly up to ~130°C (Moore and Reynolds, 1997; Schleicher et al., 2006; Williams et al., 2005). The fibrous nature suggests fault creep as the environment for their formation, but the timing of initial fracture formation remains uncertain, so that the clay-films could also represent growth along reactivated faults structures.

The large packets of illite and chlorite in the matrix show curved and damaged particle shapes with kink bands and layer terminations (Fig. 2C) that occur together with smectitic mixed-layering. The ordered and disordered 1M illite polytypes, based on SAED patterns, reflect the likely transformation of illite-smectite to smectite in the diagenetic compaction processes. Such events are well known, for example, in argillite lithologies of the Salton Sea geothermal system (Yau et al., 1987) or the Gulf Coast (Freed and Peacor, 1992).

Further investigations will be necessary to detail specific mineral reactions and swelling behavior of clays at the crystal lattice scale. Particularly important is to determine the formation mechanisms (e.g., layer-by-layer replacement vs. dissolution and neocrystallization) and fabrics of the clays, and to establish the timing of reactions in relation to the deformation history of the fault rocks.

Acknowledgements

Our SAFOD research is supported by the EarthScope Program of the U.S. National Science Foundation (NSF) and by the Deutsche Forschungsgemeinschaft (DFG).

References

Chester, F.M., Chester, J.S., Kirschner, D.L., Schulz, S.E., and Evans, J.P., 2004. Structure of large-displacement, strike-slip fault zones in the brittle continental crust. *In* Karner, G.D., Taylor, B., Driscoll N., Kohlstedt D. (Eds.), *Rheolgoy and Deformation in the Lithosphere at Continental Margins.* New York (Columbia University Press), 223–260.

Evans, J.P. and Chester, F.M., 1995. Fluid-rock interaction in faults of the San Andreas system; inferences from San Gabriel Fault rock geochemistry and microstructures. *J. Geophys. Res.*, 100:13007–13020, doi:10.1029/94JB02625.

Freed, R.L. and Peacor, D.R., 1992. Diagenesis and formation of authigenic illite-rich I/S crystals in Gulf Coast shales: TEM study of clay separates. *J. Sediment. Petrol.*, 62:220–234.

Hickman, S., Zoback, M., and Ellsworth, W., 2004. Introduction to special section: Preparing for the San Andreas Fault Observatory at Depth. *Geophys. Res. Lett.*, 31:L12S01, doi: 10.1029/2004GL020688.

Kim, J.W., Peacor, D.R., Tessier, D., and Elsass, F., 1995. A technique for maintaining texture and permanent expansion of smectite interlayers for TEM observations. *Clays and Clay Minerals*, 43:51–57, doi:10.1346/CCMN.1995.0430106.

Moore, D.W. and Reynolds, R.C. 1997. *X-ray Diffraction and the Identification and Analysis of Clay Minerals, 2nd Edition.* Oxford,U.K. (Oxford University Press).

Schleicher, A.M., van der Pluijm, B.A., Solum, J.G., and Warr, L.N., 2006. The origin and significance of clay minerals on surfaces, in fractures and in veins from SAFOD borehole samples (Parkfield, California). *Geophys. Res. Lett.*, 33: L16313, doi:10.1029/2006GL026505.

Solum, J.G. and van der Pluijm, B.A., 2004. Phyllosilicate mineral assemblages of the SAFOD Pilot Hole and comparison with an exhumed segment of the San Andreas Fault System. *Geophys. Res. Lett.*, 31:L15S19, doi:10.1029/2004GL01990.

Solum, J.G., Hickman, S.H., Lockner, D.A., Moore, D.E., van der Pluijm, B.A., Schleicher, A.M., and Evans, J.P., 2006. Mineralogical characterization of protolith and fault rocks from the SAFOD main hole. *Geophys. Res. Lett.*, 33:L21314, doi:10.1029/2006GL027285.

Warr, L.N. and Cox, S., 2001. Clay mineral transformations and weakening mechanisms along the Alpine Fault, New Zealand. *Geolog. Soc. Lond. Spec. Publ.*, 186:85–101.

Warr, L.N. and Nieto, F., 1998. Crystallite thickness and defect density of phyllosilicates in low-temperature metamorphic pelites: a TEM and XRD study of clay-minerals crystallinity-index standards. *Canad. Min.*, 36:1453–1474.

Williams, C.F., D'Alessio, M.A., Grubb, F.V., and Galanis, S.P., 2005. Heat flow in the SAFOD main hole. *Eos Trans. AGU, 86* (52), Fall Meet. Suppl., Abstract T23E-07.

Wintsch, R.P., Christoffersen, R., and Kronenberg, A., 1995. Fluid-rock reaction weakening of fault zones. *J. Geophys. Res.*, 100(B7):13021–13032.

Wu, F.T., Blatter, L., and Robertson, H., 1975. Clay gouges in the San Andreas Fault system and their possible implications. *Pure Appl. Geophys.*, 113:87–96, doi:10.1007/BF01592901.

Yau, Y.-C., Peacor, D.R., and McDowell, D., 1987. Smectite-to-illite reactions in Salton Sea shales; a transmission and analytical electron microscopy study. *J. Sediment. Res.*, 57:335–342.

Zoback, M.D., 2000. Strength of the San Andreas fault. *Nature*, 405:31–32.

Zoback, M.D., Hickman, S., and Ellsworth, W., 2005. Overview of SAFOD Phases 1 and 2: Drilling, sampling and measurements in the San Andreas Fault zone at seismogenic depth. *Eos Trans. AGU*, 86(52), Fall Meet. Suppl., Abstract T23E-01.

Authors

Anja M. Schleicher, Universität Würzburg, Geologisches Institut, Pleicherwall 1, 97070 Würzburg, Germany, e-mail: aschleic@umich.edu.

Ben A. van der Pluijm, University of Michigan, Department of Geological Sciences, 1100 University Avenue, C.C.Little Building, Ann Arbor, Mich. 48109, U.S.A.

Laurence N. Warr, Centre de Géochimie de la Surface (CNRS-ULP), 1 rue Blessig, 67084 Strasbourg, France.

John G. Solum, U.S. Geological Survey, Earthquake Hazards Team, 345 Middlefield Road, MS 977 Menlo Park, Calif. 94025, U.S.A.

45

Real-time Mud Gas Monitoring: A Technique to Obtain Information on the Composition and Distribution of Gases at Depth While Drilling

by Thomas Wiersberg and Jörg Erzinger

Several techniques have been applied to obtain information on the composition of gases and fluids in drill holes, such as downhole fluid sampling, gas and fluid analysis during pumping tests, or *in vacuo* degassing of drill core samples. All these techniques can yield valuable information on the geochemistry of gases, but they are less useful for obtaining a continuous dataset on the distribution and composition of gases at depth. To reveal such data during drilling, we have developed a technique where gas is extracted from circulating drill mud with an agitator and continuously piped into a nearby laboratory trailer for on-line analysis. In addition, we routinely determine electrical conductivity, pH, and temperature continuously during the drilling operation within the mud pit at 10-min intervals.

Hydrocarbons, helium, radon, and carbon dioxide are the most suitable gases for the real-time detection of fluid-bearing horizons such as shear zone, open fractures, sections of enhanced permeability, and occurrences of permafrost methane hydrate. Gas analysis is carried out with a quadrupole mass spectrometer for N_2, O_2, Ar, CO_2, CH_4, He, H_2, and H_2S; a gas chromatograph (GC) equipped with a flame ionization detector (FID) for hydrocarbons (CH_4, C_2H_6, C_3H_8, n-C_4H_{10}, and i-C_4H_{10}), and a custom-built Lucas cell alpha-detector for ^{222}Rn activity. When significant amounts of non-atmospheric gases are detected, off-line gas samples are automatically collected from the gas line for further investigations on isotopes (e.g. noble gases, carbon, and hydrogen) to reveal information on the origin and evolution of deep-seated fluids. The technique has been applied successfully on several International Continental Scientific Drilling Program (ICDP) projects, such as SAFOD (San Andreas Fault Observatory at Depth), Mallik, and German Continental Deep Drilling Program KTB (Erzinger et al., 2004; Wiersberg et al., 2005; Zimmer and Erzinger, 1995).

As an example, we present results from the SAFOD Main Hole, drilled in 2004–2005 within the framework of ICDP. SAFOD wants to achieve a better understanding on the processes at active plate-bounding fault systems at the location of their origin. Knowledge about the role and origin of fluids and gases associated with the San Andreas Fault (SAF) zone is relatively poor. Besides the open question on how fluids are linked with fault zone processes in general, outstanding topics include the spatial distribution of fluids at depth, and, in particular, the contribution of mantle-derived fluids to the total fluid inventory of the SAF.

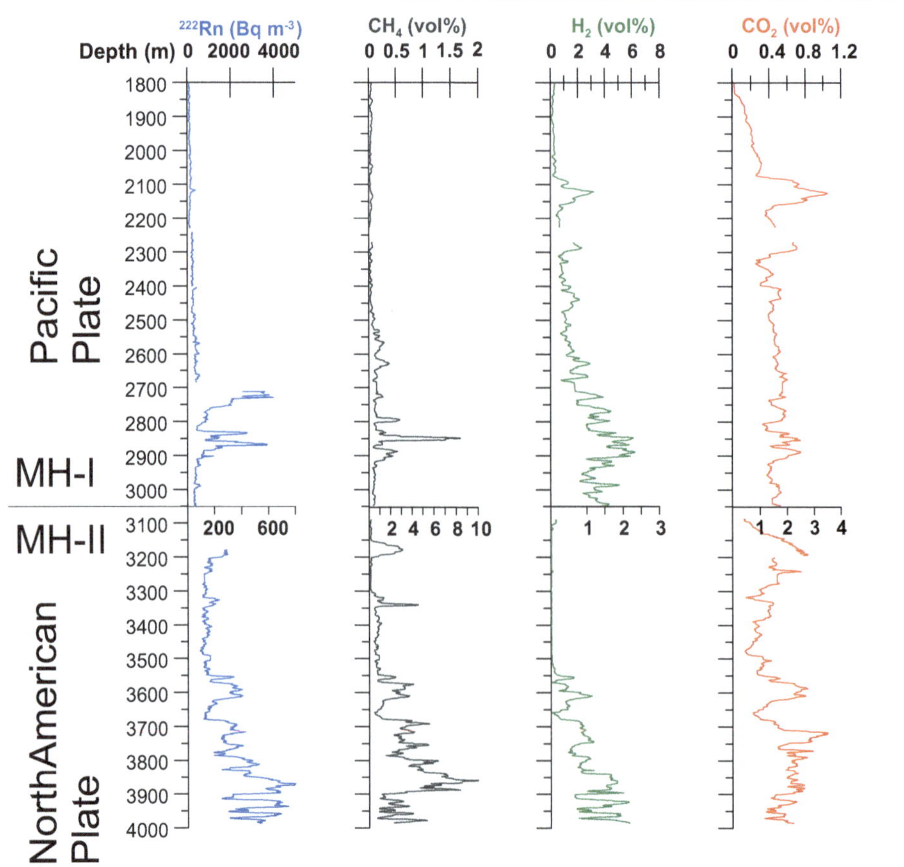

Figure 1. Mud gas profiles from the SAFOD-MH phase 1 (down to 3051 m) and phase 2 (3987 m) for the most abundant non-atmospheric gases.

The pilot hole (PH) and the main hole (MH) were both drilled at a short distance from each other. In contrast to the vertical PH, which penetrates 768 m of Quarternary and Tertiary sediments into Cretaceous granite down to 2168 m final depth, the MH was deviated towards the SAF and re-enters sedimentary strata below 1930 m, where it remains down to the bottom of the hole (3987 m). Geophysical and geological observations identify the SAF in a depth interval of ~3100–3450 m.

In the lower, sedimentary part of the SAFOD-MH, drilled in 2004 (1900–3051 m) and 2005 (3051–3980 m), the most abundant non-atmospheric gases in drill mud were hydrogen (up to 6 vol %), methane (up to 10 vol %), and carbon dioxide (up to 4 vol %). The concentration of helium remained low (>20 ppmv), whereas radon activity reached values of up to 5000 Bq m^{-3}. Two major sections could be identified where these gases were enriched in mud gas: from 2700–2900 m and from >3550 m to the bottom of the hole at 3987 m (Fig.1). A positive correlation between the numbers of beddings and fractures in these sections, identified by geophysical logging, and the radon activity implies that fluids enter the hole through open fissures. Furthermore, the radon activity indicates that the fluids are actively circulating. In both sections, radon also correlates positively with H_2, CO_2, and (in part) CH_4; however, the relative proportions of these gases are different in both sections. The upper and lower sections are enriched in ^{222}Rn and H_2, and in CO_2 and CH_4, respectively.

Based on their distinct chemical compositions, we conclude that both sections represent individual hydrologic systems. Evidence for mixing is only slight. The gas concentrations found in the drill mud between 2900 m and 3550 m depths are low, especially for ^{222}Rn. Only two sections, at 3150–3200 m and at 3340 m, show some higher gas concentrations. The spike at 3340 m, very close to the active part of SAF (3310 m), is almost exclusively composed of hydrocarbons. We assume that these are trapped in permeable sands embedded in less permeable strata, an interpretation supported by lithological data.

The air-corrected ^{3}He/^{4}He ratios of off-line mud gas samples are lower than the atmospheric ratio, but higher than the mean crustal values of ≤0.1 Ra; therefore, helium is a mixture of crustal helium with a small portion of mantle-derived helium. Down to a depth of 3051 m, air-corrected ^{3}He/^{4}He ratios of >0.45 Ra show that the contribution of mantle-derived helium is small (approx. 5%) on the Pacific plate, which composes the southwestern portion of the SAF. The ^{3}He/^{4}He ratios become significantly higher at depths below 3500 m on the North American plate, where ^{3}He/^{4}He ratios of 0.9 R_a were found, revealing a mantle contribution of 10%–12% for the total helium budget (Wiersberg and Erzinger, 2007).

Based on our observations, we conclude that the role of mantle-derived fluids in seismic processes at the SAF is small. The distinct helium isotopic compositions down to 3051 m and below 3500 m depths, with only little evidence for mixing between both hydrologic systems, again demonstrate that the SAF in some way acts as a barrier for fluid migration. Mantle-derived fluids migrate predominately through permeable country rock and leak to some extent into the SAF.

References

Erzinger, J., Wiersberg, T., and Dahms, E., 2004. Real-time mud gas logging during drilling of the SAFOD pilot hole in Parkfield, CA. *Geophys. Res. Lett.*, 31:L15S18, doi:10.1029/2003GL019395.

Wiersberg, T. and Erzinger, J., 2007. A helium isotope cross-section through the San Andreas Fault at seismogenic depth. *Geochem. Geophys. Geosyst.*, 8:Q01002, doi:10.1029/2006GC001388.

Wiersberg, T., Erzinger, J., Zimmer, M., Schicks, J., and Dahms, E., 2005. Real-time gas analysis at the Mallik 2002 gas hydrate production research well. *In* Dallimore, S.R. and Collett, T.S. (Eds.), *Scientific Results from Mallik 2002 Gas Hydrate Production Research Well Program, Mackenzie Delta, Northwest Territories, Canada.* Geological Survey of Canada, Bulletin 585.

Zimmer, M., and Erzinger, J., 1995. On the geochemistry of gases in formation and drilling fluids—results from the KTB. *Sci. Drill.*, 5:101–110.

Authors

Thomas Wiersberg and Jörg Erzinger, GeoForschungsZentrum Potsdam, Section 4.2, Telegrafenberg B323, 14473, Potsdam, Germany, e-mail: wiers@gfz-potsdam.de.

46

Low-Velocity Damage Zone on the San Andreas Fault at Depth near SAFOD Site at Parkfield Delineated by Fault-Zone Trapped Waves

by Yong-Gang Li, Peter E. Malin, and John E. Vidal

Introduction

We deployed a dense linear array of forty-five seismometers across and along the San Andreas fault (SAF) near the SAFOD site at Parkfield in 2003 to record fault-zone trapped waves generated by near-surface explosions and microearthquakes located within the fault zone. Observations and 3-D finite-difference simulations of the fault-zone trapped waves generated by microearthquakes at different depths show a ~100–200-m-wide low-velocity zone (LVZ) along the SAF reaching to depths of 6–7 km or more, within which shear velocities are reduced by 20%–40% on average from wall-rock velocities, with the maximum velocity reduction of 40%–50% in a 30–40-m-wide fault core zone, indicating the localization of severe damage on the SAF. The damage zone is not symmetric but extends farther on the southwest side of the main fault trace. The width and velocities of this zone delineated by trapped waves are confirmed by the results of the SAFOD drilling and logs that show high porosity and multiple slip planes in a ~200-m-wide LVZ across the SAF with velocity reduction of ~20%–30% at ~3.2 km depth, within which a ~35-m-wide severely damaged zone is seen around the main slip (Hickman et al., 2005). Recently, the downhole seismograph installed in the SAFOD mainhole at ~3 km depth recorded prominent fault-zone guided waves with long-duration dispersive wavetrains after *S*-arrivals from microearthquakes occurring below (Ellsworth and Malin, 2006). Finite-difference simulations of these waves show a ~30–40-m-wide core zone with velocities as low as 50% of the intact rocks embedded in the wider fault zone with intermediate velocity reduction, indicating that the damage zone on the SAF at Parkfield extends to the depth of at least several kilometers below 3 km.

The Fault Zone Observatory

The spatial extent of fault weakness, and the loss and recouping of strength across the earthquake cycle are critical ingredients in our understanding of fault mechanics. In order to relate present day crustal stresses and fault motions to the geological structures formed in their past earthquake histories, such as the matured San Andreas Fault (SAF), we must understand the evolution of fault systems on many spatial and time scales. Extensive field and laboratory research and numerical simulations indicate that fault zones undergo high, fluctuating stress and pervasive cracking during an earthquake (Chester et al., 1993). While we know slip is localized on faults because of their lower strength compared to the surrounding bedrock, the magnitude of strength reduction and its spatial extent at seismogenic depth are still not well constrained (Hickman and Evans, 1992).

Near Parkfield, California, many researchers have observed a LVZ surrounding the surface trace the SAF (Roecker et al., 2004; Thurber et al., 2003; Unsworth et al., 1997). This zone is from a few hundred meters to 1 km wide, and has velocity reductions of 10%–30% and *Vp/Vs* ratios on the order of 2.3. The LVZ is thought to be caused by intense fracturing during earthquakes, brecciation, liquid-saturation, and possibly high pore-fluid pressures near the fault core. Recent results from drilling logs at the Parkfield San Andreas Fault Observatory at Depth (SAFOD) show a LVZ more than one hundred meters wide at ~3 km depth, indicating a severely damage zone associated with the SAF (Hickman et al., 2005). Pore fluids arising from depth appear to hold a complex relationship with this damage zone, with its outer portions appearing to be more permeable than its core (Lockner et al., 2000).

Highly damaged fault rocks along the SAF at Parkfield create a LVZ that can trap seismic waves (Ben-Zion, 1998; Li et al., 1990). Using fault-zone trapped waves generated by explosions and microearthquakes, and recorded at surface seismic arrays at Parkfield, Li et al. (1997, 2004) have delineated a ~150–200-m-wide low-velocity waveguide along the SAF, appearing to extend to seismogenic depths beneath

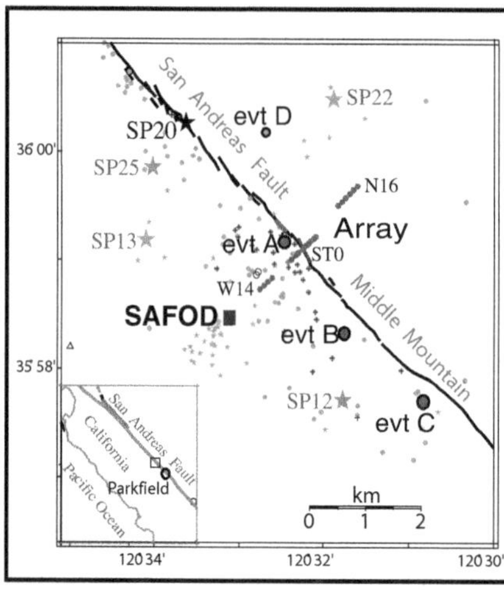

Figure 1. Map shows locations of linear seismic arrays of forty-five PASSCAL RT130s (circles along the line) deployed across and along the San Andreas fault near the SAFOD site (filled square), five explosions (stars) detonated by USGS, small shots (stars), and microearthquakes (dots) recorded at our array in the fall of 2003. Station STO in the array was located on the main fault trace. Dots denote thirty-seven earthquakes at different depths showing fault-zone trapped waves used for measurements shown in Fig. 3. Event A is a M2.2 SAFOD target event occurring at ~3 km depth on Julian date R293 in 2003. Events A, C, and D are deeper earthquakes; waveforms from them are shown in Fig. 2.

Middle Mountain, within which the shear-velocities are reduced by 20%–50%. This distinct zone is interpreted as being a damage zone along the SAF that accumulated in the recurrence of ruptures in major earthquakes over geological time.

Recently, prominent fault-zone guided waves have been recorded in the SAFOD mainhole at ~3 km depth, suggesting that a SAF-related low-velocity waveguide extends to deeper levels (Ellsworth and Malin, 2006). In this article, we discuss the fault-zone trapped waves recorded at the linear seismic array deployed across the SAF near the SAFOD site in 2003 and at the seismometers installed in the SAFOD mainhole. The data from local microearthquakes occurring at different depths provide better constraints on the depth extent of the damage structure of the SAF. These results help extend the direct measurements of fault-zone properties in the SAFOD mainhole to a 3-D structural image of the SAF zone.

Data and Results

Coordinated by SAFOD PIs, in the fall of 2003 we deployed a dense linear seismic array of forty-five PASSCAL RT130 seismometers across and along the SAF near the SAFOD site ~15 km NW of Parkfield to record fault-zone trapped (guided) waves for site characterization before drilling (Fig. 1). We recorded ~100 local earthquakes occurring at depths between 2 km and 12 km and five explosions in the fan-geometry detonated by the USGS researchers. The fault-zone trapped waves (FZTWs) generated by microearthquakes and explosions are used to delineate the rock damage extent and physical properties along the SAF at seismogenic depths.

Figure 2 exhibits seismograms recorded at cross-fault array for three microearthquakes near the SAFOD site, showing prominent FZTWs with large amplitudes and long wavetrains after *S*-arrivals at stations close to the SAF main trace in a width range of ~150–200 m for events A and C occurring within the fault zone. In contrast, trapped waves are not clear at the same stations for event D occurring ~1 km away from the fault zone, and stations located out of the fault zone register brief body waves for these three events, indicating the existence of a LVZ to form a waveguide along the SAF. It is seen that the LVZ extends farther on the southwest side of the main fault trace on which station ST0 was located. We noted some seismic energy trapped within a branch fault (BF) which may connect to the main fault at depth. The FZTWs partitioned from the main fault to this branch fault have been recorded at the borehole seismograph installed in the SAFOD mainhole (Malin et al., 2006). The FZTWs from event C occurring at ~11 km depth show much longer wavetrains than those from the shallower event A at the 3 km depth,

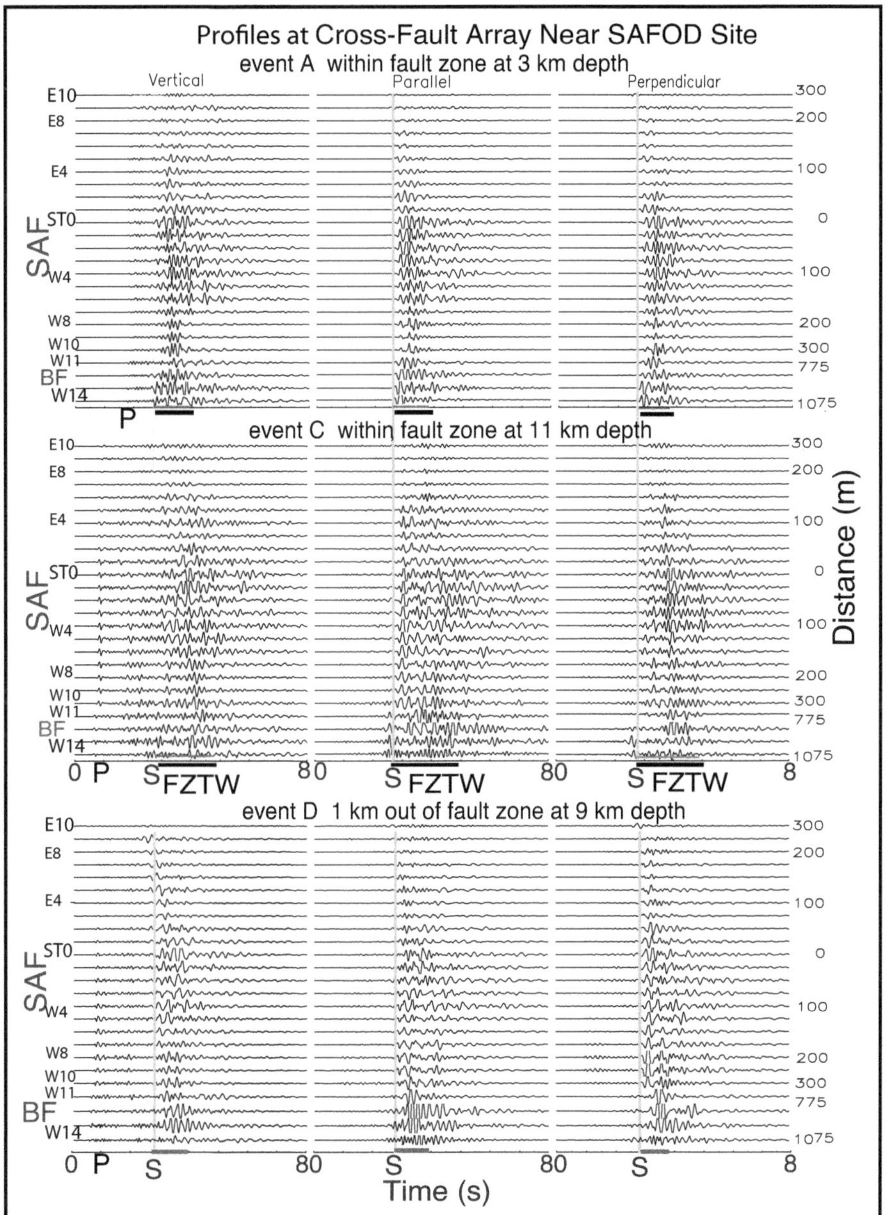

Figure 2. Three-component seismograms at the cross-fault array for three microearthquakes occurring near the SAFOD site. Seismograms have been <6-Hz filtered and plotted in a fixed amplitude scale in each plot. Prominent fault-zone trapped waves (FZTWs) with large amplitudes and long wavetrains after S-arrivals from the on-fault events A and C arrive at stations between E6 and W4 in a ~200-m range across the SAF. Vertical lines are aligned with S arrivals. Horizontal bars denote wavetrain lengths of FZTWs after S-arrivals. The FZTW wavetrains from event C occurring at ~11 km is ~2.2 s long, but is ~1.2 s for event A at ~3 km. The wavetrain for the off-fault event D occurring at the similar depth of event C is much shorter, indicating that the low-velocity waveguide on the SAF likely extends down to the depth at least 3 km. Some seismic energy is trapped within the branch fault (BF).

indicating that the low-velocity waveguide on the SAF likely extends to seismogenic depths deeper than 3 km. In contrast, the same stations registered much shorter wavetrains after *S*-arrivals for the off-fault event D at ~9 km depth compared to those from the on-fault event C, although these two events occurred at a similar depth and distance from the array. In order to examine the depth extension of the LVZ on the SAF, we used the data from thirty-seven local earthquakes located within the fault zone at different depths with the raypath incidence angles to the array smaller than 30° from vertical (Fig. 3A). Figure 3B shows seismograms and envelopes at station ST0 located on the main fault trace for eleven on-fault events at different depths near the SAFOD site. *S*-arrivals for these events are aligned at 2 s. The length of fault-zone trapped wavetrains following *S*-arrivals progressively increases from ~1.2 s to ~2.2 s as the event depths increase from 2.6 km to 11.7 km. In contrast, much shorter wavetrains after *S*-arrivals with flat changes in length are registered at the same station for eleven other events located away from the fault zone in a similar depth range. Figure 3C shows the measured wavetrain lengths of FZTWs registered at stations within the fault zone for thirty-seven on-fault events and thirteen off-fault events at depths between 2 km and 12 km. The lengths of FZTWs for on-fault events increase from 1.0 s to 2.2 s as the depth increases from ~2 km to ~12 km, but shorter wavetrains with flat depth-dependent changes are measured at the same stations for off-fault events. Stations located out of the fault zone registered much short wavetrains after *S*-arrivals for all these events. These observations indicate that the low-velocity waveguide formed by the damage zone on the SAF extends across seismogenic depths to at least ~7 km, although the velocity reduction (damage magnitude) within the zone becomes smaller with depth due to the larger confined stress at greater depths.

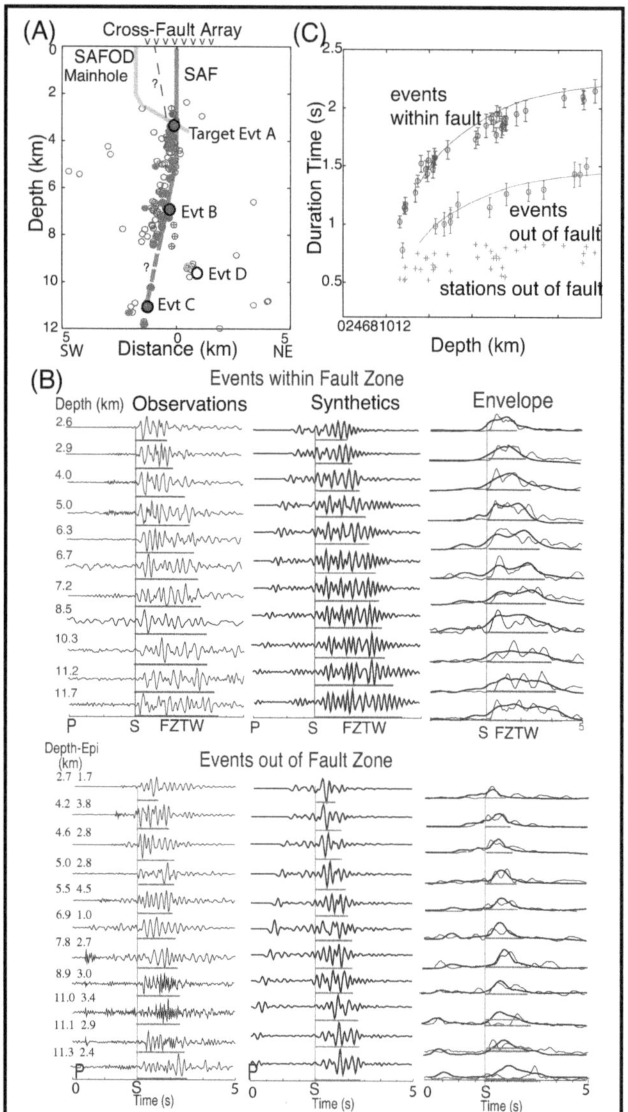

Figure 3. [A] Vertical section across the SAF near the SAFOD mainhole (gray line) shows locations of microearthquakes recorded at our seismic array in 2003. Solid (open) circles denote events showing fault-zone trapped waves (FZTWs) with long (short) wavetrains. The deep part of the SAF (thick line) dips toward SW at a steep angle. Waveforms of events A, C and D are shown in Fig. 2. The branch fault (thin dashed lines) may connect to the main fault at depth. [B] Top: Vertical-component seismograms and envelopes at station ST0 on the SAF main trace for eleven on-fault earthquakes at different depths show an increase in wavetrain length (solid horizontal bar) of FZTWs as event depths increase. S-arrivals for these events are aligned at the same time. Bottom: The same plot but for eleven off-fault events in a similar range of depths shows much shorter FZTW wavetrains and flat changes with event depths. The finite-difference synthetic seismograms and envelopes at station ST0 for these events using the model in Fig. 5A are in general agreement with observations. Event epicenters are 2 km from station ST0 in computation. [C] The measured FZTW wavetrain lengths versus focal depths for on-fault (black circles) and off-fault (gray circles) events were recorded at stations within and out of the fault zone. Each data point is averaged from measurements at four stations for the event. Error bars are standard deviations. Curves are polynomial fits to the data..

In the fall of 2005, a borehole seismograph was placed down the SAFOD mainhole at a depth of ~3 km, where the SAFOD drilling and logging records suggest the active SAF is present (Hickman et al., 2005). This downhole seismograph recorded prominent fault-zone guided waves from the deeper events (Ellsworth and Malin, 2006). For example, Figure 4 shows FZTWs with large amplitudes and long dispersive wavetrains generated by an aftershock of the M6 Parkfield earthquake at ~6 km depth within the fault zone, indicating the existence of a continuous low-velocity fault core zone between this event and the borehole seismometer. Thus, the damage zone on the SAF must extend at least several kilometers below the 3 km depth.

Based on our observations of fault-zone trapped waves at the surface array and borehole seismometer in the SAFOD mainhole, we constructed a velocity and *Q* model across the SAF near the SAFOD site (Fig. 5A). The wall-rock velocities are constrained by tomography profiles at Parkfield (Roecker et al., 2004; Thurber et al., 2003). Using a 3-D FD code (Graves, 1996), we simulated FZTWs generated by explosions to determine the shallow 1–2 km fault zone structure, and then simulated trapped waves from earthquakes at different depths and epicentral distances to obtain a model of the SAF with depth-variable structure at seismogenic depths. The width and velocities of the damage zone at 3 km depth in this model are constrained by the logging measurements in the SAFOD mainhole (Fig. 5B) (Hickman et al., 2005). For examples, Fig. 4B shows 3-D finite-difference synthetic seismograms for the SAFOD drilling target event occurring ~3 km near the array site using the model in Fig. 5A. Figure 4C exhibits synthetic seismograms at the SAFOD mainhole seismograph located at 3 km depth within the fault core zone for a deep event at 6 km with the epicenter 3 km SW of the site using the same model. We obtained a good fit of synthetics to observations, suggesting that this model is applicable for interpretation of the damage structure on the SAF at Parkfield to at least 6 km. Using the fault-zone model in Fig. 5A, we also simulated seismograms at station ST0 on the main fault trace for eleven on-fault earthquakes and eleven off-fault earthquakes at depths between 3.6 km and 11.2 km for comparison with observed seismograms at this station (Fig. 3). The lengths of synthetic FZTW wavetrains increase with the event depth for on-fault events, which generally agrees with observations. On the other hand, the synthetic seismograms for off-fault events showed much shorter wavetrains after *S*-arrivals and less change in the wavetrain length with depth. These modeling results further indicate that the low-velocity waveguide (damage zone) on the SAF at Parkfield likely extends across the seismogenic depth of at least 6–7 km.

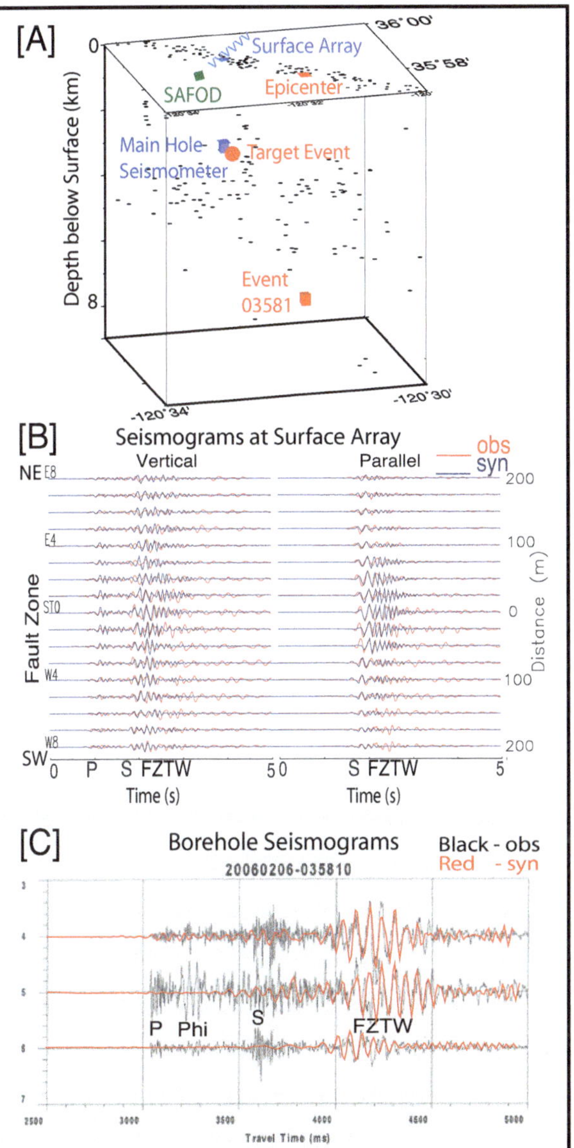

Figure 4. [A] 3-D view of locations of aftershocks (black dots) of the 2004 M6 Parkfield earthquake recorded at the SAFOD mainhole seismograph (blue square). The red square denotes an example event (03581) occurring at ~6 km depth within the fault zone with its epicenter ~3 km SE of the surface array (blue Vs) shows prominent FZTWs at the borehole seismograph (see Fig. 4C). The red dot denotes the SAFOD drilling target event. [B] Observed and 3-D finite-difference synthetic seismograms at the surface array across the SAF for the SAFOD drilling target event at ~3 km depth using the velocity model in Fig. 5A. A double-couple source is located within the fault core zone. Seismograms have been <8-Hz filtered and are plotted using a fixed amplitude scale for all traces. [C] Three-component seismograms recorded at the seismometer with the 15-Hz sensor in the SAFOD mainhole at ~3 km for a Parkfield aftershock at ~6 km (see Fig. 4A) contain direct P- and S-waves followed by lower frequency FZTWs after ~0.5 s and again after ~1 s. The large dispersed phase between the P and S waves is labeled by "Phi", referring to its likely origin in the general leaky mode type wave motion associated with special low velocity waveguides (Ellsworth and Malin, 2006). 3-D finite-difference synthetic seismograms using the velocity model in Fig. 5A match the FZTWs recorded at the borehole seismometer.

Conclusion

Observations and modeling of fault-zone trapped waves recorded at dense linear seismic arrays across the SAF at the surface and at the downhole seismograph installed in the SAFOD mainhole show the existence of a distinct low-velocity damage zone along the SAF at Parkfield, within which seismic velocities are reduced by 30%–50% from wall-rock velocities. The damage zone likely extends to the depth of at least 6–7 km. This zone is thought to be caused by intense fracturing during earthquakes, brecciation, liquid-saturation, and possibly high pore-fluid pressure nears the fault. The damage zone is asymmetric, broader on the southwest side of the main fault trace. The asymmetry may imply that the fault zone has a significant cumulative damage due to previous large earthquakes on the SAF. When a fault ruptures, it may preferentially damage the already weakened rocks in the zone, even though those rocks are not symmetrically distributed on either side of the main slip plane (Chester et al., 1993). Alternately, greater damage may be inflicted in the extensional quadrant than the compressional quadrant near the propagating crack tip (Andrews, 2005).

Although the structural model for the SAF at Parkfield delineated by FZTWs (Fig. 5A) explains part of the data, it is still a simple one. The true structure in three dimensions may be more complicated. It will be further studied using more data from deep events recorded at surface and downhole seismic arrays to document variations in the extent and magnitude of rock damage along the fault strike and with depth, as well as their relationship with the rupture distribution and stress variations over multiple length and time scales. This study contributes essential information towards further understanding of faulting mechanics and earthquake hazards at matured faults like the San Andreas fault.

Acknowledgments

This study is supported by EarthScope Grant EAR0342277, USGS Grant NEHRP20060160, and the SCEC. Special thanks to S. Hickman, W. Ellsworth, and M. Zoback for their

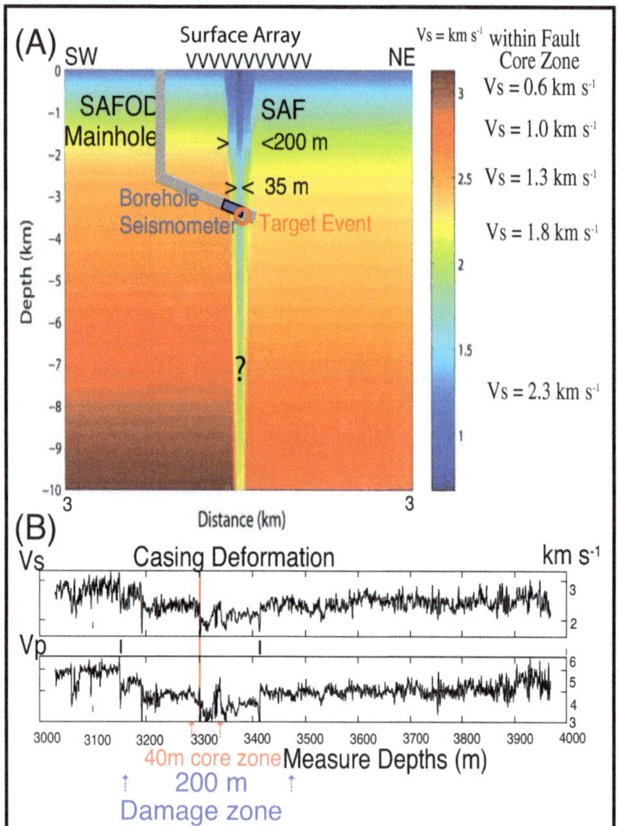

Figure 5. [A] Fault-normal depth section showing the S-wave velocity model across the SAF at the SAFOD site. The velocities within the 100–200-m-wide waveguide on the SAF at seismogenic depths and surrounding rocks were found by 3-D finite-difference fits to the FZTWs generated by explosions and earthquakes at different depths. The model suggests that, on average, a fault zone consists of two vertical layers, a 30–40-m-wide fault core and a surrounding 100–200-m-wide damage zone. S velocities within the damage zone are reduced by 25%–35% from wall-rock velocities. The maximum reduction occurs in the fault core and can be as large as 50%. The red circle denotes the SAFOD drilling target event. The SAFOD mainhole seismogram was installed at ~3 km depth in 2005 (Ellsworth and Malin, 2006). [B] Logging velocities in the SAFOD mainhole show a ~200-m-wide low-velocity damage zone on the SAF at ~3 km depth, in which a ~30–40-m-wide core zone with lowest velocities is embedded. The red line denotes the casing deformation caused in the 2004 M6 Parkfield earthquake (Hickman et al., 2005).

coordination, and E. Cochran, C. Thurber, S. Roecker, M. Rymer, R. Catchings, A. Snyder, L. Powell, B. Nadeau, N. Boness, and D. McPhee for their collaborations in our experiments at Parkfield. The author was supported by the ICDP to attend the ICDP-IODP Fault-Zone Drilling Workshop in Japan in 2006.

References

Andrews, D.J., 2005. Rupture dynamics with energy loss outside the slip zone. *J. Geophys. Res.*, 110:B01307, doi:0.1029.

Ben-Zion, Y., 1998. Properties of seismic fault zone waves and their utility for imaging low-velocity structure. *J. Geophys. Res.*, 103:12567–12585.

Chester, F., Evans, J., and Biegel, R., 1993. Internal structure and weakening mechanisms of the San Andreas fault. *J. Geophys. Res.*, 98:771–786.

Ellsworth, W.L. and Malin, P.E., 2006. A first observation of fault guided PSV-waves at SAFOD and its implications for fault characteristics. *EOS, Trans. AGU*, 87(52), Abstract T23E-02.

Hickman, S.H. and Evans, B., 1992. Growth of grain contacts in halite by solution-transfer: Implications for diagenesis, lithification, and strength recovery. *In* Evans, B., and Wong, T.-F. (Eds.), *Fault Mechanics and Transport Properties of Rocks*, San Diego, Calif., (Academic Press), 253–280.

Hickman, S.H., Zoback, M.D., and Ellsworth, W.L., 2005. Structure and composition of the San Andreas fault zone at Parkfield: Initial results from SAFOD Phase 1 and 2. *EOS, Trans. AGU*, 83(47):237.

Li, Y.G., Ellsworth, W., Thurber, C., Malin, P., and Aki, K., 1997. Observations of fault-zone trapped waves excited by explosions at the San Andreas fault California. *Bull. Seism. Soc. Am.* 87:210–221.

Li, Y.G., Leary, P., Aki, K., and Malin, P., 1990. Seismic trapped modes in Oroville and San Andreas fault zones. *Science*, 249:763–766.

Li, Y.G., Vidale, J.E., and Cochran, S.E., 2004. Low-velocity damaged structure on the San Andreas fault at Parkfield from fault-zone trapped waves. *Geophys. Res. Lett.* 31:L12S06.

Lockner, D., Naka, H., Tanaka, H., Ito, H., and Ikeda, R., 2000. Permeability and strength of core samples from the Nojima fault of the 1995 Kobe earthquake. *USGS Open file, Report 00-129*, 147–152.

Malin, P.M., Shalev, E., Balven, H., and Lewis-Kenedi, C., 2006. Structure of the San Andreas Fault at SAFOD from P-wave tomography and fault-guided wave mapping. *Geophys. Res. Lett.*, 33:L13314, doi: 10.1029/2006GL025973.

Roecker, S., Thurber, C., and D. McPhee. 2004. Joint inversion of gravity and arrival time data from Parkfield: New constraints on structure and hypocenter locations near the SAFOD drill site. *Geophys. Res. Lett.*, 31:L12S04, doi: 10.1029/2003GL019396.

Thurber, C., Roecker, S., Roberts, K., Gold, M., Powell, L., and Rittger, K., 2003. Earthquake location and 3-D fault zone structure along the creeping section of the San Andreas fault near Parkfield, CA: Preparing for SAFOD. *Geophys. Res. Lett.* 30:1112–1115.

Unsworth, M., Malin, P., Egbert, G., and Booker, J., 1997. Internal structure of San Andreas fault at Parkfield, CA. *Geology*, 356–362.

Authors

Yong-Gang Li, Zumberge Hall, Department of Earth Sciences, University of Southern California, 3651 Trousdale Parkway, Los Angeles, Calif. 90089, U.S.A., e-mail: ygli@usc.edu.

Peter E. Malin, Division of Earth and Ocean Sciences, 109A Old Chemistry Box 90227, Duke University, Durham, N.C., 27708, U.S.A.

John E. Vidale, Department of Earth and Space Sciences, University of California, 595 Charles Young Drive East, Box 951567, Los Angeles, Calif., 90095-1567, U.S.A.

47

High-resolution Imaging of Fault Zone Structures with Seismic Fault Zone Waves

by Yehuda Ben-Zion, Zhigang Peng, Michael Lewis, and Jeff McGuire

Large fault zone (FZ) structures with damaged rocks and material discontinuity interfaces can generate several indicative wave propagation signals. High crack density may produce prominent scattering and non-linear effects. A preferred crack orientation can lead to shear wave splitting. A lithology contrast can produce FZ head waves that propagate along the material interface with the velocity and motion polarity of the faster medium. A coherent low velocity layer may generate FZ trapped waves. These signals can be used to obtain high-resolution imaging of the subsurface structure of fault zones, and to track possible temporal evolution of FZ material properties. Several results have emerged from recent systematic analyses of such signals. The trapped waves are generated typically by ~100-m-wide layers that extend only to ~3–4 km depth and are characterized by 30%–50% velocity reduction and strong attenuation (Ben-Zion et al., 2003; Lewis et al., 2005; Peng et al., 2003). The trapping structures appear to be surrounded by broader anisotropic and scattering zones limited primarily to the shallow crust (Fig. 1). Results associated with anisotropy and scattering around the North Anatolian fault using repeating earthquake clusters do not show precursory temporal evolution (Peng and Ben-Zion, 2004, 2005, 2006). The anisotropy results show small co-seismic changes, while the scattering results show larger co-seismic changes and post-seismic logarithmic recovery. The temporal changes probably reflect damage evolution in the top few hundred meters of the crust. Systematic analyses of head waves along several sections of the San Andreas fault reveal material interfaces that extend to the bottom of the seismogenic zone (Ben-Zion and Malin, 1991; McGuire and Ben-Zion, 2005). Joint arrival time inversions of direct and FZ head waves and waveform modeling imply (Fig. 2) velocity contrasts of 20% or more in the top 3 km and lower contrasts of 5%–15% in the deeper section (Ben-Zion et al., 1992; Lewis et al., 2007). In several places, analyses of trapped and head waves indicate that the shallow damaged layers are asymmetric across the fault (Lewis et al., 2005, 2007). The observed damage asymmetry may reflect preferred propagation direction of earthquake ruptures (Ben-Zion and Shi, 2005).

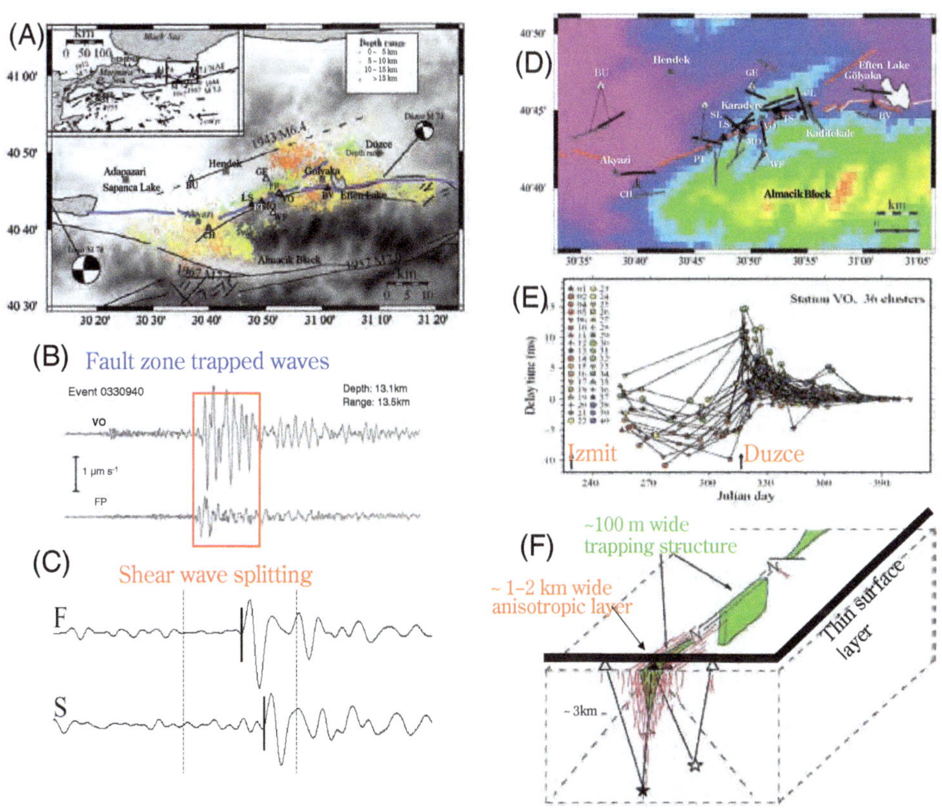

Figure 1. [A] Hypocentral distribution of ~18,000 earthquakes recorded by the PASSCAL seismic experiment along the Karadere-Düzce branch of the north Anatolian fault. [B] An example of FZ waveforms recorded by stations on (VO) and off (FP) the fault. [C] An example of rotated horizontal seismograms showing split shear waves. [D] A summary plot of average splitting parameters (bars) in our study area. The bars are oriented parallel to the average fast direction and scaled by the average delay time. [E] Median delay times for the early S-coda waves plotted against the earthquake occurrence times for the vertical-component seismograms generated by 36 repeating clusters and recorded at station VO. [F] A schematic summary of the fault zone model around the study area. Panels [A] and [B] are from Ben-Zion et al., (2003). Panels [C] and [D] are from Peng and Ben-Zion (2004, 2005). Panel [E] is from Peng and Ben-Zion (2006).

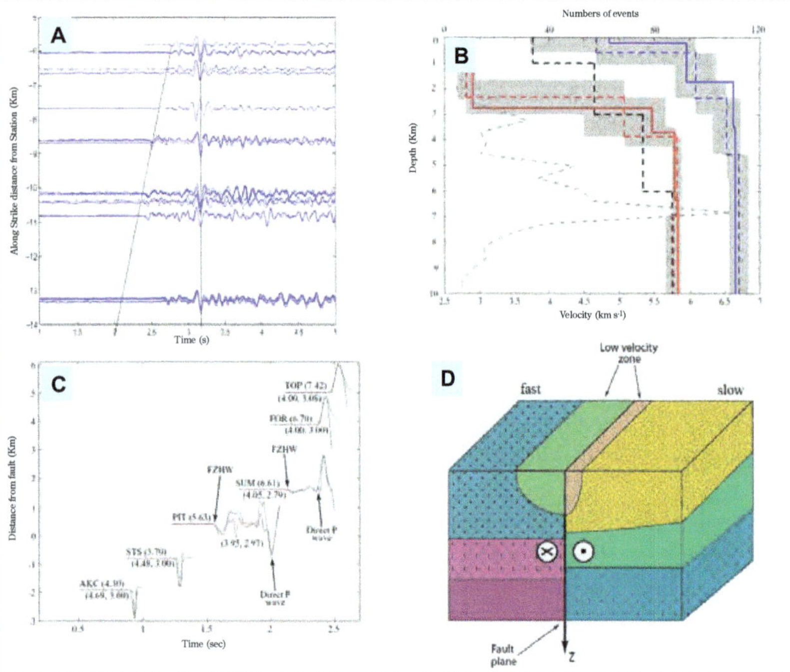

Figure 2. [A] Vertical component P waveforms at Northern California Seismic Network (NCSN) station BHR northeast of the San Andreas Fault (SAF) south of Hollister with direct body wave (vertical line) and first arriving fault zone head wave (diagonal line). [B] Inversion results of P wave velocity contrast across the SAF versus depth from arrival times of head and direct waves. The solid red and blue lines are the best-fitting velocity profiles for the slow and fast sides of the fault, respectively, out of 10 inversion runs. The dashed red and blue lines are the average depth and velocity of each layer, respectively, from the 10 inversions, and the grey shaded areas around the means represent the standard deviation of the depth and velocity of that layer. The dashed black line is the initial velocity and depth model used in the inversion. The dashed gray line (top horizontal axis) gives the number of events as a function of depth. [C] Example of synthetic waveform fits (red) for observed seismograms (blue) for stations on the fast and slow sides of the SAF. Some of the direct P and head wave arrivals are indicated by arrows with corresponding labels. The average P wave velocities in km s^{-1} of the crustal block used to generate the synthetic seismograms are given in parentheses under the waveforms. [D] A schematic diagram of the inferred velocity structure of the SAF south of Hollister, consisting of two layered quarter-spaces, joined along a sharp material interface, and a shallow asymmetric low velocity zone around the fault. Panel [A] is from McGuire and Ben-Zion (2005). Panels [B]-[D] are from Lewis et al., (2007).

Acknowledgments

The discussed studies were supported by the NSF, USGS and SCEC.

References

Ben-Zion, Y. and Malin, P., 1991. San Andreas fault zone head waves near Parkfield, California. *Science*, 251:1592–1594, doi:10.1126/science.251.5001.1592.

Ben-Zion, Y. and Shi, Z., 2005. Dynamic rupture on a material interface with spontaneous generation of plastic strain in the bulk. *Earth Planet. Sci. Lett.*, 236:486–496, doi: 10.1016/j.epsl.2005.03.025.

Ben-Zion, Y., Katz, S., and Leary, P., 1992. Joint inversion of fault zone head waves and direct P arrivals for crustal structure near major faults. *J. Geophys. Res.*, 97:1943–1951.

Ben-Zion, Y., Peng, Z., Okaya, D., Seeber, L., Armbruster, L.G., Ozer, N., Michael, A.J., Baris, S., and Aktar, M., 2003. A shallow fault zone structure illuminated by trapped waves in the Karadere-Duzce branch of the North Anatolian Fault, western Turkey. *Geophys. J. Int.*, 152:699–717, doi:10.1046/j.1365-246X.2003.01870.x.

Lewis, M.A, Ben-Zion, Y., and McGuire, J., 2007. Imaging the deep structure of the San Andreas Fault south of Hollister with joint analysis of fault-zone head and direct *P* arrivals. *Geophys. J. Int.*, 169, 1028–1042, doi:10.111/j.1365-246X.2006.03319.x.

Lewis, M.A, Peng, Z., Ben-Zion, Y., and Vernon, F., 2005. Shallow seismic trapping structure in the San Jacinto fault zone. *Geophys. J. Int.*, 162:867–881, doi:10.1111/j.1365-246X.2005.02684.x.

McGuire, J. and Ben-Zion, Y., 2005. High-resolution imaging of the Bear Valley section of the San Andreas Fault at seismogenic depths with fault-zone head waves and relocated seismicity. *Geophys. J. Int.*, 163:152–164, doi:10.1111/j.1365-246X.2005.02703.x.

Peng, Z. and Ben-Zion, Y., 2004. Systematic analysis of crustal anisotropy along the Karadere-Düzce branch of the north Anatolian fault. *Geophys. J. Int.*, 159:253–274, doi:10.1111/j.1365-246X.2004.02379.x.

Peng, Z. and Ben-Zion, Y., 2005. Spatio-temporal variations of crustal anisotropy from similar events in aftershocks of the 1999 M7.4 İzmit and M7.1 Düzce, Turkey, earthquake sequences. *Geophys. J. Int.*, 160(3):1027–1043, doi:10.1111/j.1365-246X.2005.02569.x.

Peng, Z. and Ben-Zion, Y., 2006. Temporal changes of shallow seismic velocity around the Karadere-Duzce branch of the north Anatolian fault and strong ground motion. *Pure Appl. Geophys.*, 163:567–600, doi:10.1007/s00024-005-0034-6.

Peng, Z., Ben-Zion, Y., Michael, A.J., and Zhu, L., 2003. Quantitative analysis of seismic trapped waves in the rupture zone of the 1992 Landers, California earthquake: Evidence for a shallow trapping structure. *Geophys. J. Int.*, 155:1021–1041, doi:10.1111/j.1365-246X.2003.02109.x.

Authors

Yehuda Ben-Zion, Department of Earth Science, University of Southern California, Los Angeles, 90089, Calif., U.S.A., e-mail: benzion@usc.edu.

Zhigang Peng, School of Earth and Atmospheric Sciences, Georgia Institute of Technology, Atlanta, Ga., 30332-0340, U.S.A.

Michael Lewis, Department of Earth Science, University of Southern California, Los Angeles, 90089, Calif., U.S.A.

Jeff McGuire, Department of Geology and Geophysics, Woods Hole Oceanographic Institution, Woods Hole, Mass., 02543, U.S.A.

Modern and Ancient Out-of-Sequence Thrust in the Nankai Trough: Insight from Laboratory-Derived Properties and Seismic Data

by Takeshi Tsuji, Hidekazu Tokuyama, Gaku Kimura, and Shinya Okamoto

Introduction

At convergent plate margins, characteristics of out-of-sequence thrusts (OOSTs) are important for understanding the nature of earthquake mechanisms and deformation of the accretionary prism. To understand the characteristics of OOSTs, Integrated Ocean Drilling Program (IODP) plans to drill through the seismogenic OOST in the Nankai Trough. In this phase, however, the seismogenic OOST is too deep to penetrate, so its properties have only been estimated from seismic data. Seismic reflection data from the Nankai Trough off the Kii Peninsula in southwest Japan (Figs. 1 and 2) image a strong negative polarity OOST (or splay fault) reflection branching from the major plate boundary fault (Park et al., 2002; Tsuru et al., 2005). This OOST might have ruptured during the 1944 Tonankai earthquake and associated tsunami (Tanioka and Satake, 2001). The negative polarity reflection of the OOST has been interpreted to indicate elevated fluid pressure in the fault zone (Park et al., 2002). From seismic data alone, the reflection polarity is highly useful information for estimating fault zone properties. However, polarity of deep seismic reflections is also affected by acoustic dispersion, and the fracture zone causes wavelet tuning (Costain and Coruh, 2004). Therefore, full characterization of OOST solely using reflection polarity is not possible. To investigate characteristics of a seismogenic OOST, we determined acoustic properties of discrete samples obtained from the fossil OOST outcrop in Nobeoka, and compared them with those of an active Kumano OOST imaged on seismic profile off the Kii peninsula. Because the Nobeoka OOST is interpreted as a fossil OOST, preserves *in situ* structure, and crops out (Kondo et al., 2005), it is ideal for evaluating characteristics of a seismogenic OOST.

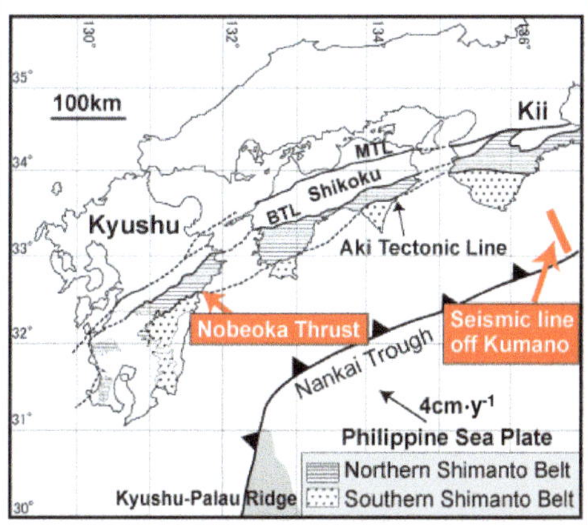

Figure 1. Locations of the Nobeoka OOST and the seismic line off Kumano.

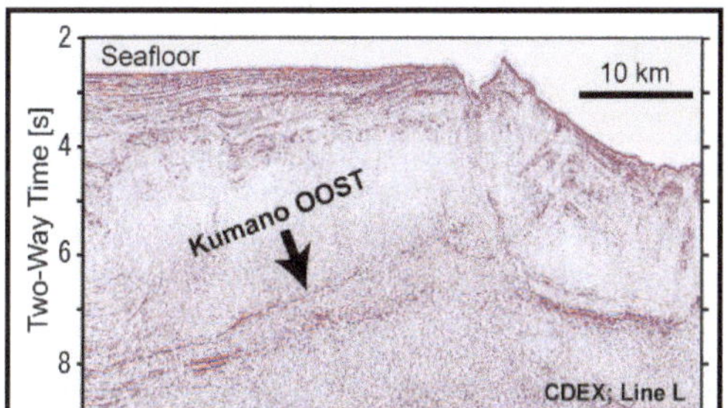

Figure 2. Out-of-sequence thrust on seismic reflection profile off Kumano. This profile crosses the planned IODP site.

Results and Discussions

P-wave and S-wave velocities of discrete (minicore-shaped) samples were measured in dry conditions because in saturated conditions fluid dispersion masks pressure effects (Tsuji et al., 2006). We observed anisotropy of velocity in the hanging wall of the Nobeoka OOST attributed to foliation of politic phyllite (Fig. 3). Foliation-normal P-wave velocities in the hanging wall are ~700 m s^{-1} slower than the foliation-parallel velocities. In contrast, the footwall is composed of brittle, deformed, chaotic shales and fine sandstones, and velocities in the footwall are lower than those in the hanging wall. Furthermore, velocities increase just above the fault core (gray zones in Fig. 3), where velocity anisotropy at atmospheric pressure is weak and velocity increase with pressure is less apparent. These characteristics may be caused by the filling of most cracks with quartz, which is observed just above the fault core in the Nobeoka OOST. The shear-related dewatering and consolidation may also affect elastic moduli just above the fault core.

Comparison of the acoustic properties of the fossil Nobeoka OOST (determined from outcrop samples) and the active Kumano OOST (calculated from seismic reflection data) is warranted by the two thrusts' similar tectonic settings and lithologies. Weathering has affected properties of the Nobeoka OOST, but overall results from the hanging wall and the footwall should reflect trends of *in situ* properties. Furthermore, for the comparison of the fossil Nobeoka and active Kumano OOSTs, we calculated water-saturated P-wave velocities from velocities determined under dry conditions via Gassmann's relation. If we assume an average bulk density of 2.5 g cm^{-3}, a seafloor depth of 2 km, an OOST depth of 7 km, and hydrostatic conditions, effective pressure in the Kumano OOST is calculated to be ~73 MPa. Assuming that velocities at 55 MPa represent those at *in situ* effective pressures, the contrast between foliation-normal P-wave velocities in hanging wall samples and footwall samples is >600 m s^{-1}. The change in P-wave velocities across the fault core of the Nobeoka OOST (<600 m s^{-1}) is larger than that estimated from the reflection amplitude of the Kumano OOST (Tsuru et al., 2005). Therefore, Kumano OOST reflection amplitudes can be explained only by the difference in matrix elastic moduli of the Nobeoka OOST.

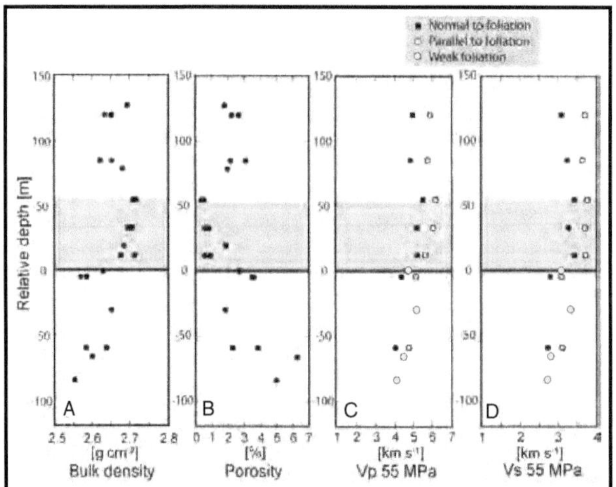

Figure 3. Depth profiles of [A] bulk densities, [B] porosities, [C] P-wave velocities at 55 MPa, and [D] S-wave velocities at 55 MPa (Tsuji et al., 2006). Vertical axis represents depth relative to the location of fault core. Black solid dots, open dots, and gray dots represent acoustic properties normal to the foliation, parallel to the foliation, and weak foliation, respectively. Black lines and gray zones represent locations of the fault core and the high velocity interval, respectively.

To compare Poisson's ratios of the Nobeoka and the active Kumano OOSTs, we modeled amplitude variation with offset (AVO) using the average P-wave and S-wave velocities of Nobeoka samples at 55 MPa and saturated conditions. The angle-dependent reflection coefficient (thick line in Fig. 4), obtained by the Zoeppritz equation compared to that of the active Kumano OOST obtained by a two-ship, long-offset survey (dots in Fig. 4; Tsuru et al., 2005), shows a far offset reflection coefficient lower than that of the active Kumano OOST. If Poisson's ratios in the footwall are higher (σ=~0.25) than those of the Nobeoka OOST, the AVO response of the Nobeoka OOST coincides well with that of the active Kumano OOST. Because Poisson's ratio becomes higher with increasing pore fluid pressure (Dvorkin et al., 1999), high fluid pressure in the footwall likely explains the AVO response of the active Kumano OOST.

Acknowledgements

When we measured velocities of the outcrop discrete samples, F. Kono, T. Saeki, and H. Mochinaga (JOGMEC) provided devices to measure the velocities. Furthermore, we thank J. Ashi (ORI, Univ. Tokyo), G. Moore (CDEX, JAMSTEC), C. Moore (University of California, Santa Cruz), T. Tsuru (Cosmo Oil), and M. Coffin (ORI, University of Tokyo) for valuable discussions. Seismic reflection data (Fig. 2) are courtesy of CDEX, JAMSTEC.

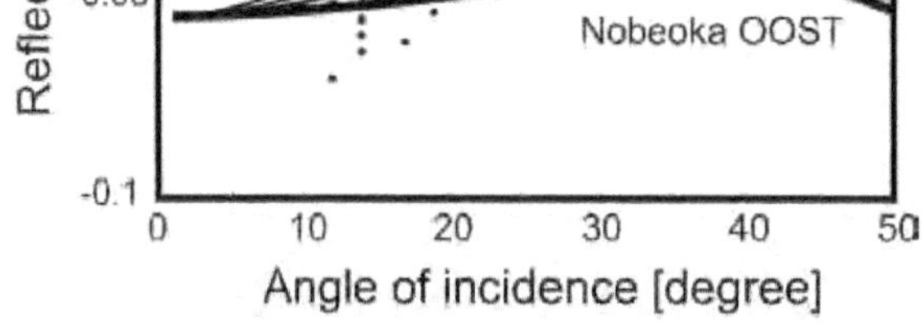

Figure 4. Comparison of angle-dependent reflection coefficients of discrete Nobeoka OOST samples (thick line) (Tsuji et al., 2006); the active Kumano OOST from seismic reflection data (solid dots) (Zone B in Tsuru et al., 2005). Thin lines represent reflection coefficients parameterized by Poisson's ratio. Poisson's ratios in the footwall of the Nobeoka OOST are lower than that of the active Kumano OOST, suggesting high fluid pressures in the footwall of the active Kumano OOST.

References

Costain, J.K. and Çoruh, C., 2004. Basic theory of exploration seismology with Mathematica notebooks and examples on CD-ROM. *Handbook of Geophysical Exploration, Seismic Exploration, Volume 1*, Amsterdam (Elsevier).

Dvorkin, J., Mavko, G., and Nur, A., 1999. Overpressure detection from compressional- and shear-wave data. *Geophys. Res. Lett.*, 26:3417–3420, doi:10.1029/1999GL008382.

Kondo, H., Kimura, G., Masago, H., Ohmori-Ikehara, K., Kitamura, Y., Ikesawa, E., Sakaguchi, A. Yamaguchi, A., and Okamoto, S., 2005. Deformation and fluid flow of a major out-of-sequence thrust located at seismogenic depth in an accretionary complex: Nobeoka Thrust in the Shimanto Belt, Kyusyu, Japan. *Tectonics*, 24: TC6008, doi: 10.1029/2004TC001655.

Park, J.-O., Tsuru, T., Kodaira, S., Cummins, P.R., and Kaneda, Y., 2002. Splay fault branching along the Nankai subduction zone. *Science*, 297:1157–1160, doi:10.1126/science.1074111.

Tanioka, Y. and Satake, K., 2001. Detailed coseismic slip distribution of the 1944 Tonankai earthquake estimated from tsunami waveforms. *Geophys. Res. Lett.*, 28:1075–1078, doi:10.1029/2000GL012284.

Tsuji, T., Kimura, G., Okamoto, S., Kono, F., Mochinaga, H., Saeki, T., and Tokuyama, H., 2006. Modern and ancient seismogenic out-of-sequence thrusts in the Nankai accretionary prism: Comparison of laboratory-derived physical properties and seismic reflection data. *Geophys. Res. Lett.*, 33:L18309, doi:10.1029 /2006GL027025.

Tsuru, T., Miura, S., Park, J.-O., Ito, A., Fujie, G., Kaneda, Y., No, T., Katayama, T., and Kasahara, J., 2005. Variation of physical properties beneath a fault observed by a two-ship seismic survey off southwest Japan. *J. Geophys. Res.*, 110:B05405, doi: 10.1029/2004JB003036.

Authors

Takeshi Tsuji, Ocean Research Institute, The University of Tokyo, 1-15-1 Minamidai, Nakano-ku, Tokyo 164-8639, Japan, e-mail: tsuji@ori.u-tokyo.ac.jp.

Hidekazu Tokuyama, Ocean Research Institute, The University of Tokyo, 1-15-1 Minamidai, Nakano-ku, Tokyo 164-8639, Japan.

Gaku Kimura, **Shinya Okamoto** Department of Earth and Planetary Sciences, The University of Tokyo, Hongou 7-3-1, Bunkyo-ku, Tokyo, 113-0033, Japan.

49

Comparison between Three Out-of-Sequence Thrusts from Japan and Alaska: Implications for Nankai Drilling Targets from the Rock Record

by Christie D. Rowe

Increasing recognition has been given to the importance of thrust faults of the "out-of-sequence" geometry as part of the seismogenic plate boundary at subduction zones. The "Good Friday Earthquake" (Prince William Sound, Alaska, 1964) produced a significant tsunami and substantial uplift of a few islands inboard of the trench. Plafker (1972) suggested this could be explained if the rupture propagated up the subduction plate boundary and reached the seafloor by slip on a higher angle splay fault, or out-of-sequence thrust (OOST). Similar tsunami generation mechanisms have been suggested for the December 2004 Sumatra earthquake and tsunami for the area directly offshore Banda Aceh (Simon Day and Steven Ward, personal communications). A large-scale OOST is a primary drilling target of the planned Nankai riser drilling project (Kimura et al., 2003).

The Nobeoka Thrust in the Shimanto Complex (Kyushu Island, Japan), very well described by Kondo et al. (2005) and Okamoto et al. (2006), is an excellent locality to study OOSTs. The Nobeoka Thrust juxtaposes two belts of the Shimanto Complex-accreted rocks with a substantial thermal gap (~70°C), suggesting total throw is 8.6–14.4 km (Kondo et al., 2005). The hanging wall and footwall are comprised of marine sedimentary rocks of Eocene age. Metamorphism of the hanging wall includes ductile deformation of quartz overprinting earlier brittle structures. The footwall contains only evidence of brittle deformation, further emphasizing the thermal gap. The hanging wall damage zone is thin, crowned by a pseudotachylyte-bearing fault surface only a few meters above the main contact. The fault core is comprised of about 10–20 cm of dense, subtly foliated cataclasite (A. Yamaguchi, personal communication).

The Uganik Thrust of the Kodiak Accretionary Complex (Afognak Island, Alaska) exhibits remarkable similarity to the Nobeoka Thrust. The thermal gap is approximately 30°C–50°C (C. Rowe, J. Hower, and O. Beyssac; unpublished data). The hanging wall is early- to mid-Cretaceous Uyak Complex, comprised of a greenstone-argillite-chert mélange with rare ductile overprinting of brittle structures (Byrne, 1985). The footwall (Kodiak Formation) is black slate and sandstone with a strong characteristic planar flaggy cleavage which is overprinted by tight folding and cleavage-parallel shear fabrics formed during subduction (Sample, 1986). The hanging wall exhibits a highly deformed, stretched, folded damage zone within 1–2 m of the fault. Mechanisms of deformation include both ductile and granular flow. The fault core is extremely narrow, about 0.5 m of intensely veined S/C cataclasites. The footwall contains several densely veined subsidiary shear zones 0.5–2 m wide which have experienced many more episodes of brittle deformation than the surrounding footwall rocks. The concentration of quartz veins is elevated to 10–50 percent above background in these zones. These subsidiary shear zones persist for approximately 85 m below the Uganik Thrust, decreasing in size and frequency with distance from the thrust surface.

The Isthmus Bay Fault is a small offset (tens to hundreds of meters) thrust fault within the Ghost Rocks Formation (a mélange complex) of Kodiak Island, Alaska. Although the hanging wall and footwall of the fault are indistinguishable in the conditions of maximum burial—225°C–250°C; 280–320 MPa (Vrolijk et al., 1988)—the fault-filling veins formed at significantly shallower conditions (140°C; C. Rowe, unpublished data), suggesting that fault activation followed a period of exhumation. The hanging wall massive sandstone is cut by steeply dipping quartz and calcite veins, possibly related to rupture of the thrust. The footwall is comprised of stratally disrupted turbidites, which are cut by the thrust at a high angle to the fabric. The footwall damage zone is identifiable by a dense network of quartz and calcite veins, and occasional small volumes of vein-supported hydraulic breccias. This zone was observed about 15 m structurally below the fault. The fault core is a thick (8–13 cm) multiply cemented breccia; sandstone breccia is cemented by calcite, which was re-brecciated and cemented by quartz. This fault core suggests strong coupling between brittle deformation and substantial fluid flow in faults of the shallow portion of the accretionary prism.

All three OOSTs display extreme asymmetry, with the footwall damage zones two orders of magnitude thicker than the hanging wall damage zones. Evidence of substantial hydrofracture and veining are pervasive in the footwall damage zone and nearly absent from the hanging wall damage zone. This is true when the hanging wall is measurably higher metamorphic grade (Nobeoka Thrust and Uganik Thrust) and when the hanging wall and footwall are of the same grade (Isthmus Bay Thrust). In the case of the deeper, greater offset OOSTs (Nobeoka and Uganik), the fault core itself is dense cataclasite and unlikely to act as a conduit to up-dip fluid flow. However, the shallower OOST (Isthmus Bay) contains a fault core dominated by veins and hydrofracture brecciation. Therefore, it is possible that at shallower structural levels, the fault core itself may be the primary conduit for up-dip fluid advection.

When OOSTs are seismically imaged, a strong negative polarity reflector is usually used to identify the fault. This is most likely caused by the presence of aqueous fluids in or near the fault. In the case of these three ancient OOSTs, the most likely structural reservoir for fluids is the highly fractured and veined footwall damage zone. This imaged area is structurally below the actual fault core and below the measured thermal gap where most offset must necessarily occur. As a drilling target, it is important to search for the core of seismogenic OOST structurally above the imaged fault.

References

Byrne, T., 1985. The Uyak Complex: A brittle-ductile shear zone. *GSA Abstracts with Programs*, 19(6):440.

Kimura, G., Tobin, H., and Group, N.W., 2003. *NanTroSEIZE: The Nankai Trough Seismogenic Zone Experiment Complex Drilling Project*. Socorro, NM (New Mexico Tech), 33.

Kondo, H., Kimura, G., Masago, H., Ohmori-Ikehara, K., Kitamura, Y., Ikesawa, E., Sakaguchi, A., and Okamoto, S., 2005. Deformation and fluid flow of a major out-of-sequence thrust located at seismogenic depth in an accretionary complex: Nobeoka Thrust in the Shimanto Belt, Kyushu, Japan. *Tectonics*, 24(6):TC6008, doi:10.1029/2004TC001655.

Okamoto, S., Kimura, G., Takizawa, S., and Yamaguchi, H., 2006. Earthquake fault rock indicating a coupled lubrication mechanism. *e-Earth*, 1:23–28.

Plafker, G., 1972. Alaskan earthquake of 1964 and the Chilean earthquake of 1960: Implications for arc tectonics. *J. Geophys. Res.*, 77:901–925.

Sample, J.C., 1986. *Structure, Tectonics, and Sedimentology of the Kodiak Formation, Kodiak and Adjacent Islands, Alaska*. University of California - Santa Cruz, Calif. (University of California - Santa Cruz), 156 pp.

Vrolijk, P., Myers, G., and Moore, J.C., 1988. Warm fluid migration along tectonic melanges in the Kodiak accretionary complex, Alaska. *J. Geophys. Res.*, 93:10313–10324.

Author

Christie D. Rowe, Earth & Planetary Science, University of California, Santa Cruz, 1156 High Street, Santa Cruz, Calif. 95064, U.S.A., e-mail: crowe@pmc.ucsc.edu.

Seismology inside the Fault Zone: Applications to Fault-Zone Properties and Rupture Dynamics

by William L. Ellsworth, Peter E. Malin, Kazutoshi Imanishi, Steven W. Roecker, Robert Nadeau, Volker Oye, Clifford H. Thurber, Felix Waldhauser, Namoi L. Boness, Stephen H. Hickman, and Mark D. Zoback

A central goal of seismology is to understand the physics of earthquakes and other sources of seismic waves in the Earth. We would like to understand how dynamic instabilities are nucleated, how they evolve in space and time, and how they come to rest. To achieve this goal, we need observations that are truly broadband with respect to source process time scales. Because the high-frequency limit of a seismogram directly controls the spatial scale at which we can resolve these processes, the requirement for "broadband" means bandwidth that is sufficient to record the shortest pulse produced by the physical system (a delta function being the ultimate broadband signal). Although there is considerable uncertainty at present about the upper frequency limit needed to capture dynamic processes, it is clearly well above the frequency range of standard seismological instrumentation (typically 30–40 Hz for 100 sample-per-second data). Even when instruments are capable of observing frequencies above 1 KHz, they must be sited close enough to the source to overcome the attenuation of the high frequency waves due to scattering and anelastic loss during propagation from the source. The natural solution to this problem is to emplace the instrumentation within the near-field of the source in boreholes and deep mines. This paper presents a review of some recent results from three deep (>2 km) boreholes in California. The three boreholes considered are the Long Valley Exploratory Well, the San Andreas Fault Observatory at Depth (SAFOD) Pilot Hole, and the SAFOD Main Hole.

The Long Valley Exploratory Well (LVEW) is a 3.0-km-deep research drill hole located near the center of Long Valley caldera in eastern California. The well was drilled in a series of stages beginning in 1989, and completed to a total depth of 2996 m in 1998 (Sackett et al., 1999) as part of the International Continental Drilling Program (ICDP) Long Valley Coring Project. Prior to the final stage of drilling, a 3-component 10-Hz seismometer, installed at a depth of 2050 m from September 1997 through May 1998, recorded tens of thousands of local events during the 1997–1998 seismic crisis in Long Valley Caldera (Prejean and Ellsworth, 2001). The value of recording at depth is clearly evident in the comparison of near-source recordings made at the surface and deep underground (Fig. 1).

In the fall of 2002, the well was converted into a deep geophysical observatory which now has a 3-component 4.5-Hz seismometer installed at a depth of 2600 m. Earthquakes as small as M2.5 have been observed at distances as close as 300 m from the seismometer (Fig. 2). Currently, the data sample rate is limited to 500 sps, which limits the highest resolvable frequency to 200 Hz. All data are available through the Northern California Earthquake Data Center (http://www.ncedc.org/). While this sample rate is adequate to detect even such small events, it is not adequate to resolve the event corner frequency. Increasing the sample rate alone, however, will not guarantee that source processes can be measured. The attenuation of seismic waves, even over distances of hundreds of meters, presents a formidable challenge to the modeler. The trade-off between earthquake source parameters and attenuation makes it difficult to separate path effects from source effects. Ide et al. (2003) overcame this problem by employing a spectral ratio method to re-interpret some of the earthquakes analyzed earlier by Prejean and Ellsworth (2001). The reanalysis demonstrated how propagation effects can contaminate source parameters interpreted from individual event recordings, even in the high frequency environment of the deep borehole.

The importance of wave attenuation and scattering even in the high Q environment of a deep borehole was further reinforced by seismic data collected by the 32-level, 1240-m-long array installed in the 2.1-km SAFOD Pilot Hole (Chavarria et al., 2004). This vertical borehole was drilled in 2002 as an ICDP project and part of the pre-Earth Scope site

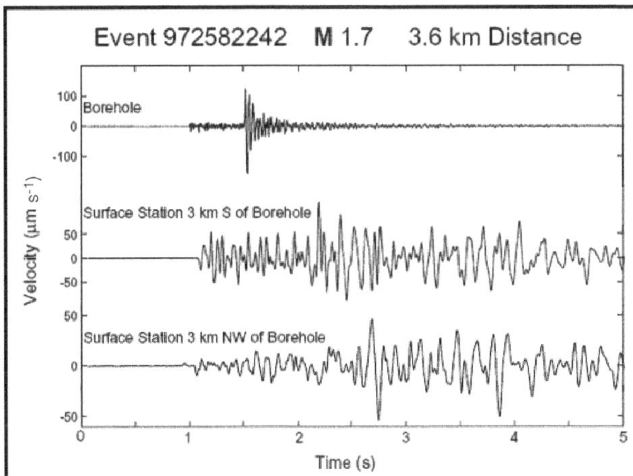

Figure 1. Comparison of vertical component seismograms recorded in LVEW at a depth of 2050 m with vertical records from two conventional surface seismic stations located near the borehole. Note the impulsive nature of the P- and S- wave arrivals, the high frequency content, and relative lack of scattering in the coda in the borehole records with respect to the surface records (from Prejean and Ellsworth, 2001).

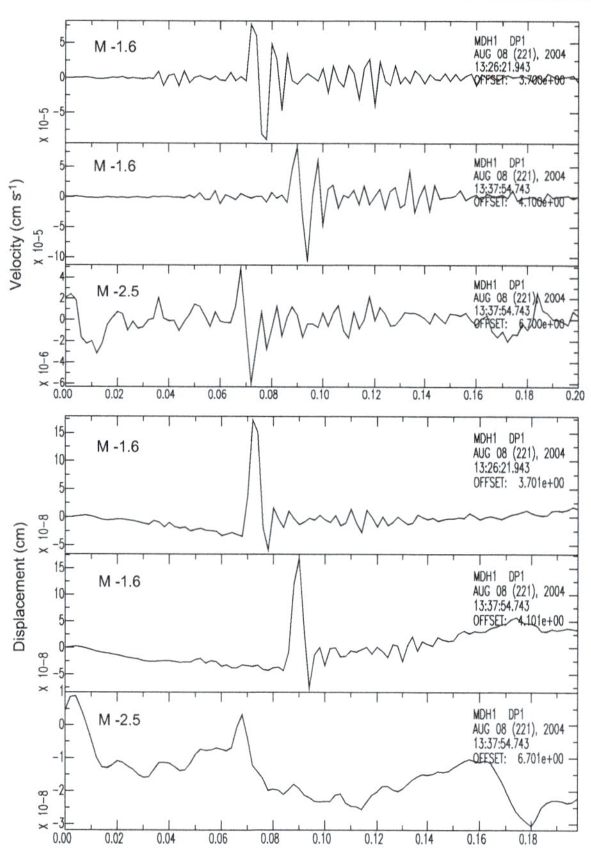

Figure 2. Seismograms of microearthquakes located approximately 300 m from the instruments in the LVEW at a depth of 2600 m. Although the seismogram sampling rate of 500 sps is too low to capture any details of the source process, the seismic moment of the events can still be easily measured from the area under the displacement pulse.

investigation program at SAFOD. Installed in granite between 856 m and 2096 m below ground level, the array recorded at sample rates between 500 sps and 1000 sps from September 2002 to April 2005. All data are available through the Northern California Earthquake Data Center. The effects of propagation on the seismograms are clearly evident in Fig. 3, which shows the seismograms of a microearthquake located about 2.5 km from the array. Note the loss of high frequency content as the wave propagates up the borehole.

Imanishi and Ellsworth (2006) applied a multi-window spectral ratio (MWSR) to seismograms of earthquake multiplets recorded in the SAFOD Pilot Hole. This procedure extended the method of Ide et al. (2003) by stacking multiple time windows of the direct body wave and its early coda to suppress noise and improve the recovery of the spectral ratio (Fig. 4). Results of their study indicated that there is no scale-dependence to static stress drop or apparent stress for earthquakes on the San Andreas Fault between M0.2 and M2.1.

The SAFOD Main Hole was drilled through the San Andreas Fault to a total vertical depth of 3.1 km in 2005. It provides a portal into the inner workings of a major plate boundary fault that makes it possible to conduct spatially extensive and long-duration observations of active tectonic processes within the actively deforming core of the San Andreas Fault.

To meet the scientific and technical challenges of building the observatory, borehole instrumentation systems developed for use in the petroleum industry and by the academic community in other deep research boreholes have been deployed in the SAFOD pilot hole and main hole over the past three years. These systems included 15-Hz omni-directional and 4.5-Hz gimbaled seismometers, micro-electro-mechanical accelerometers, tiltmeters, sigma-delta digitizers, and a fiber optic interferometric strainmeter. A 1200-m-long, 3-component 80-level clamped seismic array was also operated in the main hole for 2 weeks of recording in May of 2005, collecting continuous seismic data at 4000 sps.

Some of the observational highlights included capturing one of the SAFOD target repeating earthquakes in the near-field at a distance of 420 m, with accelerations of up to 200 cm s-1 and a static displacement of a few microns (Fig. 5). Numerous other local events were observed over the summer by the tilt and seismic instruments in the pilot hole, some of which produced

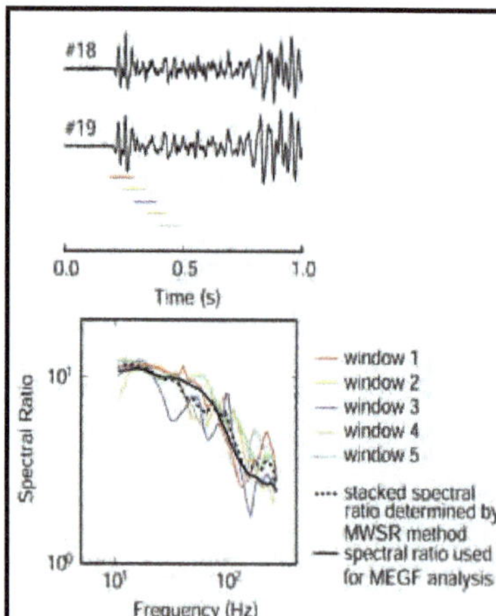

Figure 4. Example of MWSR method for measuring the spectral ratio of two earthquakes (from Imanishi and Ellsworth, 2006). Each colored spectral ratio is derived from the corresponding time window indicated on the seismograms. The averaged spectra in the stack more closely resemble the theoretical spectral ratio for the omega-squared model than the individual spectra. This is because the cancellation of path effects even for earthquakes at nearly the same location is not exact, and multiple windows suppress the uncorrelated noise introduced by multipaths between source and receiver.

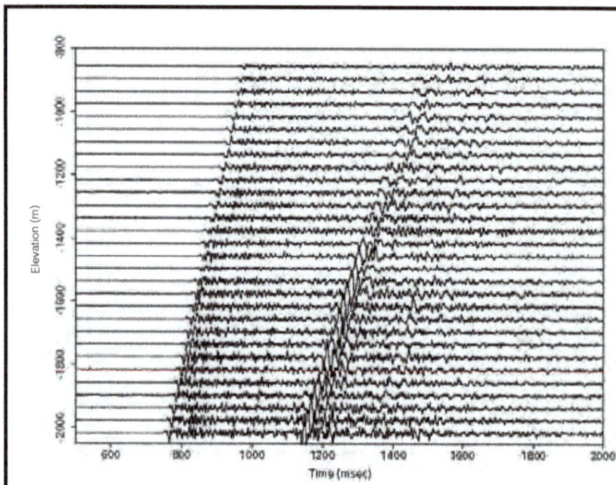

Figure 3. Local earthquake recorded on the SAFOD Pilot Hole Array illustrating the effect of wave propagation on the frequency content of body waves.

strain offsets of several nanostrain on the fiber optic strainmeter.

An exciting discovery made in the SAFOD Main Hole was a new type of seismic wave: a fault zone guided P-wave (Fig. 6). This phase arrives between the P- and S-waves, is normally dispersed, and has a clear Airy phase. It was recorded by 15-Hz seismometers located 2650 m below ground, 3270 m along the inclined borehole, and approximately 40 m from the center of a major fault that the borehole crossed. This fault has caused deformation of the borehole casing, indicating that it is one of the active stands of the San Andreas system. The existence of this phase depends on unique relationships between the P and S velocities of the fault and surrounding rocks, as well as their thicknesses. Its propagation speed, low frequency cut-off, and Airy phase frequency thus give special insights into the structure and multi-stranding of the San Andreas Fault zone.

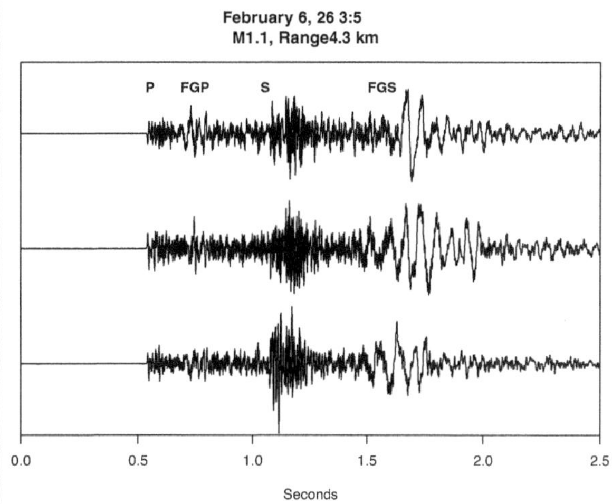

Figure 6. Microearthquake recorded on a seismometer installed in the SAFOD Main Hole located inside the San Andreas fault damage zone about 40 m from an actively slipping trace of the fault. The three velocity seismograms are radial, transverse vertical and transverse horizontal to the source (top to bottom). The high frequency P and S body waves arrive at 0.53 s and 1.07 s, respectively. The dispersed waves following S are fault zone guided S waves (FZGS), which are well known from surface observations. The lower frequency arrival following P is a fault zone guided P wave (FZGP). This is the first known observation of this phase.

In summary, observations of earthquakes at very short distances in the three borehole observatories in California demonstrate the many advantages of observing earthquake processes at short distances in deep wells. Near-source recordings are essential for the study of earthquake source processes, particularly the nucleation and early growth of dynamic rupture. SAFOD data will continue to provide a benchmark for theoretical and numerical models of earthquake processes and laboratory rock experiments, and complementary data to near-source observations of large magnitude earthquakes. Numerical simulations of the earthquake on rate-and-state faults have now advanced to the point that they can produce numerically accurate models that span the slow earthquake loading cycle and dynamic rupture on fault patches of comparable dimension to the SAFOD target earthquakes. We should anticipate a productive interchange between observations, theories, models and laboratory experiments, as we begin to gather data in the near field of the target earthquakes.

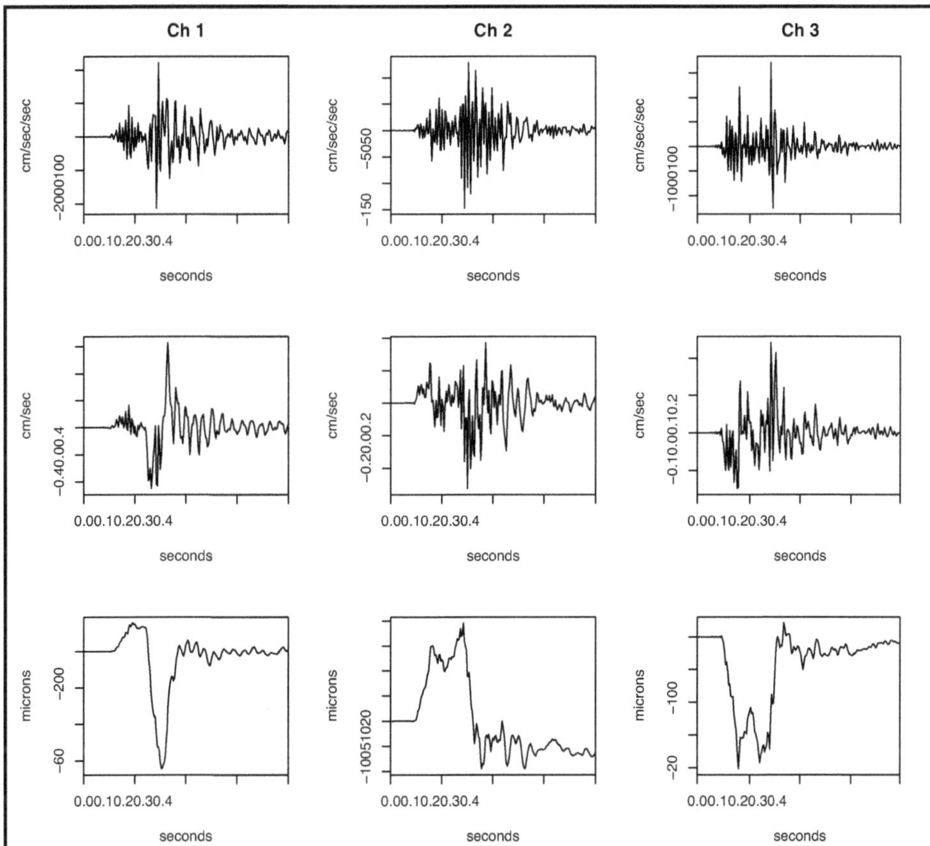

Figure 5. Seismograms of the M1.9 "L.A." target earthquake of 23 January 2003 recorded in the SAFOD Main Hole by a micro-machined accelerometer (Colibrys SF-1500). The original accelerograms (top row) were integrated to velocity (middle row) and displacement (without filtering). At this distance, the near-field contributions to the displacement seismogram become important and are clearly evident in these records.

References

Chavarria, A., Malin, P.E., and Shalev, E., 2004. The SAFOD Pilot Hole seismic array: Wave propagation effects as a function of sensor depth and source location. *Geophys. Res. Lett.*, 31:L12S07, doi:10.1029/2003GL019382.

Ellsworth, W.L. and Malin, P.E., 2006. A first observation of fault guided PSV-waves at SAFOD with implications for fault characteristics. *EOS Trans. Am. Geophys. U.* 87(52): Abstract T23E-02.

Ide, S., Beroza, G.C., Prejean, S.G., and Ellsworth, W.L., 2003. Apparent break in earthquake scaling due to path and site effects on deep borehole recordings. *J. Geophys. Res.*, 108(B5), 2271, doi:10.1029/2001JB001617.

Imanishi, K., and Ellsworth, W.L., 2006. Source scaling relationships of microearthquakes at Parkfield, CA, determined using the SAFOD pilot hole array. *In* Abercrombie, R.E., McGarr, A., Di Toro, G., and Kanamori, H. (Eds.), *Earthquakes: Radiated Energy and the Physics of Faulting, Geophys. Monogr. 170,* Washington D.C. (American. Geophysical Union), 81–80.

Prejean, S.G. and Ellsworth, W.L., 2001. Observations of earthquake source parameters at 2 km depth in Long Valley Caldera, eastern California. *Bull. Seismolog. Soc. Am.*, 91: 165–177, doi:10.1785/0120000079.

Sackett, P.C., McConnell, V.S., Roach, A.L., Priest, S.S., and Sass, J. H., 1999. Long Valley Coring Project, 1998 – Preliminary Stratigraphy and Images of Recovered Core. *U.S. Geological Survey Open-File Report 99-158.*

Authors

William L. Ellsworth, 3A-109, U.S. Geological Survey (USGS), 345 Middlefield Road Mail Stop 977, Menlo Park, Calif. 94025, U.S.A., e-mail: ellsworth@usgs.gov

Peter E. Malin, 109A Old Chemistry Box 90227, Duke University, Durham, N.C. 27708, U.S.A., e-mail: malin@duke.edu

Kazutoshi Imanishi, Geological Survey of Japan, Institute of Geology and Geoinformation (AIST), Tsukuba Central 7, 1-1-1 Higashi, Tsukuba, 305-8567, Japan.

Steven W. Roecker, Department of Earth and Environmental Sciences, Rensselaer Polytechnic Institute, Science Center 1S06, Troy, N.Y. 12180, U.S.A.

Robert Nadeau, Berkeley Seismological Lab., 207 McCone Hall, University of California Berkeley, Calif. 94720-4760, U.S.A.

Volker Oye, Norwegian Seismic Array (NORSAR), P.O. Box 53, N-2027 Kjeller, Norway.

Clifford H. Thurber, Department of Geology and Geophysics, University of Wisconsin, 1215 West Dayton Street, Madison, Wis. 53706, U.S.A.

Felix Waldhauser, LDEO-Seismology Geology and Tectonophysics, 210 Seismology, 61 Route 9W – P.O. Box 1000, Palisades, N.Y. 10964, U.S.A.

Namoi L. Boness, Chevron, Energy Technology Centers (ETC), San Ramon, Calif. 94583, U.S.A.

Stephen H. Hickman, USGS, 345 Middlefield Road, Menlo Park, Calif. 94025, U.S.A.

Mark D. Zoback, Department of Geophysics, Stanford University, Mitchell Bldg., Stanford, Calif. 94305, U.S.A.

Related Web Link

http://www.ncedc.org/

Monitoring of Rock Mass Behavior at the Closest Proximity to Hypocenters in South African Gold Mines

by Hiroshi Ogasawara and the Research Group for Semi-controlled Earthquake-Generation Experiments in South African Deep Gold Mines

To bridge the scale gap between a laboratory scale and a large earthquake scale, we need experiences in an intermediate scale at the smallest distance from the hypocenters (Fig. 1). However, the closest monitoring is not always easy, and the San Andreas Fault Observatory at Depth (SAFOD) is an exceptional opportunity to obtain experiences from natural small earthquakes (Hickman et al., 2004). Tremors in deep mines resemble natural small earthquakes, being also good to take a close look at the hypocenters (Spottiswoode and McGarr, 1975). Recently, modern monitoring is carried out with wider dynamic range and frequency band (Mendecki, 1997). The Research Group for the Semi-controlled Earthquake-Generation Experiments in South African Deep Gold Mines (SeeSA) have been co-operating with ISS International Ltd. and South African gold mines to study rock mass behavior (Iio, 1995; Iio and Fukao, 1992; Ishii and the Research Group for Semi-Controlled Earthquake Generation Experiment at South African Gold Mines, 1996; Nicolaysen, 1992; Ogasawara et al., 2001, 2002a, 2005a, 2005b; Sumitomo, 1998). We have deployed eight experimental sites at four gold mines at depths of a few km (Fig. 2; Ogasawara et al., 2002a, 2005a, 2005b), and at one of these we are collaborating with ICDP-DAFSAM/NSF-NELSAM project (PI: Ze'ev Reches).

In 1996, at the Western Deep Levels South Mine (currently Mponeng mine; Fig. 2), we deployed an array of nine triaxial accelerometers within 100 m of potential hypocenters of M>2 earthquakes, successfully delineating the finest details of small earthquakes. The relationship between seismic moment and corner frequency was the same as that for medium or large natural earthquakes (Ogasawara et al., 2002b). The source processes of some M~1 or smaller earthquakes were as complicated as large, natural earthquakes (Yamada et al., 2005).

During 2000–2003 at the Bambanani mine (Figs. 2 and 3), we successfully recorded the rock-mass behavior in the entire life span of M2.4 and M2.5 earthquakes (Mw2.9 and Mw2.0, respectively, in Fig. 4a; Ishii et al., 2000; Ogasawara et al., 2005a; Takeuchi, 2005). We installed an Ishii strainmeter (Ishii et al., 1997) within 100 m of the seismic sources and recorded the data continuously with resolutions of 25-Hz and 24-bit. The initial significant change was by an Mw0.2 event in September 2001 within a few tens of meters. The largest event (Mw3.0) within 250 m of the strainmeter took place in 2002, however, not close enough to cause the largest strain change. The largest strain changes were associated with two M2 events within 100 m of the strainmeter in 2003 (yellow circles in Fig.3, upper). The strain change was as large as 100 microstrain, corresponding to several MPa. Such a large change is recorded only at the closest distance from the seismic fault. Channel 1 subparallel to the maximum principal stress was contracting until the M2 sequence and then was released by the sequence.

However no accelerating strain preceded the hundreds of co-seismic steps for the catalogued seismic events, even for those of ~100 microstrain (stress drop ~7 MPa) (Ogasawara et al., 2005a; Takeuchi, 2005).

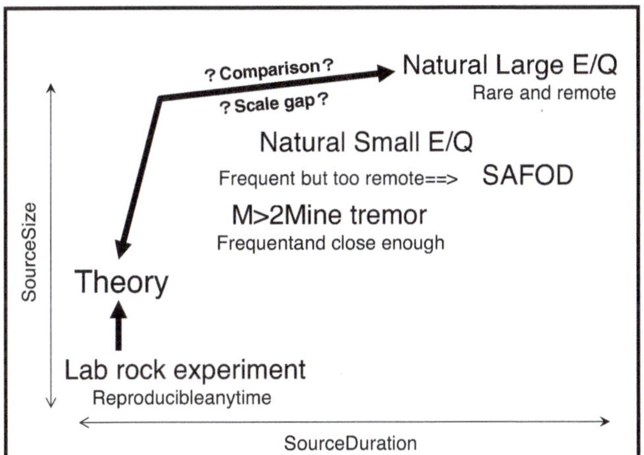

Figure 1. The role of intermediate scale experience.

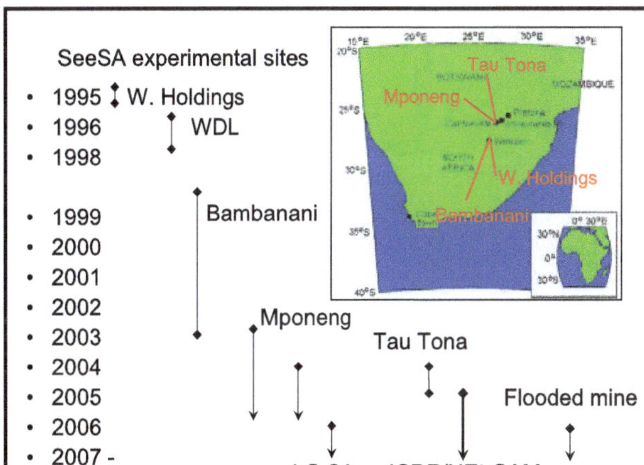

Figure 2. The history of SeeSA experimental sites.

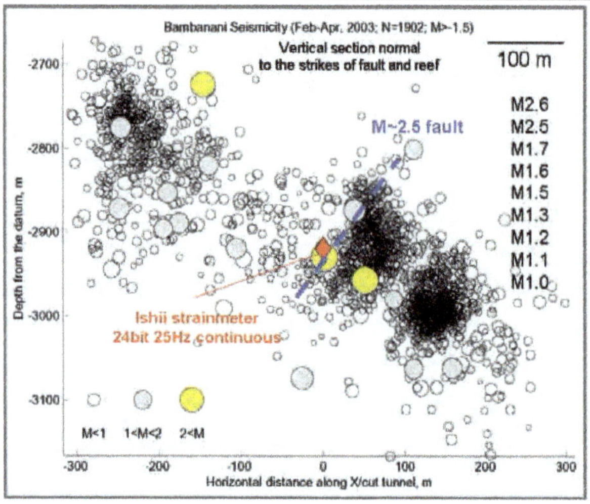

Figure 3. Configuration of an Ishii strainmeter, the M2.5 fault and the seismicity (Feb–Apr 2003) at the site at the Bambanani mine, Welkom.

Thoroughly going through the strain recordings, Yamamoto et al. (2006) found that the Ishii strainmeter picks up seismic events smaller than M= -1 (much more than those catalogued by the mine's seismic network). They also found hundreds of the aftershocks having variable senses of strain steps and postseismic changes. This suggests the strainmeter was located within the aftershock area.

The step response to the catalogued earthquake accompanied overshoot and ringing (Fig. 4b; hereinafter, dynamic

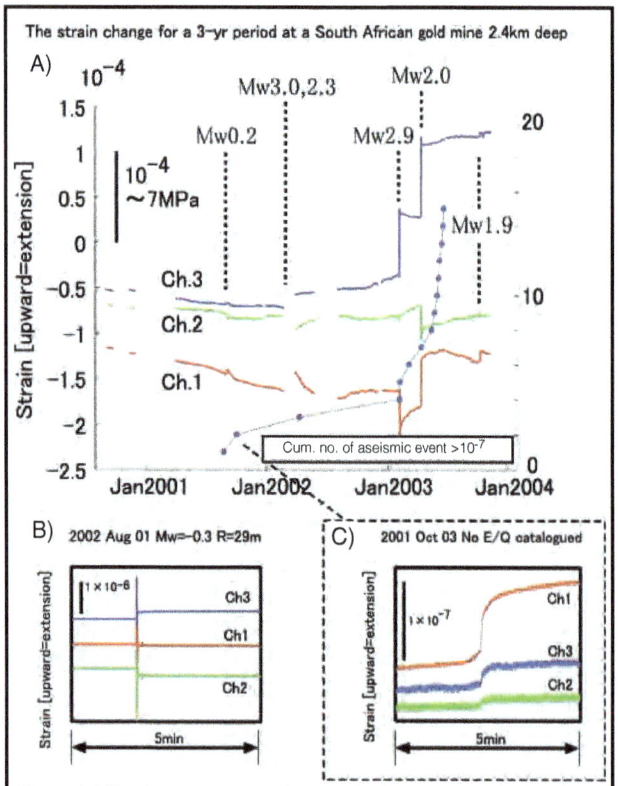

Figure 4. [A] a three-year strain changes (red, green, and blue lines) and cumulative number of aseismic events (LDR events; connected blue dots). [B] A typical response to a catalogued seismic event. Note the dynamic response (overshoot and ringing) associated with the strain step. [C] A typical example of slow step with clear forerunner (after Naoi et al., 2006 b).

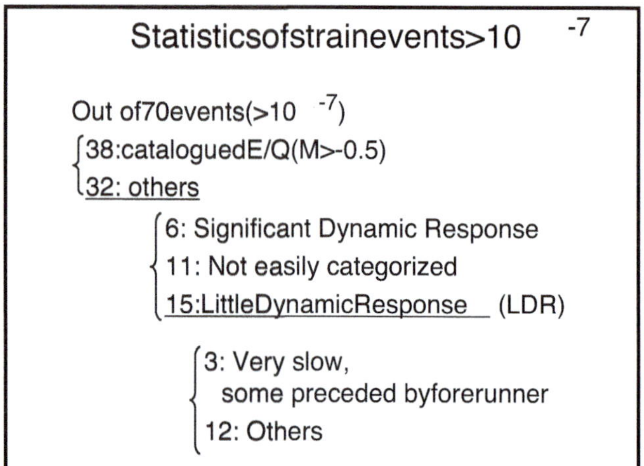

Figure 5. The statistics of the strain events (data after Naoi et al., 2006a).

response). However, Naoi et al. (2006a) found that slow strain-steps as small as 0.1 microstrain were also frequently recorded with little dynamic response (LDR). They found that the step durations were variable and significantly longer than normal seismic steps (Fig. 4c; the moments possibly correspond to M=-1 or smaller seismic events). Interestingly, most LDR events took place after the large strain change and few very slow events were preceded by the significant accelerating strain (Figs. 4c and 5).

The entire lifespan of a large natural earthquake includes loading, foreshocks, main rupture, aftershocks, and postseismic deformation. Slow events are also associated. The forerunning, accelerating deformation is not often observed, however.

For -1<M<2 mine tremors, we again observed almost the same set of phenomena. Additionally, we could find the clear forerunners associated only with the small, slow strain event. We were able to see them because we were able to install strainmeters within seismic sources.

Monitoring the smaller subsets of the phenomena (gray area in Fig. 6) could be interesting. To get the time and magnitude range closer to those for AE, we start the strain

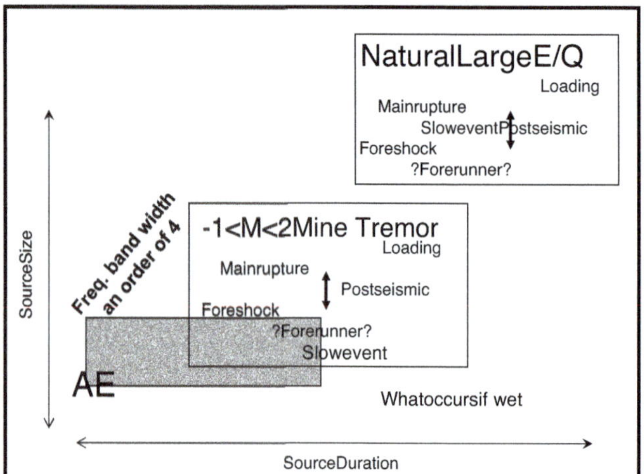

Figure 6. Comparison of sets of phenomena observed.

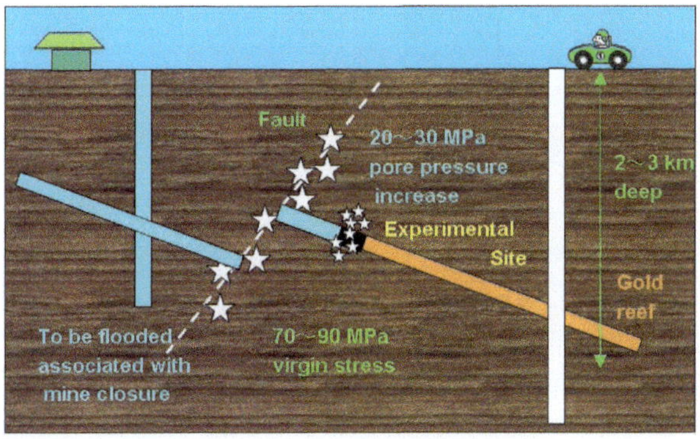

Figure 7. A schematic illustration of monitoring the seismicity (stars) associated with flooding from the adjacent working mine.

monitoring with much higher sampling rate at new sites. One of these is carried out collaterally with Nakatani et al., (2006) who attempted to monitor the wide dynamic/frequency range of fracturing process from AE (up to 200 kHz) to seismic events at a potential seismic fault.

At the previous experimental sites, the rock mass was not saturated with water. We also have to learn what behavior can be seen under wet conditions, and along these lines, the flooded South African gold mines are very unique cases to provide us with good experimental sites. We can investigate 1) the effects of flooding and the corresponding rising water levels on the stability of faults and other geological features, 2) the effects of seismicity on inter-mine water plugs and mine barriers pillars, and 3) seismic damage risks to neighboring mines in areas in which mines are mature.

These investigations are very crucial because the behavior of highly stressed rock mass during flooding is presumably analogous to that during natural earthquake swarms or stable or unstable slip at natural great earthquake hypocenters under water-saturated conditions. We cannot access inside the earthquake swarm in Japan, but we can do it in South Africa.

References

Hickman, S., Zoback, M., and Ellsworth, W., 2004. Introduction to special section: Preparing for the San Andreas Fault Observatory at Depth. *Geophys. Res. Lett.*, 31:L12S01, doi:10.1029/2004GL020688.

Iio, Y., 1995. Earthquake prediction in South African gold mines. *Iwanami Science* 65:279–281 (in Japanese).

Iio, Y. and Fukao, Y., 1992. Call for joining international experiment in South African gold mines. *Newsletter, Seismol. Soc. Jpn.*, 3(6):8–9 (in Japanese).

Ishii, H. and the Research Group for Semi-controlled Earthquake Generation Experiments at South African Gold Mines, 1996. Semi-controlled experiment in a gold mine in South Africa (5) — strain observation by borehole strain meters. *Abstracts. 1996 Japan Earth Planet. Sci. Joint Meeting:* A58 (in Japanese).

Ishii, H., Nakao, S., Ogasawara, H., and the Research Group for Semi-controlled Earthquake Generation Experiment at South African Gold Mine, 2000. Continuous monitoring of strain by 25 Hz 24 bit sampling using Ishii's borehole strainmeter on a hypocentral fault where an M > 3 event is expected - semi-controlled experiment in a gold mine in South Africa (25). *Abstracts. Seism. Soc. Jpn.*, 2:A58 (in Japanese).

Ishii, H., Yamauchi, T., and Kusumoto, F., 1997. Development of high sensitivity bore hole strain meters and application for rock mechanics and earthquake prediction study. *Rock Stress.* Rotterdam (Balkema), 253–258.

Mendecki, A.J. (Ed.), 1997. *Seismic Monitoring in Mines.* London (Chapman and Hall), 262 pages.

Nakatani, M., Carstens, R., Stanchits, S., Philipp, J., van Aswegen, G., Ide, S., Miyake, H., Yabe, Y., Ogasawara, H., Dresen, G., Manthei, G., Ward, T., and Sumitomo Norihiko International Research Group for Semi-controlled Earthquake Generation Experiment at South African Gold Mine, 2006. An ultra-accessible fault of an impending earthquake at a 3.5 km depth in a South African gold mine. *Abstract, the Japan Geoscience Union Meeting 2006*, S109-P004.

Naoi, M., Ogasawara, H., Takeuchi, J., Yamamoto, A., Shimoda, N., Morishita, K., Ishii, H., Nakao, S., van Aswegen, G., Mendecki, A.J., Lenegan, P., Ebrahim-Trolloope, R., and Iio, Y., 2006a. Small slow-strain steps and their forerunners observed in gold mine in South Africa. *Geophys. Res. Lett.*, 33:L12304, doi: 10.1029/2006GL026507.

Naoi, M. Ogasawara, H., Takeuchi, J., Shimoda, N., Morishita, K., Ishii, H., Nakao, S., Mendecki, A.J., van Aswegen, G., Ebrahim-Trollope, R. and Iio, Y., 2006b. *Newsletter, Seismol. Soc. Jpn.*, 18(1), cover page (in Japanese).

Nicolaysen, L.O., 1992. International semi-controlled experiment on seismic events: a review of the background and proposal. *Newsletter, Seismol. Soc. Jpn,*. 3(6):9–27.

Ogasawara, H. Miwa, T., and The Research Group for Semi-controlled Experiment in South African Gold Mines, 2002b. Microearthquake scaling relationship using near-source, redundant, wide-dynamic-range accelerograms in a South African deep gold mine. *In* Ogasawara, H., et al. (Eds.), *Seismogenic Process Monitoring*, Rotterdam (Balkema), 151–164.

Ogasawara, H. and The Research Group for Semi-controlled Experiments in South African Gold Mines, 2002a. Semi-controlled earthquake generation experiment in South African gold mine. A review of the semi-controlled earthquake generation experiment in South African Deep Gold Mines (1992-2001). *In* Ogasawara, H., et al. (Eds.), *Seismogenic Process Monitoring*, Rotterdam (Balkema), 119–150.

Ogasawara, H., Sato, S., Nishi, S., Sumitomo, N., Ishii, H., Iio, Y., Nakao, S., Ando, M., Takano, M., Nagai, N., Ohkura, T., Kawakata, H., Satoh, T., Kusunose, K., Cho, A., Mendecki, A., Cichowicz, A., Green, R.W.E., and Kataka, M.O., 2001. Semi-controlled seismogenic experiments in South African deep gold mines. *Proc. 5th Internat. Symp. Rockburst and Seismicity Mines*, 293–300.

Ogasawara, H., Takeuchi, J., Shimoda, N., Ishii, H., Nakao, S., van Aswegen, G., Mendecki, A.J., Cichowicz, A., Ebrahim-Trollope, R., Kawakata, H., Iio, Y., Ohkura, T., Ando, M., and the Research Group for Semi-controlled Earthquake-Generation Experiments in South African deep gold mines,

2005a. High-resolution strain monitoring during M~2 events in a South African deep gold mine in close proximity to hypocentres. *Proc. 6th Internat. Symp. Rockburst and Seismicity in Mines*, 385–391.

Ogasawara, H., Takeuchi, J., Shiimoda, N., Nakatani, M., Kato, A., Iio,Y., Kawakata, H., Yamada, T., Yamauchi, T., Ishii, H., Satoh, T., Kusunose, K., Otsuki, K., Kita, S., Nakao, S., Ward, A.K, McGill, R., Murphy, S.K., Mendecki, A.J., van Aswegen, G., and the Research Group for Semi-controlled Earthquake Generation Experiments in South African Deep Gold Mines, 2005b. Multidisciplinary Monitoring of the Entire Life Span of an Earthquake and its Practical Strategy in South African Gold Mines, *Proc. 6th Internat. Symp. Rockburst and Seismicity in Mines*, 393–398.

Spottiswoode, S.M. and McGarr, A., 1975. Source parameters of tremors in a deep-level gold mine. *Bull. Seismol. Soc. Amer.*, 65:93–112.

Sumitomo, N., 1998. International joint research: semi-controlled earthquake generation experiment. *Chikyu Monthly*, 229:391–394 (in Japanese).

Takeuchi, J., 2005. A study of strain changes associated with earthquakes observed within 100m in Bambanani mine, South Africa (in Japanese). M.S. thesis, Ritsumeikan University, Kusatsu.

Yamada, T., Mori, J.J., Ide, S., Kawakata, H., Iio, Y., and Ogasawara , H., 2005. Radiation efficiency and apparent stress of small earthquakes in a South African gold mine, *J. Geophys. Res.*, 110, B01305, doi:10.1029/2004JB003221.

Yamamoto, A., Ogasawara, H., Takeuchi, J., Shimoda, N., Naoi, M., Morishita, K., Ishii, H., Nakao, S., Mendecki, A., van Aswegen, G., Lenegan, P., Ebrahim-Trollope, S., and Sumitomo Norihiko International Research Group for Semi-controlled Earthquake Generation Experiment at South African Gold Mine, 2006. The aftershocks for ultra-micro-earthquakes seen in 25Hz continuous recordings with Ishii strainmeter within M2 source area. *Abstract, the Japan Geoscience Union Meeting 2006*, S109-P004.

Authors

Hiroshi Ogasawara, Faculty of Science and Engineering, Ritsumeikan University, 1-1-1 Noji Higashi, Kusatsu, 525-8577, Japan, e-mail: ogasawar@se.ritsumei.ac.jp.

The Research Group for Semi-controlled Earthquake-Generation Experiments in South African Deep Gold Mines, members from Ritsumeikan University, the University of Tokyo, Kyoto University, Nagoya University, Tohoku University, Tono Research Institute of Earthquake Science (TRIES) Japan, National Institute of Advanced Industrial Science and Technology (AIST) Japan, Kanazawa University, ISS International Ltd., Open House Management Solutions Ltd (OHMS), Seismogen CC., Geohydroseis CC., Gesellschaft für Materialprüfung und Geophysik (GMuG), GeoForschungsZentrum Potsdam (GFZ), Western Deep Levels, South mine, Bambanani mine, Tau Tona mine, Buffelsfontein mine, East Rand Proprietary mine.

Dynamics of Earthquake Faulting in Subduction Zones: Inference from Pseudotachylytes and Ultracataclasites in an Ancient Accretionary Complex

by Kohtaro Ujiie

The fault rocks in ancient accretionary complexes exhumed from seismogenic depths may provide an invaluable opportunity to examine the mechanisms and mechanics of seismic slip in subduction thrusts and splay faults. In order to understand the dynamics of earthquake faulting in subduction zones, we analyzed pseudotachylytes and ultracataclasites from the Shimanto accretionary complex in southwest Japan.

Pseudotachylytes

The following conclusions were derived from the analysis of pseudotachylytes (Ujiie et al., 2007a):

1. At seismogenic depths in the subduction zone, a cataclastic thrust zone develops at the top of the mélange, and the coseismic slip is concentrated into a narrow zone less than a few millimeters thick.
2. The pseudotachylyte displays a fragment-laden, glass-supported texture resulting from rapid cooling of the frictional melt (Fig. 1), which is consistent with a very short cooling time of the melt layer calculated using thermal modeling. The rapid cooling of the melt layer is due to its narrow thickness, resulting in the fast healing of the coseismic slip zone by the solidified melt layer.
3. The pseudotachylyte is derived from the frictional melting of an illite-rich ultracataclasite layer. The variation in the volume fraction of unmelted grains in the pseudotachylyte matrix primarily represents the difference in the initial volume fraction of illite in the ultracataclasite layer prior to frictional melting. The minimum melting temperature is 1100°C, which is ~850–920°C greater than the maximum temperatures recorded in the host rocks.
4. The viscosity and shear resistance of the melt layer are very low (Fig. 2); therefore, the dynamic weakness of the fault, acceleration of seismic slip, and propagation of instability can possibly occur during an earthquake. This would contribute, at least locally, to the efficiency with which stored strain energy is released and, hence, to the earthquake magnitude in subduction zones.
5. Such frictional melting of the illite-rich slip zone may be applicable to subduction thrusts and faults in other accretionary complexes. The melting of the illite-rich slip zone is likely to form a hydrous melt layer, possibly leading to a high water content in these pseudotachylytes.

Ultracataclasites

The new observations of ultracataclasites within the subduction thrust and their possible relevance to the dynamic slip processes are summarized below (Ujiie et al., 2007b).

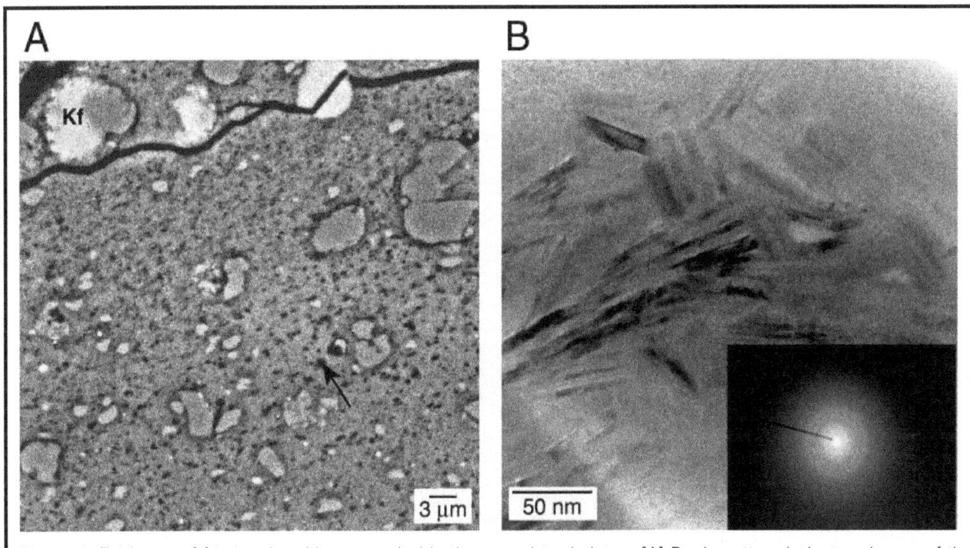

Figure 1. Evidence of frictional melting recorded in the pseudotachylytes. [A] Backscattered electron image of the pseudotachylyte, characterized by vesicles (black arrow) and embayed grains in the homogeneous matrix. Kf: K-feldspar. [B] Transmission electron micrograph of pseudotachylyte (bright-field image). Note the presence of euhedral microcrystals in a glassy matrix. Inset shows the diffraction pattern of the microcrystals.

The thrust faults in the Shimanto accretionary complex of southwest Japan imbricate the mélange derived from sedimentary rocks and basalt, which represent a duplex underplating associated with the downward step of the subduction thrust at seismogenic depths (4–6 km deep, 130–150°C). One well exposed duplex-fault zone consists mainly of foliated cataclasites and ultracataclasites derived from basalt. The fault zone structure suggests slip localization along the 2–20-cm-thick ultracataclasite layer at the

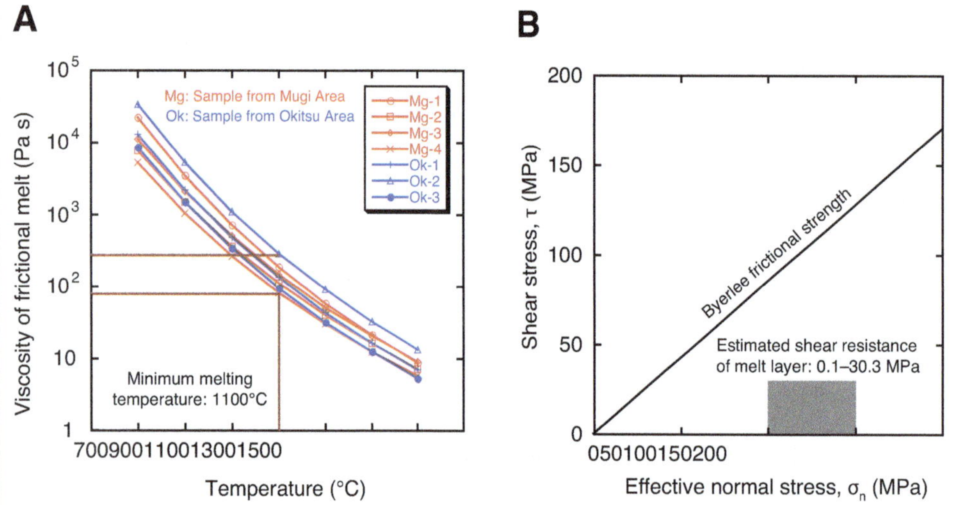

Figure 2. Estimated viscosity and shear resistance of the melt layer. [A] Viscosity versus temperature for seven samples from the Mugi and Okitsu areas. [B] Shear stress versus effective normal stress for the melt layer, compared with Byerlee's frictional strength .

boundary between different lithologies. The ultracataclasite layer is marked by fluid-related processes, as shown by the alteration along the ultracataclasite and the concentration of veins near the ultracataclasite. New observations of the ultracataclasite layer have revealed that fluidization could occur during the localized slip.

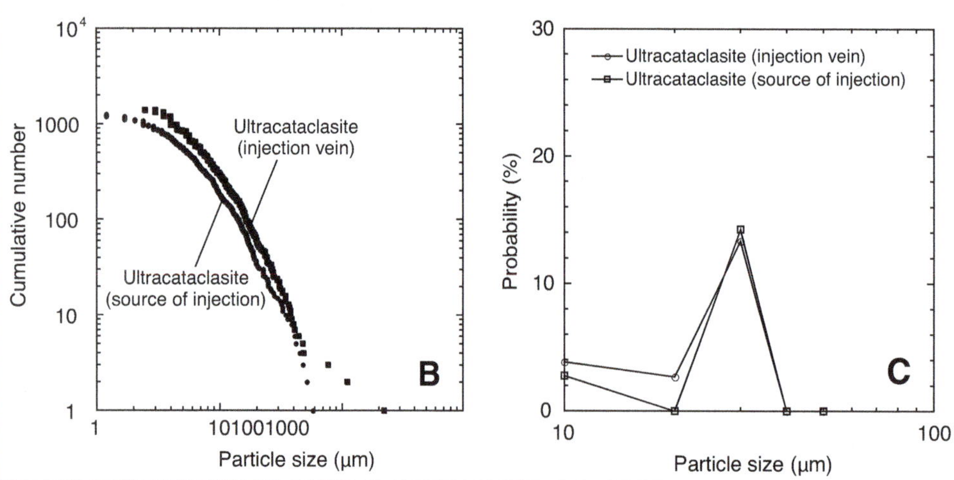

Figure 3. Evidence of fluidization. [A] Injection structures. The double arrows indicate the ultracataclasite layer. The upper boundary of the ultracataclasite is irregular due to the development of injection structures. [B] Particle size distributions from the injection vein and the source of injection. [C] Detection probability of the fragmented counterparts as a function of the particle size from the injection vein and the source of injection.

Fluidization is suggested by the injections of granular material into the mélange without the sorting of finely-grained particles, and by the quantitative data that few particles in the injected material have fragmented counterparts (Fig. 3). The fluidization during localized slip under fluid-infiltrated conditions is likely due to thermal pressurization, and injection structures may represent hydraulic fracturing normal to the least principal stress.

References

Ujiie, K., Yamaguchi, H., Sakaguchi, A., and Toh, S., 2007a. Pseudotachylytes in an ancient accretionary complex and implications for melt lubrication during subduction zone earthquakes. *J. Struct. Geol.*, 29:599–613, doi:10.1016/j.jsg.2006.10.012.

Ujiie, K., Yamaguchi, A., Kimura, G., and Toh, S., 2007b. Fluidization of granular material in a subduction thrust at seismogenic depths. *Earth Planet. Sci. Lett.*, 259:307–318, doi:10.1016/j.epsl.2007.04.049.

Author

Kohtaro Ujiie, Institute for Research on Earth Evolution (IFREE), Japan Agency for Marine-Earth Science and Technology (JAMSTEC), 3173-25 Showa-machi, Yokohama-city, Kanagawa 236-0001, Japan, e-mail: ujiiek@jamstec.go.jp.

Precise Temperature Measurements and Earthquake Heat Associated with the 1999 Chi-Chi, Taiwan Earthquake

by Yasuyuki Kano, Jim Mori, Ryo Fujio, Takashi Yanagidani, Setsuro Nakao, Hisao Ito, Osamu Matsubayashi, and Kuo-Fong Ma

Introduction

The frictional heat generated during earthquake faulting is thought to be the largest part (80% to 90%) of the total seismic energy budget, and geophysicists have long discussed the level of heat that should be observable (Brune et al., 1969; Lachenbruch and Sass, 1980; Scholz, 2002; Terada, 1930). Precise temperature measurements across the fault immediately after an earthquake can provide the most unambiguous answer; however, there has never been a significant near-fault temperature change observed for any previous large earthquake that can be attributed to the frictional heating. This is because there has been no appropriate site for temperature measurements around a fault at depth just after an earthquake. The most promising way to reach the fault zone in order to observe the frictional heat is to drill a borehole to the area where large slip occurs. There have been several drilling projects to reach deep areas of the fault zone, such as the Taiwan Chelungpu fault Drilling Project (TCDP), San Andreas Fault Observatory at Depth (SAFOD), and the planned NanTroSEIZE project. We reported the first successful temperature measurement of deep fault zone boreholes that was drilled by TCDP at the Chelungpu fault, Taiwan (Kano et al., 2006). An observation of a temperature increase, and thus an estimate of the heat generated, provides information about the frictional strength during faulting and the level of driving stress for an earthquake. These are key unknown values of important parameters that are necessary for understanding the physical process of earthquake ruptures.

In this paper we outline the results of the precise temperature measurement in TCDP Hole A as an attempt to directly measure the frictional heat produced by an earthquake. Then, we present the importance of measuring the thermal property of rocks around the fault in addition to the precise temperature measurement—the effect of thermal conductivity of the material on the temperature gradient proposed by Matsubayashi et al. (2005) is not negligible on the results of the precise temperature measurement. The details of our temperature measurement in the Chelungpu fault and its interpretation are presented in Kano et al. (2006).

Figure 1. The precise temperature instrument that was installed in the well head of TCDP Hole A.

Precise Measurements at TCDP Hole A

Following the 21 September 1999 Chi-Chi earthquake, the TCDP bored two holes which penetrated the fault at depths of about 1100 m (Ma et al., 2006) near the town of DaKeng in the northern part of the rupture zone. During the earthquake, this area had large surface rupture, and a fault displacement of about 8 m is estimated from seismic data. The boreholes provided the rare opportunity to make temperature measurements in a fault zone with large slip from a recent earthquake. The precise temperature observations were carried out in one of the boreholes (Hole A) during September 2005, six years following the earthquake. The borehole is cased with steel pipe so that there is no water flow between the borehole and surrounding rock, enabling much more stable temperature measurements.

In order to obtain a high-resolution (0.003°C) temperature profile, we developed a borehole instrument containing two quartz oscillator thermometers, separated by 3 m. The instrument was slowly lowered (about 1.0 m min^{-1}) and raised (about 0.4 m min^{-1}) in the borehole between the depths of 900 m and 1250 m, producing four independent temperature profiles across the fault zone, during 20 and 21 September (Fig. 1). The continuous recording of temperature at 10 s intervals produced 5–15 readings per meter.

All the temperature profiles show small temperature signatures, which are a small temperature increase above the geothermal gradient in the region of the fault zone at the depth of 1111 m. Kano et al. (2006) simply modeled this temperature signature as a residual temperature anomaly originating in frictional heat produced by fault slip.

Modeling of Temperature Anomaly

The temperature signature produced by frictional heat is modeled by heat conduction from the fault plane as a heat source. Here we simply modeled the signature assuming one-dimensional heat conduction, in which heat produced on a thin plane is conducted to the direction perpendicular to the plane (Officer, 1974). We assumed that all the heat generated was consumed to increase the temperature of thin fault surface. The temperature change at the point that x m away from the fault surface after t s is

$$T(x,t) = \frac{S}{2\sqrt{\pi\alpha t}} e^{-x^2/4\alpha t} \quad (1)$$

where α is the heat diffusivity of the media surrounding the fault, and S is the strength of the heat source, which is heat (product of shear stress, τ, and fault slip, u) divided by specific heat, c, and the density, ρ, of the medium with units measured in K m:

$$S = \frac{\tau \cdot u}{c \cdot \rho} \quad (2)$$

Figure 2A shows the spatial variation of temperature signature after five years and ten years from the earthquake (calculated for the parameters u = 6 m, τ = 1.1 MPa, c = 1140 J kg^{-1} K^{-1}, ρ= 2500 kg m^{-3}, and α = 3.4 x 10-7 m^2 s^{-1}). The amplitude becomes smaller with an increase of t, and the shape of the signature becomes broader. The amplitude of the peak of the temperature anomaly is only 0.06 K after five years. Figure 2B shows the temporal variation of temperature signature on the fault and at the point 1 m and 10 m apart from the fault plane.

Spatial Variation of Thermal Conductivity

The temperature profile is strongly affected by the variations of thermal conductivity of the surrounding material. Kano et al. (2006) interpreted the observed temperature profile assuming the background temperature gradient is constant. They ignored the effect of spatial variation of thermal conductivity on the temperature gradient; however, the spatial variation of thermal conductivity around the Chelungpu fault is large enough to affect the interpretation of the temperature anomaly (Matsubayashi et al., 2005). Using thermal conductivity data in Hole A, Tanaka et al. (2007) reinterpreted the temperature data by Kano et al. (2006) and suggested that the temperature signature detected by Kano et al. (2006) might be a result of thermal conductivity fluctuations, rather than a residual heat from frictional faulting. Tanaka et al. (2007) measured the thermal conductivity with core soaked in epoxy resin to fill the cracks. Doing this introduces error in the measurement of thermal conductivity because of the low value of thermal conductivity of epoxy resin compared to rocks. The effect of epoxy resin is not well evaluated. Their data, in addition, is not sampled around fault zone, which brings error to the prediction of temperature variation. To extract the heat signature produced by fault slip, we had to make measurements of the spatial variation of material thermal conductivity and examine its effect.

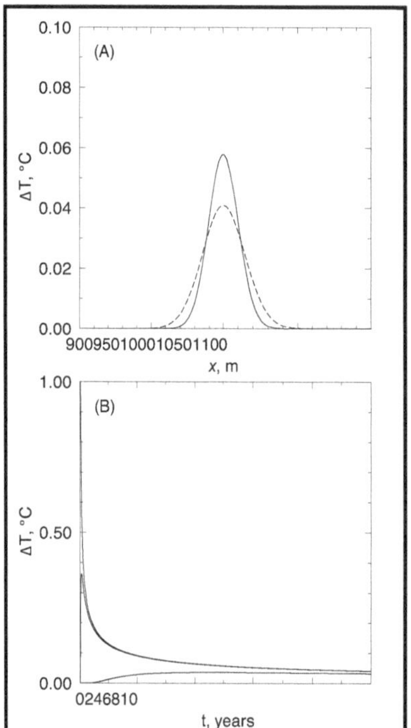

Figure 2. [A] Spatial and [B] temporal variations of temperature signature calculated for u = 6 m.

The relationship between temperature gradient (dT/dz) and thermal conductivity, κ, at depth, z, is

$$\frac{dT}{dz} = \frac{q}{\kappa(z)} \quad (3)$$

where q is heat flow. Assuming constant q, we can predict a temperature variation produced only by the spatial variation of κ,

$$T(z) = q \int \frac{1}{\kappa(z)} dz \quad (4)$$

This means that the temperature gradient produced by constant heat flow is not constant in the medium that has spatial variation of thermal conductivity. Here we predict the background temperature gradient in the temperature profile from the thermal conductivity measured using the core samples of Hole B (Matsubayashi et al., 2005). They measured the thermal conductivity every 0.04–1.5 m for the cores in which water content is carefully preserved before the measurement. We assume that the thermal conductivity, and thus lithology and porosity, in Hole A and Hole B are continuous with 26-m depth difference. Figure 3 shows the predicted temperature profile assuming a constant heat flow of 60 mW m^{-2} together with the observed temperature anomaly. The depth of Hole B is shifted by 26 m, which is the difference between the 1111 m fault zone detected in Hole A and the 1137 m fault zone in Hole B. To calibrate the response delay included in the observed temperature anomaly that is caused by thermal inertia of the temperature instrument, observed and predicted temperature anomalies are low-pass filtered (40 m). We can see peaks in the predicted temperature anomaly (1110 m, 1150 m, and 1190 m) compared to the observed temperature anomaly (1110 m and 1190 m). Those peaks in the corrected temperature observations reflect variations of thermal conductivity that are caused by differences of rock type and porosity around the fault zone.

The peak in the predicted temperature profile is smaller than the observed temperature anomaly around 1111 m, which is modeled as the heat signature of fault slip in Kano et al. (2006), and is considered to be a temperature increase superimposed on background temperature fluctuation. The contribution of residual frictional heat produced by the fault slip is then smaller than the peak modeled in Kano et al. (2006). Thus, their estimation of heat produced gives an upper bound of heat generated by fault slip. The apparent coefficient of friction is estimated to be 0.04–0.08. Laboratory determinations of the static coefficient of friction are generally quite high, 0.6 to 0.7 (Byerlee, 1978) or 0.35 to 0.5 for shales (Morrow et al., 1992), and would produce much higher amounts of heat if these values are used for the dynamic coefficient of friction.

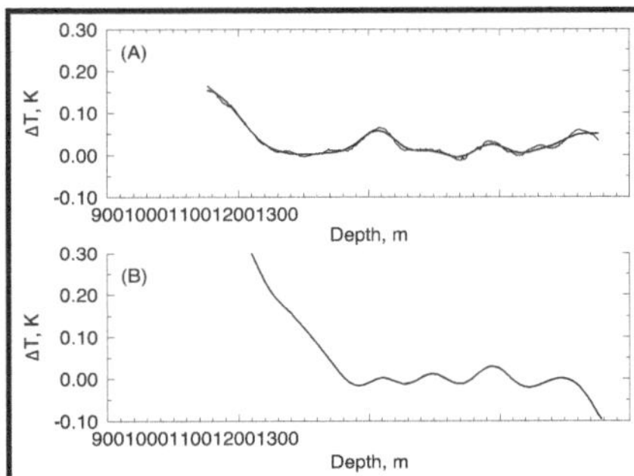

Figure 3. [A] Observed temperature anomaly in Hole A and [B] temperature anomaly predicted from the thermal conductivity measurement of core from Hole B. The depth of Hole B is shifted by 26 m, which is the difference between the 1111 m fault zone detected in Hole A and the 1137 m fault zone in Hole B. Thick lines shows the temperature anomalies that are low pass filtered with cutoff length of 40 m.

The assumption that the thermal conductivities in Hole A and Hole B are continuous is not necessarily true, since the lithology and thermal property may be three-dimensionally inhomogeneous. The best way to remove the background temperature profile is to use thermal conductivity data from Hole A to predict the temperature fluctuations contained in the temperature measurements from Hole A. There is, however, applicable thermal conductivity data measured only from the core of Hole B. To model the residual frictional heat more precisely, we need an appropriate depth correction between Hole A and Hole B.

Summary

Measuring temperature around the fault zones is a way to obtain knowledge of frictional heat produced during earthquakes and thus the energy budget, which are key unknown parameters that are necessary for understanding the physical process of earthquake ruptures. The spatial variations of material thermal conductivity may be another factor that affects the temperature signature, which is sometimes very similar to the temperature signature produced by the frictional heat of fault slip. To obtain the correct background temperature profile, we need to correct the observed temperature signature using thermal conductivities of the formation. When we make temperature measurement to seek residual frictional heat along the fault, it is important to measure material thermal properties that have enough resolution to remove the background temperature fluctuations. In our particular case, we do not have enough thermal conductivity data from Hole A to completely calibrate the temperature signature. The calculated heat in our present results was an upper bound, and it implies a very low level of dynamic friction during faulting for this region of large slip. The low level of friction we obtained needs to be confirmed for other events, and, if verified, indicates that low friction mechanisms are needed to explain the dynamic rupture process of large earthquakes.

Acknowledgements

This work was partially supported by the Kyoto University 21st Century Center of Excellence (KAGI21) and other Grants in Aid for Scientific Research (16253003) from the Ministry of Education, Culture, Sports, Science, and Technology of Japan and the National Science Council of Taiwan.

References

Brune, J., Henyey, T., and Roy, R., 1969. Heat flow, stress, and rate of slip along the San Andreas fault, California. *J. Geophys. Res.*, 74:3821–3827.

Byerlee, J.D., 1978. Friction of rocks. *Pure Appl. Geophys.*, 116:615–626, doi:10.1007/BF00876528.

Kano, Y., Mori, J., Fujio, R., Ito., H., Yanagidani, T., Nakao, S., and Ma, K.-F., 2006. Heat signature on the Chelungpu Fault associated with the 1999 Chi-Chi, Taiwan earthquake. *Geophys. Res. Lett.*, 33:L14306, doi:10.1029/2006GL026733.

Lachenbruch, A.H. and Sass, J.H., 1980. Heat flow and energetics of the San Andreas fault zone. *J. Geophys. Res.*, 85:6185–6222.

Ma, K.-F., Song, S.-R., Tanaka, H., Wang, C.-Y., Hung, J.-H., Tsai, Y.-B., Mori, J., Song, Y.-F., Yeh, E.-C., Sone, H., Kuo, L.-W., and Wu, H.-Y., 2006. Slip zone and energetics of a large earthquake: Results from the Taiwan Chelungpu-fault Drilling Project (TCDP). *Nature*, 444:473–476, doi:10.1038.

Matsubayashi, O., Lin, W., Hirono, T., Song, S., and Hung, J., 2005. Characterization of unfractured wall rocks of TCDP Hole-B by combination of thermal-property and TDR measurements in laboratory. *Eos Trans. AGU*, 86(52), Fall Meet. Suppl., Abstract T51A-1321.

Morrow, C., Radney, B., Byerlee, J., 1992. Frictional strength and the effective pressure law of montmorillonite and illite clays. *In* Evans, B., and Wong, T.F. (Eds.), *Fault Mechanics and Transport Properties of Rocks*. San Diego, Calif. (Academic Press), 69–88.

Officer, C.B., 1974. *Introduction to Theoretical Geophysics*, New York (Springer-Verlag).

Scholz, C.H., 2002. *The Mechanisms of Earthquake Faulting, 2nd Edition*, New York (Cambridge University Press), 158–167.

Tanaka, H., Chen, W.M., Kawabata, K., and Urata, N., 2007. Thermal properties across the Chelungpu fault zone and evaluations of positive thermal anomaly on the slip zones: Are these residuals of heat from faulting? *Geophys. Res. Lett.*, 34: L01309, doi:10.1029/2006GL028153.

Terada, T., 1930. On the heat generated by the deformation of the Earth crust. *Bull. Earthquake Res. Inst.*, 8:377–383.

Authors

Yasuyuki Kano, Jim Mori, Ryo Fujio, Takashi Yanagidani, and Setsuro Nakao, Disaster Prevention Research Institute, Kyoto University, Gokasyo, Uji, 611-0011, Japan., e-mail: kano@rcep.dpri.kyoto-u.ac.jp.

Hisao Ito, Japan Agency for Marine Earth Science and Technology, 3173-25, Showa-machi, Kanazawa-ku, Yokohama, Kanazawa, 236-0001, Japan.

Osamu Matsubayashi, National Institute of Advanced Industrial Science and Technology, No.7, Higashi 1-1-1, Tsukuba, 305-8567, Japan.

Kuo-Fong Ma, National Central University, No.300, Jhongda Road, Jhongli City, Taoyuan County, 32001, Taiwan (R.O.C.).

54

The Coseismic Fault Weakening Processes Inferred from Frictional and Transport Properties of Fault Rocks, TCDP

by Hiroki Sone, Hiroyuki Noda, and Toshihiko Shimamoto

Faults weaken during coseismic slip, and many factors affect the weakening process. Among these are the frictional properties and permeability of fault-consisting materials. The Taiwan Chelungpu-fault Drilling Project recovered cores from 500 m to 2000 m depth, penetrating several distinct major fault planes that are likely the slip planes from the 1999 Chi-Chi Earthquake (Ma et al., 2006). We measured transport properties of fault rocks and conducted high-velocity friction experiments using drill core samples from two fault zones at 1111 m and 1153 m depths (FZA1111 and FZA1153), and attempted to reconstruct the coseismic fault weakening processes due to the change in frictional coefficient and the effect of thermal pressurization (Sibson, 1973).

Gas permeability of the fault rock samples were measured in a tri-axial apparatus using a steady-flow method with nitrogen gas for several pore pressure gradients (0.1–2.0 MPa) at effective pressures up to 140 MPa. Results show that permeability values range between 10–14 and 10–18 m^2, and also that permeability is 1–2 orders lower in FZA1111 than in FZA1153. Data under various pore pressure gradients allowed us to subtract the Klinkenberg effect for gas permeability, and thus we estimate the water permeability to be several factors lower. High-velocity frictional experiments were conducted at a normal stress of 0.7 MPa, using a high-velocity ring-shear friction apparatus. Velocity was continuously controlled so that the velocity history reproduces the coseismic slip history derived from a study based on waveform inversions (Ji et al., 2003). The frictional coefficient decreased with acceleration and slip down to 0.1–0.2, and then recovered as the fault slip decelerated to a level below its initial friction. This indicates that a velocity weakening behavior was observed as the fault weakened by slip weakening as well.

The coseismic fault weakening process was modeled using these laboratory-derived physical properties based on equations of thermal pressurization by Lachenbruch (1980). The overall features of the resulting fault strength curve matched well with fault traction curves from inversion studies (see Fig. 1). Also the introduction of thermal pressurization seemed to enhance the difference between laboratory- and inversion-derived curves, in terms of both stress drop and weakening distance, Dc. However, temporal resolutions of waveform inversions are limited by the frequencies of the waveforms used, and thus the stress drops and weakening distances are not directly comparable. Recent studies show that inversion results preserve information on the seismological fracture energies accurately, but not the stress drop values and weakening distances (Guatteri and Spudich, 2000). We determined the apparent seismological fracture energies from the laboratory derived curve, but the physical interpretation of such manipulation was indefinite. It is important to understand which features are comparable, and which are not, between the laboratory-determined fault strength curves and traction curves derived from seismological studies.

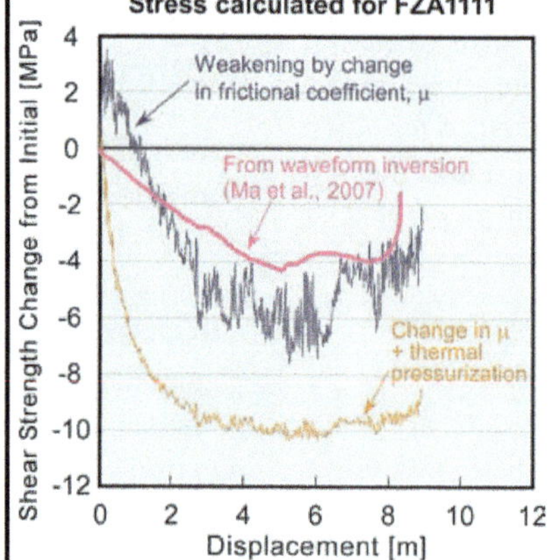

Figure 1. Comparison of the modeled and observed fault shear traction evolution.

References

Guatteri, M. and Spudich, P., 2000. What can strong-motion data tell us about slip-weakening fault-friction law? *Bull. Seismol. Soc. Am.*, 90:98–116, doi:10.1785/0119990053.

Ji, C., Helmberger, D. V., Wald, D. J. and Ma, K. F., 2003 Slip history and dynamic implications of the 1999 Chi-Chi, Taiwan, earthquake. *J. Geophys. Res.* 108(B9): 2412, doi:10.1029/2002JB001764.

Lachenbruch, A.H., 1980. Frictional heating, fluid pressure, and the resistance to fault motion. *J. Geophys. Res.*, 85:6097–6112.

Ma, K.F., Tanaka, H., Song, S.R., Wang, C.Y., Hung, J.H., Tsai, Y.B., Mori, J., Song, Y.F., Yeh, E.C., Soh, W., Sone, H., Kuo, L.W., and Wu, H.Y., 2006. Slip zone and energetics of a large earthquake from the Taiwan Chelungpu-fault drilling project. *Nature*, 444:473–476, doi: 10.1038.

Sibson, R.H., 1973. Interactions between temperature and pore fluid pressure during an earthquake faulting and a mechanism for partial or total stress relief. *Nature*, 243:66–68.

Authors

Hiroki Sone, Hiroyuki Noda, and Toshihiko Shimamoto, Department of Geology and Mineralogy, Division of Earth and Planetary Sciences, Graduate School of Science, Kyoto University, Kitashirakawa Oiwakecho, Sakyo-ku, Kyoto, 606-8502, Japan. e-mail: shima@kueps.kyoto-u.ac.jp

55

Drilling and Completion Challenges in Fault Zones: Lessons Learned from ICDP Projects

by Bernhard Prevedel

Introduction

Scientific drilling projects conducted within the framework of the International Continental Scientific Drilling Program (ICDP) and the German Continental Deep Drilling Program (KTB) represent important milestones in fault zone (FZ) drilling. This abstract intends to highlight the technical problems and potential solution pathways in such drilling programs as well as their business implications in today's hydrocarbon hype. Special attention will be paid to the fact that scientific wells today are no longer seen as pure drilling events, but rather as long-term downhole observatory installations, requiring specialized completion technology and advanced planning before drilling.

Lessons from the KTB and ICDP Projects

During the 9.1 km ultra deep KTB drilling program in Germany several fault zones in the drilled amphibolite facies crystalline rock suite were truncated. Severe drilling problems occurred primarily at a depth between 6.9 and 7.3 km depth in a main fault zone characterized by cataclasis, secondary mineralization, fluid inflow and a pronounced seismic reflectivity of the whole steeply dipping fault zone. Main contributors to the difficulties were stress-induced breakouts that increased rapidly with time especially in the central part of the fault to up to ten times the nominal drilled volume, as observed through repeated caliper logs. The other factor creating drilling setbacks were caliper reduction in a zone consisting of chlorite gneisses at 8300 m depth. Both issues finally made additional cementing, sidetrack drilling, or protracted casing setting necessary, and led to a time loss and a cost overrun of approximately 20%.

Similar breakout phenomena were observed in the drilling of main hole of the San Andreas Fault Zone Observatory at Depth (SAFOD), where the 8"hole section developed cave-outs of 20" and more in the lower 60 degree deviated section of the well from approximately 3.0 to 4.0 km, a depth range where the active trace of the San Andreas Fault is suspected.

Another observation during the drilling of fault zones is sudden changes in formation fluid pressure. For example, during the drilling of the 1000-m-deep Corinth borehole, a pressure surge of approximately 10 bars was observed at a depth of 760 m when penetrating the fault.

A selection of the most important lessons learned from previous drilling efforts in fault zones includes; 1) careful planning with multi-disciplinary industry and academia advisory panels is necessary, including extended well-bore stability planning; 2) use of a top drive plus vertical steering system with downhole motors is advised, to ensure operational speed and precision in drilling, and also in logging (including use of downhole MWD/LWD systems); 3) expandable casings or casing-drilling systems for loose and unstable formations or extended casing scheme with additional casings needs to be planned beforehand; 4) enhanced pumping and mud carrying capacity of the drilling mud to balance under- or overpressures is necessary; 5) wireline coring techniques for successful and rapid sample gain are advantageous; 6) synergy between research drilling projects and programs need to be developed; and 7) designs of scientific wells for life are necessary.

The last point represents a significant paradigm change from the way we regarded scientific well drilling in the past. Today such wells are planned and designed for long life, involving permanent monitoring arrays and completion technologies, very similar to the ones employed in the oil and gas industry.

Downhole Observatories

Completion technology is a term from the oil field industry that stands for the equipment installed in a well after the drilling phase, enabling the production of either (i) hydrocarbons in oil and gas or (ii) the acquisition of data from research wells. Although the objectives are different, the equipment designs and investment efforts are almost the same.

Long-term downhole data acquisition today is ensured from completed scientific wells by means of permanent monitoring technology. They can in principle be categorized in two basic embodiments:

Type-1: Outside the casing cemented cables and sensors (outside casing arrays = OCA), which are primarily based on (i) fiber-optic sensors and cables for distributed temperature, pressure, rock strain and acceleration measurements, as well as (ii) analogue sensors on copper cables for electric conductivity, seismic and pressure measurements.

The main characteristics for this entirely analogue sensor technology are the requirement for a pressure-isolating wellhead outlet design, its system ruggedness combined with highest redundancy and no external power requirement.

Type-2: Inside the casing semi-permanently installed tools (inside casing arrays = ICA), which are instruments run primarily on fiber/electrical armoured cable, like digital 3-component seismometer, tiltmeter, accelerometer and other logging equivalent sensors.

Main features of this primarily digital technology are its persistent problems of cable survival in long-term deployments, sensitivity to gas attack on elastomers, poor sensor coupling to the borehole wall, limited overall system redundancy, but good stability of the sensors in a high temperature environment.

Present Drilling Market

One key indicator of today's market development is the utilization of the worldwide rig fleet, both for marine and land drilling units. Their day-rates rose since 1994 by a factor of six to seven, with a rig utilization today consistently hitting 100%. As with rig rates, the service industry increased prices recently for equipment and services and further tightened delivery conditions. On an average, prices rose +18% in the last 6 months and delivery times for critical well equipment increased from half a year to one year and in some instances even to 17 months.

Due to the lack of drilling equipment suitable for scientific drilling and coring, the GFZ (GeoForschungsZentrum) Potsdam has decided to design and invest in its own proprietary scientific drilling rig, named Innova Rig. This totally new and fully hydraulic rig concept, built by a tunnel bore manufacturing works in southern Germany will be delivered by the end of 2007. The rig will be operated by a joint industry cooperation company for the drilling of both commercial as well as research projects. GFZ will be the owner of the rig while the drilling contractor partner will be in charge of staffing and daily operations.

Table 1. Technical specifications of Innova Rig:

Drilling depth	3500–5000 m
Regular hook-load	3500 kN
Nominal rotary speed	220 rpm (max.)
Rotary torque	40–75 kNm
Wire-line coring speed	500 rpm (max.)
Wire-line coring torque	12–18 kNm
Tripping speed	500 m/hr
Hoist cylinder stroke	22 m
Drive power upper limit	4000 kw
Rid weight (approx.)	370 tons
Mud pumps	3 x 1000kW
WL coring pump	1 x 350 kW
Mud pressure	350 bar (max.)
Mud tank system	240 m^3
General set	3 x 1540 kVA
Rack capacity	>7000 m
Wire-line coring winch	5500 m
Coring line	d=12.7 mm

The technical highlights of Innova Rig will be its special hybrid rig design and on-board equipment aiming especially at scientific drilling and integrating well proven drilling techniques around a variable application of rotary drilling and coring with the least possible environmental impact. Based on its container-module design, the hook-load will be directly supported by substructure, and rig moves will be greatly reduced in time due to optimized modular truck loads.

The entire energy management of all rig modules is designed form a holistic viewpoint, with highest safety standards due to full automation, integrated noise attenuation design and reduced footprint of the entire drill site. The hoist and feed system are fully hydraulic (no drilling line), constructed to handle doubles including a pipe manipulator to store the tubulars outside the rig on the pipe rack including a torque unit on the catwalk. (See Table 1.)

With this equipment planned, built, and operated for science, a land equivalent to the Integrated Ocean Drilling Program (IODP) research drilling vessels will be at hand for research projects and increase efficiency and hence reduce the high costs of such endeavors. Due to the dual-use concept in industry and academia however, the Innova Rig can be set aside for commercial industry operations and therefore maximize rig utilization and return of investment.

Concluding Remarks

Fault-zone and seismogenic zone drilling projects are complex and must be considered in their planning as well as operation phase as high-tier drilling operations. Due to the present boom in hydrocarbon and mineral exploration, the existing drilling equipment is very short in supply and quite expensive. The special engineering and equipment needs for scientific fault-zone drilling and monitoring instrumentation are not available off the shelf. Therefore, an extended close communication and interaction with research and industry experts are crucial. The establishment of advisory panels, the enhancement of exchange between IODP and ICDP and the joint development of procedures and engineering processes will be important steps to ensure safe fault-zone drilling and stable long-term monitoring observatories at depth in the future.

Author

Bernhard Prevedel, ICDP/OSG (Operational Support Group), GFZ-Potsdam, Telegrafenberg A34, D-14473, Potsdam, Germany, e-mail: prevedel@gfz-potsdam.de.

56

Drilling Challenge for the NanTro Fault Zone Drilling by *Chikyu*

by Kazuhiro Higuchi and Shomei Kobayashi

The *Chikyu* is a riser-equipped, dynamically positioned, scientific drill ship owned and operated by the Japan Agency for Marine-Earth Science and Technology (JAMSTEC). *Chikyu* was delivered in 2005 and is preparing for the IODP scientific drilling operation (Curewitz et al., 2006; Taira, 2005). For specifications, see Table 1.

To access scientific targets, deeper than ever before beneath the deep-sea floor, the target was set to 7000 m, in water depths initially up to 2500 m, eventually as deep as 4000 m (Taira, 2005). To drill deeper, weighted mud is required to keep holes stable within the pressure window higher than pore pressure and lower than fracture pressure. For the circulation of mud and transportation of gas/cuttings, specially designed risers are prepared for *Chikyu*. The high pressure BOP (Blow Out Preventer) aboard *Chikyu* is designed to control the borehole pressure balanced with mud hydrostatic pressure, even in occasional under-balanced conditions. Also, high-strength drill pipes are necessary to access deeper targets to endure high torque and high tensile stress. These drill pipes are designed to have large inside diameters to pass through the wireline retrievable core barrel. Several sizes of cemented casings are required to drill through various pressure/stress zones or fractures.

Although *Chikyu* has various state of the art equipment as described above for the deeper scientific drilling, well planned engineering is the key to achieving the scientific targets without encountering drilling hazards. Potential drilling hazards might be posed by sea and weather conditions, weak formation, shallow gas, low fracture gradients, abnormal pressure, high stress, fractured faults, high temperatures, or human error (Fig. 1).

The drilling engineer's responsibility is to accomplish the objective of the hole, safely, efficiently and economically. The objective might be the access to the target position, depth or formation, data from wireline logging or LWD, samples of core/cuttings/gas/water, testing, or the deployment of monitoring systems.

Table 1: *Chikyu* Specifications

Overall Length	210 m
Breadth (Mid)	38 m
Depth (Mid)	16.2 m
Draft	9.2 m
Derrick Height	107 m
Drillstring Length	10,000 m
Accommodation	150
Speed	~Symbol 10kt

Keys to the success of any drilling are to have accurate pressure, temperature and geology estimation, a good understanding of what might happen from offset wells, a choice of field-proven reliable tools and procedures, a simple design, preparation for back-up tools and contingency plans, and good communication with related people.

NanTroSEIZE (Tobin and Kinoshita, 2006) will commence from 2007 to drill a complex splay fault zone. The challenge lies in the lack of offset well data and difficulties regarding predictions of pressure, temperature, and stress. NanTroSEIZE drilling is the first scientific drilling by *Chikyu* and the first access to a large active splay fault of a plate-boundary thrust.

As scientific success is not achieved without drilling success, cooperation and communication between drilling engineers and scientists are essential points to consider.

References

Curewitz, D., Kuramoto, S., and Kawamura, Y., 2006. Drilling vessel *Chikyu*: Status, capabilities, and current operations. *Sci. Drill.*, 3:52–53.

Taira, A., 2005. The *Chikyu*: Meeting the challenges of a new scientific drilling era. *Sci. Drill.*, 1:32–33.

Tobin, H. and Kinoshita, M., 2006. NanTroSEIZE: The IODP Nankai Trough seismogenic zone experiment. *Sci. Drill.*, 2:23–27.

Authors

Kazuhiro Higuchi and Shomei Kobayashi, Center for Deep Earth Exploration (CDEX), Japan Agency for Marine-Earth Science and Technology (JAMSTEC), 3173-25 Showa-machi, Kanazawa-ku, Yokohama, Kanagawa 236-0001, Japan, e-mail: higuchik@jamstec.go.jp.

Figure 1. Potential drilling hazards.

57

Coring Techniques for Scientific Investigation of Faults

by Dennis L. Nielson

Introduction

Faulted and fractured rocks are difficult to core because of differential hardness, angular unsupported fragments, under- or over-pressured fluids, and overall fault architecture. Often, the most scientifically interesting parts of the fault are the most difficult to recover. The scientific objectives of the drilling and the techniques required (sampling, logging, monitoring), as well as the depth and geometry of the target, will determine the drilling methods used.

Operational Issues

The character of fault zones often leads to drilling problems. Due to the abrasive nature of faults, the bit may break or more likely wear on the face or inner and outer gauges. Bits may also plug with a piece of core, or their waterways may become blocked by clay or rock fragments. Cores may jam in the bit or core barrel due to small shifts along fractures during the drilling process. Bits can be custom-designed for specific drilling conditions, and bit life and penetration rate may be improved by size and placement of diamonds, hardness of matrix, and location and geometry of waterways.

Hole stability is a critical factor in operational efficiency; it can be controlled by the circulation of mud whose composition and circulation rate can be optimized. Coring in fault zones often encounters different fluid pressure within short distances. Formation overpressure (greater that hydrostatic) leads to the flow of fluids into the hole, and underpressure results in the loss of circulation in permeable rock. Under conditions where circulation is lost, drilling mud flows into the formation rather than returning to the surface. Cuttings can build up in the hole and stick to the drill pipe. This condition tends to be more serious in rotary drilling than coring, where it is common to core ahead without mud returning to the surface. Lightweight injected grout can be used to shut off lost circulation zones and stabilize fractured rock.

A fault can influence directional control, hole inclination, and doglegs (rapid changes in direction). In general, holes tend to curve into a fault, particularly when the core hole and the fault are both at steep angles. Additionally, the drill bit can become "trapped" within a fault and wander down a soft component, unable to escape. In most cases it is best to plan crossing the fault at as high an angle as possible.

The location of a fault in the subsurface is often uncertain, and intersecting a steeply dipping structure with a vertical borehole is problematic (Are we there yet?). Many coring systems are designed to drill holes at an angle as well as vertically. Directional coring from a borehole is also possible either using a wedge or whip-stock or downhole mud motors. Any of these options allows multiple intersections of a fault to take place from one borehole (multi-laterals).

Core Quality Control

Core is the principal product of most scientific drilling, and core quality is of utmost concern. High-quality core often requires additional effort, time, and cost. Ideally, the core is a continuous and representative sample of the fault; however, faults are generally broken formations (unless cemented by hydrothermal minerals) and can contain materials of dramatically different hardness. This increases the risk that pieces will be lost or that softer parts of the core may be washed away by the drilling fluid. Once the core is cut, mechanical and thermal stress relief are inevitable and result in the formation of secondary fractures.

Improper core handling can also lead to quality control problems, and it is best to have a well formulated protocol before the drilling phase of the project commences. This is especially important when different people are responsible for handling the core. Problems can result during removal of core from the core barrel, and during boxing, labeling, and transportation.

Bore Hole Quality Control

The process of drilling changes the borehole environment from *in situ* conditions. Mechanical disruption results from imposing a hole where there was previously solid rock, and borehole breakouts occur as stress is released. Circulation of drilling mud fills the hole with a fluid that is different from formation fluids in temperature, pressure, and composition. Separate zones that originally had distinct temperature, pressure, and chemistry mix in the borehole. The return to pre-drilling temperatures may require some time (or may never take place). Solids (mud, cement, lost circulation material [LCM]) have probably invaded the rock along faults, fractures, and intergranular spaces. Fault heterogeneity can mean that the contamination itself may not be uniform. Chemical and physical tracers can be introduced to assess contamination of the core by the drilling fluid.

Drilling Rigs and Coring Techniques

Several different types of drilling rigs are available, and they must be chosen to conform to the objectives of the investigation. Rotary rigs are designed to drill a hole with a high penetration rate, and they are principally used in petroleum and groundwater applications. Rotary drilling uses high weight on bit (WOB) and low rotation rate (rpm) in order to crush the rock. High mud volumes are needed to clear the crushed rock from the borehole. Core can be collected by a rotary rig using a conventional barrel, but the quality is subject to the operating parameters of high WOB and low rpm. In addition, the entire drill string must be tripped to recover core from the conventional barrel. This process can be time-consuming, and conventional coring is normally used for spot coring within a zone of interest.

Diamond coring rigs are designed to collect core samples. They do this by cutting the rock with a diamond bit with relatively low and precise control of the WOB and with high rotation rates. Generally, the penetration rate for coring is less than with rotary methods, and the cost/meter can be relatively high. However, the sample is generally superior to that collected by a rotary rig using a conventional barrel. Core is generally collected by wireline systems where the core barrel is retrieved to the surface using a wire while the drilling rods remain in the hole. Wireline coring was developed for mining applications, and coring equipment comes in standard sizes (PQ = 85 mm, HQ = 63.5 mm; NQ = 47.6 mm), although the labels are not particularly intuitive. Larger sizes may be more beneficial in the sampling of faulted rock. A variation on this technique uses a triple tube where the core is collected into a metal or plastic liner that is placed within the core barrel. The core is slightly smaller (PQ_3 = 83.0 mm, HQ_3 = 61.1 mm), but there is also much less sample disruption.

Slim-hole wireline coring systems are designed to core a telescoped hole where smaller standard sizes are designed to core through larger sizes (PQ>HQ>NQ). Therefore, if penetration is stopped, rods can be left in the hole as casing, and the next smaller size system can be used to continue coring.

Because of the need for the characteristics of both rotary and diamond coring, several hybrid rigs have been constructed. These rigs can drill rotary and set casing and then switch to wireline diamond coring for the collection of core samples. Examples of this type of rig include the DOSECC (Drilling, Observation and Sampling of the Earth's Continental Crust) Hybrid Coring System (DHCS) that has been used to core the Long Valley, Hawaii, and Chicxulub Scientific Drilling Projects. The DHCS can attach to most large rotary rigs, thereby limiting its shipping costs. ICDP's new Innova Rig is a self-contained rig that performs both coring and rotary functions.

Concluding Statement

Faults often have complex histories and, as a result, they show three-dimensional architecture with intricate cross- cutting features. A borehole (core sample) is small, and by itself is largely one-dimensional. Because of the complexity of faults and the quality control issues discussed above, it is advantageous to incorporate logging to support the interpretation of the core. In particular, imaging logs are of great value for their ability to orient fractures in the core and, in some cases, to document material that was lost during the coring process. Temperature logs are also of great value in determining the location and characteristics of zones where fluids are either entering or exiting the borehole. Multiple completions drilled directionally from the same borehole are useful for documenting fault heterogeneity.

Author

Dennis L. Nielson, DOSECC, Inc., P.O. Box 58857, Salt Lake City, Utah 84158-0857, U.S.A., e-mail: dnielson@dosecc.org.

58

Core Handling and Real-Time Non-Destructive Characterization at the Kochi Core Center: An Example of Core Analysis from the Chelungpu Fault

by Weiren Lin, Tetsuro Hirono, En-Chao Yeh, Wataru Tanikawa, and Wonn Soh

Abstract

As an example of core analysis carried out in active fault drilling programs, we report the procedures of core handling on the drilling site and non-destructive characterization in the laboratory. This analysis was employed on the core samples taken from Hole B of the Taiwan Chelungpu-fault Drilling Project (TCDP), which penetrated through the active fault that slipped during the 1999 Chi-Chi, Taiwan earthquake. We show results of the non-destructive physical property measurements carried out at the Kochi Core Center (KCC), Japan. Distinct anomalies of lower bulk density and higher magnetic susceptibility were recognized in all three fault zones encountered in Hole B. To keep the core samples in good condition before they are used for various analyses is crucial. In addition, careful planning for core handling and core analyses is necessary for successful investigations.

Introduction

An enormous and damaging earthquake (Mw7.6) occurred in west-central Taiwan on 21 September 1999 due to the convergence between the Philippine Sea and Eurasian plates (Shin and Teng, 2001). Its epicenter was located at the vicinity of the country town of Chi-Chi, and the hypocenter was at about 10 km depth. Abundant teleseismic observation data of the earthquake revealed that the slip displacement and slip velocity increased to as much as 8 m and 300 cm s^{-1}, respectively, but the high-frequency acceleration decreased when the rupture propagated from south to north. For this reason, it was suggested that the fault at the northern segment was lubricated during rupturing (Ma et al., 2003). In order to solve questions about the mechanism of earthquake generation and rupture propagation of the fault, the Taiwan Chelungpu-fault Drilling Project (TCDP) was undertaken (Ma et al., 2006). The drilling site is located in the northern segment where a large surface coseismic displacement occurred.

We jointed the TCDP and transported all the cores of Hole B taken from the depth range 950–1350 m to the KCC, Japan to conduct continuous non-destructive measurements. From the excellent scientific results, original papers were published (Hirono et al., 2006a, 2006b), while others have been submitted and some are still under preparation. Here, we only report the procedures of core handling and non-destructive characterization and show several valuable examples of the measurement results.

Core Handling on the Drilling Site

In order to keep the retrieved core samples in good condition, we paid close attention to i) prevent moisture change while the core was undergoing non-destructive measurement, and until the working half cores were split for individual sampling, ii) avoid contact with oxygen, and iii) keep them cool (but not less than 0°C) to minimize possible chemical reactions and/or biological activity. The most

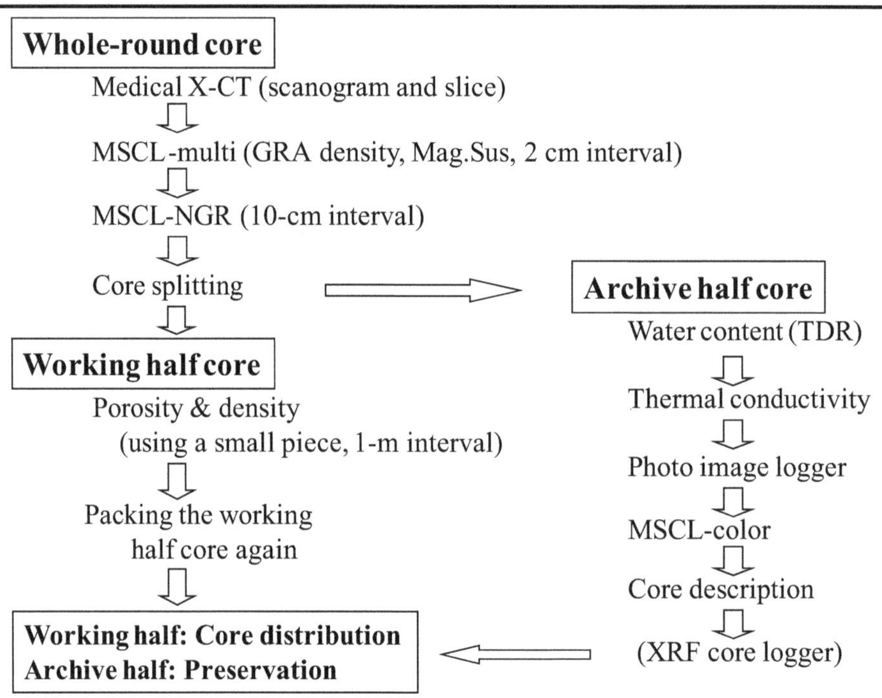

Figure 1. Flow of the non-destructive measurements using all the core samples taken from Hole B, TCDP. MSCL: Multi sensor core logger; GRA: gamma ray attenuation; NGR: natural gamma ray; TDR: time domain reflectometry; XRF: x-ray fluorescence.

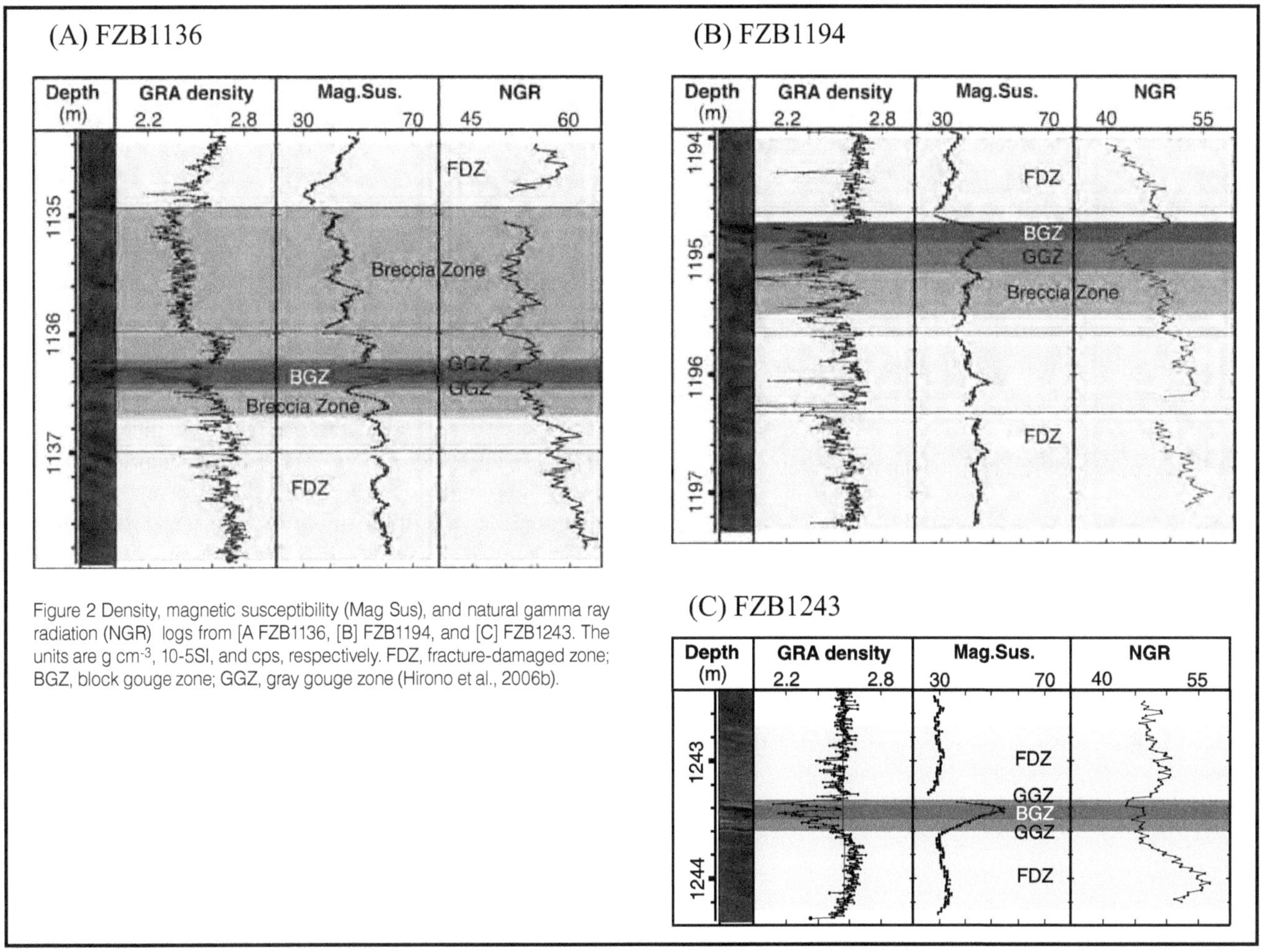

Figure 2 Density, magnetic susceptibility (Mag Sus), and natural gamma ray radiation (NGR) logs from [A FZB1136, [B] FZB1194, and [C] FZB1243. The units are g cm^{-3}, 10-5SI, and cps, respectively. FDZ, fracture-damaged zone; BGZ, block gouge zone; GGZ, gray gouge zone (Hirono et al., 2006b).

important thing for on-site core handling is packing the core samples as quickly as possible after core recovery. Needless to say, to prevent mechanical damaging during the planned shipping or land transporting to the KCC, Japan, we also filled any gaps between core and the wooden core box with cushioning materials. The flow chart of the routine works of core handling for all the cores of Hole B on the drilling site is as follows:

- Wash cores gently and carefully
- Cut core into an appropriate length to fit the size of core box
- Measure core length
- Take core picture for every 1-m with a scale and a color bar
- Jacket core with a tray half pipe and cover with shrink wrap
- Put core with a wet sponge into an aluminum pack with a very effective sealing ability.
- Replace air in the pack with N2 gas
- Remove extra N2 gas from the pack
- Seal the aluminum pack
- Put core in core box with cushioning materials
- Store the core box in a refrigeration container at 4°C even during shipping and land transporting.

Of course, we paid special attention to fault zone cores in handling them more quickly and carefully.

In order to determine *in situ* stress orientations and to estimate magnitudes, we did anelastic strain recovery (ASR) measurement using a few cores by the same method as Lin et al (2006). Because the anelastic strain recovers immediately from the *in situ* stress released by drilling, the measurements have to be conducted as quickly as possible after retrieving the core sample. Therefore, we did it at the on-site laboratory on the drilling site.

Non-Destructive Measurements at the KCC

Figure 1 shows the work flow for non-destructive measurements on the core of Hole B, TCDP conducted at the KCC. First, an x-ray CT image is taken, while the core sample is still in the aluminum wrapping (ideally without any change in moisture and without oxidation). Next the aluminum packaging was opened and subsequent measurements with the MSCL-multi (Multi-Sensor Core Logger), core splitting and the measurement of thermal conductivity were performed in an optimized process in order to minimize waiting time and avoid moisture vaporization. The physical properties obtained from the measurements are GRA (gamma ray attenuation), wet-bulk density, magnetic susceptibility, NGR (natural gamma ray), porosity, dry-bulk density, volumetric water content (by Time Domain Reflectometry, TDR indirect measurement), thermal conductivity, and color indexes (L*, a*, b* which correspond to brightness, chromaticity of red-green and blue-yellow, respectively). Moreover, x-ray CT scanograms, slice images on the whole-round core, and optical pictures on split surfaces are available. Unfortunately, the accuracy and reliability of P-wave velocity and electrical resistivity data by the MSCL were low, so that the data need further calibration and discussion before interpretation. In addition, mainly due to a problem with poor smoothness of the split core surface, the analyses to identify chemical elements by using XRF (x-ray fluorescence) core logger are also pending.

Examples of Measurement Results

(1) MSCL measurement results

The Hole B drilling penetrated the Chelungpu fault and recovered core samples ranging from about 950 m to 1350 m in vertical depth. Three fault zones—FZB1136 (fault zone at about 1136 m depth in Hole B), FZB1194, and FZB1243—were recognized in the core samples on-site as series within the Chelungpu fault system. The results of MSCL measurements (see Fig. 2) revealed distinct anomalies, lower wet-bulk densities, and higher magnetic susceptibilities within black gouge zones in all three fault zones (Hirono et al., 2006b). Higher magnetic susceptibilities can indicate that they have experienced intense shearing and/or frictional heating. The non-destructive continuous physical property measurements can provide important preliminary knowledge for understanding the faulting mechanism of the 1999 Chi-Chi earthquake.

(2) An ASR measurement result

We showed an example of the application of the ASR method for stress measurement on the drill core of the TCDP hole penetrated into the active Chelungpu fault. The anelastic strains of a drill core specimen in nine directions, including six independent directions, were measured using wire strain gauges after having retrieved the core from the borehole to ground level (Lin et al., 2007). The core specimen was sandstone taken from a depth of 592 m in Hole A. Acquired anelastic strains (Fig. 3A) were extensions – they reached several hundred microstrains, which is a level high enough to ensure satisfactory measurement accuracy. These strains were used for a three-dimensional analysis to determine the orientations and estimation of the magnitudes of the principal *in situ* stresses. The orientations determined by the strain data are shown in Fig. 3B. The estimated magnitudes of the maximum, intermediate, and minimum principal stresses are 14.6 MPa, 12.6 MPa, and 12.1 MPa, respectively. The obtained results can be considered as valid; consequently, it can be said that the anelastic strain recovery measurement is well suited to the task of directly determining the orientations of principal *in situ* stresses and of estimating the magnitude of the stresses at great depth. The results of theses stress measurements suggest both the orientation and magnitude of current stress at the vicinity of the Chelungpu fault in TCDP holes might be influenced by the fault rupturing.

Summary

As an example of core analysis carried out for the active fault drilling program, we reported the procedures of core handling and non-destructive characterization employed for the core samples taken from Hole B of the Taiwan Chelungpu-fault Drilling Project. Then, we showed two examples of the results; one is the non-destructive physical property measurements carried out at the KCC, Japan. Distinct anomalies of lower bulk density and higher magnetic susceptibilities were recognized in all three fault zones encountered in Hole B. Another example is carried out on the drilling site. Anelastic strain was measured by using a core sample after the *in situ* stress was released. It shows the anelastic strain result was satisfactory and can be used for determining three-dimensional principal stress orientations and estimating their magnitudes. Needless to say, the studies using core samples are important for scientific drilling, specifically for active fault drilling programs. However, it is more important

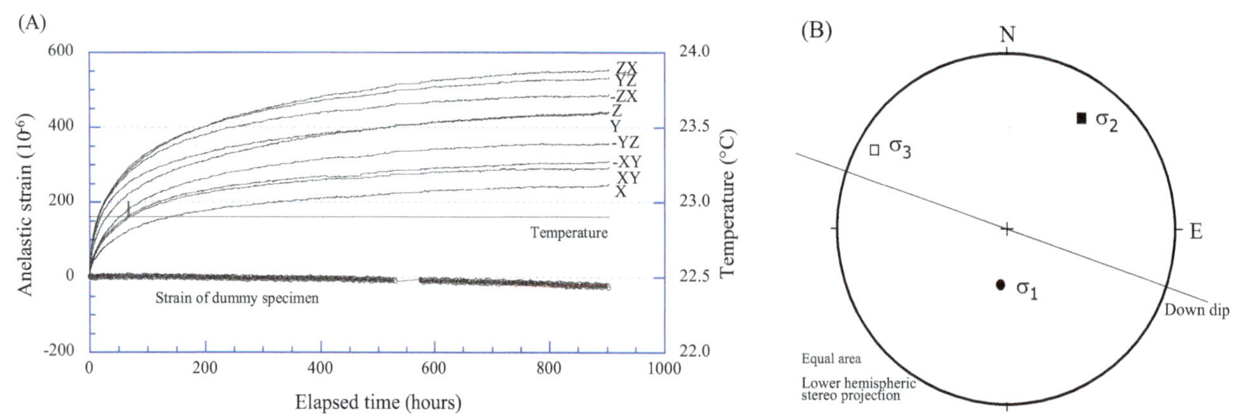

Figure 3. [A] An example of anelastic strain of a sandstone core sample from a depth of about 600 m in Hole A. [B] Three-dimensional *in situ* stress orientation determined by the ASR data shown in [A] (Lin et al., 2007).

to keep cores in the good conditions before they are used for various analyses. Careful planning for core handling and core analysis is necessary for successful investigations.

Acknowledgements

First, we thank the principal investigators, Y.-B. Tsai, C.-Y. Wang, K.-F. Ma, S.-R. Song, and J.-H. Hung of TCDP for giving us the opportunity to use all the cores of Hole-B for the measurements at KCC. We also thank the working group in Japan including Y. Hashimoto, H. Sone, L-W. Kuo, O. Matsubayashi, K. Aoike, H. Ito, M. Kinoshita, M. Murayama, and other colleagues from CDEX/JAMSTEC and Marine Work Japan, Ltd., and many students from Kochi University. We are very grateful for much kind cooperation of B. Lin and Y. Wei and others of the drilling company, as well as Taiwanese assistants and students from NCU and NTU during the handling of cores on site.

References

Hirono, T., Ikehara, M., Otsuki, K., Mishima, T., Sakaguchi, M., Soh, W., Omori, M., Lin, W., Yeh, E.-C., Tanikawa, W., and Wang, C.-Y., 2006a. Evidence of frictional melting from disk-shaped black material, discovered within the Taiwan Chelungpu fault system. *Geophys. Res. Lett.*, 33:L19311, doi:10.1029/2006GL027329.

Hirono, T., Lin, W., Yeh, E.-C., Soh, W., Hashimoto, Y., Sone, H., Matsubayashi, O., Aoike, K., Ito, H., Kinoshita, M., Murayama, M., Song, S.-R., Ma, K.-F., Hung, J.-H., Wang, C.-Y., and Tsai, Y.-B., 2006b. High magnetic susceptibility of fault gouge within Taiwan Chelungpu fault: Nondestructive continuous measurements of physical and chemical properties in fault rocks recovered from Hole B, TCDP. *Geophys. Res. Lett.*, 33:L15303, doi:10.1029/2006GL026133.

Lin, W., Kwasniewski, M., Imamura, T., and Matsuki, K., 2006. Determination of three-dimensional *in situ* stresses from anelastic strain recovery measurement of cores at great depth. *Tectonophysics*, 426:221–238, doi:10.1016/j.tecto.2006.02.019.

Lin, W., Yeh, E.-C., Ito, H., Hirono, T., Soh, W., Wang, C.-Y., Ma, K.-F., Hung, J.H., and Song, S.-R., 2007. Preliminary results of stress measurement by using drill cores of TCDP Hole-A: An application of anelastic strain recovery method to three-dimensional *in situ* stress determination, *Terr. Atm. Ocean Sci.*, 18:379–393, doi:10.3319/TAO.2007.18.2.379(TCDP).

Ma, K.-F., Brodsky, E.E., Mori, J., Ji, C., Song, T.-R.A., and Kanamori, H., 2003. Evidence for fault lubrication during the 1999 Chi-Chi, Taiwan, earthquake (Mw7.6), *Geophys. Res. Lett.*, 30:1244–1247, doi:10.1029/2002GL015380.

Ma, K-F, Tanaka, H., Song, S.-R., Wang, C.-Y., Hung, J.-H., Tsai, Y.-B., Mori, J., Song, Y.-F., Yeh, E.-C., Soh, W., Sone, H., Kuo, L.-W., and Wu, H.-Y., 2006. Slip zone and energetics of a large earthquake from the Taiwan Chelungpu-fault drilling project. *Nature*, 444:473–476, doi:10.1038/nature05253.

Shin, T-C. and Teng, T.-L., 2001. An overview of the 1999 Chi-Chi, Taiwan, Earthquake. *Bull. Seismol. Soc. Am.*, 91:895–913, doi:10.1785/0120000738.

Authors

Weiren Lin, Wataru Tanikawa, and Wonn Soh, Kochi Institute for Core Sample Research, Japan Agency for Marine-Earth Science and Technology, B200 Monobe, Nankoku, Kochi, 783-8502 Japan, e-mail: lin@jamstec.go.jp.

Tetsuro Hirono, Department of Earth and Space Science, Graduate School of Science, Osaka University, Toyonaka 560-0043, Osaka, Japan.

En-Chao Yeh, National Taiwan University, No.1, Dec. 4, Roosevelt Road, Taipei, 106147, Taiwan (R.O.C.).

59

Third-Party Borehole Seismic Experiments during the Ocean Drilling Program

by Ralph A. Stephen, Stephen A. Swift, S. Thompson Bolmer, and Hartley Hoskins

Introduction

The first borehole seismic experiments on DSDP and ODP were two-ship Oblique Seismic Experiments (Stephen, 1979; Stephen, et al., 1979, 1980; Swift, et al., 1988). By recording on the drill ship and shooting explosives out to ranges of 8 km, the upper 1.5 km of the upper crust (Layer 2) adjacent to the borehole could be imaged (Fig. 1; Stephen and Harding, 1983). Azimuthal anisotropy (Stephen, 1981, 1985) and lateral heterogeneity (Stephen, 1988; Swift and Stephen, 1989) could also be studied by shooting circles of shots at a fixed range from the borehole.

Western Atlantic South of Bermuda (DSDP Hole 418, Leg 102)

Third-party borehole seismic experiments on the Ocean Drilling Program began with an Oblique Seismic Experiment at Site 418 in the Western Atlantic south of Bermuda on Leg 102. The experiment confirmed the velocity structure of upper Layer 2 including azimuthal anisotropy and azimuth-dependent scattering. It is interesting to note that travel-time and amplitude data from two Oblique Seismic Experiments on 110-ma crust in the slow-spreading western North Atlantic and fast-spreading northwestern Pacific show that compressional velocities within layer 2 are, within experimental error, identical (Kong et al., 1985).

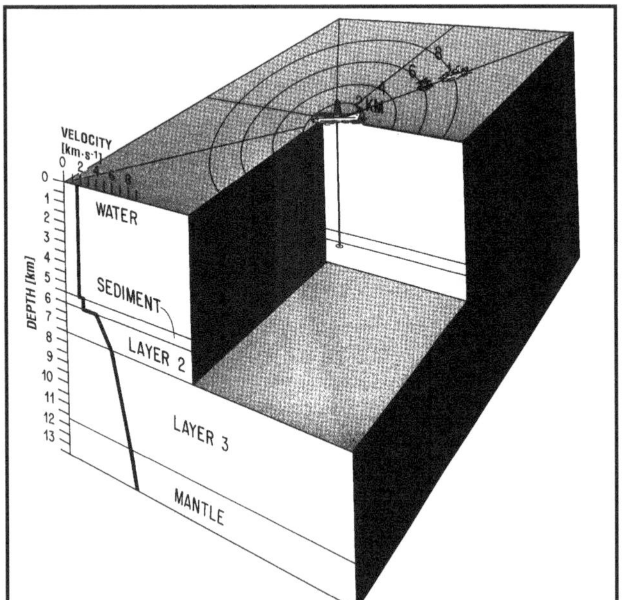

Figure 1. In an Oblique Seismic Experiment the drill ship (top center) maintains its position over the borehole and monitors the downhole geophone while a shooting ship sails concentric circles and radial lines. A generalized oceanic crustal structure and velocity-depth function are shown. The object at 8km is the shooting ship. The object at 6km is intended to represent the splash from a shot in the water (from Little and Stephen, 1985).

Southwest Indian Ridge (Hole 735B, Leg 118)

Normal incidence Vertical Seismic Profiles (VSP) were carried out on ODP Legs 104, 109, and 111 before the VSP at Hole 735B on Leg 118 on the Southwest Indian Ridge. This experiment measured velocities corresponding to Layer 3 which was consistent with the gabbroic petrology of the cores. Anomalously high attenuation was also observed, prompting the hypothesis that the gabbro cored may not actually represent the bulk of Layer 3 material.

Argo Abyssal Plain (Hole 765, Leg 123)

In thick sedimentary sequences VSPs can be very useful in correlating the drilling results with the seismic reflection profiling results. This was demonstrated on ODP Leg 123 which drilled in the Argo Abyssal Plain (Leg 123 Shipboard Scientific Party, 1990a; b). VSPs were also carried out on ODP Legs 127/128, 129, 131, and 146.

Costa Rica Rift (Hole 504B, Legs 111 and 148)

The VSP data acquired at Hole 504B in the eastern equatorial Pacific on Leg 148 helped to constrain the velocity-depth structure at the site (Swift, et al., 1996, 1998a, 1998b) and showed that upper Layer 3 at this site, at a depth of over 2 km into the crust, did not consist of gabbros but rather consisted of the lower portions of the sheeted dykes (Detrick et al., 1993). VSPs were also carried out on Leg 156.

Gas Hydrates on the Blake Ridge (Holes 994, 995, and 997; Leg 164)

Both offset and normal incidence VSPs were run on Leg 164 to study the seismic velocity structure of gas hydrates on the Blake Ridge, offshore South Carolina. Seismic velocities measured in three drill holes through the gas hydrate deposit indicated that substantial free gas exists to at least 250 m beneath the bottom-simulating reflector (BSR; Holbrook

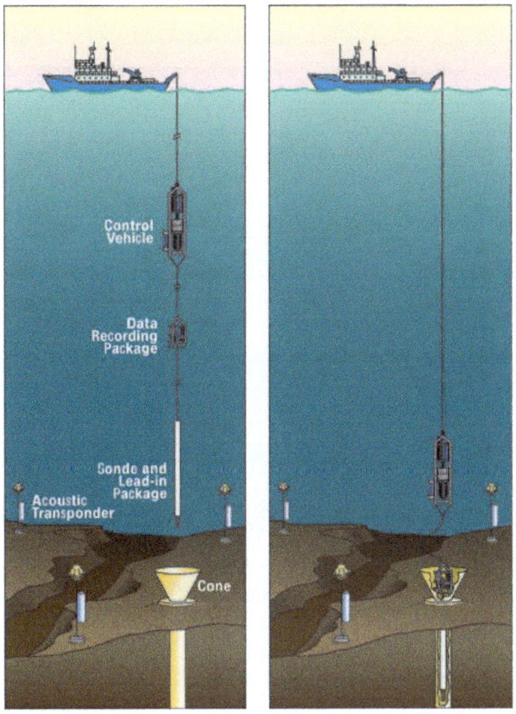

Figure 2. In the Wireline Re-entry System, a borehole sonde, data recording package, and control vehicle are suspended from a conventional research vessel on coaxial or fiber-optic cable (left). The control vehicle, navigated within a network of acoustic transponders, is used to guide the sonde into the borehole. The sonde is lowered into the borehole until the data recording package lands in the re-entry cone (right). When the seafloor and subseafloor systems are operating properly, the control vehicle is disconnected from the data recoding package and is recovered back onboard the ship. The borehole system acquires data autonomously for up to a year until the system is recovered by grappling. From Stephen, et al., 2003.

et al., 1996). Both methane hydrate and free gas exist even where a clear BSR is absent. The low reflectance, or blanking, above the BSR is caused by lithologic homogeneity of the sediments rather than by hydrate cementation.

Seismology with the Wireline Re-entry System

The ODP era also saw the development of systems for re-entering boreholes from conventional research vessels and submersibles after the drill ship left the site (Fig. 2; Legrand et al., 1989; Montagner et al., 1994a; Spiess et al., 1992). Borehole seismic experiments and installations that used wireline re-entry technology were carried out in DSDP Holes 534 (Blake-Bahama Basin;Bradley et al., 1997; Stephen et al., 1994) and 396 (Mid-Atlantic Ridge at 23degrees north; Montagner et al., 1994a) and in ODP Hole 843B (south of Oahu; Stephen et al., 2003). The latter experiment (the Ocean Seismic Network Pilot Experiment) carried out a test of three configurations of broadband seafloor seismic installation in preparation for extending the Global Seismic Network to the deep ocean.

Broadband Borehole Seismology

A new innovation on ODP was the deployment of broadband seismometers in boreholes (Suyehiro et al., 2006). Whereas the conventional VSPs and offset VSPs mentioned above operate in the frequency range of 1–100 Hz (similar to the band used for reflection and refraction seismology), broadband seismometers are used in earthquake seismology and operate in the frequency range of 0.001–10 Hz. The first broadband borehole seismometer test was carried out from the drill ship on Leg 128 in the Japan Sea in 1989 (Kanazawa et al., 1992; Suyehiro et al., 1992). The first submersible-assisted broadband borehole seismometer test was carried out on the Mid-Atlantic Ridge (Montagner et al., 1994a, 1994b). On the Ocean Seismic Network Pilot Experiment a broadband borehole sensor in ODP Hole 843B was compared with a shallow buried and a seafloor seismometer for a period of over three months (Stephen et al., 2003). Subsequently, four permanent broadband borehole seismic observatories were installed in the western Pacific and Japan Trench on Legs 186, 191 and 195 (Araki 1999; Araki et al. 2004; Suyehiro et al., 2002). Three other boreholes have been drilled specifically for borehole seismic installations: at the Ninety-east Ridge, near the Hawaii-2 Observatory, and in the equatorial Pacific.

Acknowledgements

RAS would like to thank the Earthquake Research Institute at the University of Tokyo for a six-month Visiting Professorship during which this synthesis was carried out.

References

Araki, E., 1999. Geophysical nature of broadband seismic signals in deep oceans. PhD thesis, University of Tokyo, Tokyo.

Araki, E., Shinohara, M., Sacks, S., Linde, A., Kanazawa, T., Shiobara, H., Mikada, H., and Suyehiro, S., 2004. Improvement of seismic observation in the ocean by use of seafloor boreholes. *Bull. Seism. Soc. Am.*, 94:678–690.

Bradley, C.R., Stephen, R.A., Dorman, L.M., and Orcutt, J.A., 1997. Very low frequency (0.2-10.0Hz) seismoacoustic noise below the seafloor. *J. Geophys. Res.*, 102(11):703–711,718, doi:10.1029/96JB3183.

Detrick, R.S., Collins, J.A., Stephen, R.A., and Swift, S.A., 1994. *In situ* evidence for the nature of the seismic layer 2/3 boundary in oceanic crust, Nature, 370:288–290, doi:10.1038/370288a0.

Holbrook, W.S., Hoskins, H., Wood, W.T., Stephen, R.A., Lizarralde, D., and Leg 164 Science Party, 1996. Methane hydrate and free gas on the Blake Ridge from vertical seismic profiling. *Science*, 273:1840–1843, doi: 10.1126/science.273.5283.1840.

Kanazawa, T., Suyehiro, K., Hirata, N., and Shinohara, M., 1992. Performance of the ocean broadband downhole seismometer at site 794. *In* K. Tamaki, Suyehiro, K., Allan, J., McWilliams, M., et al. (Eds.), *Proceedings of the Ocean Drilling Program, Scientific Results 127/128 Part 2*, College Station, Texas (Ocean Drilling Project), 1157–1171.

Kong, L., Brocher, T.M., and Stephen, R.A., 1985. Spreading rate independence of oceanic seismic layer 2. *Geophys. Res. Lett.*, 12:219–222.

Leg 123 Shipboard Scientific Party, 1990a. Explanatory Notes. *In* Ludden, J.N., Gradstein, F.M., et al., (Eds.), *Proceedings of the Ocean Drilling Program, Initial Report.* College Station, Texas (Ocean Drilling Program), 27–59.

Leg 123 Shipboard Scientific Party, 1990b. Site 765., *In* Ludden, J.N., Gradstein, F.M., et al (Eds.), *Proceedings of the Ocean Drilling Program, Initial Report.* College Station, Texas (Ocean Drilling Program), 63–267.

Legrand, J., Echardour, A., Floc'h, H., Floury, L., Gieskes, J., Harmegnies, F., Loaec, G., Pozzi, J.-P., Raer, Y., and Stephen, R., 1989. Campagne FARE: Wireline reentry of DSDP Hole 396B using the NADIA system. *EOS Trans. Am. Geophys. Union*, 70:729–730, 741.

Little, S.A. and Stephen, R.A., 1985. Costa Rica Rift borehole seismic experiment, Deep Sea Drilling Project Hole 504B, Leg 92. *Init. Repts. DSDP,* 83:517–528.

Montagner, J.P., Karczewski, J.-F., Romanowicz, B., Bouaricha, S., Lognonne, P., Roult, G., Stutzmann, E., Thirot, J.-L., Brion, J., Dole, B., Fouassier, D., Koenig, J.-C., Savary, J., Floury, L., Dupond, J., Echardour, A., and Floc'h, H., 1994a. The French Pilot Experiment OFM-SISMOBS: first scientific results on noise level and event detection. *Physics of the Earth and Planetary Interiors*, 84:321–336, doi: 10.1016/003 1/9201(94)90050-7.

Montagner, J.P., Romanowicz, B., and Karczewski, J.-F., 1994b. A first step toward an oceanic geophysical observatory. *EOS Trans. Am. Geophys. Union*, 75:150–151, 154.

Spiess, F.N., Boegeman, D.E., and Lowenstein, C., 1992. First ocean-research-ship-supported fly-in re-entry to a deep ocean drill hole. *Mar. Tech. Soc. J.*, 26:3–10.

Stephen, R.A., 1979. The oblique seismic experiment in oceanic crust - equipment and technique. *Mar. Geophys. Res.*, 4:213–226, doi: 10.1007/BF00286406.

Stephen, R.A., 1981. Seismic anisotropy observed in upper oceanic crust. *Geophys. Res. Lett.*, 8:865–868.

Stephen, R.A., 1985. Seismic anisotropy in the upper oceanic crust. *J. Geophys. Res.*, 90:(11):383–311, 396.

Stephen, R.A., 1988. Lateral heterogeneity in the upper oceanic crust at DSDP Site 504. *J. Geophys. Res.*, 93:6571–6584.

Stephen, R.A., and Harding, A.J., 1983. Travel time analysis of borehole seismic data. *J. Geophys. Res.*, 88:8289–8298.

Stephen, R.A., Koelsch, D., Berteaux, H., Bocconcelli, A., Bolmer, S., Cretin, J., Etourmy, N., Fabre, A., Goldsborough, R., Gould, M., Kery, S., Laurent, J., Omnes, G., Peal, K., Swift, S., Turpening, R., and Zani, C., 1994. The Seafloor Borehole Array Seismic System (SEABASS) and VLF Ambient Noise. *Mar. Geophys. Res.*, 16:243–286, doi: 10.1007/BF01224745.

Stephen, R.A., Louden, K.E., and Matthews, D.H., 1979. The oblique seismic experiment on Deep Sea Drilling Project leg 52, *In* Donnelly, T., Francheteau, J., Bryan, W., Robinson, P., Flower, M., and Salisbury, M. (Eds.), *Initial Reports of the Deep Sea Drilling Project, 51-53,* Washington, DC (U.S. Government Printing Office), 675–704.

Stephen, R.A., Louden, K., and Matthews, D.H., 1980. The oblique seismic experiment on DSDP Leg 52. *Geophys. J. R. Astron. Soc.*, 60:289–300.

Stephen, R.A., Spiess, F.N., Collins, J.A., Hildebrand, J.A., Orcutt, J. A., Peal, K.R., Vernon, F.L., and Wooding, F.B., 2003. Ocean seismic network pilot experiment. *Geochem. Geophys. Geosyst.*, 4:1092, doi: 1010.1029/2002GC000485.

Suyehiro, K., Araki, E., Shinohara, M., and Kanazawa, T., 2002. Deep sea borehole observatories ready and capturing seismic waves in the Western Pacific. *EOS Transactions (Supplement)*, 83:621, 624–625, doi: 10.1029/2002EO000420.

Suyehiro, K., Kanazawa, T., Hirata, N., Shinohara, M. and Kinoshita, H., 1992. Broadband downhole digital seismometer experiment at Site 794: A technical paper, *In* Tamaki, K., Suyehiro, K., Allan, J., McWilliams, M., et al. (Eds.), *Proceedings of the Ocean Drilling Program, Scientific Results*, College Station, Texas (Ocean Drilling Program), 1061–1073.

Suyehiro, K., Montagner, J.-P., Stephen, R.A., Araki, E., Kanazawa, T., Orcutt, J., Romanowicz, B., Sacks, S., and Shinohara, M., 2006. Ocean seismic observatories. *Oceanography*, 19:144–149.

Swift, S.A., and Stephen, R.A., 1989. Lateral heterogeneity in the seismic structure of upper oceanic crust, Western North Atlantic. *J. Geophys. Res.*, 94:9303–9322.

Swift, S.A., Hoskins, H., and Stephen, R.A., 1996. Vertical seismic profile into upper oceanic crust at Hole 504B, *In* Alt, J.C., Kinoshita, H., Stokking, L.B., Michael, P.J., et al. (Eds.), *Proceedings of the Ocean Drilling Program, Scientific Results*, Washington, DC (U.S. Government Printing Office), 339–347.

Swift, S.A., Kent, G.M., Detrick, R.S., Collins, J.A., and Stephen, R.A., 1998a. Oceanic basement structure, sediment thickness, and heat flow near Hole 504B. *J. Geophys. Res.*, 103:15377–15391.

Swift, S.A., Lizarralde, D., Hoskins, H., and Stephen, R.A., 1998b. Seismic attenuation in upper oceanic crust at Hole 504B. *J. Geophys. Res.*, 103:27193–27206.

Swift, S.A., Stephen, R.A., and Hoskins, H., 1988. Structure of upper oceanic crust from an Oblique Seismic Experiment at site 418A, Western North Atlantic, *In* Salisbury, M.H., Scott, J.H., et al. (Eds.), *Proceedings of the Ocean Drilling Program, Scientific Results*, Wasthington, DC (U.S. Government Printing Office), 97–113.

Authors

Ralph A. Stephen, Department of Geology and Geophysics, Woods Hole Oceanographic Institution, Clark South 282, MS#24, Woods Hole, Mass. 02543, U.S.A., e-mail: rstephen@whoi.edu

Stephen A. Swift, Department of Geology and Geophysics, Woods Hole Oceanographic Institution, Clark South 286A, MS#24, Woods Hole, Mass. 02543, U.S.A.

S. Thompson Bolmer, Department of Geology and Geophysics, Woods Hole Oceanographic Institution, Clark South 286B, MS#24, Woods Hole, Mass. 02543, U.S.A.

Hartley Hoskins, Computer and Information Services, Department of Geology and Geophysics, Woods Hole Oceanographic Institution, Woods Hole, Mass. 02543, U.S.A.

60

Borehole Seismic and Strain Observatories in Seafloor Settings—Experiences after ODP Legs 186, 191, 195 and Future Plans

by Eiichiro Araki and Kiyoshi Suyehiro

Borehole Seismo-Geodetic Observatories

In the period of Ocean Drilling Program Legs 186, 191, and 195, we successfully deployed a set of seismo-geodetic sensors in deep seafloor boreholes in four locations (JT1, JT2, WP1, and WP2). The JT1 and JT2 borehole observatories are located in the landward slope of the Japan Trench, where we expect seismic activity beneath the observatories. These borehole observatories gave us good platforms to look at the dynamic process on the plate boundary in the seafloor.

The WP1 and WP2 are considered a part of a global seismic network of observatories currently distributed mostly on land and ocean islands. For permanent seismic observation in the seafloor setting, installation of broadband seismic sensors in the seafloor borehole was recommended, and candidate locations were chosen (Suyehiro et al., 2006). The WP1 and WP2 sites are one of the first borehole observatories to implement such a permanent seafloor observatory.

We called these seafloor borehole observatories "NEREID" ('Neath Seafloor Equipment for Recording Earth's Internal Deformation). The NEREID observatories are characterized by several key features that are thought to acquire optimum performance from the borehole sensors.

Firstly, sensors are cemented at the bottom of the borehole. This is very important because long-period seismic and tilt/strain sensors are sensitive to fluid flow around them, so it is necessary to couple them to the ground very well. Cementing sensors at the bottom of the borehole was able to fix sensors rigidly in place, and eliminated fluid around the sensors. Cementing is necessary especially for strainmeters, because the strain of the ground should be transmitted to the sensing cylinder of the strainmeter by some means. On land, it has been widely practiced for that purpose.

Secondly, separate cables connect borehole sensors uphole for power feeding and seafloor data recording. This is to deal with possible failure of the sensors, cables, and connectors in the borehole during and after the installation in the borehole. It worked very well, and we were able to replace failed sensors after the installation without affecting operation of other sensors.

Sensors deployed in these observatories were configured to target specifications needs in each location of the observatories. For JT1 and JT2 observatories, a tiltmeter (AG510), a strainmeter (dilatometer and three component types), and broadband seismometers (CMG1T, PMD) are chosen to mainly target the earthquakes and slow deformation processes below the observatory. The WP1 and WP2 observatories are important for monitoring global seismic activity, so two identical broadband seismometers (Guralp CMG1T) were installed for redundancy.

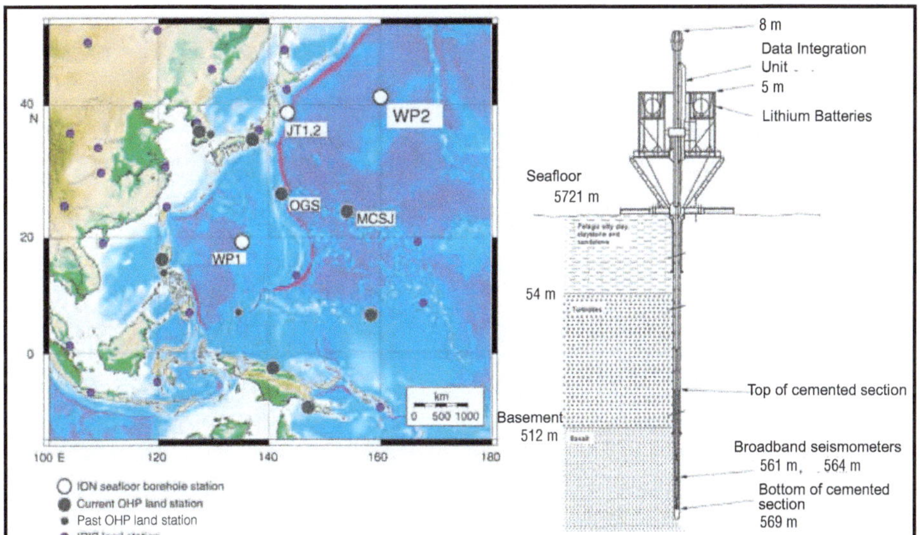

Figure 1. Location of NEREID borehole observatories (left: white circles), and schematic section of the NEREID observation system at WP1 (right).

Data from borehole sensors are recorded in the seafloor. Electric power necessary for these sensors is provided by a stack of seawater batteries or lithium batteries stored in a titanium sphere. Data are recovered by ROV by retrieving the data recorder because the amount of the recorded seismic data was as large as 10GB per year.

We conducted visits to each observatory for data recovery and maintenance once a year or more. From most of the observatories,

Table 1. Period of data obtained from the NEREID observatories since 1999 (JT1, JT2, WP1, WP2)

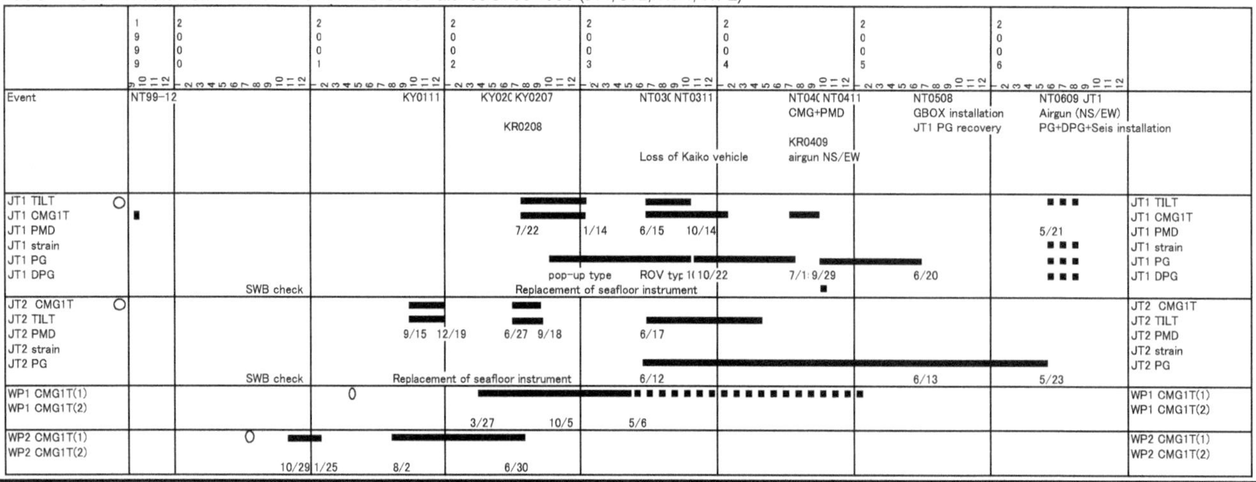

data for more than a year were recovered so far (Table 1). The observatories are still functional more than 5 years post-installation.

Improvement of Seismic and Tilt Measurement

The seismic data from the NEREID borehole observatories exhibited excellent performance with low background noise. The JT1 borehole seismometer showed less horizontal noise than a seismometer in the seafloor by 80 dB (Araki et al., 2004) in a 1000-second period. As a result, we were able to see effects of long period ocean gravity waves tilting the ground (at JT1 borehole 1.2 km below seafloor). This environment in the borehole enabled us to detect smaller teleseismic events (Shinohara et al., 2006), resulting in better understanding of seismic structure in the deep solid earth, such as structure of the transition zone of the Earth's mantle (Suetsugu et al., 2005).

Effects of earth tide were also seen in the seismic records from JT1, JT2, and WP2 boreholes. The observed tidal effect can be explained by deformation of ground due to tidal force on the solid earth. Broadband seismometers in the seafloor also show tidal effects, but this is usually the effect from tidal motion of water on the seafloor.

Furthermore, up to a year period, we investigated the stability of boreholes from tilt records. JT2 borehole tilt drifted at approximately 7 microradians per year. JT1 borehole showed a similar drift rate. JT1 and JT2 boreholes are located at a seismogenic plate boundary with different background seismicity. We sought slow events in the tilt record, as well as local earthquakes, to evaluate the activity of subducting plate boundary. No slow event was identified during the initial observation period in these boreholes.

Borehole seismometers are also very useful for observing local seismic activity. Figure 2 compares a small local earthquake's records from JT1 NEREID borehole observatory and nearby seafloor OBSs (Ocean Bottom Seismographs). The OBS records exhibit monotonic response around 10 Hz especially in the horizontal component. This prevents precise analysis of the hypocenter and mechanism of earthquakes. Borehole records do not show such a response for all components; therefore, installing seismometers in boreholes will be very important for the study of dynamic processes of earthquakes beneath the seafloor.

Future Plans

We continue observations in existing long-term observatories. Activities in these borehole observatories include not only data recovery but also renovation of seafloor equipment to allow very long-term (more than several years) continuous observations by tiltmeter.

On the basis of present experiences of seismo-geodetic observation in the existing ODP seafloor boreholes, we plan on future long-term borehole observation in the IODP era. We proposed to implement scientific long-term borehole observatories in the Nankai Trough south of Japan as a part of NanTroSEIZE project. The area where Nankai Trough drilling is conducted covers a whole subduction process of oceanic plate, before the subduction to seismo-genesis of mega-thrust earthquake of magnitude greater than 8.

The plan of Nankai Trough borehole observatory includes CORKs (see Becker and Davis, 2005) and a long-term obser-

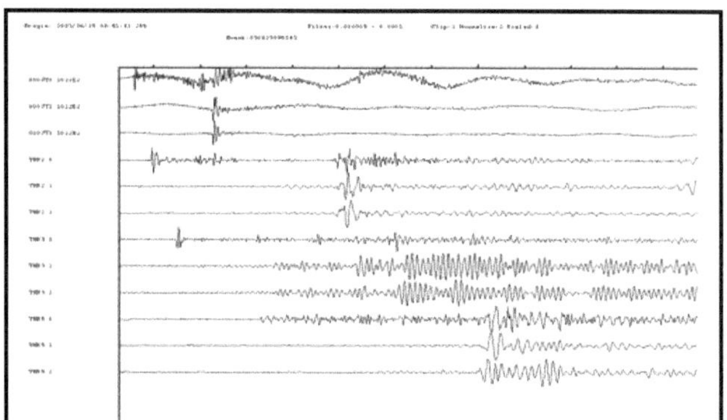

Figure 2. Local seismic records from the JT1 borehole seismometer (top three traces for vertical and horizontal components) and those from OBS around the JT1 borehole observatories (three stations [THK] each with three components).

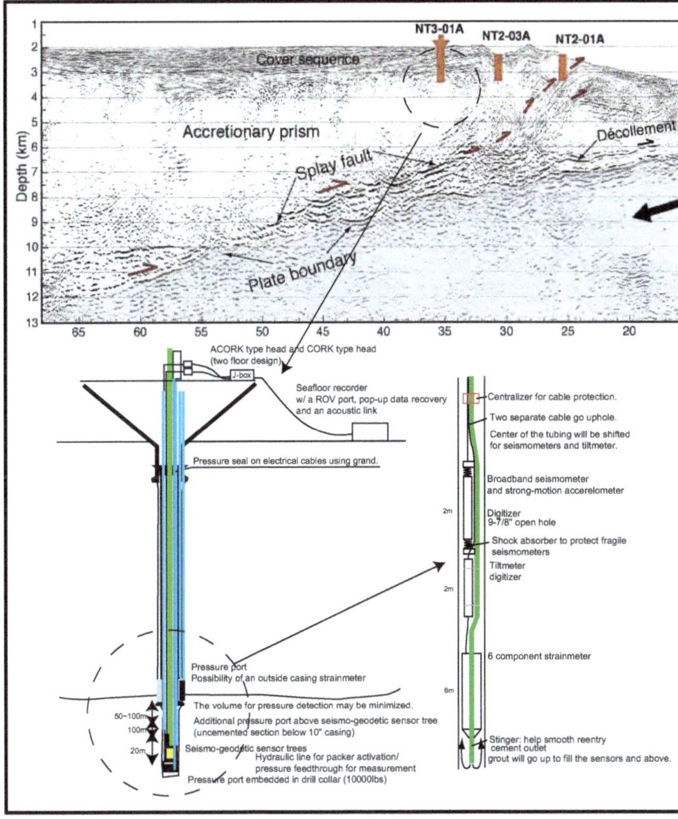

Figure 3. Schematic plan of the future borehole observatory in the Nankai Trough. Top: cross-section of borehole location, penetration depths with location of splay fault and the plate boundary to the Philippine Sea Plate (PSP). Only the boreholes drilled by non-riser drilling platform are displayed. Bottom left: Cross-sectional schematic of observation system of NT3-1 planned observatory. Bottom right: arrangement of bottom hole sensors (three component strainmeter, tiltmeter, broadband seismometer, and strong motion accelerometers).

vatory in a deep borehole of 3.5 km below the seafloor that requires a riser drilling platform. In addition to these observatories, we also plan to establish a seismo-geodetic-hydrological observatory that integrates the present A-CORK system for hydrological measurement and NEREID system for seismo-geodetic measurement. Figure 3 shows a schematic diagram of the observation system. At the bottom of the borehole drilled by a non-riser platform, we plan to install a three-component strainmeter, a tiltmeter, a strong motion accelerometer, and a broadband seismometer. We will also install a hydraulic tube below the seismic sensors to measure pore-pressure. By integrating various observation parameters obtained from the borehole observatory, we expect to obtain a better understanding of dynamics in the plate boundary of the Nankai Trough.

Understanding the seismogenesis of mega-thrust earthquakes, which is one of the major targets of the Nankai Trough drilling project (NanTroSEIZE), cannot be achieved solely by activities during drilling. Continuous long-term observations in these borehole observatories should play a key role in documenting and understanding ongoing subduction processes. Sufficient infrastructures and science-technology commitment are required to realize continuous long-term observations to achieve scientific goals. JAMSTEC will establish a scientific seismic network (DONET) around the area of the Nankai Trough drilling by 2010. The network utilizes submarine fiber optic cable to feed power and transmit data from instruments installed in the seafloor. We also plan to connect the borehole observatories to the DONET submarine cable network to assure long-term realtime observations using the IODP boreholes.

References

Araki, E., Shinohara, M., Sacks, S., Linde, A., Kanazawa, T., Shiobara, H., Mikada, H., and Suyehiro, K., 2004. Improvement of seismic observation in the ocean by use of seafloor boreholes. *Bull. Seismol. Soc. Am.*, 94:(2)678–690, doi: 10.1785/0120020088.

Becker, K. and Davis, E.E., 2005. A review of CORK designs and operations during the Ocean Drilling Program. *In* Fisher, A.T., Urabe, T., Klaus, A., and the Expedition 301 Scientists (Eds.) *Proceedings of the Integrated Ocean Drilling Program, Volume 301*, College Station, Texas (Integrated Ocean Drilling Program Management International, Inc.), 104, doi:10.2204/iodp.proc.301.104.2005.

Shinohara, M., Araki, E., Kanazawa, T., Suyehiro, K., Mochizuki, M., Yamada, T., Nakahigashi, K., Kaiho, Y., and Fukao, Y., 2006. Deep-sea borehole seismological observatories in the Western Pacific: temporal variation of seismic noise level and event detection. *Ann. Geophys.*, 49(2/3):625–641.

Suetsugu, D., Shinohara, M., Araki, E., Kanazawa, T., Suyehiro, K., Yamada, T., Nakahigashi, K., Shiobara, H., Sugioka, H., Kawai, K., and Fukao, Y., 2005. Mantle discontinuity depths beneath the West Philippine Basin from receiver function analysis of deep-sea borehole and seafloor broadband waveforms. *Bull. Seismol. Soc. Am.*, 95(5):1947–1956.

Suyehiro, K., Montagner, J.-P., Stephen, R.A., Araki, E., Kanazawa, T., Orcutt, J., Romanowicz, B., Sacks, S., and Shinohara, M., 2006. Ocean seismic observatories. *Oceanogr.*, 19(4):144–149.

Authors

Eiichiro Araki, Japan Agency for Marine-Earth Science and Technology (JAMSTEC), 2-15 Natsushima-cho, Yokosuka, 237-0061, Japan, e-mail: araki@jamstec.go.jp.
Kiyoshi Suyehiro, Japan Agency for Marine-Earth Science and Technology (JAMSTEC), 2-15 Natsushima-cho, Yokosuka, 237-0061, Japan.

BABHY—A New Strategy of Hydrofracturing for Deep Stress Measurements

by Takatoshi Ito, Kentaro Omura, and Hisao Ito

Paradox

Hydraulic fracturing in a vertical borehole induces fractures that will be vertical and normal to the minimum horizontal stress S_h (parallel to the maximum horizontal stress S_H), if there is no influence of natural fractures. The induced fractures close with venting and open with re-pressurization. At those times, there appear two kinds of critical borehole pressure, the reopening pressure P_r and the shut-in pressure P_s, which characterize the variation of borehole pressure during the test. The conventional theory tells us that those two pressures are related to the two stress components of S_H and S_h as follows (Haimson and Cornet, 2003):

$$P_r = 3S_h - S_H - P_p \qquad (1)$$

$$P_s = S_h \qquad (2)$$

Note that P_p is pore pressure in the fracture before opening. Those two equations give the principle for the two values of S_H and S_h to be determined from the two measured pressures of P_r and P_s.

The interpretation of Eq. (2) for P_s is supported by considerable experimental and theoretical works. On the other hand, if the interpretation of Eq. (1) for P_r is also correct, the measured values of P_r and P_s should change independently in response to the combination of S_H and S_h which will vary site by site. However, the data of field tests so far indicate that incidences where the measured reopening pressure lies close to the shut-in pressure (i.e., $P_r = P_s$) are far more numerous than can reasonably be expected (Lee and Haimson, 1989). This strange phenomenon could happen if the crust is in a stress condition of $(S_H - P_p)/(S_h - P_p) = 2$. Nevertheless, it is hard to consider that such a condition has been held everywhere in the crust. It may be more reasonable to consider that, contrary to the conventional theory, the measured reopening pressure does not coincide with the "true" reopening pressure (the borehole pressure at which the fracture truly begins to open from its mouth at the borehole wall) and that pressure takes the same value with the shut-in pressure (i.e., with S_h). If this is true, we could estimate with hydraulic fracturing only the minimum component of stress S_h but not the maximum component of stress S_H, which is the most desired concern in the stress measurement. Furthermore, a serious problem may occur should such a large error be included in the estimates of maximum stress S_H based on the reopening pressure so far.

True and Apparent Reopening Pressures

In order to explain the paradox described above, we have to take into account two factors which have been ignored in all conventional theory. Those factors are (i) residual aperture of fracture, and (ii) hydraulic compliance of test systems C. The C corresponds to an amount of fluid required to elevate fluid pressure in a test system by a unit magnitude, and it can be represented equivalently as $C = ßV_{eff}$, where $ß$ is the fluid compressibility, and V_{eff} is the effective system volume. While the details of explanation have been described by Ito et al. (1999, 2005, 2006), they can be outlined briefly as follows.

The residual fracture aperture causes pressure penetration into the fracture prior to opening. Evidence of this has already been shown by laboratory studies (Cornet, 1982; Durham and Bonner, 1994; Zoback et al., 1977). The pressure penetration will be almost wholly transmitted to the fracture surface since the net area of contact of the two surfaces is usually a small fraction of their nominal area. Thus, the third component in Eq. (1) should be borehole pressure rather than P_p. The borehole pressure at fracture opening is defined as P_r, and so substituting P_p with P_r in Eq. (1) yields

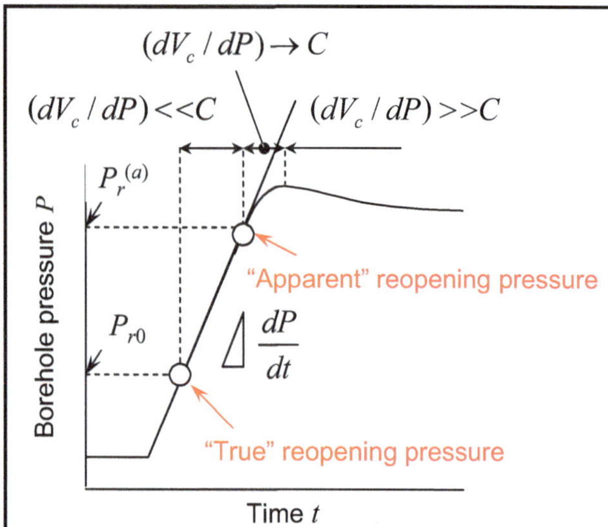

Figure 1. Effect of the system compliance C and fracture opening on borehole pressure variation.

$$P_r \equiv P_{r0} = \frac{1}{2}(3S_h - S_H) \qquad (3)$$

Thus, the effect of including pressure penetration into the fracture prior to opening is to reduce the reopening pressure by a factor of almost two from the value expected using conventional theory (the reduction is precisely two when pore pressure is negligible). We will refer the borehole pressure given by Eq. (3) to the true reopening pressure P_{r0}.

On the other hand, the influence of the compliance C on fracture opening is more problematic; it is concerned with the correct identification of the true reopening pressure from the borehole pressure P minus time t curves. Note that the reopening pressure is usually detected as the borehole pressure P at which the P - t curve is seen to deviate from linearity (Fig. 1). The effect of fracture opening on the borehole pressure variation can be expressed as follows (Ito et al., 1999):

$$\frac{dP}{dt} = \frac{Q}{(dV_c / dP) + C} \qquad (4)$$

where dV_c is the change in pressurized fluid volume due to fracture opening. Since the flow rate Q and the system compliance C are constant, Eq. (4) indicates that deviations of the P - t curve from linearity are governed by changes in the value of dV_c/dP and its relative value with respect to C. That is, prior to fracture opening, dV_c/dP is zero and the borehole pressure P increases linearly with t. After fracture opening, dV_c/dP becomes greater than zero, and the P - t curve will deviate from linearity to some degree or other. However, even if a flexible hydraulic tube with small ID (less than 10 mm) is used to convey fracturing fluid from a pump to a test section in a borehole, C is considerably large. As a result, at the early stage of fracture opening, dV_c/dP should be very small compared with C so that no detectable change occurs on the P - t curve, as shown schematically in Fig 1. When P reaches a level of S_h, the stress acting normally to the fracture surface becomes almost equal to or less than the value of S_h anywhere. Such a balanced stress condition leads to the criticality that the fracture aperture increases abruptly with a small increment in borehole pressure. As a result, dV_c/dP becomes a larger value compared with C, and finally the P - t curve begins to deviate from the initial linear trend. The same process occurs regardless of the S_H value. Thus, we provide an explanation as to why incidences where the apparent (or measured) reopening pressure coincides with P_s (i.e., the minimum stress S_h are so common as described above). We will denote hereafter the apparent reopening pressure as $P_r^{(a)}$.

A Strategy for the Maximum Stress Measurements

Thus the strange observation of $P_r^{(a)} = P_s$ in field tests arises because the compliance of typical hydraulic fracturing systems is far larger than that of fracture until P reaches a level of minimum stress S_h. However, it should be recalled here that the compliance in concern is that of the volume between the flow meter and the fracture mouth. Taking this into account, if the flow meter is placed as close as possible to the test interval, as illustrated in Fig. 2, the system compliance C can be reduced drastically, and a more objective measure of flow entering the fracture can be obtained. In this case, it is not a matter of course what kind of tubing (drill pipe)—flexible tube or stainless pipe with small ID—is used to convey fracturing fluid from a pump to a test interval. To demonstrate this idea, we developed a test system with a downhole flow meter (Ito et al., 2002). The system is basically the same as the conventional one except that the transducer to measure flow rate of injection is installed at the top of a straddle packer tool. Due to this modification, we succeeded in reducing the system compliance C drastically. The straddle packer tool is conveyed in boreholes on 6-conductor wireline. A single high-pressure hose is used to supply pressure from a hydraulic pump at the ground surface to the packer elements and the straddle interval so that a switch valve controllable from the surface is attached to the straddle packer. The system is designed to use in a borehole with 101 mm (HQ size) diameter at depths up to 1 km.

However, such a modification as above is still not sufficient to achieve the stress measurement at depths more than 1 km because of the following reasons.

(a) The stress measurement at deep depths cannot be done, of course, without deep boreholes, which generally have a large diameter, and accordingly the straddle packer tool needs to be large. The large size of the straddle packer tool leads to an increase in the system compliance C.

(b) For monitoring and recording flow rate and pressure during tests of the transducer installed on top of the straddle packer tool, the transducers should be connected with a data acquisition system placed at the ground surface by wires. To do this, it is appropriate to convey the straddle packer tool in

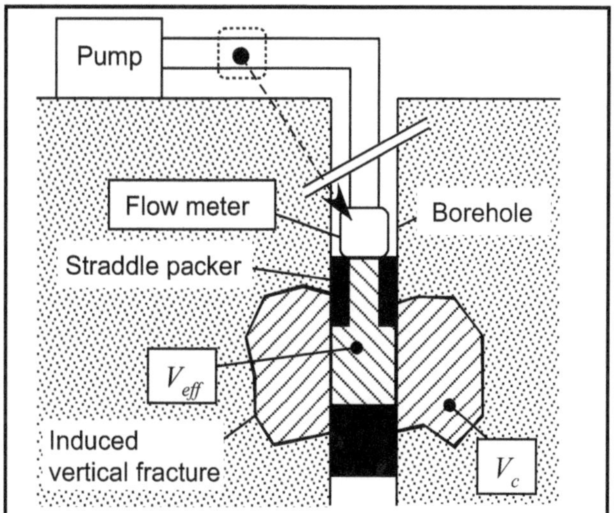

Figure 2. A concept to reduce the system compliance C (i.e., the volume, Veff) by placing the flow meter just above the straddle packer tool.

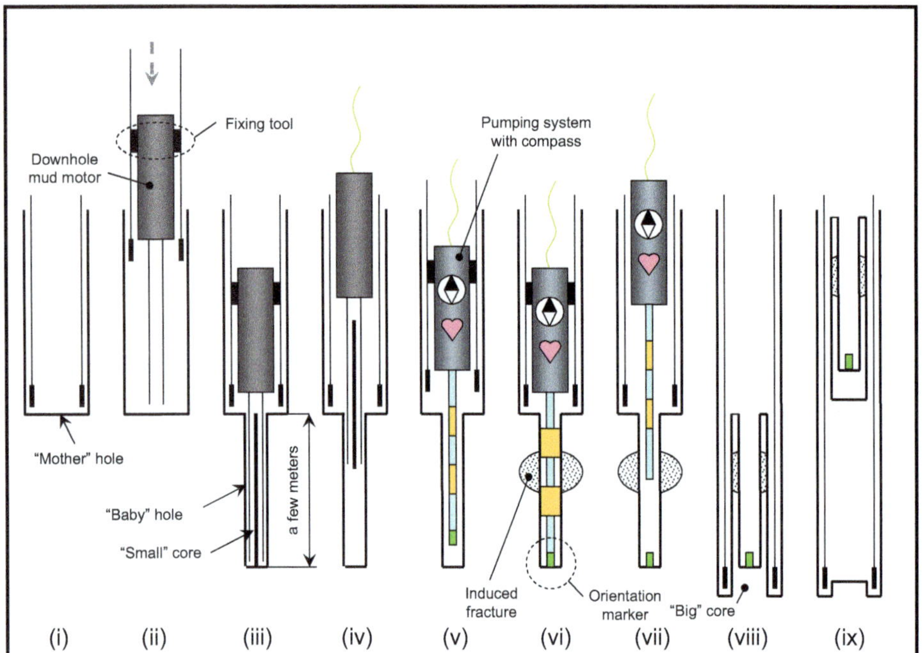

Figure 3. Proposed new strategy, BABHY, to achieve stress measurements by hydraulic fracturing at depths of more than 1 km, and its procedures. The procedure consists of three parts as follows: (i–iv), drilling the baby hole; (v–vii), *in situ* testing of hydraulic fracturing; and (viii–ix), extending the mother hole and retrieving the large core.

boreholes on wireline. The use of wireline is also effective to save the time for the tool running in boreholes. However, as the depth of measurement becomes greater, so does the risk for the tool to become stuck in the borehole. The financial risk in losing the advantages of wireline logging is very severe. For this reason, the straddle packer tool has generally been conveyed so far on drill pipe in the case of deep measurement, but the use of drill pipe makes it hard to arrange the wires connecting the downhole transducers and the surface data acquisition system.

Realistic Proposal: BABHY

Such a dilemma could be solved by a new strategy appropriate for the stress measurement at deep depths, as shown in Fig. 3. There are two components used in the strategy: (i) the compact drilling tool with a built-in mud motor, and (ii) the straddle packer tool with a pump and a digital compass. Each is conveyed in drill pipe on wireline. The compact drilling tool is used to drill an additional hole, several tens of millimeters in diameter and a few meters in length, at the bottom of an original borehole, and the hydraulic fracturing is carried out in the drilled hole by using the small packer tool. The additional hole and the original borehole are referred to the "baby" hole and the "mother" hole, respectively. The procedure can be outlined as follows.

i) Set drill pipe with coring bit in the mother hole.

ii) Lower the compact drilling tool in drill pipe on wireline and fix it onto the drill pipe.

iii) Pump drilling mud through the drill pipe to drive the mud motor in the compact drilling tool, and drill the baby hole at the bottom of the mother hole.

iv) Retrieve the compact drilling tool and the small core, inspect pre-existing fractures in the core, and determine the depth of test section(s) in the baby hole.

v) Lower the straddle packer tool in the drill pipe on wireline, and fix it onto the drill pipe.

vi) Lower the drill pipe slightly to squeeze the packer element for isolating the test interval, and pressurize the test interval to induce axial fractures by using the pump installed in the tool. During the test, the pressure and flow rate of injected fluid and the tool orientation are monitored by the transducers installed in the tool and transmitted through wireline to the data acquisition system at the surface.

vii) Retrieve the straddle packer tool while leaving the orientation marker at the bottom of the baby hole.

viii) Lower and set a core barrel, and drill out the test section for getting the big core.

ix) Retrieve the big core, and inspect the fractures induced by pressurization in step (vi). The fracture orientation can be determined from the orientation marker in the core and the tool orientation recorded in step (vi).

Subsequently, the reopening pressure $P_r^{(a)}$ and the shut-in pressure P_s will be detected from the records of pressure and flow rate during the test. Finally, the stress magnitudes S_H and S_h will be estimated from those detected pressures based on Eqs. (2) and (3), assuming $P_r^{(a)} = P_{r0}$, since the system compliance is to be sufficiently small, and the stress orientation will be estimated from the fracture orientation

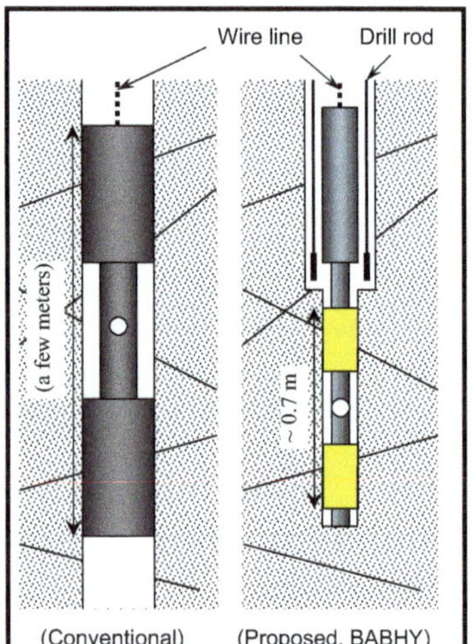

Figure 4. The compact size of the BABHY system makes it easy to find the test section free from pre-existing fractures.

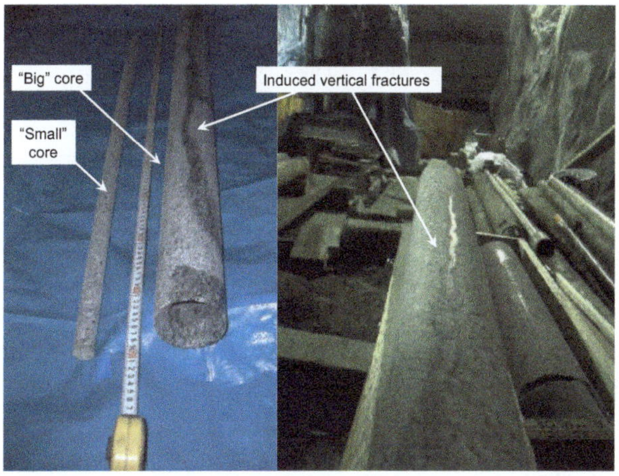

Figure 5. "Small" core (AQ size) retrieved by the baby hole drilling and "big" core (PQ size) retrieved by the over-coring. Hydraulically-induced fractures appeared in the big core.

detected in step (ix). We call this strategy Baby Borehole Hydrofracturing, BABHY for short.

This strategy will allow us to improve many defects in the conventional method as follows. It is easy to reduce the system compliance sufficiently because of a very compact size of the straddle packer tool (Fig. 4). The test section being free from pre-existing fractures can be chosen with certainty by the inspection of the small core, and then the straddle packer tool can be adjusted as the pressurized interval to be located at the chosen test section rightly. Note that the axial length between the top of the upper packer element and the bottom of the lower packer element is very short (less than 1 m) compared with that of a few meters for the conventional tool. The shorter length will make it much easier to choose the test section. We have already completed development of the compact drilling tool, and we are now developing the straddle packer tool. They are designed to be used in the mother holes with diameters larger than 101 mm (HQ size). As a part of the development process, we carried out a field test to confirm the procedure of step (vi) in particular (installing the straddle packer in the baby hole and carrying out hydraulic fracturing) at the Kamioka mine in Japan. For this test, we used a vertical borehole with 123 mm (PQ size) diameter and about 30 m deep drilled from the floor of a chamber at a depth of about 500 m from the ground surface. The baby hole with 47 mm (AQ size) diameter and 1 m length was drilled at the bottom of the borehole. The test succeeded quite well so that a pair of typical fractures in axial direction was induced, and the shape and orientation of the induced fractures were clearly detected from the large retrieved core (Fig. 5).

References

Cornet, F.H., 1982. Analysis of injection tests for *in situ* stress determination. *In* Zoback, M.D., Haimson, B.C., and Jacobson, M.L. (Eds.), *Proceedings of Workshop XVII, Workshop on Hydraulic Fracturing Stress Measurement.* Menlo Park, Calif. (The Survey), 414–443.

Durham, W.B. and Bonner, B.P., 1994. Self-propping and fluid flow in slightly offset joints at high effective pressures. *J. Geophys. Res.*, 99:9391–9399.

Haimson, B.C. and Cornet, F.H., 2003. ISRM suggested methods for rock stress estimation – Part 3: hydraulic fracturing (HF) and / or hydraulic testing of pre-existing fractures (HTPF). *Int. J. Rock Mech. Min. Sci.*, 40:1011–1020.

Ito, T., Evans, K., Kawai, K., and Hayashi, K., 1999. Hydraulic fracture reopening pressure and the estimation of maximum horizontal stress. *Int. J. Rock Mech. Min. Sci. & Geomech. Abstr.* 36:811–826.

Ito, T., Igarashi, A., Ito, H., and Sano, O., 2006. Crucial effect of system compliance on the maximum stress estimation in the hydrofracturing method: Theoretical considerations and field-test verification. *Earth Planets Space*, 58:963–971.

Ito, T., Kato, H., Karino, H., and Hayashi, K., 2002. Hydrofrac stress measurements on true reopening pressure and development a system for the measurements at depths up to 1000 m. *In* Choi, S.-Y., et al (Eds.), *Proceedings of 2002 ISRM Regional Symposium on Rock Engineering Problems And Approaches in Underground Construction*, 93–100.

Ito, T., Igarashi, A., Ito, H., and Sano, O., 2005. Problem for the maximum stress estimation by hydraulic fracturing method and its potential solution. *Proceedings of 40th U.S. Rock Mechanics Symposium*, Anchorage, Alaska: ARMA/USRMS 05-862 (CD-ROM).

Lee, M.Y. and Haimson, B.C., 1989. Statistical evaluation of hydraulic fracturing stress measurement parameters. *Int. J. Rock Mech. Min. Sci. & Geomech. Abstr.*, 26:447–456, doi::10.1016/0148-9062(89)91420-4.

Zoback, M.D., Rummel, F., Jung, R., and Raleigh, C.B., 1977. Laboratory hydraulic fracturing experiments in intact and pre-fractured rock. *Int. J. Rock Mech. Min. Sci. & Geomech. Abstr.*, 14:49–58, doi:10.1016/0148-9062(77)90196-6.

Authors

Takatoshi Ito, Institute of Fluid Science, Tohoku University, 2-1-1 Katahira Aoba-ku Sendai, 980-8577, Japan, e-mail: ito@ifs.tohoku.ac.jp.

Kentaro Omura, National Research Institute for Earth Science and Disaster Prevention, Tennodai 3-1, Tsukuba, Ibaraki 305-0006, Japan.

Hisao Ito, Japan Agency for Marine-Earth Science Technology (JAMSTEC), 3173-25 Showa-machi, Kanazawa-ku Yokohama, Kanazawa, 236-0001, Japan.

62

Long-term Monitoring in Deep Boreholes in the Nankai Subduction Zone

by Hisao Ito

Introduction

IODP NanTroSEIZE is planning to construct observatories in deep boreholes in the Nankai subduction zone to monitor interseismic behavior at and above the updip limit of seismogenic zone. The deep observatories will be at NT2-03 and NT3-01; NT2-03 will be 3.5-km deep below sea floor to penetrate several splay faults, and NT3-01 will be 6-km deep to penetrate the splay faults and plate boundary fault.
Scientific needs for long term *in situ* monitoring at and near splay faults and plate boundary fault include detailed analysis of low frequency/slow events, strain partioning during interseismic period, and near-source hydrological/seismic/geodetic observations.

Scientific Needs

A deep borehole observatory has advantages in surface noise reduction (Araki et al., 2004), high frequency recording with little attenuation, observation in the close vicinity of active processes, and installing a vertical sensor array along the borehole to improve the depth resolution (Chavarria et al., 2003).

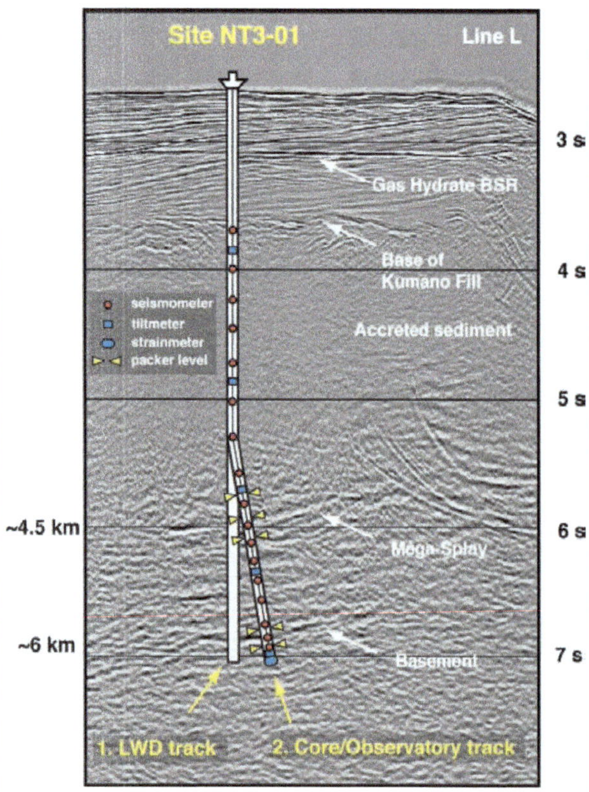

Figure 1. Schemtic observatory plan (Tobin and Kinoshita, 2006).

Drilling through the updip limit of the fault and establishing borehole observatories is important to understand the following issues:

- Crustal deformation throughout the hanging wall to the fault zone
- How the strain caused by the backstop and subducting plate affect the fault plane strain
- Existence of any slip across the fault
- Is the fault locked? Is the fault weak?
- Correlation with asperity inferred from 3-D seismics
- Importance of mesoscopic structure for dynamic rupture
- Slow deformation in fault zone before dynamic rupture
- Structure of plate boundary/splay fault
- Occurrence of dynamic rupture in thin slip zones (main slip) and slow ductile deformation in thick fault zones.

Answers to the following questions may also be provided from a borehole observatory:

- Can both (thin slip zone and thick fault zone) occur within "asperity"?
- How thin or thick are the zones?
- How do they rupture coseismically ?
- How do they deform in an interseismic period?
- How are they deforming now ?
- How different are they in plate boundary / splay fault?

NanTroSEIZE scientists proposed an observatory plan (Harold and Kinoshita, 2006; Shinohara et al., 2003). One of the most essential parts of the basic idea is a distributed, multi-level multi-sensor system (Fig. 1), with seismometer, tiltmeter, strainmeter, and pressure sensors.

Possible Plan

Based on this basic concept, we have been working on a possible plan for sensors, a downhole telemetry system, a sensor/downhole telemetry system interface, and system installation. We understand that distributed sensors are essential. Currently, strain and pressure sensors need to be set at the bottom of the hole, and tilt/seismic/temperature monitors may be set at bottom- and mid-hole sections (Fig. 2).

Major Technical Challenges

The following major technical issues for developing a deep borehole observatory in IODP are recognized:

1. Great depths telemetry, sensor, etc.
2. High temperatures (80°C–100°C for 3.5 km, 170°C–180°C for 6 km)
3. Long-term reliability and stability
4. Coupling of the sensors to the for mation/casing
5. Geodetic/seismic and pore pressure measurements at multiple intervals (multi-packer vs. multi-hole?)
6. Vertical drilling and core sampling at the fault interval
7. Broadband high dynamic range and high resolution recording
8. Deployment
9. Telemetry system through Christmas tree/wellhead system of riser drill holes
10. Real time monitoring through seafloor cables

Borehole Telemetry System

The system should be reliable and work for many years at high temperatures. At the same time, the system should be able to deliver data that scientists need. The system should be able to be deployed practically and maintained with reasonable efforts. The fundamental requirements may be summarized as follows: reliability, long mean-time-before-failure (MTBF), redundancy, protection from failure, high dynamic range, continuous recording, low power consumption, and limited number of connections through pressure controlled well head/Christmas tree.

The system consists of three sections: 1) a downhole module array that digitizes seismic signals continuously and transmits the data to a recorder on sea floor, 2) a subsea recorder that receives data from downhole and stores data, and 3) a communications unit that sends commands from sea surface to the subsea recorder to check the status of the downhole systems and receives QC data (Fig. 3).

The downhole module can also send data from other sensors (e.g., pressure, temperature, and strain monitors) interfaced to auxiliary channels at a lower sampling rate.

Sensor Development

Sensors/sensor housing/connectors should be developed to work under high temperatures and great depth/hostile installation conditions.

We must carefully consider sensor installation in deep boreholes and at multi-levels. Strainmeters have to be cemented in open hole sections of the borehole. Tiltmeters

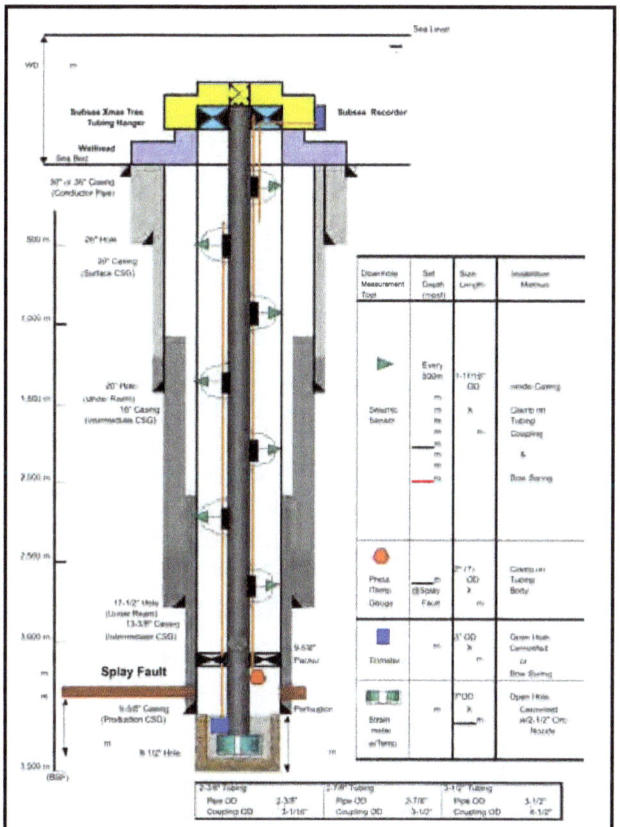

Figure 2. Schematic diagram of multi-level multi-sensor sensor configuration for NT3-02.

and seismometers might be cemented, however for multi-level installations we need to consider installing them by other methods (e.g., by locking arm or bow spring).

Pore pressure monitoring is useful as a proxy of strain and pore pressure variation along the fault (Davis et al., 2006; Kano and Yanagidani, 2006). To be useful as a proxy for strain, the compliance of the pressure measurement system needs to be extremely small. For monitoring pore pressure of the fault, the gault must be isolated from other sections of the borehole either by packer or cementing.

Links to Shallow Boreholes/ Land Stations

The NanTroSEIZE deep borehole observatories, which will deploy vertical array sensor systems, will be linked to a sea floor cable network and NanTroSEIZE shallow borehole observatories to be integrated to three dimensional ocean observatories. The IODP NanTroSEIZE observatories will be situated at/around the updip limit. Ideally, the IODP NanTroSEIZE observatories would also be integrated with on-land observatories, which would be above the asperities.

Possible Application to Other Projects

The telemetry system and sensors that have been developed for NanTroSEIZE will be applied in other deep boreholes, such as Costa Rica Seismogenesis Project CRISP (Ranero et al., this issue), and also The Kanto Asperity

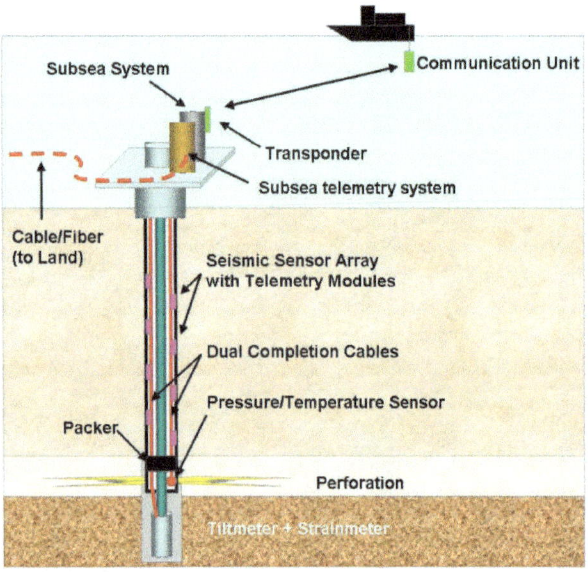

Figure 3. Long Term Borehole Monitoring System.

Project (KAP): KAP is a linked offshore/onshore regional geodetic and seismic network covering the asperity and non-asperity areas of the Kanto Region (Kobayashi et al., 2007). In the KAP, it is proposed to monitor strain, tilt, uplift and seismicity in the area underlain by asperities. This will show the relationship between plate motion, strain accumulation, and earthquake mode.

References

Araki, E., Shinohara, M., Sacks, S., Linde, A., Kanazawa, T., Shiobara, H., Mikada, H., and Suyehiro, K., 2004. Improvement of seismic observation in the ocean by use of seafloor boreholes. *Bull. Seismol. Soc. Amer.*, 94:678–690, doi:10.1785/0120020088.

Chavarria, J.A., Malin, P.E., and Shalev, E. 2004. The SAFOD Pilot Hole seismic array: Wave propagation effects as a function of sensor depth and source location. *Geophys. Res. Lett.*, 31: L12S07, doi:10.1029/2003GL019382.

Davis, E., Becker, K., Wang, K., Obara, K., Ito, Y., and Kinoshita, M., 2006. A discrete episode of seismic and aseismic deformation of the Nankai trough subduction zone accretionary prism and incoming Philippine Sea plate. *Earth Planet. Sci. Lett.*, 242:73–84, doi:10.1016/j.epsl.2005.11.054.

Kano, Y. and Yanagidani, T. 2006. Broadband hydroseismograms observed by closed borehole wells in the Kamioka mine, central Japan: Response of pore pressure to seismic waves from 0.05 to 2 Hz. *J. Geophys. Res.*, 111:B03410, doi:10.1029/2005JB003656.

Kobayashi, R., Nakao, S., Nishimura, T., Curewitz, D., Matsu'ura, M., Sagiya, T., Satake, K., Ito, T., Stein, R., Malin, P., Shalev, E., Stephen, R., Sato, T., Mochizuki, K., Hori, T., Kamiya, S. and Naka, S., 2007. Kanto Asperity Project network: a wide geodetic and seismic network covering Kanto asperity and non-asperity regions, IODP CDP proposal #723, http://www.iodp.org/index.php?option=com_docman&task=doc_download&gid=1550.

Shinohara, M., Araki, E., Kamata, M., Kinoshita, M., Kyo, N., Kuroki, K., Kosuge, Y., Kobayashi, S., Konno, S., Goto, T., Saito, S., Suzuki, M., Takahashi, T., Tadokoro, K., Tsunogai, U., Tezuka, K., Nanba, K., Nishi, M., Hino, R., Mikada, H., Morita, N., Yoshida, C., and Ito, H., 2003. Long-term monitoring using deep seafloor boreholes penetrating the seismogenic zone. *Bull. Earthquake Res. Inst. Univ. Tokyo*, 78:205–218.

Tobin, H.J., and Kinoshita, M., 2006. NanTroSEIZE: The IODP Nankai Trough Seismogenic Zone Experiment. *Sci. Drill.*, 2:23–27.

Tobin, H. and Kinoshita, M., 2007. Report on NanTroSEIZE Long-Term Observatories Workshop, July 17-19, 2006, San Jose and Parkfield, California.

Author

Hisao Ito, Center for Deep Earth Exploration (CDEX), Japan Agency for Marine-Earth Science and Technology (JAMSTEC), 3173-25 Showa-machi, Kanazawa-ku, Yokohama, Kanagawa 236-0001, Japan, e-mail: hisaoito@jamstec.go.jp.

Last Name	First Name	Country
Ando	Ryosuke	U.S.A.
Araki	Eiichiro	Japan
Ashi	Juichiro	Japan
Babaie	Hassan	U.S.A.
Behrmann	Jan	Germany
Ben-Zion	Yehuda	U.S.A.
Boullier	Anne-Marie	France
Brodsky	Emily	U.S.A.
Byrne	Tim	U.S.A.
Chester	Frederick M.	U.S.A.
Cornet	Francois H.	France
Davis	Earl	Canada
Doan	Mai-Lin	U.S.A.
Dresen	Georg	Germany
Ellsowrth	William L.	U.S.A.
Geraud	Yves	France
Guerin	Gilles	U.S.A.
Hashimoto	Yoshitaka	Japan
Hickman	Steve	U.S.A.
Higuchi	Kazuhiro	Japan
Hori	Takane	Japan
Hung	Jih-Hao	Taiwan
Hyndman	Roy D.	Canada
Ide	Satoshi	Japan
Ito	Hisao	Japan
Ito	Takatoshi	Japan
Johnson	Malcolm	U.S.A.
Kanagawa	Kyuichi	Japan
Kano	Yasuyuki	Japan
Kiang	Tan Hoon	Singapore
Kimura	Gaku	Japan
Kinoshita	Masataka	Japan
Kirschner	David L.	U.S.A.
Kopf	Achim	Germany
Kryc	Kelly	U.S.A.
Kusaba	Yoko	Japan
Li	Yong-Gang	U.S.A.
Lin	Weiren	Japan
Lockner	David	U.S.A.
Ma	Kuo-Fong	Taiwan
Malin	Peter	U.S.A.
Marone	Chris	U.S.A.

Last Name	First Name	Country
Moore	Gregory	Japan
Moore	J. Casey	U.S.A.
Mori	Jim	Japan
Nadeau	Robert M.	U.S.A.
Nielsen	Dennis	U.S.A.
Nikaizo	Akira	Japan
Nishigami	Kin'ya	Japan
Ogasawara	Hiroshi	Japan
Okamoto	Shin'ya	Japan
Omura	Kentaro	Japan
Park	Jin-oh	Japan
Prevedel	Bernhard	U.S.A.
Rack	Frank R.	U.S.A.
Ranero	César R.	Spain
Reches	Ze'ev	U.S.A.
Roecker	Steven	U.S.A.
Rowe	Christen D.	U.S.A.
Saffer	Demian	U.S.A.
Saito	Saneatsu	Japan
Sakaguchi	Arito	Japan
Shibata	Ihiro	Japan
Shimamoto	Toshihiko	Japan
Solum	John	U.S.A.
Sone	Hiroki	Japan
Stephen	Ralph	U.S.A.
Suyehiro	Kiyoshi	Japan
Tsuji	Takeshi	Japan
Talwani	Manik	U.S.A.
Tanikawa	Wataru	Japan
Tezuka	Kazuhiko	Japan
Thu	Moe Kyaw	Japan
Tobin	Harold	U.S.A.
Townend	John	New Zealand
Ujiie	Kohtaro	Japan
Underwood	Michael	U.S.A.
van der Pluijm	Ben	U.S.A.
Vanucchi	Paola	Italy
von Huene	Ronald	U.S.A.
Wiersberg	Thomas	Germany
Yamaguchi	Asuka	Japan
Zoback	Mark	U.S.A.

Field excursion to the Nobeoka thrust.

Permissions

All chapters in this book were first published in Scientific Drilling, by Copernicus Publications; hereby published with permission under the Creative Commons Attribution License or equivalent. Every chapter published in this book has been scrutinized by our experts. Their significance has been extensively debated. The topics covered herein carry significant findings which will fuel the growth of the discipline. They may even be implemented as practical applications or may be referred to as a beginning point for another development.

The contributors of this book come from diverse backgrounds, making this book a truly international effort. This book will bring forth new frontiers with its revolutionizing research information and detailed analysis of the nascent developments around the world.

We would like to thank all the contributing authors for lending their expertise to make the book truly unique. They have played a crucial role in the development of this book. Without their invaluable contributions this book wouldn't have been possible. They have made vital efforts to compile up to date information on the varied aspects of this subject to make this book a valuable addition to the collection of many professionals and students.

This book was conceptualized with the vision of imparting up-to-date information and advanced data in this field. To ensure the same, a matchless editorial board was set up. Every individual on the board went through rigorous rounds of assessment to prove their worth. After which they invested a large part of their time researching and compiling the most relevant data for our readers.

The editorial board has been involved in producing this book since its inception. They have spent rigorous hours researching and exploring the diverse topics which have resulted in the successful publishing of this book. They have passed on their knowledge of decades through this book. To expedite this challenging task, the publisher supported the team at every step. A small team of assistant editors was also appointed to further simplify the editing procedure and attain best results for the readers.

Apart from the editorial board, the designing team has also invested a significant amount of their time in understanding the subject and creating the most relevant covers. They scrutinized every image to scout for the most suitable representation of the subject and create an appropriate cover for the book.

The publishing team has been an ardent support to the editorial, designing and production team. Their endless efforts to recruit the best for this project, has resulted in the accomplishment of this book. They are a veteran in the field of academics and their pool of knowledge is as vast as their experience in printing. Their expertise and guidance has proved useful at every step. Their uncompromising quality standards have made this book an exceptional effort. Their encouragement from time to time has been an inspiration for everyone.

The publisher and the editorial board hope that this book will prove to be a valuable piece of knowledge for researchers, students, practitioners and scholars across the globe.

List of Contributors

Clive R. Neal
Department of Civil Engineering and Geological Sciences, 156 Fitzpatrick Hall, University of Notre Dame, Notre Dame, Ind., 46556, U.S.A

Millard F. Coffin
National Oceanography Centre, Southampton, University of Southampton Waterfront Campus, European Way, Southampton SO14 3ZH, U.K

Nicholas T. Arndt
Laboratoire de Géodynamique des Chaînes Alpines, Université de Grenoble, Grenoble, France

Robert A. Duncan
College of Oceanic and Atmospheric Sciences, 104 COAS Admin Bldg., Oregon State University, Corvallis, Ore. 97331, U.S.A

Olav Eldholm
Department of Earth Science, The University of Bergen, Allegaten 41, N-5007, Bergen, Norway

Elisabetta Erba
Dipartimento di Scienze della Terra, University of Milan, via Mangiagalli, 34, 20133 Milano, Italy

Cinzia Farnetani
Laboratoire de Dynamique des Fluides Géologiques, Institut de Physique du Globe, Boite 89, 4 pl. Jussieu, 75252 Paris, France

J. Godfrey Fitton
School of Geosciences, 210 Grant Institute, University of Edinburgh, The King's Buildings, West Mains Road, Edinburgh EH9 3JW, U.K

Stephanie P. Ingle
National Oceanography Centre, Southampton, University of Southampton Waterfront Campus, European Way, Southampton SO14 3ZH, U.K

Nao Ohkouchi
Institute for Research on Earth Evolution, Japan Agency for Marine-Earth Science Technology, 2-15 Natsushima-Cho, Yokosuka, 237-0061, Japan

Michael R. Rampino
Earth & Environmental Science Program, New York University, 1009 Silver Center, 100 Washington Square East, New York, N.Y. 10003, U.S.A

Marc K. Reichow
Department of Geology, University of Leicester, University Road, Leicester, LE1 7RH, U.K

Stephen Self
Department of Earth Sciences, The Open University, Milton Keynes MK7 6BJ, U.K

Yoshiyuki Tatsumi
Institute for Research on Earth Evolution, Japan Agency for Marine-Earth Science Technology, Natsushima-cho 2-15, Yokosuka 237-0061, Japan

Craig S. Fulthorpe
The University of Texas at Austin Institute for Geophysics, John A. and Katherine G. Jackson School of Geosciences, J.J. Pickle Research Campus, Bldg. 196 (ROC), 10100 Burnet Road (R2200), Austin, Texas 78758-4445, U.S.A

Kenneth G. Miller
Department of Earth and Planetary Sciences, Wright Labs, Rutgers, The State University of New Jersey, 610 Taylor Road, Piscataway, N.J. 08854-8066, U.S.A

André W. Droxler
Department of Earth Science, Rice University, P.O. Box 1892, Houston, Texas 77251-1892, U.S.A

Stephen P. Hesselbo
Department of Earth Sciences, University of Oxford, Parks Road, Oxford OX1 3PR, U.K

Gilbert F. Camoin
CEREGE UMR 6635 CNRS, BP 80, Europôle Méditerranéen de l'Arbois, F-13545 Aix-en-Provence cedex 4, France

Michelle A. Kominz
Department of Geosciences, Western Michigan University, 1133 Rood Hall, 1903 West Michigan Avenue, Kalamazoo, Mich., 49008, U.S.A

Fabio Florindo
Istituto Nazionale di Geofisica e Vulcanologia, Via di Vigna Murata, 605, 00143 Rome, Italy

David Harwood
ANDRILL Science Management Office, and Department of Geosciences, University of Nebraska-Lincoln, 126 Bessey Hall, Lincoln, Neb. 68588-0341, U.S.A

Richard Levy
ANDRILL Science Management Office, and Department of Geosciences, University of Nebraska-Lincoln, 126 Bessey Hall, Lincoln, Neb. 68588-0341, U.S.A

Bernhard Prevedel
GeoForschungsZentrum Potsdam, Telegrafenberg, D-14473 Potsdam, Germany

Lothar Wohlgemuth
GeoForschungsZentrum Potsdam, Telegrafenberg, D-14473 Potsdam, Germany

Jan Henninges
GeoForschungsZentrum Potsdam, Telegrafenberg, D-14473 Potsdam, Germany

Kai Krüger
GeoForschungsZentrum Potsdam, Telegrafenberg, D-14473 Potsdam, Germany

Ben Norden
GeoForschungsZentrum Potsdam, Telegrafenberg, D-14473 Potsdam, Germany

Andrea Förster
GeoForschungsZentrum Potsdam, Telegrafenberg, D-14473 Potsdam, Germany

Shin'ichi Kuramoto
JAMSTEC, Yokohama Institute of Earth Sciences, 3173-25 Showamachi, Kanazawa, Yokohama 236-0001 Japan

Daniel Curewitz
JAMSTEC, Yokohama Institute of Earth Sciences, 3173-25 Showamachi, Kanazawa, Yokohama 236-0001 Japan

Moe Kyaw Thu
JAMSTEC, Yokohama Institute of Earth Sciences, 3173-25 Showamachi, Kanazawa, Yokohama 236-0001 Japan

Hideki Masago
JAMSTEC, Yokohama Institute of Earth Sciences, 3173-25 Showamachi, Kanazawa, Yokohama 236-0001 Japan

Weiren Lin
Kochi Institute for Core Sample Research, Japan Agency for Marine-Earth Science and Technology, Nankoku 783-8502, Japan

Koji Yamamoto
Technology Research Center, Japan Oil, Gas and Metals National Corporation, Chiba, Japan

Hisao Ito
Center for Deep Earth Exploration, Japan Agency for Marine- Earth Science and Technology, Yokohama, Japan

Hideki Masago
Center for Deep Earth Exploration, Japan Agency for Marine- Earth Science and Technology, Yokohama, Japan

Yoshihisa Kawamura
Center for Deep Earth Exploration, Japan Agency for Marine- Earth Science and Technology, Yokohama, Japan

Greg Myers
IODP-MI Engineering and Operations Manager, IODP-MI Headquarters, 815 Connecticut Avenue, NW, Suite 210, Washington, DC 20006 U.S.A

Roger C. Searle
Department of Earth Sciences, Durham University, Durham, DH1 3LE, U.K

Philippe Claeys
Department of Geology, Vrije Universiteit Brussel, Pleinlaan 2, B-1050 Brussels, Belgium

Alex Deutsch
Institut für Planetologie, Westfaelische Wilhelms-Universität Münster Wilhelm-Klemm-Str. 10, D-48149 Münster, Germany

Henning Dypvik
(coordinator), Department of Geosciences, University of Oslo, P.OBox 1047, No 0316 Oslo, Norway

Frank T. Kyte
Institute of Geophysics and Planetary Physics, UCLA University of California, Los Angeles, Calif. 90095-1567, U.S.A

Takafumi Matsui
Department of Complexity Science and Engineering, University of Tokyo, Tokyo, Japan

Morten Smelror
Geological Survey of Norway, No 7491 Trondheim, Norwa

Georg Dresen
GeoForschungsZentrum Potsdam, Telegrafenberg, D-14473, Potsdam, Germany

Marco Bohnhoff
GeoForschungsZentrum Potsdam, Telegrafenberg, D-14473, Potsdam, Germany

Mustafa Aktar
Bogazici University, Kandilli Observatory and Earthquake Research Institute (KOERI), Cangelkoy, Istanbul, 81220, Turkey

Haluk Eyidogan
Istanbul Technical University, Mining Faculty, Department of Geophysics, 34469, Maslak, Istanbul, Turkey

Yongjian Huang
Research Center for Tibetan Plateau Geology, China University of Geosciences, Xueyuan Road 29, Haidian District, 100083, Beijing, People's Republic of China

Cheng-Shan Wang
Research Center for Tibetan Plateau Geology, China University of Geosciences, Xueyuan Road 29, Haidian District, 100083, Beijing, People's Republic of China

Paul E. Olsen
Lamont-Doherty Earth Observatory (LDEO) of Columbia University, 61 Route 9W, Palisades, N.Y., 10964-1000, U.S.A

Dennis V. Kent
Department of Earth & Planetary Sciences, Rutgers University, 610 Taylor Road, Piscataway, N.J. 08854, U.S.A. and Lamont-Doherty Earth Observatory, 61 Route 9W, Palisades, N.Y. 10964,, U.S.A

John W. Geissman
Department of Earth and Planetary Sciences, University of New Mexico, Albuquerque, N.M., 87131, U.S.A

Takeshi Sagiya
Research Center for Seismology, Volcanology, and Disaster Mitigation, Graduate School of Environmental Studies, Nagoya University, Furo-cho, Chikusa-ku, Nagoya 464-8601 Japan

Harold Tobin
Department of Geoscience, University of Wisconsin – Madison, 1215 West Dayton Street, Madison, Wis. 53706, U.S.A

Masataka Kinoshita
Institute for Research on Earth Evolution/JAMSTEC, Kanazawa-ku, Yokohama, Kanagawa 236-0001, Japan

Juichiro Ashi
Ocean Research Institute, The University of Tokyo, 1-15-1 Minamidai, Nakano-ku, Tokyo, 164-8639, Japan

Siegfried Lallemant
Département des Sciences de la Terre et de l'Environnement, Université de Cergy-Pontoise, Cergy-Pontoise 95031 Cedex, France

Gaku Kimura
Department of Earth and Planetary Science, Graduate School of Science, The University of Tokyo, Tokyo, Japan

Elizabeth Screaton
Department of Geology, University of Florida, P.O. Box 112120, Gainesville, Fla. 32611, U.S.A

Moe Kyaw Thu
Center for Deep Earth Exploration/JAMSTEC, 3173-25 Showa-machi, Kanazawa-ku, Yokohama, Kanagawa 236-0001, Japan

Hideki Masago
Center for Deep Earth Exploration/JAMSTEC, 3173-25 Showa-machi, Kanazawa-ku, Yokohama, Kanagawa 236-0001, Japan

Daniel Curewitz
Center for Deep Earth Exploration/JAMSTEC, 3173-25 Showa-machi, Kanazawa-ku, Yokohama, Kanagawa 236-0001, Japan
Department of Earth Sciences, Syracuse University, Syracuse, N.Y. 13244, U.S.A

Julien Thiesson
Université Pierre et Marie Curie, UMR 7619, Sisyphe, case 105, 4 place Jussieu, 75252 Paris cedex 05, France and INRA, Centre de Recherche d'Orléans, UR 0272 Science du Sol, 2163 Avenue de la Pomme de Pin, CS 40001 ARDON, 45075 ORLEANS Cedex 2, France

Emilia Huret
Andra, 1/7 rue Jean Monnet, Parc de la croix blanche, 92298 Châtenay Malabry cedex, France

Alain Tabbagh
Université Pierre et Marie Curie, UMR 7619, Sisyphe, case 105, 4 place Jussieu, 75252 Paris cedex 05, France

Bruno Galbrun
Université Pierre et Marie Curie, ISTeP, case 117, 4 place Jussieu, 75252 Paris cedex 05, France

Eugene G. Grosch
Africa Earth Observatory Network (AEON) and the Department of Geological Sciences, University of Cape Town, Rondebosch 7700, South Africa

Maarten de Wit
Africa Earth Observatory Network (AEON) and the Department of Geological Sciences, University of Cape Town, Rondebosch 7700, South Africa

Nicola McLoughlin
Department of Earth Science and Centre for Geobiology, University of Bergen, Allegaten, 41, N-5007 Bergen, Norway

Harald Furnes
Department of Earth Science and Centre for Geobiology, University of Bergen, Allegaten, 41, N-5007 Bergen, Norway

Bernd Zolitschka
Geomorphology and Polar Research (GEOPOLAR), Institute of Geography, University of Bremen, Celsiusstraße FVG-M, D-28359 Bremen, Germany

Flavio Anselmetti
Swiss Federal Institute of Aquatic Science and Technology (Eawag), Department of Surface Waters, Überlandstrasse 133, CH-8600 Dübendorf, Switzerland

Daniel Ariztegui
Section of Earth & Environmental Sciences, University of Geneva, Rue de Maraîchers 13, CH-1205 Geneva, Switzerland

Hugo Corbella
Argentine Museum of Natural History, Av. Angel Gallardo 470, Buenos Aires, Argentina

Pierre Francus
Institut National de la Recherche Scientifique, Centre Eau, Terre et Environnement, 490 rue de la couronne, Québec, QC G1K 9A9, Canada

Christian Ohlendorf
Geomorphology and Polar Research (GEOPOLAR), Institute of Geography, University of Bremen, Celsiusstraße FVG-M, D-28359 Bremen, Germany

Frank Schäbitz
Seminar for Geography and Education, University of Cologne, Gronewaldstr. 2, D-50931 Cologne, Germany

Noriaki Masui
Geomicrobiology Group, Kochi Institute for Core Sample Research, Japan Agency for Marine-Earth Science and Technology (JAMSTEC), Monobe B200, Nankoku, Kochi 783-8502, Japan

Yuki Morono
Geomicrobiology Group, Kochi Institute for Core Sample Research, Japan Agency for Marine-Earth Science and Technology (JAMSTEC), Monobe B200, Nankoku, Kochi 783-8502, Japan

Fumio Inagaki
Geomicrobiology Group, Kochi Institute for Core Sample Research, Japan Agency for Marine-Earth Science and Technology (JAMSTEC), Monobe B200, Nankoku, Kochi 783-8502, Japan

Peggy Delaney
Professor of Ocean Sciences, Ocean Sciences Department/ Institute of Marine Sciences, 1156 High Street, University of California, Santa Cruz, Calif., U.S.A

Sean Higgins
Associate Director, Ocean Drilling Programs, Consortium for Ocean Leadership, U.S. Implementing Organization, Washington, DC, U.S.A

Barry M. Freifeld
Earth Sciences Division, Lawrence Berkeley National Laboratory, MS 90-1116, One Cyclotron Road, Berkeley, Calif. 94720, U.S.A

Juergen Thurow
Department of Earth Sciences, University College London, Gower Street, London, WC1E 6BT, U.K

Larry C. Peterson
Rosenstiel School of Marine and Atmospheric Science, University of Miami, 4600 Rickenbacker Causeway, Miami, Fla., 33149, U.S.A

Ulrich Harms
International Continental Scientific Drilling Program, GFZ German Research Centre for Geosciences, Telegrafenberg, 14473, Potsdam, Germany

David A. Hodell
Department of Earth Sciences, University of Cambridge, Cambridge, CB2 3EQ, U.K

Heather Cheshire
Department of Earth Sciences, University College London, Gower Street, London, WC1E 6BT, U.K

Michael Schulz
Faculty of Geosciences/Center for Marine Environmental Research (MARUM), University of Bremen, D-28334, Bremen, Germany

Hans-Juergen Brumsack
Institute for Chemistry and Biology of the Marine Environment (ICBM), University of Oldenburg, P.O. Box 2503, D-26111, Oldenburg, Germany

Tomohisa Irino
Faculty of Environmental Earth Science, Hokkaido University, Kita 10, Nishi 5, Kita-ku, Sapporo, 060-0810, Japan

Valerie Masson-Delmotte
Laboratoire des Sciences du Climat et de l'Environnement, IPSL/CEA-CNRS UMR 1571, Gif-sur-Yvette, France

Ryuji Tada
Department of Earth and Planetary Science, Graduate School of Science, University of Tokyo, Science Bldg. No.1-734, 7-3-1 Hongo, Bunkyo-ku, Tokyo, 113-0033, Japan

Nicholas Christie-Blick
Department of Earth and Environmental Sciences and Lamont-Doherty Earth Observatory of Columbia University, Palisades, N.Y. 10964-8000, U.S.A

Mark H. Anders
Department of Earth and Environmental Sciences and Lamont-Doherty Earth Observatory of Columbia University, Palisades, N.Y. 10964-8000, U.S.A

Gianreto Manatschal
Université de Strasbourg, IPGS-EOST, 1 rue Blessig F-67 084, France

Brian P. Wernicke
Division of Geological and Planetary Sciences, California Institute of Technology, Pasadena, Calif. 91125, U.S.A

Andrew Cohen
Department of Geosciences, University of Arizona, Tucson, Ariz., U.S.A

Ramon Arrowsmith
School of Earth and Space Exploration, Arizona State University, Tempe, Ariz., U.S.A

Anna K. Behrensmeyer
Department of Paleobiology, National Museum of Natural History, Washington, DC, U.S.A

Christopher Campisano
Institute of Human Origins, Arizona State University, Tempe, Ariz., U.S.A

Craig Feibel
Departments of Geological Sciences and Anthropology, Rutgers University, New Brunswick, N.J., U.S.A

Shimeles Fisseha
Department of Earth Sciences, Addis Ababa University, Addis Ababa, Ethiopia

Roy Johnson
Department of Geosciences, University of Arizona, Tucson, Ariz., U.S.A

Zelalem Kubsa Bedaso
Department of Geology, University of South Florida, Tampa, Fla., U.S.A

Charles Lockwood
(deceased), Department of Anthropology, University of Wittwatersrand, Johannesburg, South Africa

Emma Mbua
Department of Paleontology, National Museums of Kenya, Nairobi, Kenya

Daniel Olago
Department of Geology, University of Nairobi, Nairobi, Kenya

Richard Potts
Human Origins Program, National Museum of Natural History, Washington, DC, U.S.A

Kaye Reed
Institute of Human Origins, Arizona State University, Tempe, Ariz., U.S.A

Robin Renaut
Department of Geological Sciences, University of Saskatchewan, Saskatoon, Saskatchewan, Canada

Jean-Jacques Tiercelin
CNRS, Université de Rennes, Rennes, France

Mohammed Umer
Department of Earth Sciences, Addis Ababa University, Addis Ababa, Ethiopia

Emily E. Brodsky
Deptartment of Earth and Planetary Sciences, UC Santa Cruz, 1156 High Street, Santa Cruz, Calif., 95060, U.S.A

Kuo-Fong Ma
Department of Earth Sciences, Institute of Geophysics National Central University, No.300, Jhongda Road, Jhongli City, Taoyuan County 32001, Taiwan (R.O.C.)

Jim Mori
Disaster Prevention Research Institute, Kyoto University, Gokasho, Uji, Kyoto, 611-0011, Japan

Demian M. Saffer
Department of Geosciences, The Pennsylvania State University, 310 Deike Building, University Park, Pa., 16802, U.S.A

John Townend
School of Geography, Environment, and Earth Sciences, Victoria University of Wellington, P.O. Box 600, Wellington, New Zealand

Rupert Sutherland
GNS Science, P.O. Box 30368, Lower Hutt, New Zealand

Virginia Toy
Department of Geology, University of Otago, P.O. Box 56, Dunedin, New Zealand

Harold Tobin
Associate Professor, Department of Geology and Geophysics, University of Wisconsin-Madison, 1215 West Dayton Street, Madison, Wis. 53706, U.S.A

Hisao Ito
Center for Deep Earth Exploration (CDEX) - Japan Agency for Marine-Earth Science and Technology (JAMSTEC), 3173-25 Showa-machi, Kanazawa-ku, Yokohama, Kanagawa, 236-0001, Japan

Jan Behrmann
Marine Geodynamics, IFM-GEOMAR, Wischhofstr. 1-3, 24148, Kiel, Germany

Steve Hickman
U.S. Geological Survey, MS977, 345 Middlefield Road, Menlo Park, Calif., 94025, U.S.A

Gaku Kimura
Department of Earth and Planetary Sciences, Graduate School of Science, The University of Tokyo, 7-3-1 Hongo, Bunkyo-ku, Tokyo, 113-0033, Japan

J. Casey Moore
Earth and Planetary Sciences, University of California Santa Cruz, Santa Cruz, Calif., 95064, U.S.A

Jim Mori
Disaster Prevention Research Institute, Kyoto University, Gokasho, Uji, Kyoto, 611-0011, Japan

Demian Saffer
Department of Geosciences, The Pennsylvania State University, 310 Deike Building, University Park, Pa. 16802, U.S.A

Satoshi Ide
Department of Earth and Planetary Science, University of Tokyo, 7-3-1, Hongo, Bunkyo, Tokyo, 113-0033, Japan

Ryosuke Ando
Natonal Research Institute for Earth Science and Disaster Prevention, 3-1 Tennodai, Tsukuba, Ibaraki, 305-0006, Japan

Teruo Yamashita
Earthquake Research Institute, The University of Tokyo, 1-1-1 Yayoi, Bunkyo-ku, Tokyo, Japan

Stephen Hickman
U.S. Geological Survey, 345 Middlefield Road, Mail Stop 977, Menlo Park, Calif., 94025, U.S.A

Mark Zoback
Stanford University, Mitchell Building, Room 359, Stanford, Calif.,94305-2215 U.S.A

William Ellsworth
U.S. Geological Survey, 345 Middlefield Road, Mail Stop 977, Menlo Park, Calif.,94025,U.S.A

Naomi Boness
Chevron Energy Technology Company, 6001 Bollinger Canyon Road, San Ramon, Calif., 94583, U.S.A

Peter Malin
Duke University, 109A Old Chem, Box 90227, Durham, N.C., 27708, U.S.A

Steven Roecker
Department of Earth and Environmental Sciences, Rensselaer Polytechnic Institute, Troy, N.Y., 12180, U.S.A

Clifford Thurber
Department of Geology and Geophysics, University of Wisconsin, 1215 West Dayton Street, Madison, Wis., 53706, U.S.A

Kuo-Fong Ma
Department of Earth Science and Institute of Geophysics, National Central University, No.300, Johngda Rd., Johngli City, Taoyuan County 32001, Taiwan (R.O.C.)

Hidemi Tanaka
Solid Earth Science Group, Department of Earth and Planetary Sciences, The University of Tokyo, Faculty of Science Building #1, Hongou 7-3-1, Bunkyo-ku, Tokyo, 113-0033, Japan

François Henri Cornet
Institut de Physique du Globe de Strasbourg, 5 rue René Descartes, 67000 Strasbourg, France

Harold Tobin
Department of Geology and Geophysics, University of Wisconsin, 1215 West Dayton Street, Madison, Wis., 53706 U.S.A

Masataka Kinoshita
IFREE (Institute for Frontier Research on Earth Environment), JAMSTEC (Japan Agency for Marine-Earth Science and Technology), 2-15 Natsushima-cho, Yokosuka-city, Kanagawa 237-0061, Japan

Georg Dresen
GeoForschungsZentrum Potsdam, Sektion 3.2 Deformaton und Rheologie, Telegrafenberg, D-14473, Potsdam, Germany

Mustafa Aktar
Bogazici University, Kandilli Observatory and Earthquake Research Institute (KOERI), Cangelkoy, Istanbul, 81220, Turkey

Marco Bohnhoff
GeoForschungsZentrum Potsdam, Haus D, Projectbereich 3.2, Telegrafenberg, D-14473, Potsdam, Germany

Haluk Eyidogan
Istanbul Technical University, Mining Faculty, Department of Geophysics, 34469, Maslak, Istanbul, Turkey

Achim Kopf
Research Center Ocean Margins (RCOM), University of Bremen, Leobener Strasse, 28359 Bremen, Germany

Marco Bohnhoff
GeoForschungsZentrum Potsdam, Telegrafenberg D424, 14473 Potsdam, Germany

Shin'ya Okamoto
Department of Earth and Planetary Sciences, Graduate School of Science, The University of Tokyo, 7-3-1 Hongo, Bunkyo-ku, Tokyo, 113-0033, Japan

Gaku Kimura
Department of Earth and Planetary Sciences, Graduate School of Science, The University of Tokyo, 7-3-1 Hongo, Bunkyo-ku, Tokyo, 113-0033, Japan; and Institute for Research on Earth Evolution (IFREE), Japan Agency for Marine-Earth Science and Technology, 2-15 Natsushima-cho, Yokosuka, Kanagawa, 237-0061, Japan

Asuka Yamaguchi
Department of Earth and Planetary Sciences, Graduate School of Science, The University of Tokyo, 7-3-1 Hongo, Bunkyo-ku, Tokyo, 113-0033, Japan

Haruka Yamaguchi
Department of Earth and Planetary Sciences, Graduate School of Science, The University of Tokyo, 7-3-1 Hongo, Bunkyo-ku, Tokyo, 113-0033, Japan; and Institute for Research on Earth Evolution (IFREE), Japan Agency for Marine-Earth Science and Technology, 2-15 Natsushima-cho, Yokosuka, Kanagawa, 237-0061, Japan

Yoko Kusuba
Department of Earth and Planetary Sciences, Graduate School of Science, The University of Tokyo, 7-3-1 Hongo, Bunkyo-ku, Tokyo, 113-0033, Japan

Kentaro Omura
National Research Institute for Earth Science and Disaster Prevention, Tennoudai 3-1, Tsukuba, Ibaraki, 305-0006

Anja M. Schleicher
Universität Würzburg, Geologisches Institut, Pleicherwall 1, 97070 Würzburg, Germany

Ben A. van der Pluijm
University of Michigan, Department of Geological Sciences, 1100 University Avenue, C.C.Little Building, Ann Arbor, Mich. 48109, U.S.A

Laurence N. Warr
Centre de Géochimie de la Surface (CNRS-ULP), 1 rue Blessig, 67084 Strasbourg, France

John G. Solum
U.S. Geological Survey, Earthquake Hazards Team, 345 Middlefield Road, MS 977 Menlo Park, Calif. 94025, U.S.A

Thomas Wiersberg
GeoForschungsZentrum Potsdam, Section 4.2, Telegrafenberg B323, 14473, Potsdam, Germany

Jörg Erzinger
GeoForschungsZentrum Potsdam, Section 4.2, Telegrafenberg B323, 14473, Potsdam, Germany

Yehuda Ben-Zion
Department of Earth Science, University of Southern California, Los Angeles, 90089, Calif., U.S.A

Zhigang Peng
School of Earth and Atmospheric Sciences, Georgia Institute of Technology, Atlanta, Ga., 30332-0340, U.S.A

Michael Lewis
Department of Earth Science, University of Southern California, Los Angeles, 90089, Calif., U.S.A

Jeff McGuire
Department of Geology and Geophysics, Woods Hole Oceanographic Institution, Woods Hole, Mass., 02543, U.S.A

Takeshi Tsuji
Ocean Research Institute, The University of Tokyo, 1-15-1 Minamidai, Nakano-ku, Tokyo 164-8639, Japan

Hidekazu Tokuyama
Ocean Research Institute, The University of Tokyo, 1-15-1 Minamidai, Nakano-ku, Tokyo 164-8639, Japan

Gaku Kimura
Department of Earth and Planetary Sciences, The University of Tokyo, Hongou 7-3-1, Bunkyo-ku, Tokyo, 113-0033, Japan

Shinya Okamoto
Department of Earth and Planetary Sciences, The University of Tokyo, Hongou 7-3-1, Bunkyo-ku, Tokyo, 113-0033, Japan

Christie D. Rowe
Earth & Planetary Science, University of California, Santa Cruz, 1156 High Street, Santa Cruz, Calif. 95064, U.S.A

William L. Ellsworth
3A-109, U.S. Geological Survey (USGS), 345 Middlefield Road Mail Stop 977, Menlo Park, Calif. 94025, U.S.A

Peter E. Malin
109A Old Chemistry Box 90227, Duke University, Durham, N.C. 27708, U.S.A

Kazutoshi Imanishi
Geological Survey of Japan, Institute of Geology and Geoinformation (AIST), Tsukuba Central 7, 1-1-1 Higashi, Tsukuba, 305-8567, Japan

Steven W. Roecker
Department of Earth and Environmental Sciences, Rensselaer Polytechnic Institute, Science Center 1S06, Troy, N.Y. 12180, U.S.A

Robert Nadeau
Berkeley Seismological Lab., 207 McCone Hall, University of California Berkeley, Calif. 94720-4760, U.S.A

Volker Oye
Norwegian Seismic Array (NORSAR), P.O. Box 53, N-2027 Kjeller, Norway

Clifford H. Thurber
Department of Geology and Geophysics, University of Wisconsin, 1215 West Dayton Street, Madison, Wis. 53706, U.S.A

Felix Waldhauser
LDEO-Seismology Geology and Tectonophysics, 210 Seismology, 61 Route 9W - P.O. Box 1000, Palisades, N.Y. 10964, U.S.A

Namoi L. Boness
Chevron, Energy Technology Centers (ETC), San Ramon, Calif. 94583, U.S.A

Stephen H. Hickman
USGS, 345 Middlefield Road, Menlo Park, Calif. 94025, U.S.A

Mark D. Zoback
Department of Geophysics, Stanford University, Mitchell Bldg., Stanford, Calif. 94305, U.S.A

Hiroshi Ogasawara
Faculty of Science and Engineering, Ritsumeikan University, 1-1-1 Noji Higashi, Kusatsu, 525-8577, Japan

Kohtaro Ujiie
Institute for Research on Earth Evolution (IFREE), Japan Agency for Marine-Earth Science and Technology (JAMSTEC), 3173-25 Showa-machi, Yokohama-city, Kanagawa 236-0001, Japan

Yasuyuki Kano
Disaster Prevention Research Institute, Kyoto University, Gokasyo, Uji, 611-0011, Japan

Jim Mori
Disaster Prevention Research Institute, Kyoto University, Gokasyo, Uji, 611-0011, Japan

Ryo Fujio
Disaster Prevention Research Institute, Kyoto University, Gokasyo, Uji, 611-0011, Japan

Takashi Yanagidani
Disaster Prevention Research Institute, Kyoto University, Gokasyo, Uji, 611-0011, Japan

Setsuro Nakao
Disaster Prevention Research Institute, Kyoto University, Gokasyo, Uji, 611-0011, Japan

Hisao Ito
Japan Agency for Marine Earth Science and Technology, 3173-25, Showa-machi, Kanazawa-ku, Yokohama, Kanazawa, 236-0001, Japan

Osamu Matsubayashi
National Institute of Advanced Industrial Science and Technology, No.7, Higashi 1-1-1, Tsukuba, 305-8567, Japan

Kuo-Fong Ma
National Central University, No.300, Jhongda Road, Jhongli City, Taoyuan County, 32001, Taiwan (R.O.C.)

Hiroki Sone
Department of Geology and Mineralogy, Division of Earth and Planetary Sciences, Graduate School of Science, Kyoto University, Kitashirakawa Oiwakecho, Sakyo-ku, Kyoto, 606-8502, Japan

Hiroyuki Noda
Department of Geology and Mineralogy, Division of Earth and Planetary Sciences, Graduate School of Science, Kyoto University, Kitashirakawa Oiwakecho, Sakyo-ku, Kyoto, 606-8502, Japan

Toshihiko Shimamoto
Department of Geology and Mineralogy, Division of Earth and Planetary Sciences, Graduate School of Science, Kyoto University, Kitashirakawa Oiwakecho, Sakyo-ku, Kyoto, 606-8502, Japan

Bernhard Prevedel
ICDP/OSG (Operational Support Group), GFZ-Potsdam, Telegrafenberg A34, D-14473, Potsdam, Germany

Kazuhiro Higuchi
Center for Deep Earth Exploration (CDEX), Japan Agency for Marine-Earth Science and Technology (JAMSTEC), 3173-25 Showa-machi, Kanazawa-ku, Yokohama, Kanagawa 236-0001, Japan

Shomei Kobayashi
Center for Deep Earth Exploration (CDEX), Japan Agency for Marine-Earth Science and Technology (JAMSTEC), 3173-25 Showa-machi, Kanazawa-ku, Yokohama, Kanagawa 236-0001, Japan

Dennis L. Nielson
DOSECC, Inc., P.O. Box 58857, Salt Lake City, Utah 84158-0857, U.S.A

Weiren Lin
Kochi Institute for Core Sample Research, Japan Agency for Marine-Earth Science and Technology, B200 Monobe, Nankoku, Kochi, 783-8502 Japan

Wataru Tanikawa
Kochi Institute for Core Sample Research, Japan Agency for Marine-Earth Science and Technology, B200 Monobe, Nankoku, Kochi, 783-8502 Japan

Wonn Soh
Kochi Institute for Core Sample Research, Japan Agency for Marine-Earth Science and Technology, B200 Monobe, Nankoku, Kochi, 783-8502 Japan

Tetsuro Hirono
Department of Earth and Space Science, Graduate School of Science, Osaka University, Toyonaka 560-0043, Osaka, Japan

En-Chao Yeh
National Taiwan University, No.1, Dec. 4, Roosevelt Road, Taipei, 106147, Taiwan (R.O.C.)

Ralph A. Stephen
Department of Geology and Geophysics, Woods Hole Oceanographic Institution, Clark South 282, MS#24, Woods Hole, Mass. 02543, U.S.A

Stephen A. Swift
Department of Geology and Geophysics, Woods Hole Oceanographic Institution, Clark South 286A, MS#24, Woods Hole, Mass. 02543, U.S.A

S. Thompson Bolmer
Department of Geology and Geophysics, Woods Hole Oceanographic Institution, Clark South 286B, MS#24, Woods Hole, Mass. 02543, U.S.A

Hartley Hoskins
Computer and Information Services, Department of Geology and Geophysics, Woods Hole Oceanographic Institution, Woods Hole, Mass. 02543, U.S.A

Eiichiro Araki
Japan Agency for Marine-Earth Science and Technology (JAMSTEC), 2-15 Natsushima-cho, Yokosuka, 237-0061, Japan

Kiyoshi Suyehiro
Japan Agency for Marine-Earth Science and Technology (JAMSTEC), 2-15 Natsushima-cho, Yokosuka, 237-0061, Japan

Takatoshi Ito
Institute of Fluid Science, Tohoku University, 2-1-1 Katahira Aoba-ku Sendai, 980-8577, Japan

Kentaro Omura
National Research Institute for Earth Science and Disaster Prevention, Tennodai 3-1, Tsukuba, Ibaraki 305-0006, Japan

Hisao Ito
Center for Deep Earth Exploration (CDEX), Japan Agency for Marine-Earth Science and Technology (JAMSTEC), 3173-25 Showa-machi, Kanazawa-ku, Yokohama, Kanagawa 236-0001, Japan